CHEMICAL ENGINEERING METHODS AND TECHNOLOGY

POLYMER CHAINS: STRUCTURE, PHYSICAL PROPERTIES AND INDUSTRIAL USES

Chemical Engineering Methods and Technology

Additional books in this series can be found on Nova's website under the Series tab.

Additional E-books in this series can be found on Nova's website under the E-book tab.

Polymer Science and Technology

Additional books in this series can be found on Nova's website under the Series tab.

Additional E-books in this series can be found on Nova's website under the E-book tab.

CHEMICAL ENGINEERING METHODS AND TECHNOLOGY

POLYMER CHAINS: STRUCTURE, PHYSICAL PROPERTIES AND INDUSTRIAL USES

RÉGINE PENZKOFER
AND
YASUE WU
EDITORS

Nova Science Publishers, Inc.
New York

For permission to use material from this book please contact us:
Telephone 631-231-7269; Fax 631-231-8175
Web Site: http://www.novapublishers.com

Library of Congress Cataloging-in-Publication Data

Polymer chains : structure, physical properties, and industrial uses / editors, Rigine Penzkofer and Yasue Wu.
p. cm.
Includes index.
ISBN 978-1-62100-422-6 (hardcover)
1. Polymerization. 2. Polymers. I. Penzkofer, Rigine. II. Wu, Yasue.
TP156.P6P583 2011
547'.28--dc23
2011034219

Published by Nova Science Publishers, Inc. + New York

CONTENTS

PREFACE

The physical properties of a polymer are strongly dependent on the size or length of the polymer chain. As chain length is increased, melting and boiling temperatures increase quickly. Impact resistance also tends to increase with chain length, as does the viscosity, or resistance to flow, of the polymer in its melted state. In this book, the authors present topical research in the study of the structure, physical properties and industrial uses of polymer chains. Topics discussed include the flexibility of polyheteroarylenes and the effect on several physical properties of these polymers; aliphatic polyester-based nanocomposites; bioplastic-based blends; interactions in small permeants in polymeric matrices; the role of polymer chain ends in plasma surface modification and preceramic polymer chains.

Chapter 1 - Conformational parameters for a wide variety of aromatic cyclochain polymers have been calculated assuming free rotation about virtual bonds. The flexibilility with free rotation A_{fr} is shown to be determined by the geometry of repeat unit and can vary over a wide range (10 Å - ∞). Experimental value of Kuhn segment length A_{exp} are calculated from literature data on the hydrodynamic behavior of polymeric macromolecules of this class in dilute solution. For most polymers A_{fr} is equal to experimental A_{exp} and hindered rotation is mainly due to bulky substitutions. The dependences of several physical properties such glass transition temperature, friction coefficient, elasticity modulus, dielectric constant, remaining normalized surface potentials, fire-resistance parameters on conformational rigidity of polymer chain are revealed for polyheteroarylenes. The contribution of conformational rigidity of a polymer chain to physical properties of this class of polymers is shown to be significant.

Chapter 2 - Nowadays, aliphatic polyerster-based polymers have attracted a great deal of interest as promising biodegradable materials. Nevertheless, in certain practical applications, these polymers suffer from limitations, such as its thermal stability and insufficient stiffness, low-melt viscosity, and poor mechanical properties. Inorganic nanocomposite, made by adding a small amount of inorganic material within a matrix of polymer, is an effective way of improving these drawbacks and widening its application. Such nanocomposites usually exhibit substantially improved thermal, mechanical, and physical properties in comparison to the corresponding neat polymer. They can combine the favored properties of the inorganic material (e.g., increased strength and modulus, high thermal stability and better gas barrier property) and the organic polymer matrix (e.g., flexibility, low density and processability). Furthermore, they have merits when compared with conventional polymer composites, such as lightweight, transparency, and easier processability. Consequently, organic polymer-based nanocomposites are considered as next-generation materials.

Chapter 3 - With numerous environmental problems caused by petrochemical plastics waste, poly(hydroxyalkanoates) (PHAs), which can be produced from renewable carbon resources, has attracted a great deal of attention from scientific community and industry as these microbially synthesized polyesters can be completely biodegraded to CO_2 and H_2O under aerobic and anaerobic conditions.

Poly(3-hydroxybutyrate) (PHB), discovered by Lemoigne in 1925, is the most representative member of the PHA family. PHB possesses a melting point close to that of polypropylene (PP), forms a better barrier to oxygen than poly(ethylene terephthalate) (PET) and PP, and has mechanical properties resembling those of polystyrene (PS) and PP. However, it is rigid and stiff because of the high crystallinity and larger spherulites. Furthermore, PHB is also thermally unstable during the conventional melt processing due to the narrow processing temperature window. To overcome these drawbacks of PHB, several PHB copolymers have been biosynthesized in a variety of microorganisms. Table 1-1 shows the various PHB copolymer structures.

Recently, Kaneka Corp. (Osaka, Japan) synthesized poly(3-hydroxybutyrate-*co*-3-hydroxyhexanoate) (PHB–HHx) for the first time from oils and fats using *Aeromonas* sp. OL338 and *Aeromonas* sp. FA440. This new co-polyester has characteristics similar to those of conventional polyethylene and polyester in terms of, for example, its elasticity, toughness, dyeability, and chemical digestibility. In addition, PHB–HHx has been reported to not only possess improved thermoplasticity and mechanical properties over the other PHAs , but also to have excellent compatibility with other biodegradable polymers.

Chapter 4 - High-polymeric inorganic polyphosphates are linear polymers containing a few to several hundred orthophosphate residues linked by energy-rich phosphoanhydride bonds. These inorganic compounds can be easily obtained by fusion of potassium or sodium phosphates at 700-800 °C for several hours followed by rapid cooling. These polymers are present in the cells of all living organisms from bacteria to mammals. Numerous functions of this compound in the cells, including phosphate and energy reservation, sequestration and storage of cations, formation of membrane channels, participation in phosphate transport, involvement in cell envelope formation and function, gene activity control, regulation of enzyme activities, stress response and stationary phase adaptation, are provided by the following properties: polyphosphate is an osmotically neutral form of phosphorus, the vital cell element necessary for the biosynthesis of nucleic acids and phospholipids, phosphorylation of proteins and sugars; the energy of phosphoanhydride bonds between phosphate residues is close to the energy of terminal phosphoester bond in ATP; these polyanions can form complexes both with inorganic cations and with positively charged organic compounds, modifying the properties of these compounds.

Chapter 5 - An ionomer is a polymer that is partially ionized by the metallic ion. Research involving ionomers have received substantial attention in the past decades. As originally proposed by Eisenberg, the interaction between ionic groups leads to the formation of multiplets containing a small number of ion pairs, and ionic clusters, which constitute a second phase and contain many multiplets as well as portions of the hydrocarbon chains. These ionic structures have been shown to act as strong electrostatic cross-links and enhance the mechanical properties of the polymer. The improvement in mechanical properties of the ionomer has been reported to be related to the strength of the electrostatic cross-links, which can be varied by changing the size and type of the counterion present in the ionomer. Among several types of ionomers, poly(ethylene-*co*-methacrylic acid) (EMAA) is the most widely used base polymer for metal cation ionomers in industrial applications. Structural and mechanical properties of ethylene ionomers have been extensively investigated in both solid and melt states.

Chapter 6 - This chapter discusses the interaction between small permeant states in polymer-based membranes. The permeants, in the vapor and liquid phases outside the membranes, are sorbed into different membrane systems, which include hydrophilic, hydrophobic and organic/inorganic composite films. The permeants and membranes are characterized using low temperature differential scanning calorimetry (DSC). The permeant state can be classified into freezable, intermediate, or non-freezable bound states. We use three systems to represent these various permeant-polymer interactions. In the weak interaction system of water and organic solvents in poly(dimethyl siloxane) (PDMS), the sorbed permeant exists in both freezable and non-freezable states. The fusion temperature and enthalpy of the freezable permeant resembles those of the bulk, pure solvent. At low sorption levels, the permeant molecules are present in the non-freezable bound state and the swollen PDMS film behaves like a pseudo-homogeneous system. Beyond a critical sorption level, all of the excessive permeant molecules in the PDMS are in the freezable state and the films behave like a phase-separated solution. In medium interaction systems, such as water in polyvinyl alcohol (PVA), the water distributes into freezable (with freezing at 0°C), intermediate (with sub-zero freezing temperatures), and non-freezable bound water. However, the incorporation of fumed silica (FS) nano-particles into the PVA results in a complex matrix in which FS generates dilution, shielding and suppression of hydrogen bond formation, as well as a shift in the freezable and non-freezable water equilibrium. In the strong interaction system of water in perflurosulfonic acid (Nafion®), no free water is detected. Only non-freezable bound water and intermediate water with subzero fusion temperatures are found. Mathematical models are proposed to express the permeant state distribution in these three systems. The interpretation of the corresponding plots of the solvent content of each state versus total solvent uptake is elucidated. These various micro-interactions may have different impacts on the transport properties of polymer-based membranes, hydrogels, and solid electrolytes used in separation, gas barriers, controlled drug release, fuel cells, and other applications.

Chapter 7 - Plasma treatment is a useful technique for modification of polymeric surface properties. In the plasma treatment process, two different reactions, implantation of functional groups (chemical modification) and etching of polymer surface (physical modification) occur on polymer surfaces. These reactions do not occur separately but occur simultaneously during the plasma treatment process. How to control the etching reactions on polymer surfaces is an important factor to accomplish successful surface modification by plasma.

Plasma modification reactions are restricted at a thin topmost layer of a few nanometers on polymer surfaces, and never occur at deep layers (a few micrometers or more) from the surfaces. Therefore, polymer segments located at a topmost layer of polymer surfaces are subjected to be modified by plasma. What segments of polymer chains are located at the polymer surfaces?

Polymer chain ends do not distribute homogeneously in polymer bulks, but tend to aggregate at surface layers of the polymers. Therefore, surface functionalities such as adhesion with Cu metal are mainly due to polymer chain ends aggregated at the polymer surfaces.

Chapter 8 - Copolymerization reactions are widely used for the modification of many important end use properties of common polymers. In these copolymers, the composition and the chemical heterogeneity appear as main factors to understand polymer behaviour, and have to be determined in addition to the common characterization properties for polymer systems as molecular weight, polydispersity index, melting temperature, crystallinity, etc.

In this chapter, a set of ethylene/propylene copolymers covering the whole composition range was analyzed in order to fully describe the strong composition effects on thermodynamic and conformation properties in solution. Ethylene/propylene copolymers were synthesised by using the heterogeneous catalytic system MAO/rac-$Me_2Si[2Me\text{-}Ind]_2ZrCl_2$ and the polymerization process was carried out with controlled monomers feed that lead to copolymers covering a wide composition range. Details on that synthesis process leading to controlled compositions are also included.

Chapter 9 - Preceramic polymers are compounds with a skeletal structure that does not include only carbon atoms but also others atoms such as metals and metalloids (boron and silicon). We can distinguish homo-atomic polymers with one kind atom in the main chain such as polysilanes containing silicon-silicon and hetero-atomic polymers with more than one type of atom in the main chain, mostly two types of atoms alternate along the main chain such as polycarbosilanes (Si-C), polysilazanes (Si-N), polymethylaminoboranes (B-N) and others. They are available as liquid and solid (fusible or infusible) with a large range of viscoelastic properties to generate a large variety of shapes.

Chapter 10 - A novel method for template synthesis of carbon materials with homogeneous pores has been proposed on the basis of using a high porous monolithic material from alumina nanofilaments as a template matrix. Monolithic alumina with high surface area template was obtained by direct water vapor oxidation of aluminum plates with the purpose to apply it as a template matrix. The templates were impregnated by different carbon precursors – sucrose, furfuryl alcohol, polyacrylonitrile and petroleum pitch. After polymerization, carbonization of precursors and removing the template, the porous structures of initial template and the carbon materials were characterized by analyzing the low temperature nitrogen adsorption and water vapor isotherms. Also, the surface and structural characteristics of these nanostructured materials were investigated using FTIR and Raman spectroscopy, and electron microscopy methods. The results showed that the highly developed homogeneous porous structure of the carbon materials with narrow pore size distribution were the consequences of the replication of the high porous structure of the nanofibrous template. According to the spectroscopy data the carbon nanoporous materials are rather homogeneous, and their structural features are similar to nanomaterials produced by other methods. Their properties depend on precursor properties and impregnation conditions. The materials obtained from sucrose are similar to the double wall nanotubes (DWNTs). They have an average inner diameter in the range of 0.4 – 0.6 nm and the surface BET area equal to 400 m^2/g. It was concluded from the spectroscopic data that these materials had mainly semiconductor properties and showed a strong nonlinear response to an action of infrared radiation. Impregnation of the template by furfuryl alcohol resulted in the monolithic carbon materials, which occurred to be similar to the single wall nanotubes (SWNTs) with a diameter of about 1.8 nm and the BET surface equal to 580 m^2/g, exhibiting semiconductor and metallic properties. Uniformed supermicropores with the diameters exceeding 3.5 nm were found in the carbon materials obtained by using petroleum pitch as precursor.

In: Polymer Chains
Editors: Régine Penzkofer and Yasue Wu

ISBN: 978-1-62100-422-6

Chapter 1

On the Flexibility of Polyheteroarylenes and the Effect on Several Physical Properties of These Polymers

_Inga Ronova_*
Nesmeyanov Institute of Organoelement Compounds,
Moscow, Russia

Abstract

Conformational parameters for a wide variety of aromatic cyclochain polymers have been calculated assuming free rotation about virtual bonds. The flexibilility with free rotation A_{fr} is shown to be determined by the geometry of repeat unit and can vary over a wide range (10 Å - ∞). Experimental value of Kuhn segment length A_{exp} are calculated from literature data on the hydrodynamic behavior of polymeric macromolecules of this class in dilute solution. For most polymers A_{fr} is equal to experimental A_{exp} and hindered rotation is mainly due to bulky substitutions. The dependences of several physical properties such glass transition temperature, friction coefficient, elasticity modulus, dielectric constant, remaining normalized surface potentials, fire-resistance parameters on conformational rigidity of polymer chain are revealed for polyheteroarylenes. The contribution of conformational rigidity of a polymer chain to physical properties of this class of polymers is shown to be significant.

Introduction

Polyphenylenes, aromatic polyamides, polyesters and polycarbonates, polyimides, polyoxdiazols, etc. belong to class of aromatic cyclochain polymers. Since the 60s of last century, much attention has been given to these polymers because of their special properties, which are due, in particular, to high energies of intermolecular interaction. For instance, they have good mechanical properties and can be used at comparatively high temperatures.

* E-mail: ron@ineos.ac.ru.

Additionally, several aromatic polymers, in particular *para*-linked polyamides, are characterized by high rigidity of polymer chain.

Chain stiffness, combined with strong intermolecular interactions, makes it possible to obtain polymeric materials with specified properties. Such high rigidity is not an exclusive characteristic of polyamides but can be attained by deliberate "tailoring" of chemical structure of polymer chain repeat unit.

The effect of chemical structure of the repeat unit (i.e., its chemical composition and the structure of its active groups and fragments, which provide intermolecular interactions, such as the dipole-dipole interactions, H-bonding, the ionic interactions, etc.) is widely considered in the literature. The effect of the geometry is usually not considered. Actually, consideration of pure geometric contribution presents a complex problem, because, as a rule, it is difficult to separate the effect of geometric structure from simultaneous effect of functional groups.

As is shown in [1-5], a class of the polymers exists, for which the effect of geometric structure can be separated. This class comprises the polymers with the main chain consisting of aromatic cycles and heterocycles. The cycles are connected with each other directly or through bridging amide, ester or carbonate groups, or one atom of carbon, oxygen, sulfur, etc. The conformational rigidity of such polymers (i.e., polyheteroarylenes) varies in a wide range depending on the structure of the repeat units, which consist of virtual bonds of different bond lengths and angles. When constructing such a chain, one can introduce any active groups providing desirable intermolecular interactions into the chain design. In other words, it is possible to construct the repeat units in such a way that for the same molecular interaction the conformational rigidity of the chains will vary in a wide range.

Attention should be drawn to a specific feature of aromatic polymers. Because the rotation of the repeat units with respect to each other is almost free, the conformational rigidity of the polymer chain does not depend on temperature [6], and all the properties associated with temperature changes should correlate unambiguously with the conformational rigidity of a polymer chain.

I. Conformational Parameters with Free Rotation

Before considering basic matters, we give several definitions from conformational analysis, using the treatment we will employ further.

The *virtual bond* is rigid section of chain approximated by a straight line, about which rotation occur. In particular, it can be a valence bond or bond can contain both valence bonds and rings.

The *contour length* of structural repeat unit l_o is the sum of projections of virtual bonds on the direction of main chain of extended macromolecule in *trans*-conformation. The concept of contour length was introduced for aliphatic polymers having nearly the same values for all virtual bonds lengths and valence angles, which resulted in coincidence of the sum of bond projections with the symmetry axis of macromolecule (fig. 1a). For aromatic polymers, characterized by nonuniform virtual bond lengths and valence angles, the projection of the virtual bonds on macromolecular axis does not form a continuous line (Fig. 1b).

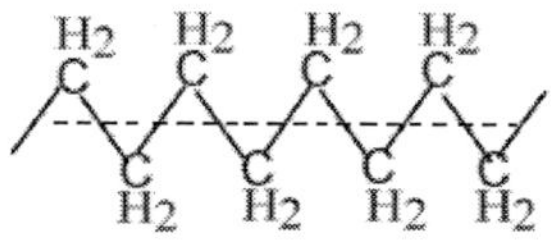

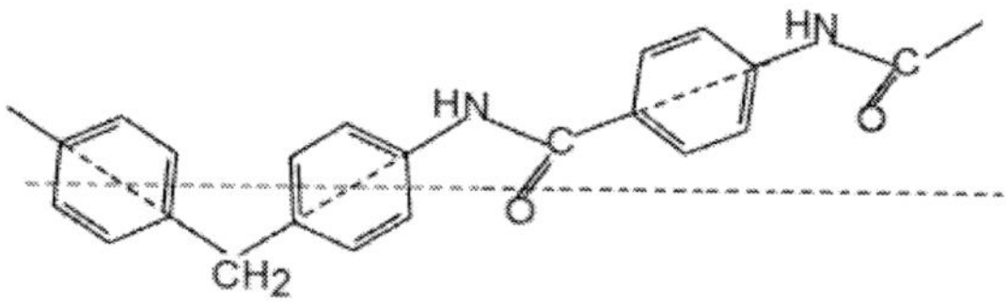

Figure 1. The definition of contour length as a sum of projections of bond lengths on the chain's direction: a aliphatic polymer, b polyheteroarylenes.

For polyheteroarylenes the contour length is defined as the length of the broken line that connects the centers of the virtual bonds (Fig. 2). In this case L does not depend on the conformation of the chain and can be regarded as a measure of the macromolecule's length.

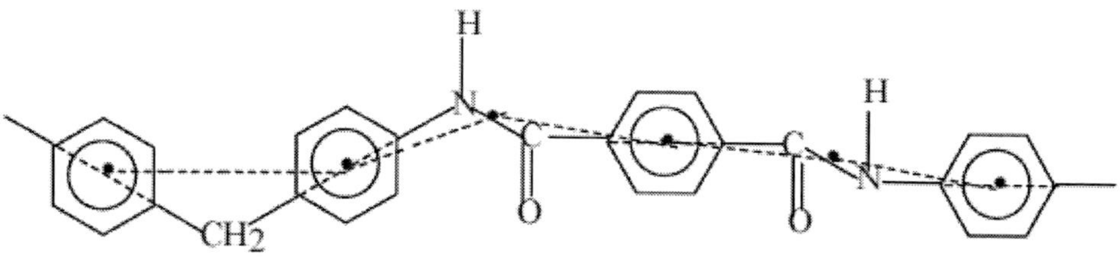

Figure 2. Contour length in a monomer unit of polyheteroarylenes.

There are two characteristic of chain rigidity. The kinetic rigidity characterized the mobility of chain units and is determined by the potential barriers to rotation about virtual bonds. The chain is kinetically rigid if, under given conditions, potential barriers to rotation are so high that they cannot be overcome by Brownian movement. Thus, the kinetic rigidity is defined by the value of potential barrier to rotation. The conformational flexibilility (for kinetically flexible chain it is called the thermodynamic or equilibrium flexibility) is characterized by the degree of coiling of macromolecule and defined by the geometric parameters of the chain – i. e. by value of virtual bonds and angles between these bonds.

These two characteristics do not necessary coincide. Thus, poly(*para*-benzamide) has a high kinetic flexibility and a low conformational one. And, on the contrary, one can image a macromolecule that under given conditions, exist as a successive series of rotation isomers separated by high potential barriers. In first case, easily realized rotations do not result in essential changes of distances between the chain links, whereas in the second case these distances depend strongly on population of rotational isomers.

It is not known, whether these two extreme case really exist, but they are plausible in principle. This example illustrates the difference between the kinetic-rigidity and conformational-rigidity concept. Variations in combinations of these properties are basis of many physical properties of polymers. Thus, the two polymers mentioned above lack high elasticity for different reasons: poly(*para*-benzamide) due to a lack of conformational flexibility, while the rigidly coiled chain lacks kinetic flexibility. Characteristics and assessment methods for these two types of flexibility are different. Here we will consider the conformational rigidity of polymers.

The conformational rigidity is usually characterized by the Kuhn statistical segment

$$A_{fr} = \lim_{n\to\infty}\left(\frac{<R^2>}{nl_o}\right), \qquad (1)$$

where $<R^2>/nl_o$ is the ratio of the average square end-to-end distance of a chain to its contour length; n is the number of the repeat units; l_o is the contour length of the repeat unit.

Other conformational parameter is characteristic ratio:

$$C_\infty = \frac{A_{fr}}{l_o} \qquad (2)$$

It shows how many monomer units Kuhn segment includes.

The third conformational parameter

$$p = C_\infty * n, \qquad (3)$$

where $\boldsymbol{p}$ is the parameter which is shown the number of aromatic rings per Kuhn segment, $\boldsymbol{n}$ is the number of aromatic rings in repeat unit of the polymer.

The hindrance parameter

$$\sigma = \sqrt{\frac{A_{exp}}{A_{fr}}}, \qquad (4),$$

specifies the deviation of the experimental value Kuhn segment from calculated for a free rotating chain.

As to polymers with aromatic rings in the chain, rotation about a bond between neighboring links is characterized by potential barriers at rotation angles 180°. If variations in these positions are symmetrical, σ is unity, independent of the height of potential barriers to rotation. Nevertheless, the knowledge of this value is useful since it is characteristic of asymmetry of vibrations.

There are many computational methods for macromolecular conformational parameters, or more precisely, for values of $<R^2>$, which enter all these parameters. For aliphatic polymers, there were methods developed that allowed for both independent and mutually dependent hindered rotation of groups with respect to one another [7]. All of these are complicated and calculation of conformational energy maps is required in each individual case. In contrast to this, aromatic polymers appear to be more amenable to calculations. Owing to their long virtual bonds, including, for instance, phenyl rings in *para* position (Fig. 3), and large angles between virtual bonds, distances between groups at the ends of neighboring virtual bonds exceed the sum of Van der Waals radii for atoms constituting these groups. Hence, from this point of view there is nothing to hinder rotation about the virtual bonds. However, the interaction between an aromatic ring and substitution in position *ortho* to neighboring link can produce hindrances to rotation. An analysis of internal rotation about virtual bonds containing cyclic groups (phenyl) united by bridging atoms or groups X (where X= -O-, $-CH_2-$, -(C=O)-) or by diatomic groups of the main chain (-CO—NH-, -CO—O-) carry out by a semiempirical method (EHT) by Birstein [8, 9]. In accordance with the benzene ring symmetry, several symmetrically situated energy minima, whose positions were determined by the chemical nature of bridging groups, were shown to exist in conformational space. Averaged values of cosines of internal rotation angles were also found to be zero. From the thermodynamic viewpoint, this means a free rotation. The calculations also showed that the *trans* configuration was preferred for bridging amide and ester bonds. Analogous results were obtained by Tonelly [10-12].

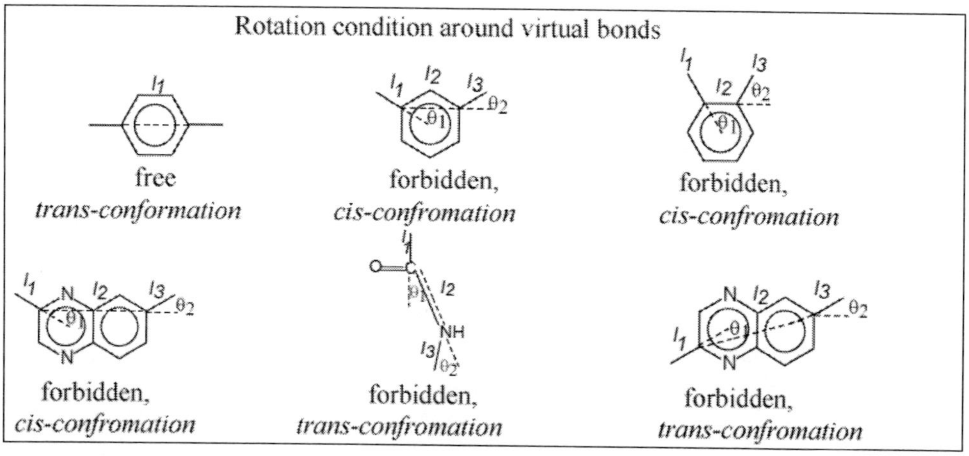

Figure 3. Construction of virtual bonds in cycles and heterocycles, conformation, and rotation conditions.

There are several methods for calculating mean-square end-t-end distances $<R^2>$ for cyclochain polymers. They are not quite equivalent because of different approximations. Here were calculated the Kuhn segments by the Monte-Carlo method. We used the Volkenstein rotational isomeric state approximation [13] by consideration of only discrete values of rotation angles and the Flory approximation [7] by the assumption that rotation around virtual bonds are independent. Tonelly [10-12]. With these assumptions, one can write coordinates of any vector in the reference frame associated with the first vector as

$$V_j=T_1....T_{j-2}T_{j-1}l_j$$

where T is the Flory matrix

$$T_j=\begin{bmatrix} \cos\theta_j ....\sin\theta_j ....0 \\ \sin\theta_j \cos\varphi_j ...-\cos\theta_j \cos\varphi_j ...\sin\varphi_j \\ \sin\theta_j \sin\varphi_j ...-\cos\theta_j \sin\varphi_j ...-\cos\varphi_j \end{bmatrix}$$

and (π-θ_j) are the angles between virtual bonds. On constructing the polymer chain, the angles θ_j recur periodically with the period depending on the number of virtual bonds N in the repeat structural unit. The index j runs from 1 to n·N.

The values of the rotation angles φ_j were chosen in one of the following two ways depending on the nature of a bond: (i) they were determined by a Monte Carlo procedure in the assumption of a uniform distribution within the interval (0,2π); (ii) φ was constant, i.e. any rotation about bond was forbidden. The ensemble averages $\langle R^2/n \cdot N \rangle$ were obtained by generating on a computer a set of independent chain sequences (in each sequence n ran from 1 to 2500) and averaging over the set of the values of $R^2/n \cdot N$ relating to the same n. For each of the polymers the value of n=n' beyond which the average $\langle R^2/n \cdot N \rangle$ as a function of n tended to converge was found. Usually n' was of the order of 300. [1].

In calculations carried out by this method, free rotation about virtual bonds was forbidden for the following cases: (1) if the virtual bond passes through a ring and is not the symmetry axis of this ring, (2) length of bonds -CO—NH- and -CO—O- are less than the sum of covalent radii and partial double bond character, which forbids rotation about these bonds (figure 3).

The model of the monomer unit is built in the molecular editor and its geometry is refined by the quantum-mechanical AM1 method [14]. In the Table 1 are listed conformational parameters for free rotation calculated by the Monte Carlo for different groups of cyclochain polymers.

The first thing that is seen from the data of Table I is the fact that changes in chain flexibility can be controlled by choosing the structure of the skeleton fragment of the repeat unit. The conformational flexibility varies over a wide range values of the Kuhn segment are between 15 and 1500 *Å* In many cases, the length of the Kuhn segment does not exceed the value for aliphatic polymers.

Table 1. Structural repeat unit and conformational parameters, calculated by method Monte Karlo

No	Structural repeat unit	l_o, Å	A_{fr}, Å
1	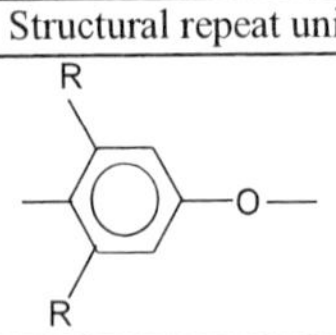	4.87	23.7
2	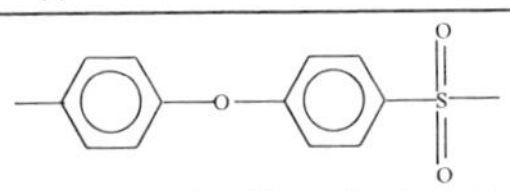	10.00	38.1
3	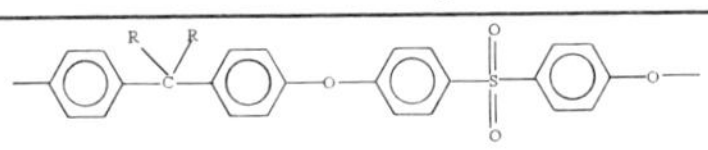	19.81	16.3
4	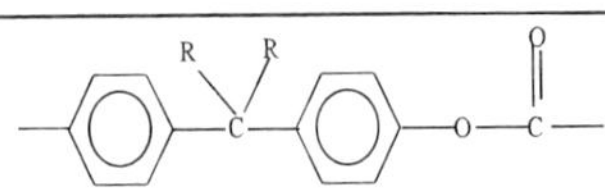	12.12	17.9
5[a]	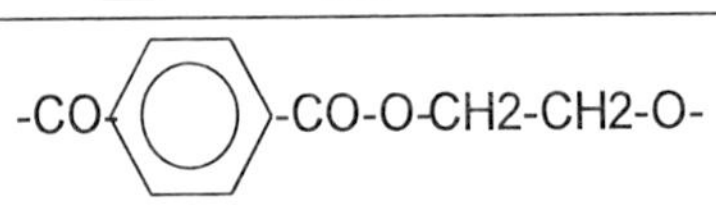	11.29	8.0

No	Structural repeat unit	l_o, Å	A_{fr}, Å
6	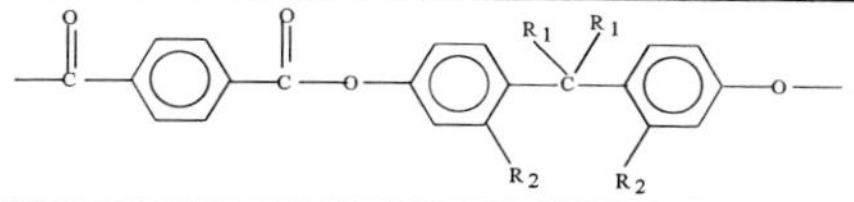	17.53	25.7
7	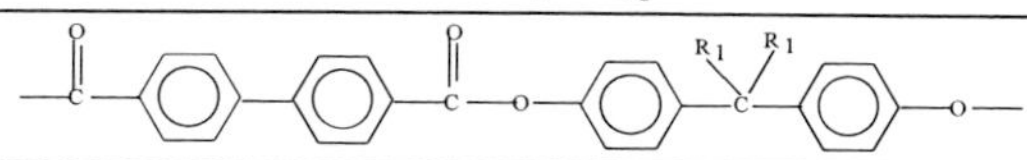	21.78	33.7
8	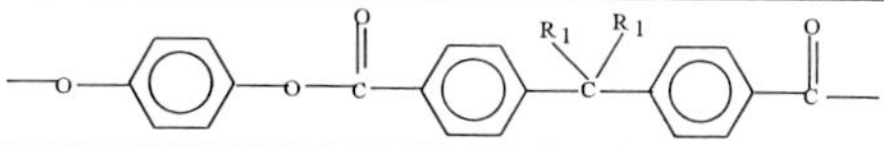	17.53	24.5
9	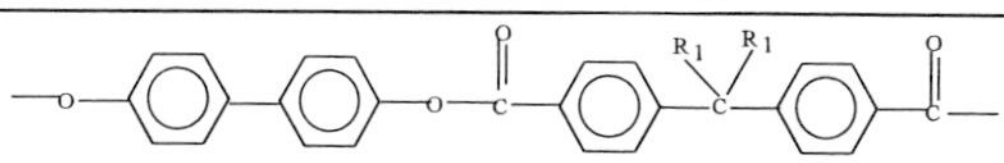	21.78	31.7
10	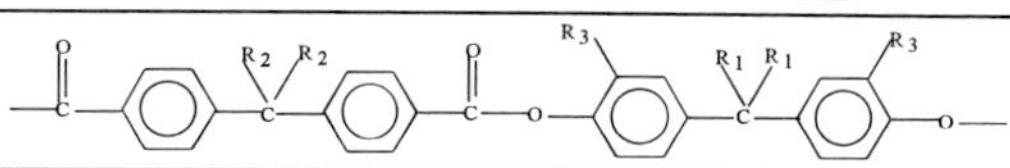	22.51	19.2
11	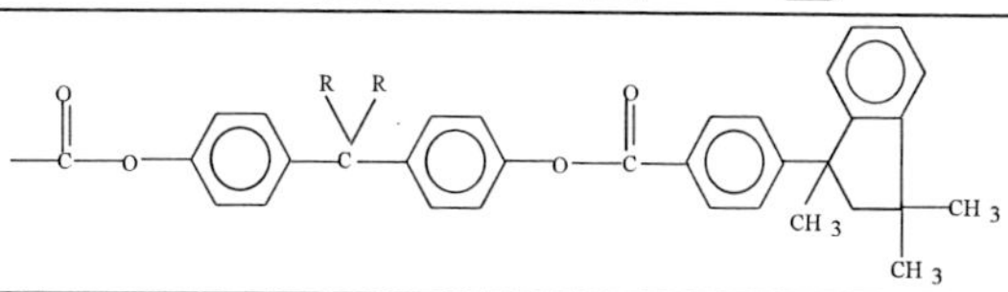	19.03	19.7
12	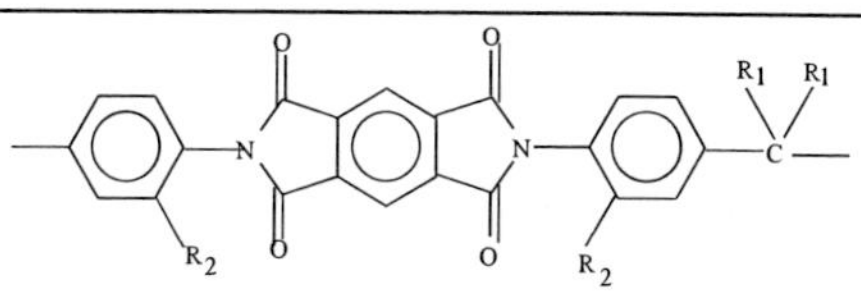	15.11	53.4
12a	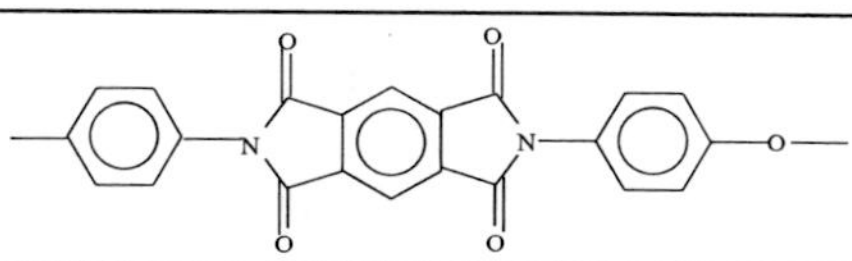	16.02	78.1
12b	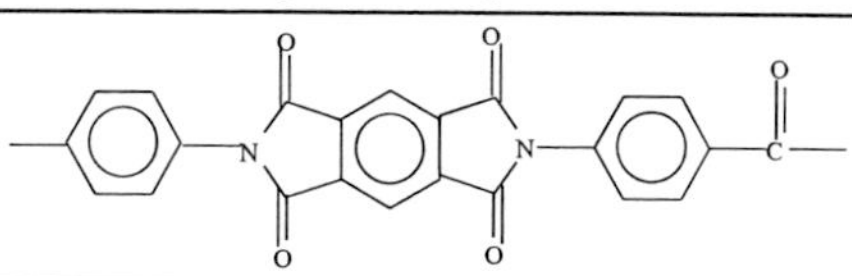	16.01	70.2
12c	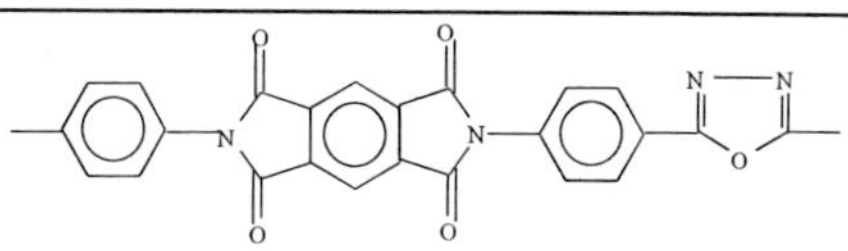	20.03	157.5
12d	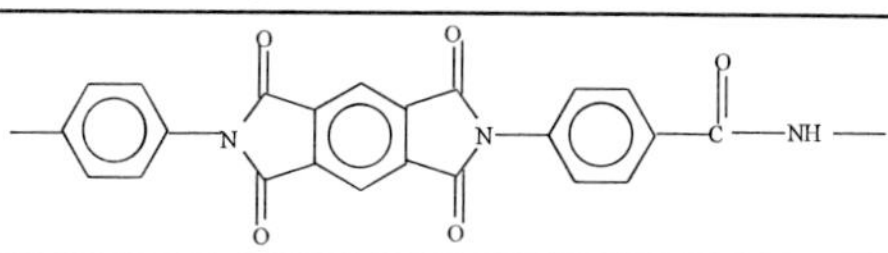	18.92	846
13	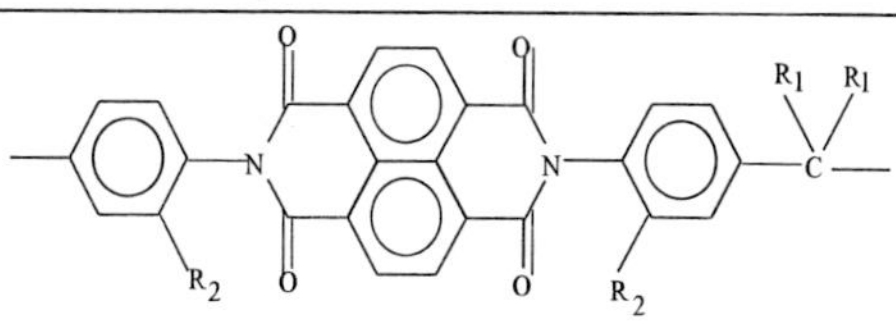	15.29	48.9
14	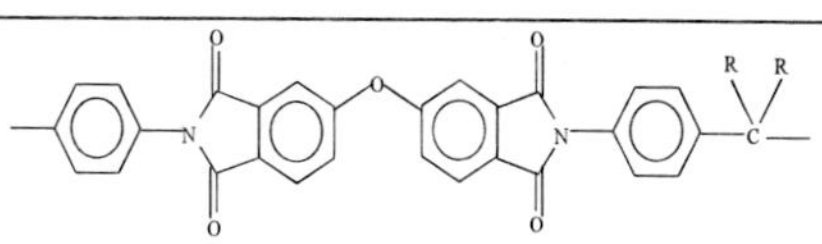	22.33	30.3

Table 1. (Continued)

No	Structural repeat unit	l_o, Å	A_{fr}, Å
15	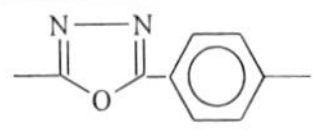	7.61	65.0
16	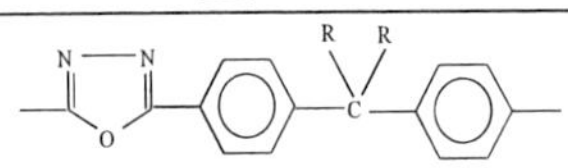	12.29	25.2
17	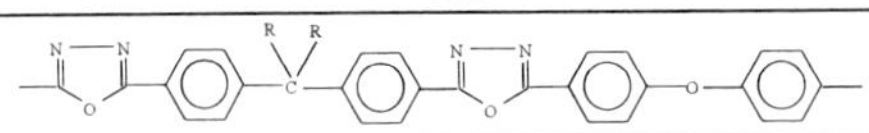	24.82	31.4
18	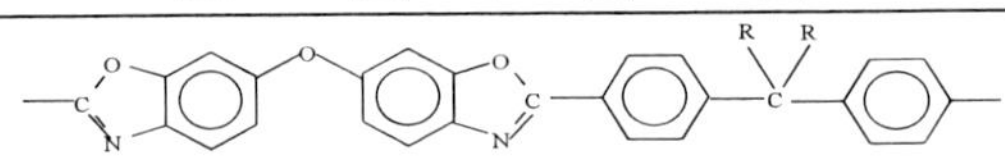	22.24	29.1
19	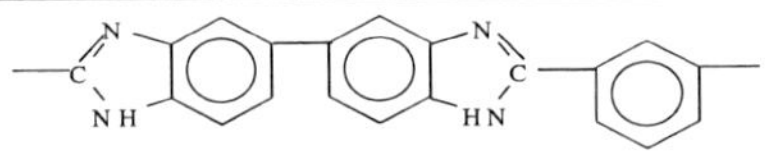	15.69	34.1
20	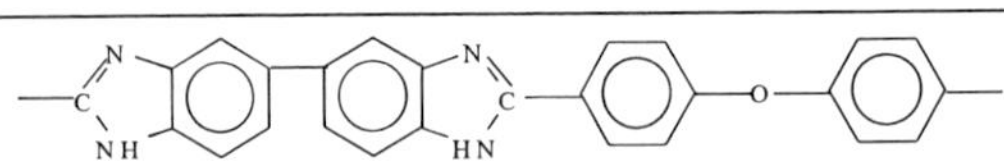	21.02	57.8
21	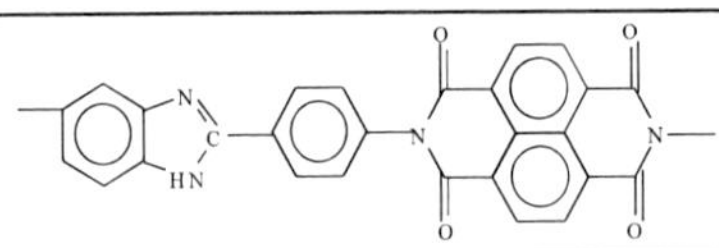	18.51	266
22	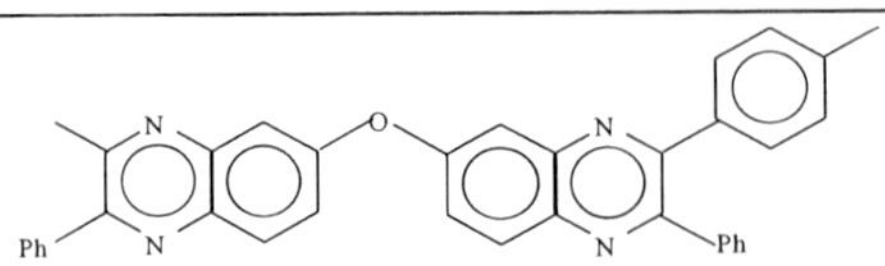	17.24	21.8
23	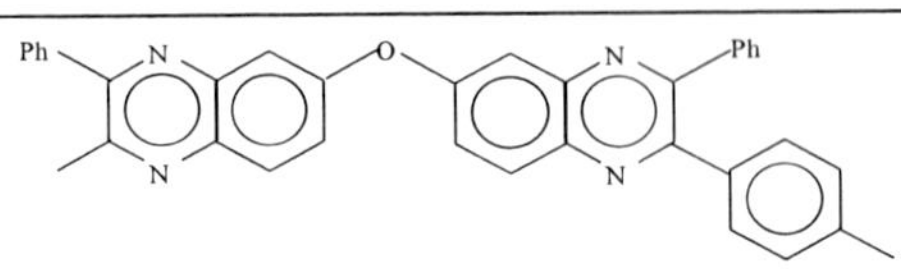	17.99	59.4
24	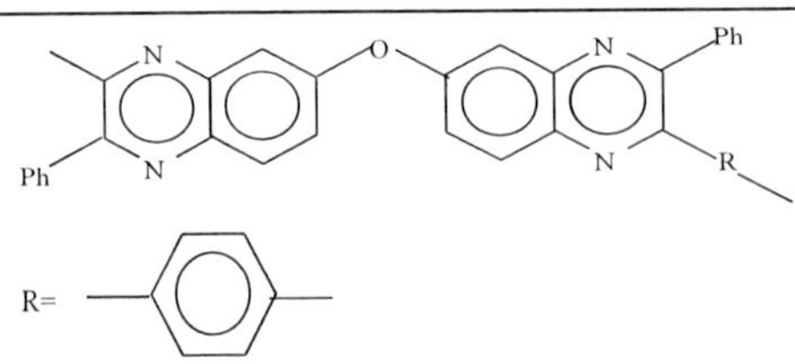	18.93	35.2
24a	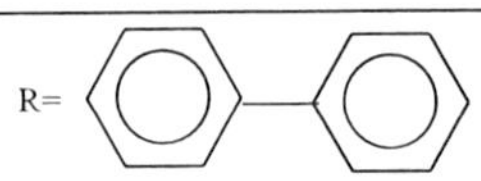	21.78	48.9
24b	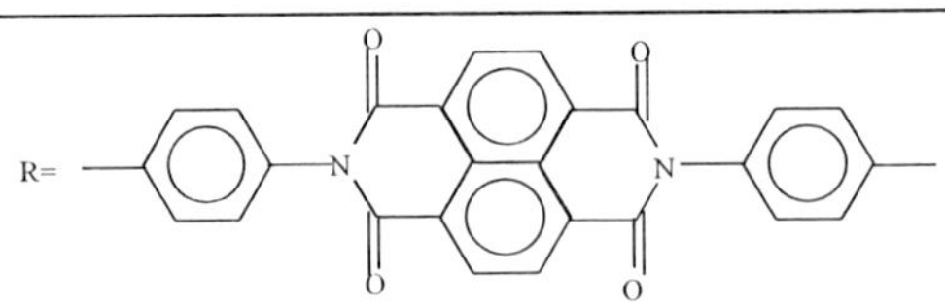	30.21	71.9
25	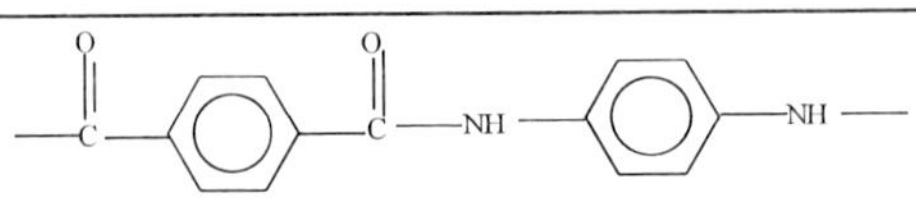	13.01	947

No	Structural repeat unit	l_o, Å	A_{fr}, Å
26	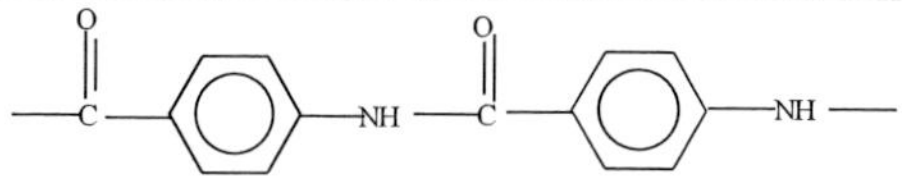	13.01	942
26a	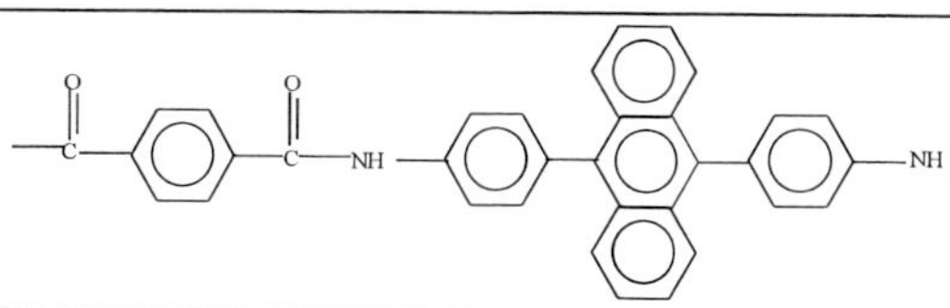	21.40	1565
27	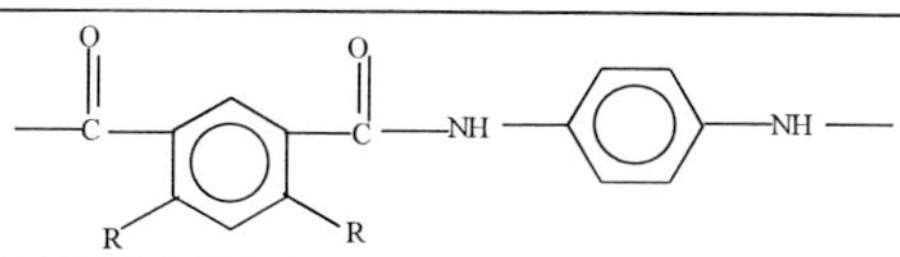	12.59	36.4
28	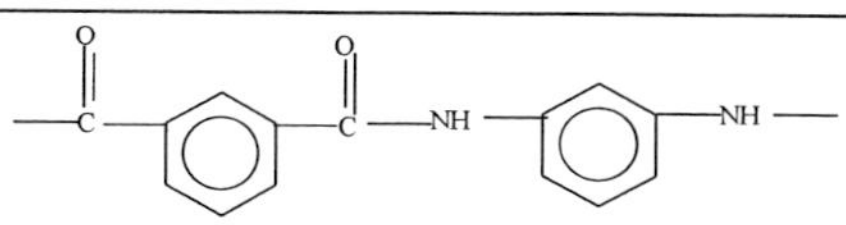	12.41	19.7
29	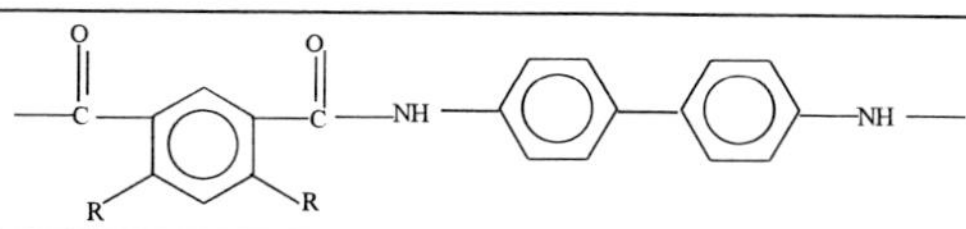	16.72	48.6
30	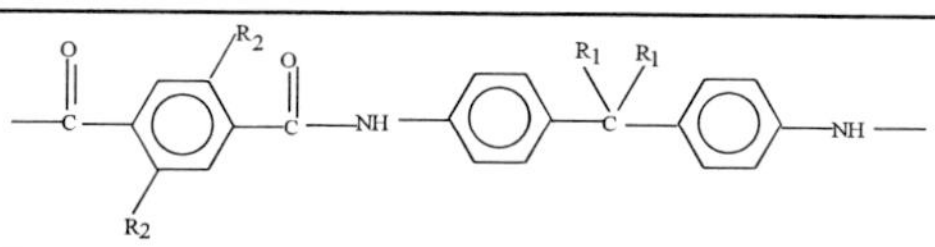	16.78	43.8
31	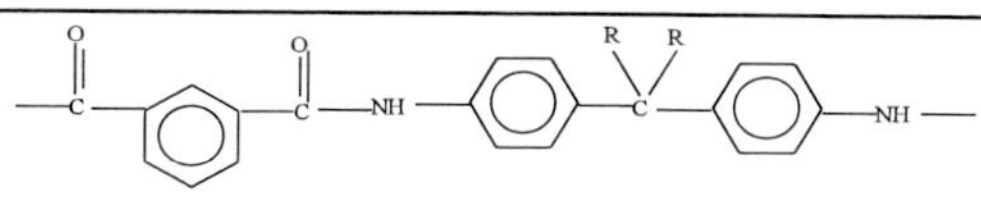	17.53	25.2
32	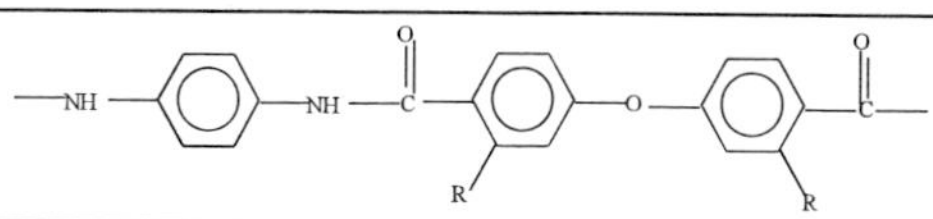	17.79	76.9
33	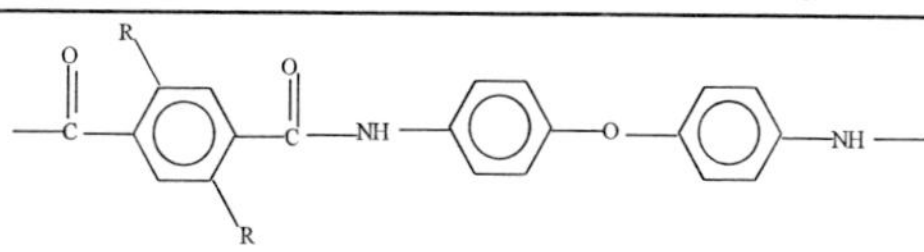	17.79	72.7
34	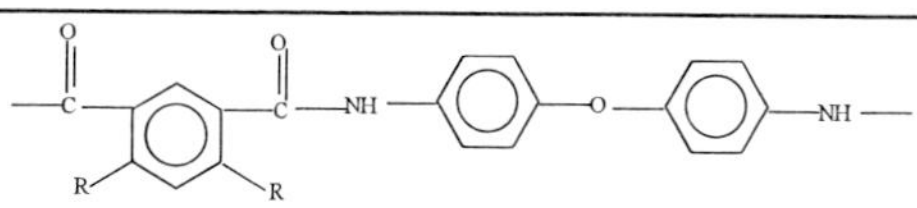	17.42	31.4
35	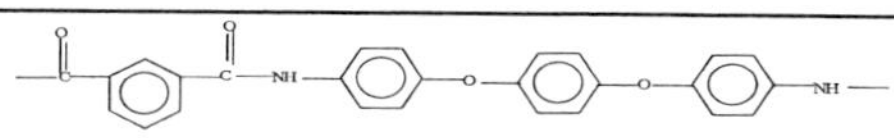	22.28	28.7
36	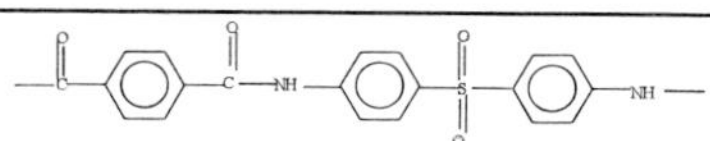	18.01	37.9
37	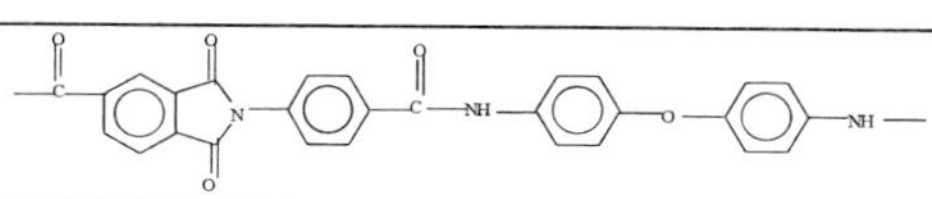	24.12	50.80
38	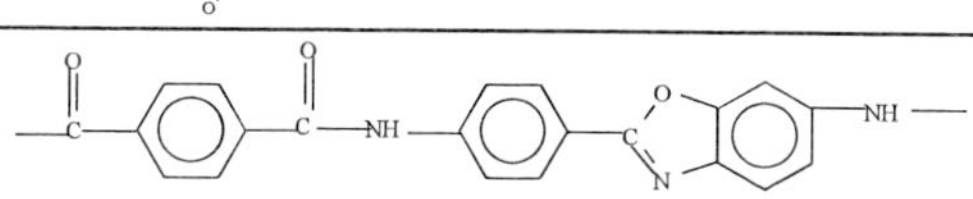	19.13	216.4

Table 1. (Continued)

No	Structural repeat unit	l_o, Å	A_{fr}, Å
39	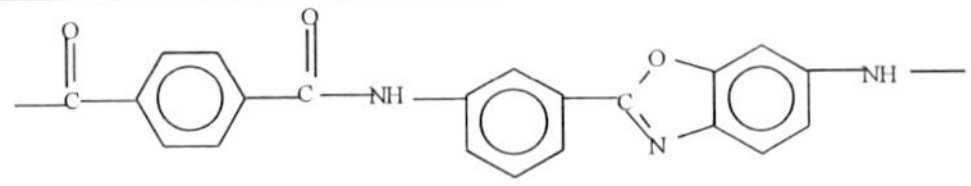	18.69	43.7
40	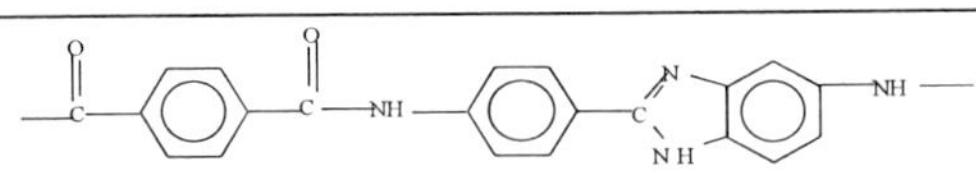	19.02	207 258
41	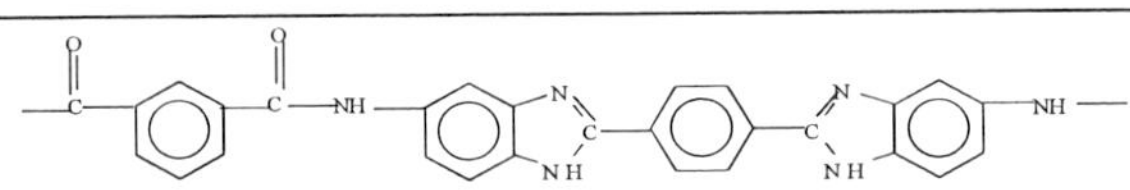	24.68	48.8
42	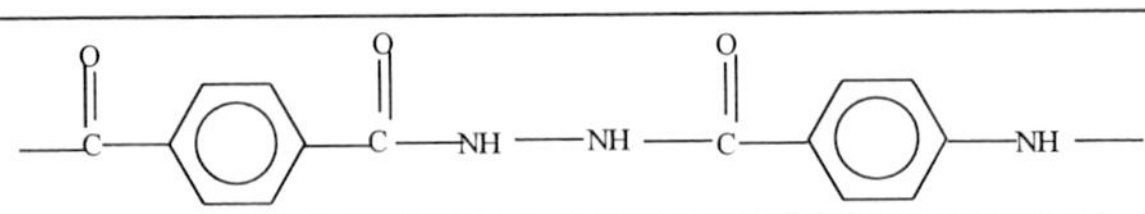	15.32	498
43	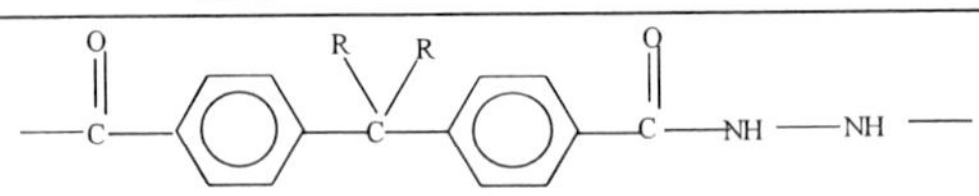	13.11	24.1
44	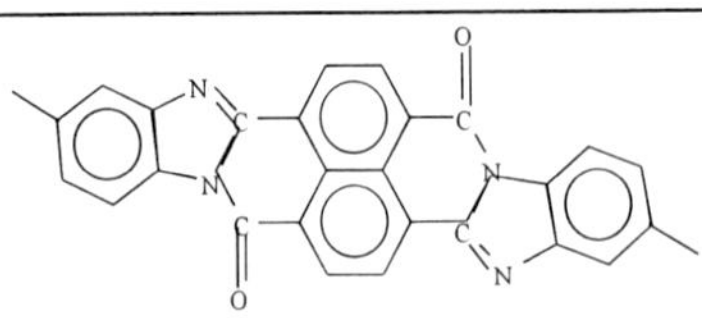	16.42	∞
45	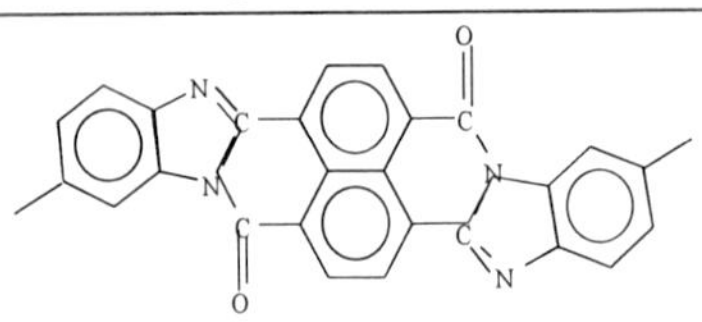	16.00	∞
46	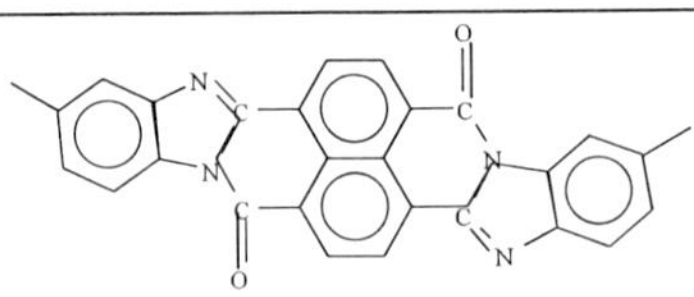	15.88	61.3
47	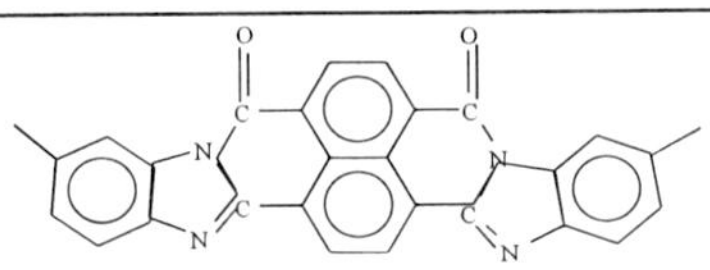	16.73	45.6
48	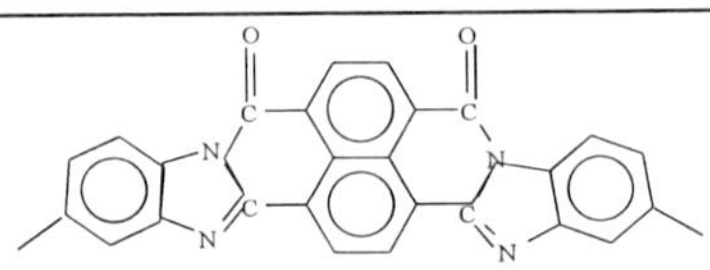	16.11	99.0
49	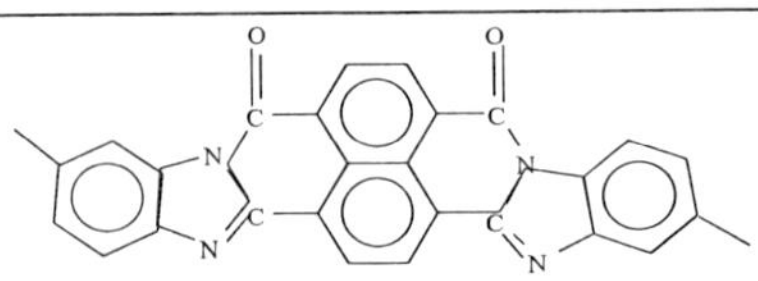	15.92	1125

No	Structural repeat unit	l_o, Å	A_{fr}, Å
50[b]	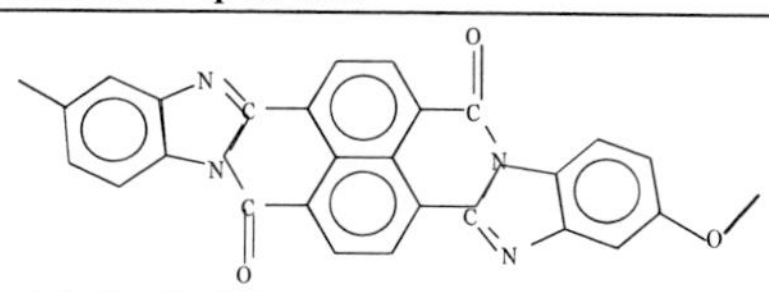	I 17.41 II 16.69 III 16.92	96.0 77.2 27.4
51	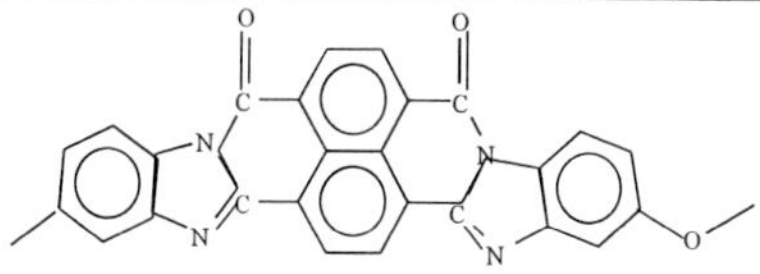	IV 16.71 V 17.09 VI 17.02	27.8 36.7 151.3
52[c]	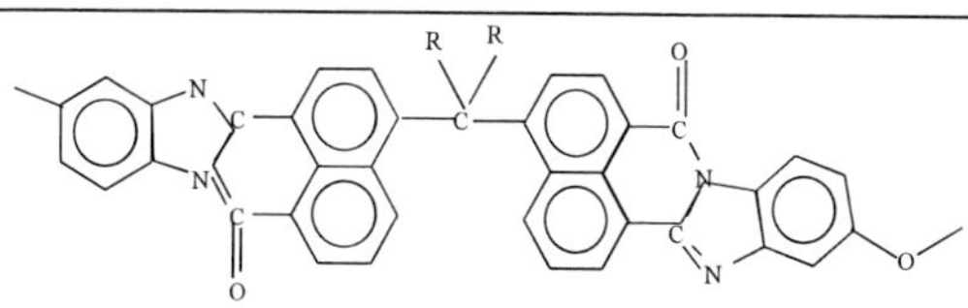	22.44 23.53 22.91	27.5 26.7 31.8

[a] A_{fr} calculated in Ref. 9.
[b] Isomers are analogous to those of polymers 44-49.
[c] Isomers are analogous to those of polymers 22-24.

Thus, the effect of the heterocycle and bridging group structure on the Kuhn segment can be observed in polyimides (polymers 12-14). Here we consider three cases. In polymer 12 the diimide fragment with one benzene ring (diimide of pyromellitic acid) forms a virtual bond containing two additional rings (phenylenes) in the *para* position (Figure 4a). The polymer 13 include naphtailenimid fragment instead dimide piromellitic acid (Figure 4b). In the third case a bridging atom is situated between fragments containing the imide rings so that the whole fragment is described by four virtual bonds (polymer 14, Figure 4c). Polymers 12, 13 and 14 demonstrate a decrease in the polymer rigidity depending on the structure of the imide fragment. One can observe the effect of bridging groups bonding the imide fragments on polymers 12,12a-d). The sp^3 hybridized carbon atom in the chain alters the direction of the chain by 68°, the oxygen atom by 56°, the oxadiazol cycle alters the direction of the chain by 40°, the amide group merely by 9°, and the chain rigidity A_{fr} increases from 53 to 846 Å.

The effect of benzene rings in *para* and *meta* positions in the chain is demonstrated by a number of examples for *para* and *meta* positions by polymers 25 and 27(one ring in *meta*-position) and 28 (two rings in *meta*-position) or 33 and 34 (one ring in *meta* positions). In all these cases, the introduction of a benzene ring in the *meta* position leads to a decrease in rigidity.

It is demonstrated by polyphenylquinoxalines that an increase in bond lengths, under otherwise equal conditions, results in a marked decrease in flexibility. With polymers 24, 24a, and 24b, the virtual bond length increases from 5.6 to 18 Å and the Kuhn segment increases from 35 to 71 Å.

There are polymers with strict alternation of bonds with allowed and forbidden rotation and small variations in the axis direction. Thus, the amide group causes little change in the chain direction (δ = 66°-57° =9°); and in polymers 25, 26, and 26a the rigidity is large and increases with increasing virtual bond length. If angles between virtual bonds adjacent to both ends of a virtual bond with forbidden rotation are equal (δ = 0°). Then, according to a definition by Tsvetkov, there is a conformation of the "crankshaft" type with equal angles that can in no case form a coil. For such polymers the Kuhn segment concept cannot be applied.

Polymers 44 and45 (isomers I and II) serve as examples. In such cases we denote the Kuhn segment as ∞. However, if a bond with *trans* conformation and forbidden rotation is exchanged for a *cis* bond, a rotation about which is also forbidden, the Kuhn segment assumes quite finite values (polymers 44-49; isomers III - V).

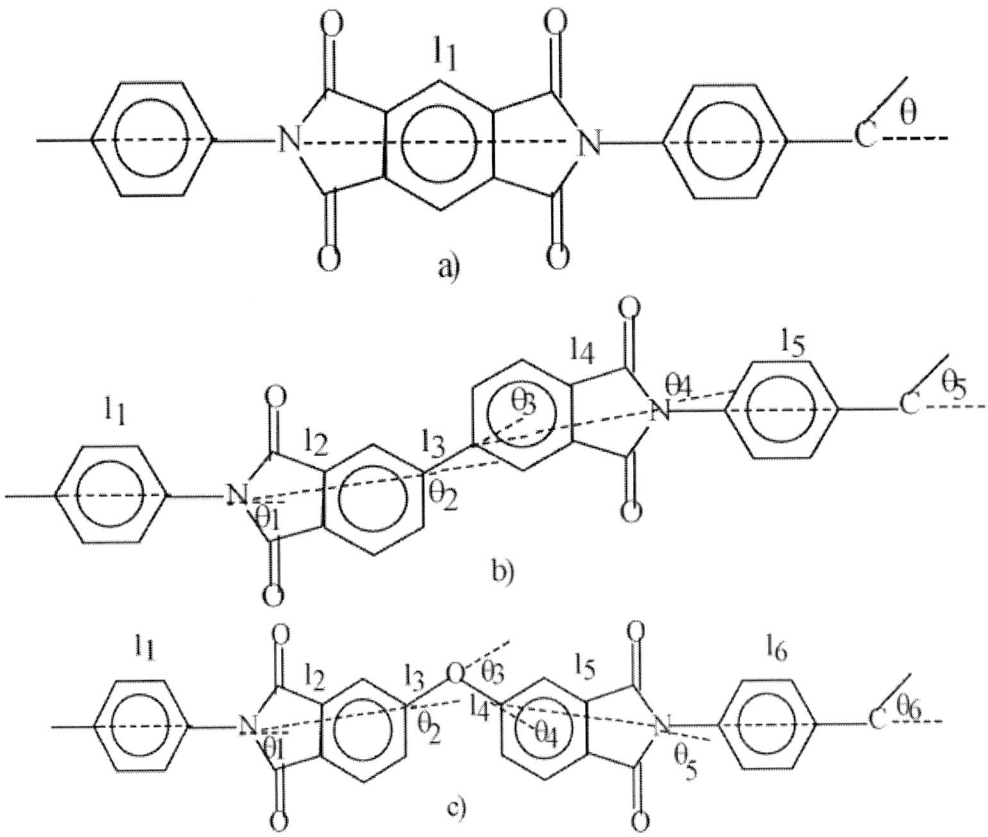

Figure 4. Construction of virtual bonds for three types of imide heterocycle structure.

For a number of polymers, linkages of heterocycles are possible so that three isomers can be obtained for polymer 22-24, six for 44-49, six for 50, 51 and three for polymer 52. It is interesting to note that conformational parameters for polymer 44-49 were computed by Monte Carlo method. The calculations were carried out for the repeat unit containing six isomers, whose contributions were assumed to be equal. If the Kuhn segment for this polymer is calculated using the formula $1/A_{sum}=\Sigma 1\ /n_i A_{fr}$, where n_i is the number of isomers, the segment Kuhn obtained is also 81.2 Å. For six isomers of polymers 50, 51 the Kuhn segment is 43.6 Å; and for three isomers of polymer 22 24.is 32.9 Å; and for three isomers of polymer 52 - 28.5Å. Thus, in the assuming of a free rotation, there is a possibility to calculate the Kuhn segment and other conformation parameters for any number of virtual bonds in a structural unit of a polymer.

II. Assement of Conformational Parameters from Hydrodynamic Data

Unperturbed dimensions of the macromolecular coil can be obtained by direct measurements of the angular dependence of light scattering, small-angle X-ray scattering, and flow birefrigerence. However, most simple, accessible methods for assessing conformational parameters are indirect—e.g., hydro- dynamic properties of dilute solutions: intrinsic viscosity [η], sedimentation constant s_0, and diffusion constant D_0. Uncertainties in determination of conformational parameters by these methods arise in the choice of the model of the macromolecular coil in solution since this is not always unambiguous. The crucial factor is the L/A ratio, i.e., the molecular mass range and chain rigidity. The shape of an isolated macromolecule changes with increasing molecular mass from rod to a coil. The behavior of the macromolecule in solution can be described by that of a coil generated by a persistent chain, of a semipermeable (partially draining) Gaussian coil (SGC), and eventually, of an impermeable (nondraining) Gaussian coil (IGC) depending on rigidity and molecular weight.

According to Flory [15, 16], the behavior of flexible macromolecules in ideal solutions can be simulated by the IGC. With the following ratio between the characteristic viscosity in a θ solvent and the Kuhn segment:

$$[\eta]=\Phi_0\frac{\left(\langle R^2\rangle\right)^{3/2}}{M} \quad or \quad \frac{[\eta]M_L^{3/2}}{M^{1/2}}=\Phi_0 A^{3/2} \tag{5}$$

where Φ_0 is a hydrodynamic coefficient largely dependent on chain geometry and polydispersity. For linear Gaussian chains in θ solvent, Φ_0 = 2.87 • 10^{21}. In the case of kinetic friction, the Kuhn segment is calculated from the kinetic friction coefficient

$$f_0=\frac{M(1-\nu\rho_0)}{N_A s_0}=\frac{kT}{D}$$

using the equation

$$f_0=P_0\eta_0\left(\langle R^2\rangle\right)^{1/3} \tag{6}$$

where η_0 is the solvent viscosity and P_0 is coefficient analogous to Φ_0. For IGC in an ideal solvent, P_0 is equal to 5.11; υ is the partial specific volume of the dissolved polymer, and ρ_0 is the solvent density.

Conformational parameters of a macromolecule calculated using eqs. (5) and (6) in a η_0 solvent are inherent dimensions of the chain due only its flexibility. Usually it is difficult to find an ideal solvent, where the second virial coefficient A_2 is zero. For aromatic polymers, only a very limited number of solvents are available. In good solvents long-range interactions appear and as a result of this the mean- square end-to-end distance $\langle R^2 \rangle$ increases as compared to $\langle R_o^2 \rangle$ in a ϑ solvent. Determination of unperturbed parameters of the macromolecular coil in nonideal solvents is based on measurements of [η] or s_0 and D_0 of polymer fractions in good solvents and extrapolation of the data obtained to low molecular weight. For IGC in a nonideal solution, the Fixman-Stockmayer extrapolation [17] from viscosimetric data and that of Cowie [18] from sedimentation data are mostly used:

$$\frac{[\eta]M_L}{(L)^{1/2}} = \Phi_\infty A^{3/2} + 0.51\Phi_\infty BM_L^2 L^{1/2} \tag{7}$$

$$\frac{L^{1/2}M_L}{[s_o]} = N_A P_\infty A^{1/2} + 0.201 N_A BA^{-1} M_L^2 L^{1/2} \tag{8}$$

where $M_L = M_o/l_o$ and B is the thermodynamic interaction parameter in the polymer-solvent system, $B = \beta/M_o^2$, where $\beta\ 0$ is the excluded volume of the polymer repeat unit.

If the IGC model cannot be applied (as for rigid and/or insufficiently long chains where $L/A < 20$), the "chain of finite length" (CFL) equations, describing the approximate dependence of coil dimensions on molecular weight, are usually used. In these equations, changes in $\Phi(M)$ for viscosimetric data and in $P(M)$ for sedimentation and diffusion data are taken into account.

In this work, experimental data obtained at our laboratory as well as those reported in the literature were used. These data were processed using two models for the behavior of the macromolecular coil in solution—the Hearst-Stockmayer model (I) [19, 20]:

$$\frac{L}{[\eta]M_L} = \frac{L^{1/2}}{\Phi_o A^{3/2}} + 2.15\left(ln\frac{A}{d} - \gamma\right) \tag{9}$$

$$\frac{s_o}{M_L} = \frac{1.843}{3\pi N_A}\left[\frac{L}{A}\right]^{1/2} + \frac{1}{3\pi N_A}\left(ln\frac{A}{d} - \gamma\right) \tag{10}$$

and the Bohdanecky model (II) [21-23]:

$$\left[\frac{M^2}{[\eta]}\right]^{1/3} = A_\eta + B_\eta M^{1/2} \tag{11}$$

Where

$$A_\eta = A_o M_L \Phi_o^{-1/3}$$

$$B_\eta = B_o \Phi_o^{-1/3} \left[\frac{\langle R^2 \rangle}{M} \right]^{-1/2} \qquad (12)$$

$A_0 = 0.46 - 0.53 \log_{10} d_r$, $B_0 = 1. - 0.0367 \log_{10} d_r$, and $d_r = d/A$—the reduced hydrodynamic diameter of the chain, which is found from the ratio:

$$\frac{d_r^2}{A_o} = \frac{4\Phi_o \nu B^4}{1.215\pi A}$$

The approach developed by Bohdanecky is based on a theoretical analysis carried out by Yamakawa and Fujii [24, 25]. From approximate calculations after Bohdanecky, one can also evaluate the M_L value ($M_L = M_o/l_o$), which is a convenient criterion for the adequacy of calculation since M_o is easily calculated from the structural formula of the repeat unit. The application of this criterion is limited by values of $d_r < 0.1$ at which $log_{10}(d_r^2 / A_o = 0173 + 2.158 log_{10} d_r$. Unfortunately, eqs. (9)-(11) do not take account of the excluded volume. Therefore, calculations using hydrodynamic data measured in various solvents can lead somewhat different results. In choosing a model to compute the Kuhn segment from experimental data, its value for free rotation serves as a criterion since A_{exp} cannot be smaller than A_{fr}. This is illustrated in Figures 5 and 6.

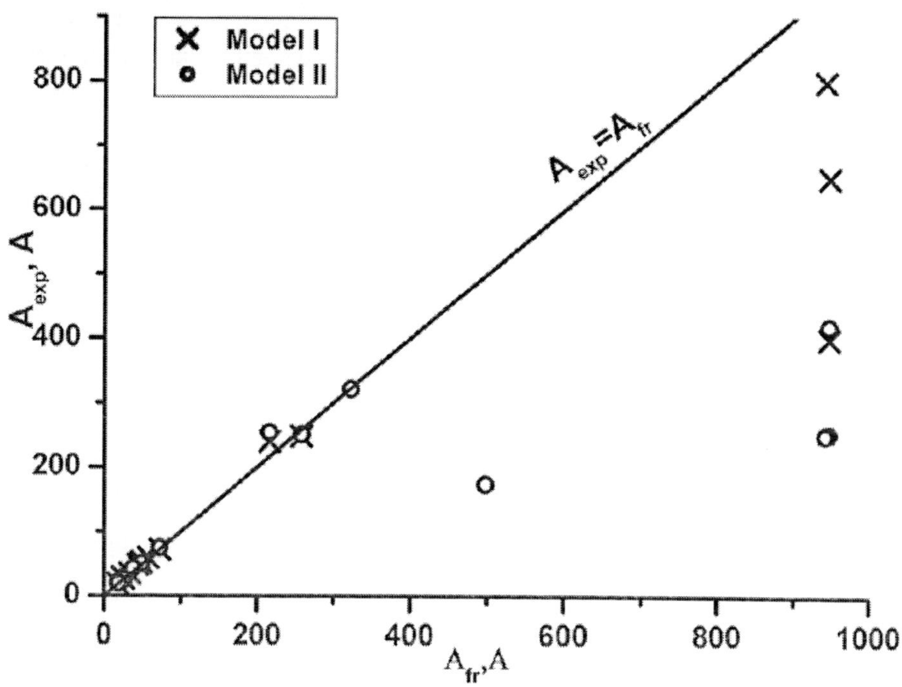

Figure 5. Dependence of A_{exp} on A_{fr}in the experimentally studied range of rigidity (**x**: model I, CFL; O: model II, CFL).

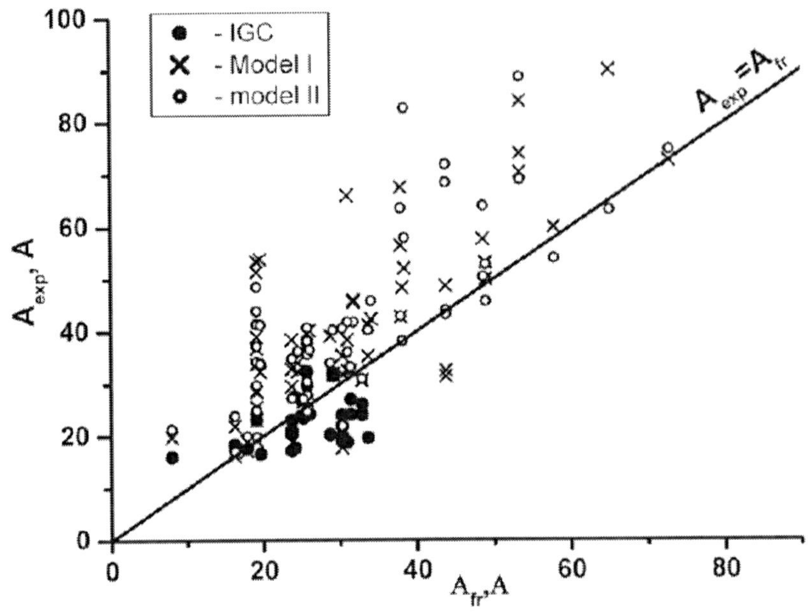

Figure 6. Detailed Dependence of A_{exp}on A_{fr}in initial rigidity range up to 100 Å (**x**: model I, CFL; O: model II, CFL; •IGC).

In Table 2, experimental data for a number of aromatic polymers studied by light scattering, sedimentation, diffusion, and viscometry are presented. For these polymers, experimental data for molecular masses, viscosities, and sedimentation coefficients reported in the literature were processed using both models. The code numbers in Table II correspond to the numbers in Table I. It should be noted at once that among the polymers under study, there are those for which the Mark-Houwink constant α is close to or equal to 0.5. For such polymers the choice of a model is hardly significant and they are well described by the IGC model. For all of them, the Kuhn segment values obtained from viscosities and sedimentation coefficients are close to A_{fr}, i.e., $\sigma = 1$. However, the computations were also carried out using both models for chains of finite length. In those cases for which the partial specific volume was not reported, $\boldsymbol{M}_{L,exp}$ and d values for model II were not calculated.

In Figure 5 there are compared A_{fr} and A_{exp} calculated from models I and II for the behavior of the macromolecular coil in solution. First of all, attention must be paid to a pronounced correlation between A_{exp} and A_{fr} over a large range of the rigidity parameter (A varies from 8 to 500 Å), A_{exp} values being slightly overestimated. Hence, it can be ascertained that the models employed for the behavior of the molecular coil in solution give a sufficiently accurate description of deviations from IGC conformation due to finite chain length ($L/A < 20$).

Table 2. Experimental value of polyheteroarilene conformational parameters*

No	Side Substituent	Method Measurement[a] M_w	Solvent[b]	Molecular Weight Range[c] M_w 10^3	A_{fr}, Å	M_L^t	IGC A_{exp}, Å	Chain of Finite Length: Model I A_{exp}, Å	Model I d, Å	Model II A_{exp}, Å	Model II M_L^e	Model II d, Å	Range σ	Refs.
1	$R=CH_3$	GPC LS O	TCC	26.5-464.0 F	23.7	24.9	17.2- 21.0	29.5	4.7	27.2	-	-	0.85- 1.12	26, 27
1	$R=C_6H_6$	LS GPC	TL Cl-B	43.8-1445.5 F	23.7	50.1	20.2- 23.0	32.8 38.4	7.2 8.5	34.8 -	- -	- -	0.92- 1.27	27, 28
2	$R=CH_3$	S D	DMFA	73.5-476.0 F	38.1	23.3	-	48.4	3.6	38.2	-	-	1.00- 1.13	29
3	$R=CH_3$	LS O	DMSO DMFA THF	18.9-91.9 F	16.3	22.3	18.2 18.3	16.2 17.0 21.9	3.8 4.0 5.0	- 17.1 23.8	- - -	- - -	0.86- 1.21	30
4	$R=CH_3$	LS O, D S	MC THF	4.3-800.0 F	17.9	21.0	17.6	17.2 19.1	3.8 3.9	17.9 -	- -	- -	0.98- 1.03	31-35
5		LS O	TFAA	26.0-118.0	8.0	17.0	16.0	19.8	3.3	21.4	-	-	1.41- 1.63	1, 36,37
6	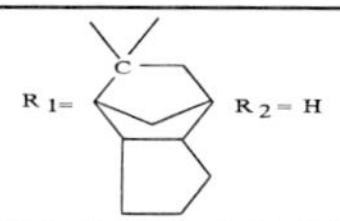	LS	TCE	8.6-246.0 F	25.7	26.4	32.3	28.3	5.0	27.8	-	-	1.04- 1.12	38
6	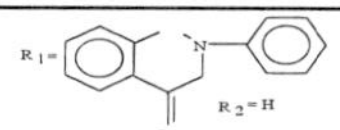	LS	TCE	8.3-67.3 F	25.7	30.1	36.0	39.7	6.5	35.6	28.7	10.6	1.18- 1.24	39,40
6	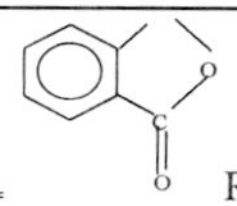 $R_1=$ $R_2=H$	LS	THF	7.0-93.0 F	25.7	25.6	-	36.4	6.4	30.4	18.9	7.7	1.09- 1.19	41-44
6	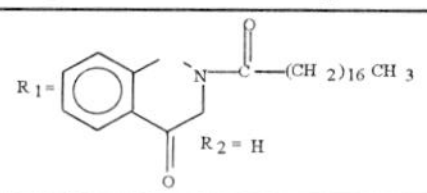	LS	THF	11.5-62.5 F	25.7	40.7	-	87.0	12.2	54.0	40.7	7.3	1.45- 1.84	40,45
6	$R_1=R_2=CH_3$	LS	TCE	9.2-110.0 F	25.7	22.1	29.7	25.2	6.0	24.7	-	-	0.98- 1.08	46
6	$R_1=CH_3$, $R_2=Cl$	LS	TCE	27.5-620.0 F	25.7	24.3	-	36.6	8.5	38.2	-	-	1.19- 1.22	47

Table 2. (Continued)

No	Side Substituent	Method Measurement[a] M_w	Solvent[b]	Molecular Weight Range[c] M_w 10^3	A_{fr}, Å	M_L^t	IGC A_{exp}, Å	Chain of Finite Length					Range σ	Refs.
7	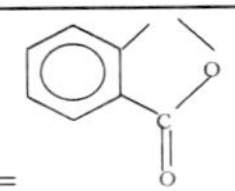 R_1=	LS	TCE THF	5.6-82.0 F	33.7	24.0	19.6	35.3 41.3	6.9 9.5	- 40.3	- 22.7	- 6.1	0.76- 1.11	44, 48
8	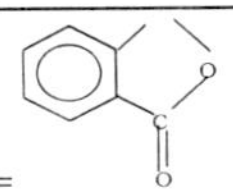 R=	LS	DCE THF	12.0-30.0 F	24.5	25.6	-	32.4 34.0	10.3 12.0	- 36.2	- -	- -	1.15- 1.21	49
9	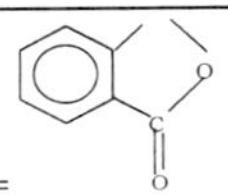 R=	LS	DCE THF	1.6-82.0 F	31.7	24.0	-	46.0 45.6	9.0 7.0	- 41.8	- 25.2	- 5.8	1.15- 1.20	41
10	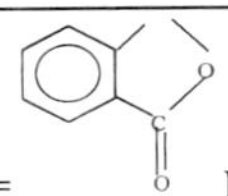 R_1=R_2= R_3=H	LS	DCE TCE	12.6-90.0 F	19.2	29.1	23.0	23.0 28.4	5.5 4.9	- 24.9	- 25.7	- 6.9	1.09- 1.22	44, 50
10	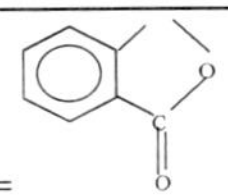 R_1=R_2= R_3=CH(CH_3)2	LS	THF TCE	11.0-58.0 F	19.2	32.6	-	36.8 39.0	8.8 9.3	37.2 41.4	- -	- -	1.38- 1.47	51
10	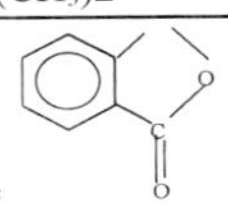 R_1=R_2= R_3=CH_2-CH_2-CH_3	LS	THF TCE	7.6-51.7 F	19.2	32.6	-	28.6 33.8	6.8 7.9	29.7 34.2	- -	- -	1.22- 1.34	51
10	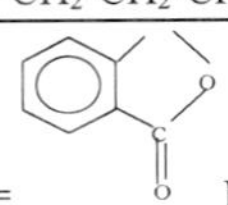 R_1= R_2=CH_3 R_3=$(CH_2)_3$-CH_3	LS	THF TCE		19.2	35.1	-	51.4 53.5	11.9 12.4	43.9 48.5	- -	- -	1.52- 1.67	51

No	Side Substituent	Method Measu rement[a] M_w	Solvent[b]	Molecular Weight Range[c] M_w 10^3	A_{fr}, Å	M_L^t	IGC A_{exp}, Å	Chain of Finite Length					Range σ	Refs.
10	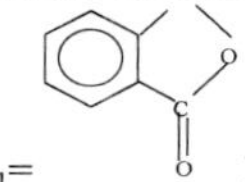 R_1= R_2=R_3=CH_3	LS	THF TCE		19.2	27.6	-	19.1 25.5	4.5 5.9	19.8 24.9	- -	- -	1.00- 1.15	51
11		LS, O GPC	THF	26.0- 290.0, F	19.7	27.8	16.6	32.3	4.8	33.8	-	-	1.28- 1.31	52-54
12	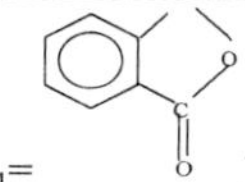 R_1= R_2=H	S	DMFA	8.0- 51.6 F	53.4	33.0	-	[η][d] 70.4 [s][d] 74.1	10.6 21.7	69.0 -	36.3 -	7.1 -	1.14- 1.18	3
12	R_1=CH_3, R_2=OH	S	DMFA	5.8-45.0 F	53.4	27.0	-	84.1	12.1	88.7	37.5	7.0	1.25- 1.34	3
13	R_1=H, R_2=Cl	D	N-MP H_2SO_4 96%	21.5- 146.6 F	48.9	30.3	-	53.1 53.2	11.5 11.4	50.0 52.8	- -	- -	1.01- 1.04	55
13	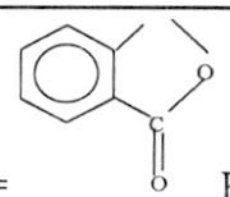 R_1= R_2=H	S	N-MP	10.0- 90.1 F	48.9	35.8	-	49.9	12.0	45.7	37.6	7.5	0.97- 1.01	56-58
14	R= 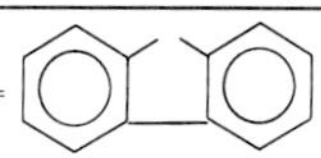	S	DMFA	20.0- 1220.0 F	30.3	27.9	[η] 21.9 [s] 24.0	31.9 35.4	4.5 9.6	40.6 -	44.0 -	8.3 -	0.85- 1.16	3
14	R_1=CH_3, R_2=C_6H_6	D S	DMFA	7.0- 280.0 F	30.3	25.2	[η] 20.1 [s] 18.9	17.7 20.4	3.7 -	21.9	27.1 -	6.7 -	0.76- 0.84	59
15		LS D	H_2SO_4 96%	61.0- 208.0	65.0	18.9	-	90.0	-	63.2	-	-	0.98- 1.18	60
16	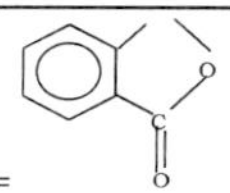 R=	S	DMAA	7.0- 580.0 F	25.2	27.9	[η] 23.4 [s] 23.5	26.6 27.2	4.8 8.8	27.1 -	27.4 -	6.8 -	0/96- 1.04	1, 61

Table 2. (Continued)

No	Side Substituent	Method Measu rement[a] M_w	Solvent[b]	Molecular Weight Range[c] M_w 10^3	A_{fr}, Å	M_L^t	IGC A_{exp}, Å	Chain of Finite Length					Range σ	Refs.
17	R= 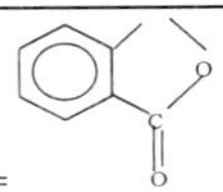	S	N-MP	9.5- 650.0 F	31.4	23.7	[η] 24.1 [s] 27.0	32.5 32.6	5.1 10.4	33.3 -	23.0 -	4.9 -	0.88- 1.03	1
18	R= 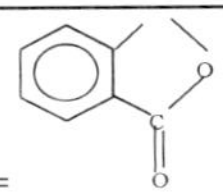	LS S	N-MP	3.5- 140.0 F	29.1	24.2	[η] 31.6 [s] 32.0	32.0 31.7	7.6 7.6	40.3 -	20.1 -	- -	1.04- 1.18	62
19		LS -O-	DMMA+ LiCl	11.0- 42.0 F	34.1	19.6	12.9	42.4	8.8	45.8	-	-	0.62- 1.16	63, 64
20		LS	DMMA	2.9- 23.0	57.8	19.0	-	60.0	8.3	54.0	-	-	0.97- 1.02	65
21		LS D	H_2SO_4 98%	6.7- 250.0	208 323	24.6	-	-	-	321.0	-	-	0.99	66, 67
22 - 24[e]	R= 	S	N-MP	10.0- 370.0 F	32.9	28.4	[η] 24.0 [s] 26.0	30.6 30.8	5.8 11.5	31.0	25.0	6.0	0.73- 0.94	68-70
25		LS GPC	H_2SO_4 98% $ClSO_3H$ +LiCl	1.6-123.0	947	18.6	185	400- 650	4.8	253.0 420	-	6.0	0.52- 0.83	71-80
26		LS GPC	H_2SO_4 98% DMAA +LiCl	4.5-72.3	942	18.6	-	800.0	-	250.0	-	8.2	0.52- 0.92	71-73, 75, 81-84
27	R=COOH	LS O	N-MP DMAA +LiCl	4.5-280	38.4	26.1	-	52.1	11.6	57.9 82.8	- 29.9	- 5.1	1.16- 1.46	85, 87
28		S D	DMAA +LiCl	4.3- 305.0 F	19.7	19.2	-	53.8	10.1	41.3	21.8	4.5	1.45- 1.65	87-89

No	Side Substituent	Method Measurement[a] M_w	Solvent[b]	Molecular Weight Range[c] M_w 10^3	A_{fr}, Å	M_L^t	IGC A_{exp}, Å	Chain of Finite Length					Range σ	Refs.
29	R=COOH	LS	DMAA DMAA+DO	10.0- 60.0	48.6	24.2	35.1	57.7	12.8	65.0 50.0	- -	- -	0.84- 1.17	90
30	$R_1=C_6H_5$, $R_2=H$	D	H_2SO_4 98% DMFA	85.0- 140.0 F	43.8	28.9	-	48.7	10	44.0 43.2	-	-	1.00- 1.05	91
30	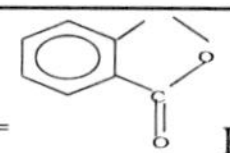 R_1= R_2=COOCH 3	S	DMFA N-MP	5.2-65.3 F	43.8	33.8	-	31.4 32.5	7.5 7.8	68.5 71.9	35.8 37.6	6.5 6.7	0.85- 1.28	92
31	R=H	S	DMAA	1.4-100.0 F	25.2	18.7	24.2	40.2	11.2	36.5	39.3	3.4	0.98- 1.24	93
32	R=COOH	LS O	DMFA	8.0-40.0	76.9	23.3	16.0	25.3	5.7	29.4	-	-	0.46- 0.62	90, 94
33	R=COOH	LS O, S	DMAA +LiCl	22.5-266.0	72.7	23.5	14/0	75.6	13.9	74.7	37.9	5.9	0.44- 1.02	95
34	R=COOH	LS S	DMAA +LiCl DMFA +LiCl	6.6-350.0	31.0	24.0	18.8	32.5 [η] 38.4 [s] 66.0	5.6 10.6 18.1	36.0 41.8 -	- 29.0 -	- 6.6 -	0.78- 1.46	85, 90, 95, 96
35	R=COOH	LS	DMAA	90-210	28.7	22.9	20.2	39.2	7.4	34.0	-	-	0.84- 1.17	90, 95
36		LS S	DMAA DMAA +LiCl H_2SO_4 98%	25.0-220.0	37.9	20.8	-	42.8 56.5 67.6	7.2 11.4 13.5	63.6 42.9 42.9	- 22.7 -	- 4.9 -	1.06- 1.34	97- 101
37		S D	DMFA	31.0-181.0	50.8	19.7	-	[η] 50.1 [s] 56.2	12.0 19.3	52.7 -	59.5 -	3.1 -	0.99- 1.05	102
38		D	H_2SO_4 98%	0.8-41.5	216.4	18.6	-	240.0	8.6	254.0	-	12.3	1.05- 1.08	103, 104
39		D	H_2SO_4 98%	1.0-42.0	43.7	19.0	-	57.3	13.5	64.6	-	15.8	1.14- 1.22	103, 105

Table 2. (Continued)

No	Side Substituent	Method Measu rement[a] M_w	Solvent[b]	Molecular Weight Range[c] M_w 10^3	A_{fr}, Å	M_L^t	IGC A_{exp}, Å	Chain of Finite Length					Range σ	Refs.
40		S D	H_2SO_4 98%	0.9-61.0 F	207.0-258.0	18.6	-	248.0	4.2	250.0	-	-		106-109
41		S D	DMAA +LiCl	6.6-190.0 F	48.8	19.0	-	85.0	19.1	79.7	-	-	1.28-1.32	110, 111
42		LS S D	DMSO DMAA TFAA	26.0-127.0	498	18.4	-	-	-	173.8	-	7.3	0.59	88, 106, 112-117
43	R= 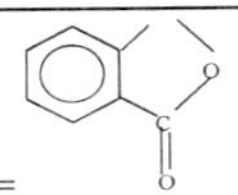	S	DMAA	10.0-150.0 F	24.1	28.3	[η] 17.7 [s] 27.5	30.6 56.8	6.1 16.5	-	-	-	0.86-1.50	1, 118
44-49[e]		D	H_2SO_4 98%	2.7-34.0	81.2	25.8	-	95.8	16.0	105.5	-	13.0		119-121
50 51[e]		D	H_2SO_4 98%	2.7-84	46.3	25.1	-	45.0	11.9	49.8	-	-		122-124
52[e]	R= 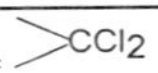	D	H_2SO_4 98%	33.0-117.0	28.5	29.2	-	29.2	8.9	32.8	-	-		119, 125

*The polymers are numbered as in Table 1

[a] - GPC: gel permeation chromatography; LS: light scattering; S: sedimentation; O: osmometry; D: diffusion.

[b] – TCC: carbon tetrachloride; TL: toluene; CB: chlorobenzene; DMFA: dimethylformamide; DMSO: dimthylsulfoxide; THF: tetrahydrofurane; TCE: tetrochlorethane; DCE: dichlorethane; DO: dioxane; TFAA: triftoroacetic acid; TFA: trifluoroacetone; MC: methylchloride; N-MP: N-methylpyrrolidone.

[c] – F: polymer fraction: notation-polymer specimens.

[d] – [η]: conformational parameters found from viscometry; [s_o]: from sedimentation.

[e] – A_{fr} corresponds to equal contributions of all isomers (Table I). Quantitative analysis for isomers in specimens has not been carried out.

A more detailed examination of the initial part of this dependence shows that in the range of molecular masses characteristic of polycondensation polymers (Table 2) the IGC model is adequate for macro molecular behavior in solution only for the polymers whose rigidity does not exceed 20 Å. For more rigid macromolecules, processing of experimental data following the IGC model results in $A_{exp} < A_{fr}$ which is unlikely. At the same time both CFL models give roughly coinciding results (Figures 5 and 6), differing little from the rigidity parameter with free rotation. The only exception, as could be expected, are several polymers with side-chain substituents. Additionally, experimental values of the statistical segment for most rigid chain poly (*para*-amides) and polyamidohydrazides appeared to be much smaller than those calculated with free rotation. These two groups will be discussed in more detail.

Since in the absence of bulk side substitutes the experimental value of a Kuhn segment A_{exp} is equal approximately to its theoretical value $\cong A_{fr}$, which has been calculated with the assumption of free rotation, so to comparative analysis of polymers one can use henceforth its value calculated in the framework of free rotation as the measure for the rate of conformation rigidity. This is especially important as an experimental determination of the Kuhn segment is connected with significant difficulties such as solvent selection (preferable a θ-solvent) and a fractionating of a polymer onto separate fractions with approximately equal molecular mass.

III. Effect of Side Groups

When The Kuhn segment is calculated assuming free rotation, the side substituents are not taken into account. In the calculations based on the assumption of free rotation, Monte Carlo method has been used according to which the random numbers for chain generation are evenly distributed in the [0, 2π] range. Taking into account the hindrance of rotation, random numbers cannot be considered as evenly distributed, because in the presence of considered substituent, only those conformations are realized in which the distances between non-bonded atoms in structural units do not exceed the sum of their van der Waals radii. In this case, the probability of rotation through angle φ around a virtual bond depends on the form of potential energy function $U(\varphi)$. To take this into account in construction of chain a distribution function was introduced over the internal rotation angles $F(\varphi)$ with weights proportional to Boltzmann factor $exp(-U(\varphi)/kT)$ [126], and random number were generated in conformity with this function. If the structural units are closely spaced, some angles of rotation are never encountered, and some others occur in rare cases and, with is most probable, in absence of steric interactions between structural units.

A number of polyarylates with various sizes and positions of side substituents, for witch experimental values for Kuhn segment are known, have chosen. Dilute solutions of these polymers have been studied previously by light scattering [127], and the results can be used to compare the calculated A_{hin} values with those found experimentally. Table 3 contains the structural units and the Kuhn segment values for the polyarylates studied, found experimentally, and calculated with allowance for hindrance. The hindrance parameters σ are also included. To consider the difference between the sizes of the side groups, we introduced the value M_L/M_{Lo}, where M_{Lo} and M_L are the ratios of the molecular weight of a polyarylate unit to its contour length in the absence of the substituent and with allowance for the molecular weight of the substituent, respectively. It can be seen from Table 3 that increasing the volume, and, hence, the weight of the substituent tends to enhance hindrance to the rotation. However, this approach does not take into account the position of the substituent in the polymer chain. The polyarylates presented in Table 3 can be divided into two groups depending on whether substituents are located in the phenyl ring of the backbone (1) or at the bridging C atom (2).

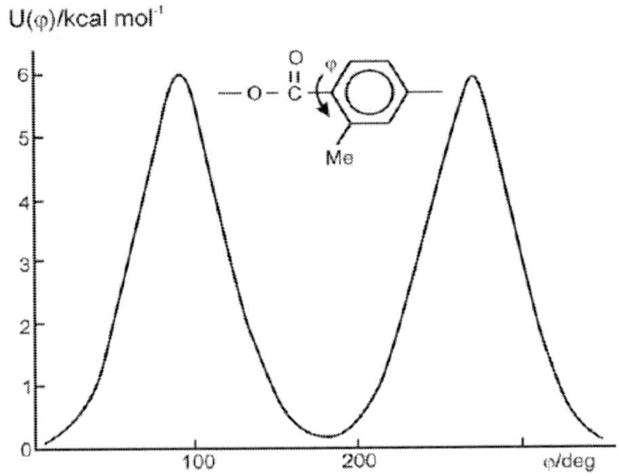

Figure 7. Dependence of the potential energy $U(\varphi)$ on the angle of rotation around the Ph_1-C virtual bond in polymers with a side groups in the backbone.

Table 3. Conformational parameters of polyarylates [127]

No	Structural repeat unit	M_L/M_{Lo}	A_{hin}, Å	A_{exp}, Å	σ
1	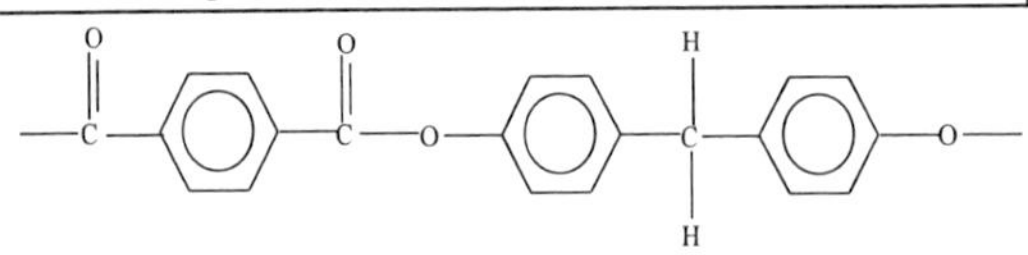 l_o=17.5 Å, A_{fr}=25.7Å .0				
2	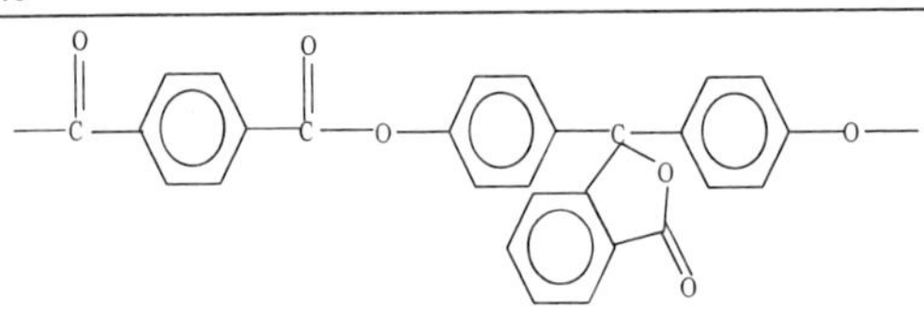	1.36	36.1	36.4(THF) 40.8(CP)	1.19-1.26
3	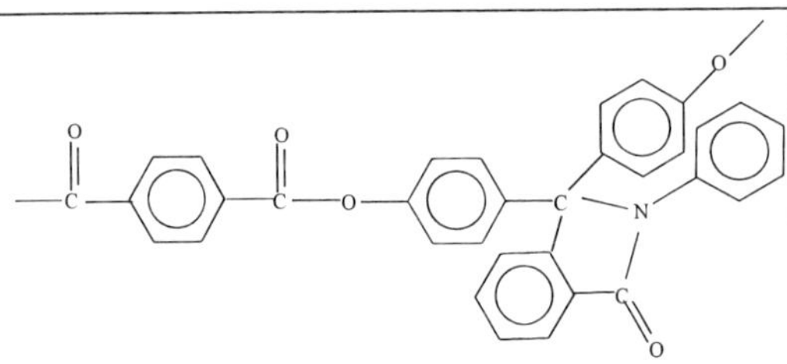	1.59	42.1	42.1(CP)	1.28

No	Structural repeat unit	M_L/M_{Lo}	A_{hin}, Å	A_{exp}, Å	σ
4	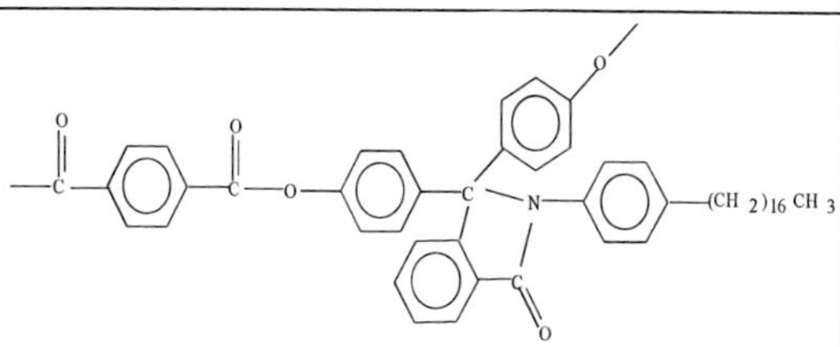	2.16	58.8	58.0 (θ) 61.4(CP)	1.50- 1.54
5	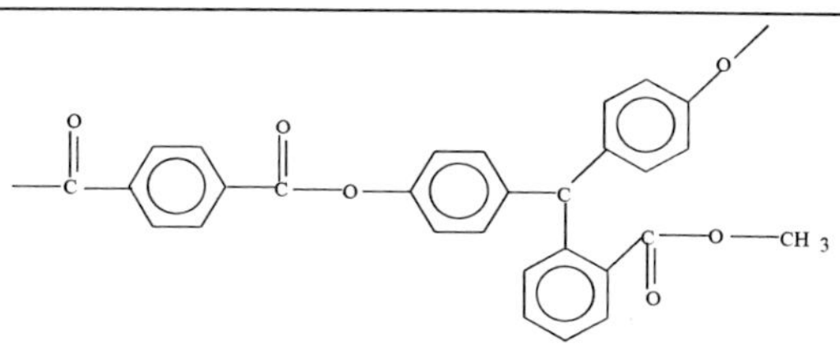	1.40	37.1	37.3(THF) 37.8(DO)	1.20- 1.21
6	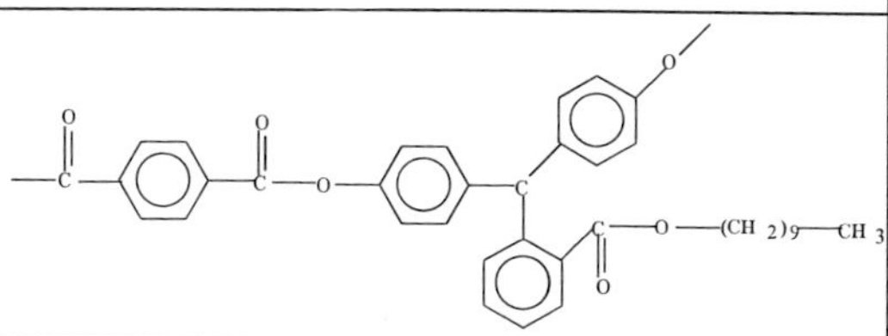	1.79	47.6	47.3(CP) 51.4THF)	1.36- 1.41
7	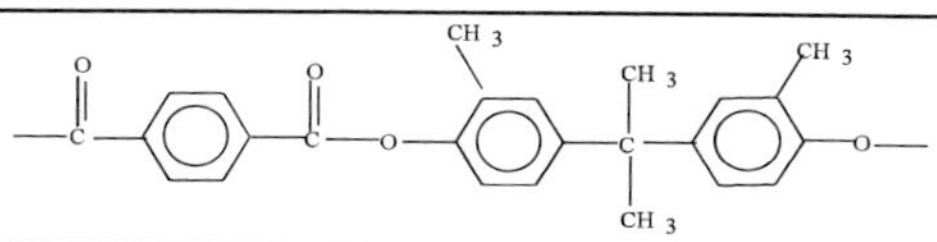	1.08	26.6	27.8(TCE)	1.04
8	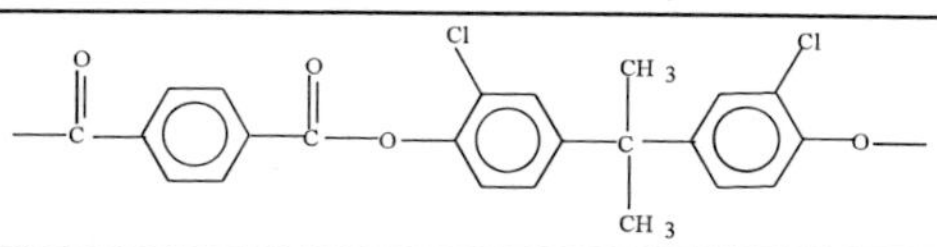	1.19	37.0	38.2(TCE)	1.22
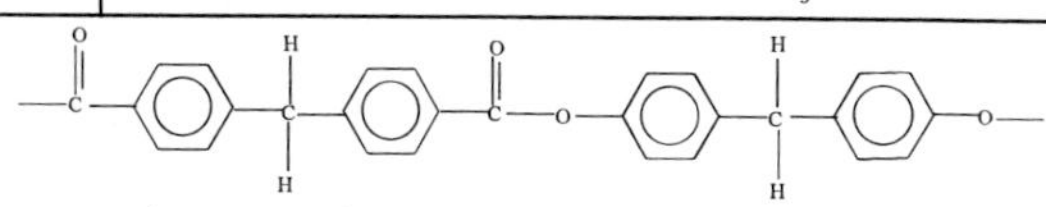 l_o=22.5 Å, A_{fr}=19.2Å					
9	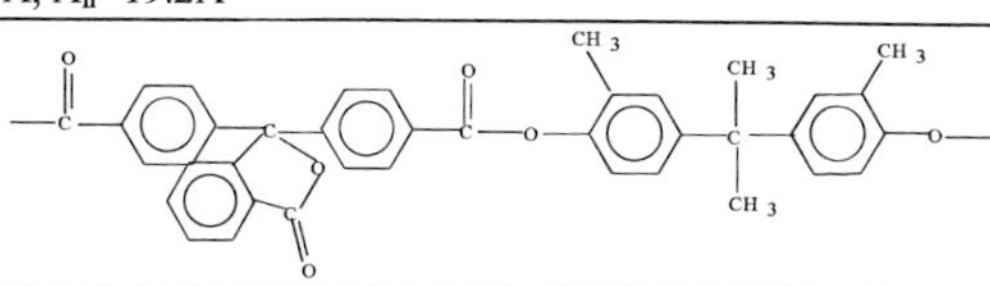	1.12	23.5	24.9(TCE)	1.15
10	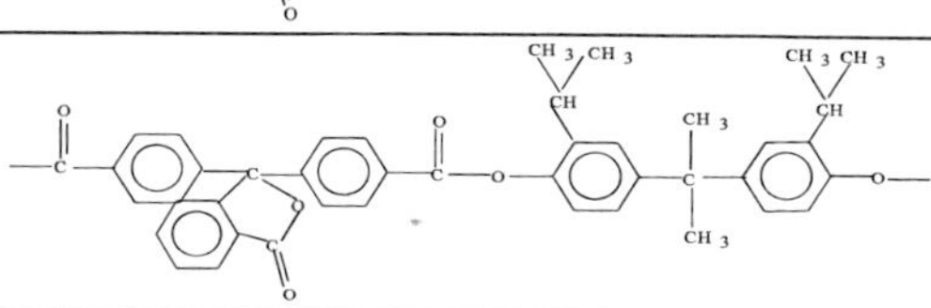	1.35	38.2	39.0(TCE) 41.4(TCE)	1.42- 1.47
11	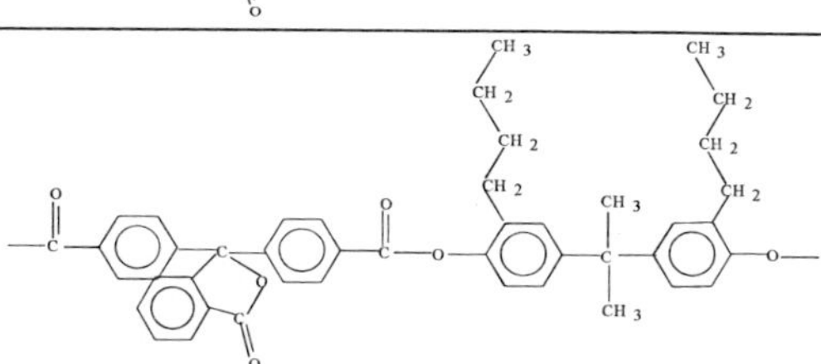	1.35	34.0	33.8(TCE) 34.2(TCE)	1.32

Table 3. (Continued)

No	Structural repeat unit	M_L/M_{Lo}	A_{hin}, Å	A_{exp}, Å	σ
12	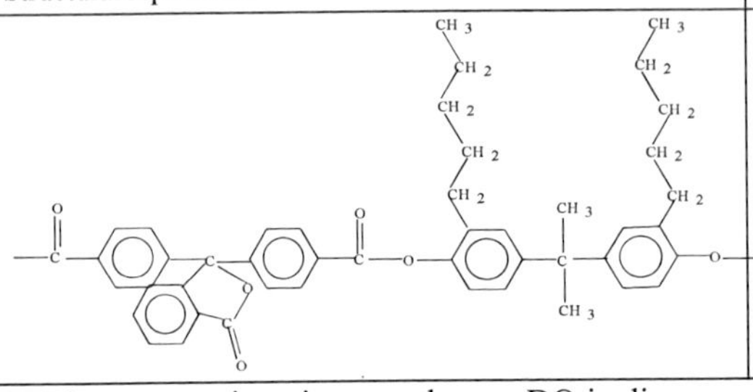	1.40	41.8	43.9(THF) 48.5(TCE)	1.51- 1.59

Note: The solvents are given in parentheses: DO is dioxane; CP is chloroform; THF is tetrahydrofuran; TCE is tetrachloroethylene, and θ is a solvent corresponding to the θ conditions.

Figure 7 shows the angular dependence $U(\varphi)$ for a polyarylate fragment containing a side substituent in the phenyl ring of the backbone. The $U(\varphi)$ function was calculated by the MM3 method [14]; the φ angle was varied from 0 to 360° with as step of 5°, the other geometric parameters of the macromolecule fragment were optimized. Since the sub substituent at the bridging C atom is asymmetrical, the potential energy function $U(\varphi_1,\varphi_2)$ for compounds of the second group depends on the angles of rotation around the Ph_1-C and Ph_2-C bonds, φ_1 , and φ_2 (scheme).

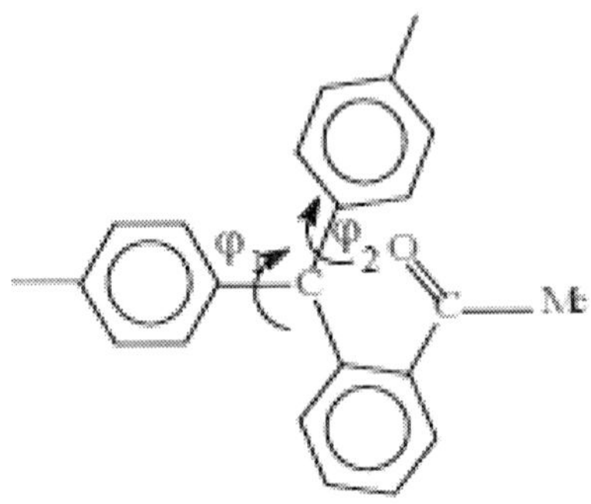

Figure 8 shows this dependence for the polyarylate fragment presented in Table 3 (polymer 5). The initial conformation with angles φ_1 and φ_2 equal to zero, corresponds to the *trans*-chain. In the initial conformation, the benzene ring of the side group attached to the bridging C atom lies in the plane perpendicular to the image plane.

For all the polymers listed in Table 3, the conformation energy have been calculated as a function of the angle of rotation, $U(\varphi)$, and the A_{hin} values. The A_{exp} values for these polymers were found by light scattering and viscosimetry from the hydrodynamic properties of dilute solutions of these polymers in poor solvents [127]. Comparison of the A_{hin}, and A_{exp} values implies that the data obtained by our method are in good agreement with the experimental results for a number of polyarylates and can be used to estimate the rigidity of polymers, for which determination of the A_{exp} values is very difficult.

For some polymers, the experimental determination of the rigidity is impossible because of the poor solubility. However, the same polymers possess unique thermal and electric properties, which should correlate with their conformational rigidity. Therefore, it is of interest to calculate the rigidity for these polymers by the above-proposed method.

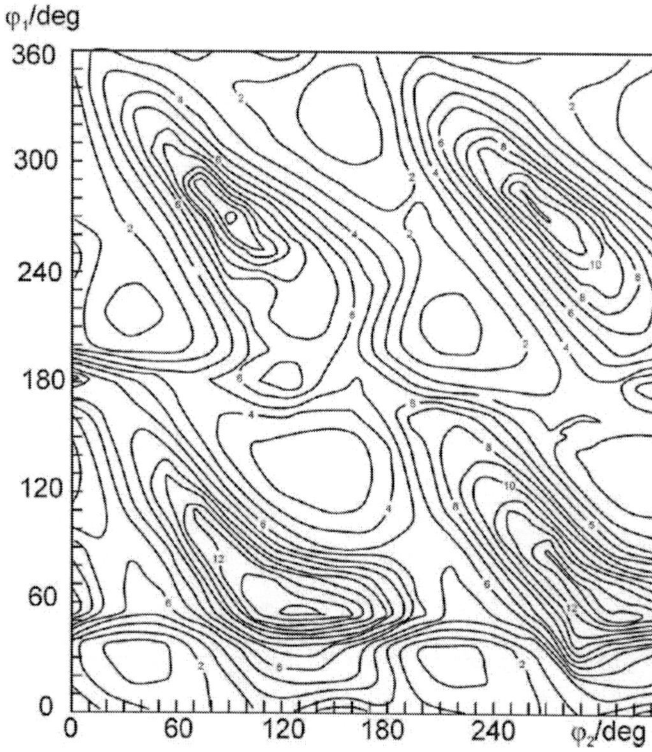

Figure 8. Dependence of potential energy ($U(\varphi_1,\varphi_2)$ kcal/mol, denoted by numerals) on angles of rotation in a fragment with a side substituent at the bridging atom of the polyarilate backbone.

Table 4. Conformational parameters of polyamides with phtalimide side substituens

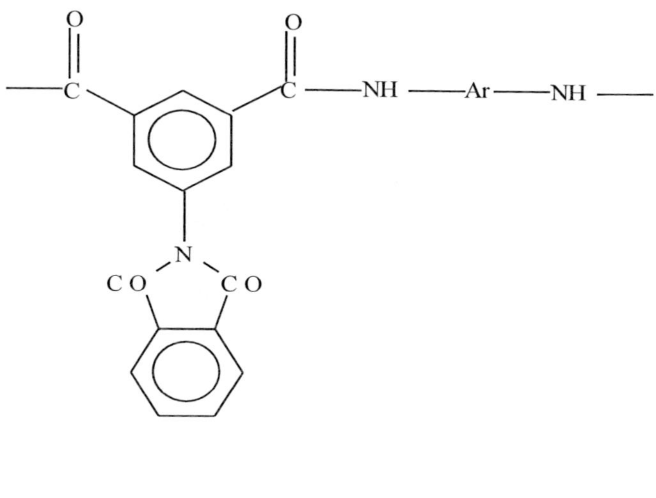

Table 4. (Continued)

No	Ar	l_o, Å	A_{fr}, Å
1	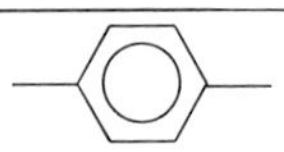	12.59	36.4
2	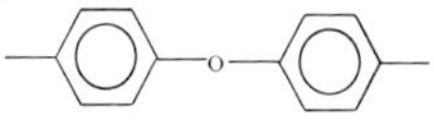	17.60	31.4
3	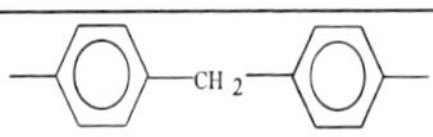	17.4	25.2
4	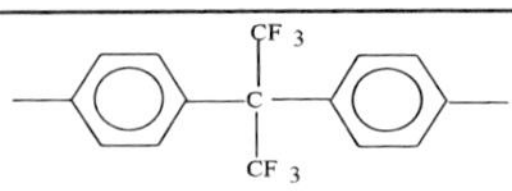	17.42	25.1
5	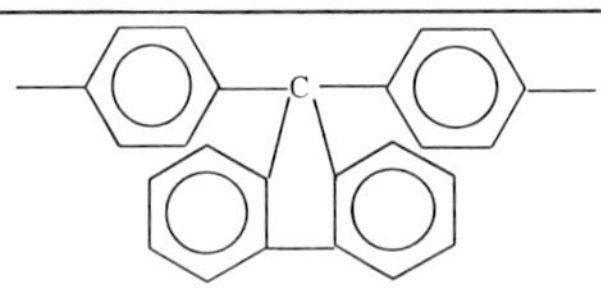	17.53	25.4
6	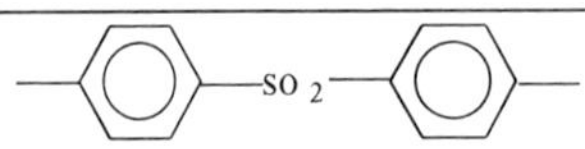	17.82	21.6
7	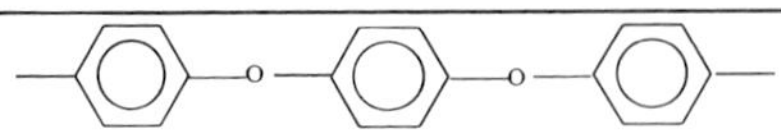	22.28	28.7
8	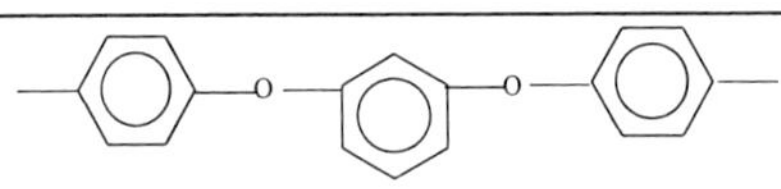	22.83	20.6
9	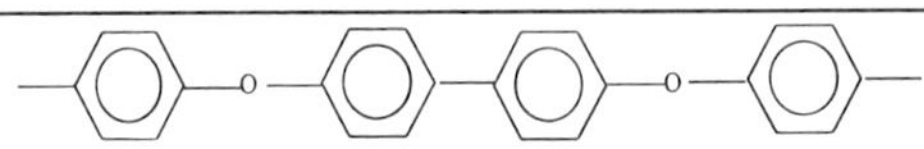	26.41	30.5
10	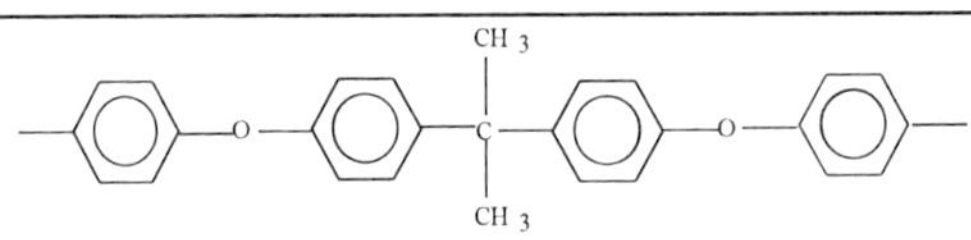	27.24	22.7

Tables 4-7 include polymers containing substituents in the backbone phenylene rings and the relevant Kuhn segments A_{fr} calculated neglecting the side groups. More detailed analysis of the geometric structures of the side substituents reveals that these polymers contain substituents of three types. The first type is a phthalimide ring in the *meta*-position to the carbonyl fragment of the backbone (see Table 4). The second type is represented by the phenyl groups in the *orto*-position in relation to the quinoxaline ring (see Tables 5 and 6), and the third type is a phenoxide side group (see Table 7).

Table 5. Conformational parameters of polymers

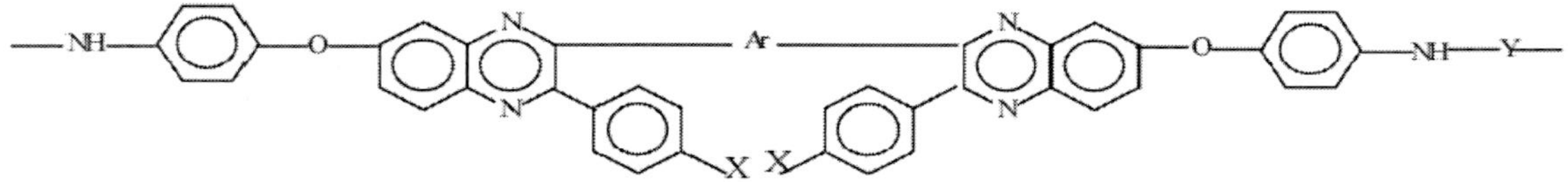

Ar$_1$= 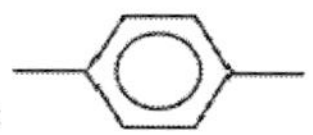 Ar$_2$= 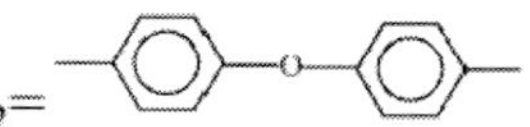 X= H, NO_2

No	Structural Repeat Unit	l_o, Å	A_{fr}, Å	A_{hin} Å		σ
				X=H	X=NO_2	
1	Y= 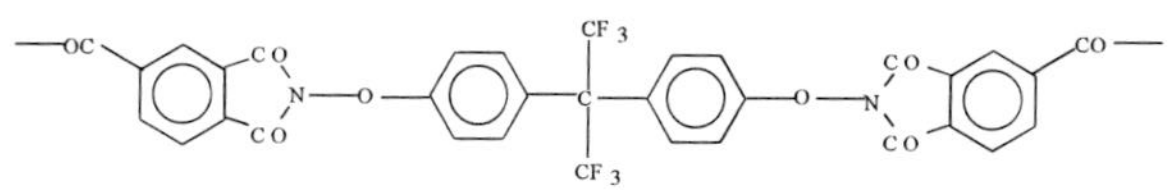 Ar$_1$ Ar$_2$ 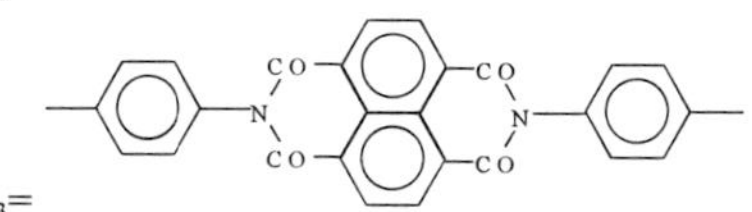 Ar$_3$=	52.4 57.3 64.9	35.6 33.9 42.6	39.2 34.6 42.6	42.1 37.2 42.6	1.10 1.18 1.02 1.10 1.00
2	Y= 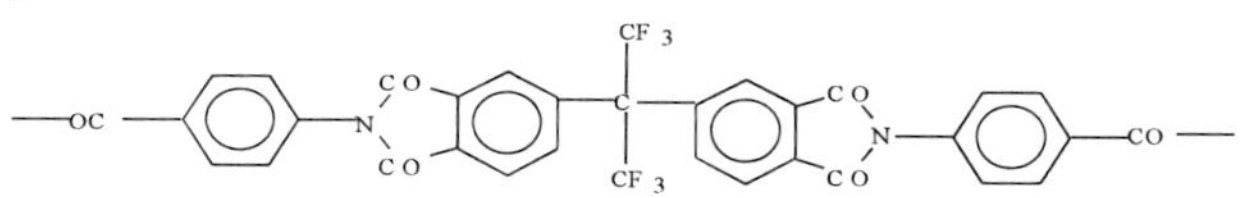 Ar$_1$ Ar$_2$	53.1 58.0	35.5 34.0	38.7 34.7	41.5 37.4	1.09 1.17 1.02 1.10
3	Y= 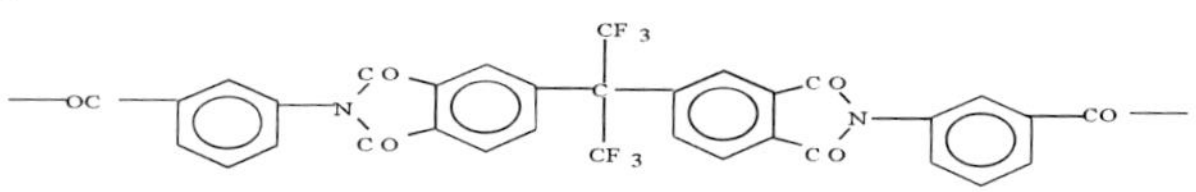 Ar$_1$	52.4	25.7	28.3	30.3	1.09 1.17

Table 5. (Continued)

No	Structural Repeat Unit	l_o, Å	A_{fr}, Å	A_{hin} Å		σ
				X=H	X=NO_2	
4	Y=					
	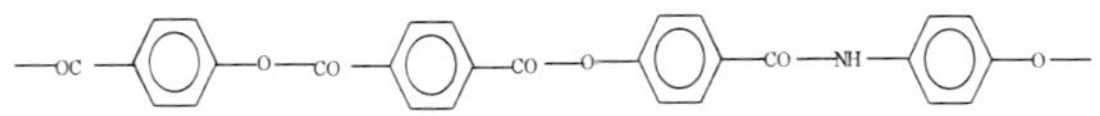					1.10
		47.9	44.3	48.8	52.3	1.18
	Ar_1					1.02
	Ar_2	52.8	43.4	44.3	47.7	1.10
5	Y=					
	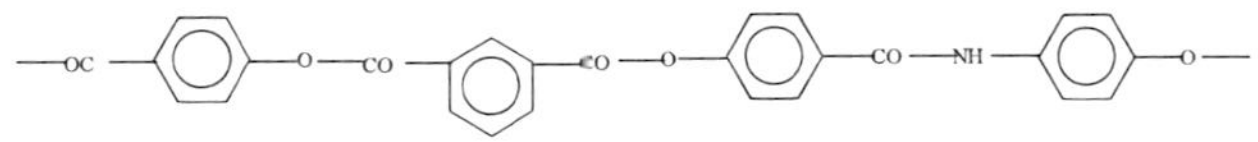					1.10
		47.7	34.6	38.1	40.5	1.17
	Ar_1					1.02
	Ar_2	52.6	33.5	34.2	36.8	1.10

The structures of the first-type side group (phthalimide ring in a polyamide chain) in *trans*- and *cis*-chains are substantially dissimilar.

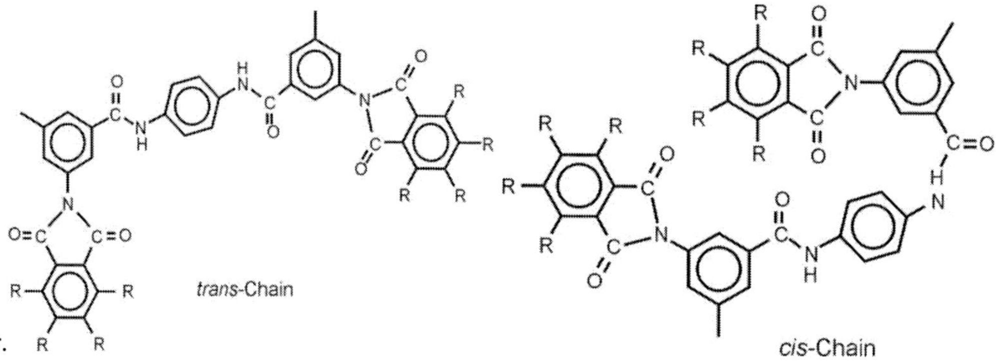

r.

In both cases, the distance between two side groups (phthalimide rings) is much greater than the sum of the van der Waals radii, and the rotation does not become hindered. The rotation around the CO—Ph bond could be hindered due to the fact that the distance between the closest carbonyl groups in the benzimide ring of the side substituent and in the amide fragment of the backbone is -2.5 Å in both chains; this is somewhat shorter than the sum of the van der Waals radii of two O atoms. However, this does not lead to substantial hindrance to rotation, because the phthalimide fragment is preferably arranged perpendicular to the plane of the benzene ring attached to it. Consequently, the distance between the amide and imide carbonyl groups increases to 3.5 Å, which is greater than the sum of the van der Waals radii of two O atoms. Thus, the introduction of a phthalimide ring as the side substituent into the polymer chain should have no effect on the Kuhn segment length of the polymers (see Table 4), and A_{fr} can serve to characterize the conformation of these polymers.

Table 6. Conformational parameters of polymers

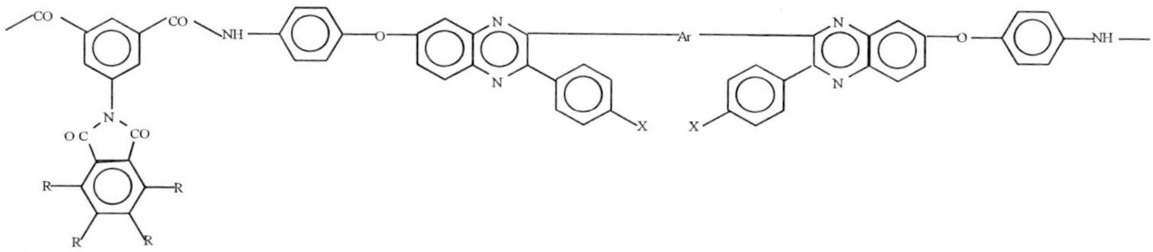

No	Structural Repeat Unit	l_o, Å	A_{fr}, Å	A_{hin} Å				σ
				R=H X=H	R=Cl X=H	R=H $X=NO_2$	R=Cl $X=NO_2$	
1		35.0	27.4	30.1	30.5	32.3	32.8	1.10 1.18
2	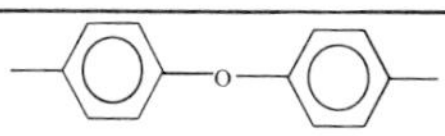	40.0	25.8	26.3	27.3	28.3	29.0	1.02 1.12
3	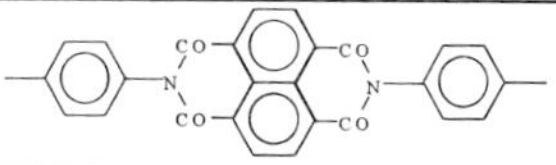	47.5	37.3	37.3	38.6	39.3	39.7	1.00 1.06

R= H, Cl and X=H, NO_2.

The side groups of the second type are phenyl rings in the quinoxaline fragment linked to the quinoxaline cycles. In this case, the hindrance to rotation largely depends on the length of the bridge between the quinoxaline cycles. Benzene ring is the shortest bridge among those considered in our study (see Table 5). In this case, the side groups closely approach one another. When the phenyl bridge between the quinoxaline fragments is replaced by a biphenyl oxide or a diphenylnaphthoyleneimide bridge, the distance between the quinoxaline rings increases, and the side groups are arranged freely without interfering with one another; even the presence of NO_2 groups in the diphenylnaphthoyleneimide fragment does not enhance the rotation hindrance. For the fragment with phenyl bridges, we calculated the dependence of the conformation en energy $U(\varphi)$ on the angle of rotation φ around a virtual bond between the phenylquinoxaline fragments (Fig. 9a). In this case, the *cis*-chain with $\varphi = 0$ was taken as the initial structure, whereas $\varphi = 180°$ was assigned to the *trans*-conformation. With 4-nitrophenyl as the side group, rotation be comes more hindered (see Fig. 9, *b).* In both cases (see Fig. 9, *a, b),* the $U(\varphi)$ curves clearly demonstrate the presence of two strong peaks (at $\varphi \approx 0$ and $\varphi \approx 180°$) and a number of weak peaks. For both nonsubstituted and substituted phenyl side groups, the first peak is rather intense (~9—11 kcal mol^{-1}). Thus, for both polymers, the occurrence of conformers in this region is relatively unlikely

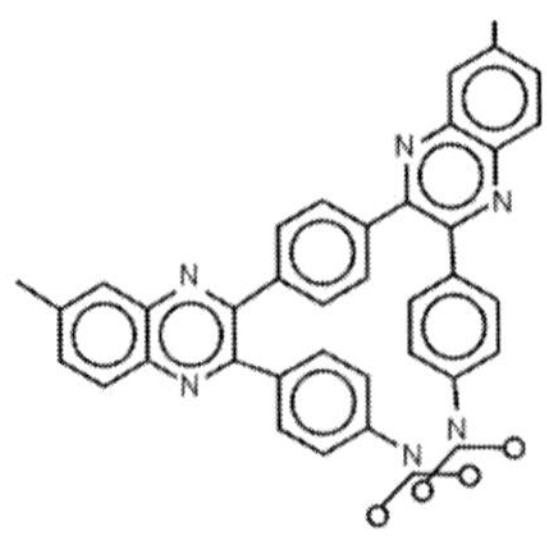

cis-Conformation

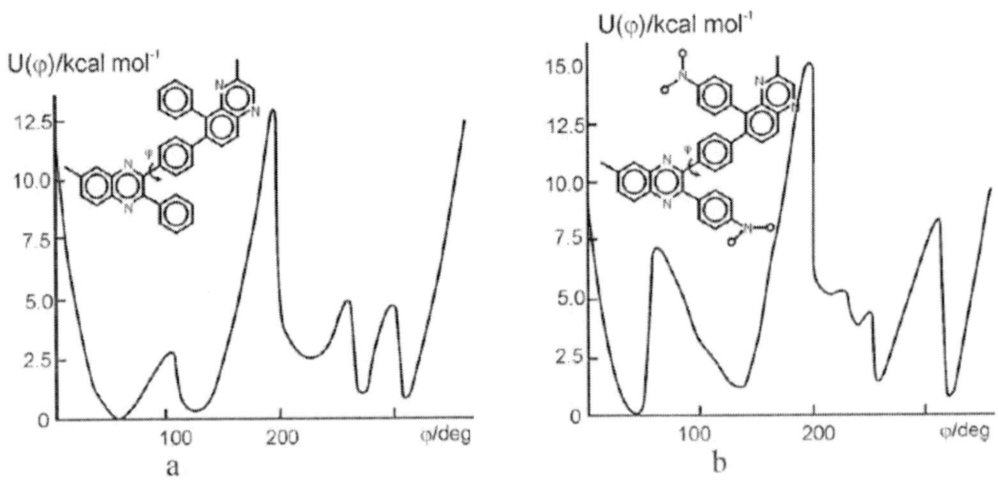

Figure 9. Dependence of the potential energy $U(\varphi)$ on the angle of rotation around the virtual bond between two quinoxaline rings with phenyl (a) and 4-nitrophenyl (b) as side groups.

Table 7. Conformational parameters of poly-1,3,4-oxadiazolamides containing phenoxyterephthalates

Structural Repeat Unit	l_o, Å	A_{fr}, Å	A_{hin} Å	σ
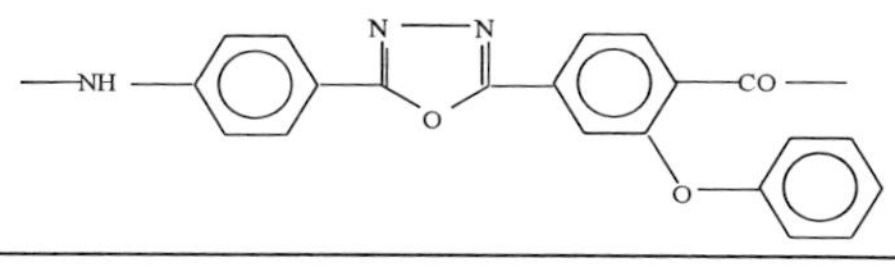	9.9	73.4	100.6	1.37
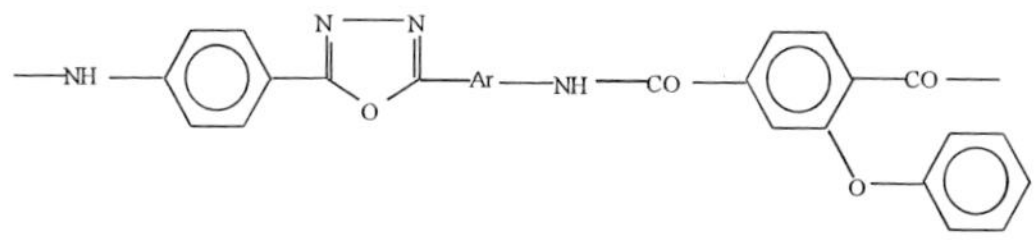				
Ar= 	20.5	143.5	185.1	1.29
Ar= 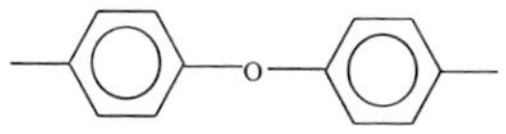	30.3	51.3	63.6	1.24
Ar= 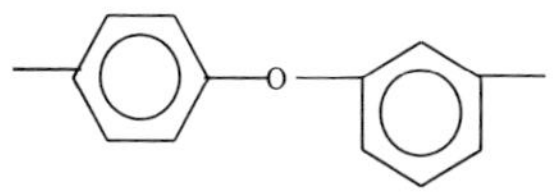	30.2	27.0	33.5	1.24

The second peak at φ≈180° (*cis*-conformation) is intense (12—15 kcal mol^{-1}) and asymmetrical in both cases. This confirms the assumption that when a chain with the *cis*-conformation contains side groups, the distance between their H atoms (especially, in the case of 4-nitrophenyl group) is much shorter than the sum of the relevant van der Waals radii. Therefore, after the angle φ = 180°±20° with respect to the *cis*-conformation has been attained, the groups cannot continue approaching each other, and there are no conformers in this region either. Thus, the set of conformers for these polymers is depleted, and the molecule becomes more stretched and more rigid. The values of the Kuhn segment A_{hin}, calculated with allowance for the hindrance to rotation, are listed in Table 5.

In all cases, the magnitude of the parameter σ confirmed our hypothesis that in the presence of a short bridge (benzene ring), the rotation hindrance is enhanced, and when 4-nitrophenyl is present as the side group, rotation is even more hindered. However, since in the case of the polymers listed in Table 5, a unit contains a large number of virtual bonds around which free rotation is possible, and only one bond around which rotation is hindered, this has only a slight effect on the final rigidity of the polymer.Table 6 includes polymers containing side groups of the first and second types. The hindrance parameters for these polyheteroarylenes are close to those for polymers containing only side groups of the second type. Table 7 also presents the conformation parameters for polymers with phenoxide substituents.

In this case, the rotation hindrance is mostly due to the short distances between the H atoms in the phenylene ring of the backbone amide fragment and in the phenylene ring of the side group; thus, the latter tends to be arranged in a plane perpendicular to the image plane. In addition, the rotation hindrance is determined by the proximity of the O atoms in the carbonyl group of the amide fragment and in the ether group of the side substituent. It can be easily seen from Fig. 10 that for the rotation around the Ph—CO bond (angle φ_1), the difference between the energies of the *cis*- and *trans*-conformations is close to 6 kcal mol^{-1}, and again the full set of conformers corresponding to free rotation in the polymer chain is not realized. For rotation around the NH—Ph bond (angle (φ_2), the angular dependence of the potential energy follows a pattern normal for polyamides [128]. It was found experimentally that a substituent in the *ortho*-position to the carbonyl group accounts for a rotation hindrance described by σ = 1.2—1.5 (Table 3). The values that we obtained lie in the same range. It can be seen from the data presented in Tables 3 —7 that the introduction of side substituents in a chain does not necessarily result in a substantial rotation hindrance. This occurs only in those cases where the side group cannot be freely arranged in the niche formed by long virtual bonds in polyheteroarylenes, and short (shorter than the sum of the van der Waals radii) contacts between the backbone and side-group atoms arise. This paragraph shows the possibility to calculate the Kuhn segment in case of bulk side substitutes taking in account a hindrance. However, this route is very labor consuming.

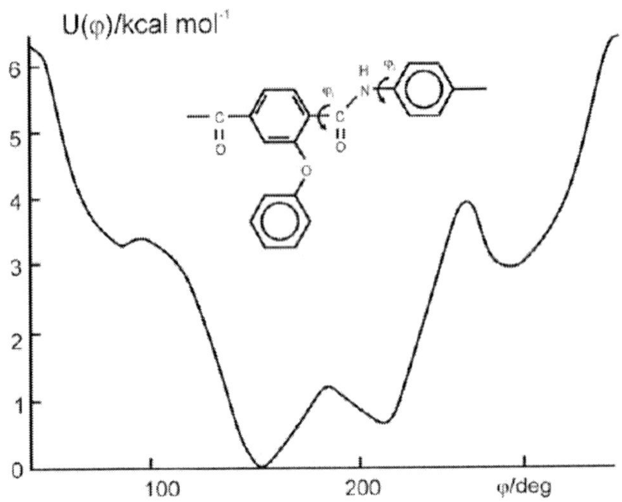

Figure 10. Dependence of the potential energy $U(\varphi)$ on the angle of rotation for fragment with a side phenoxide substituent.

IV. Rigid-Chain Polymers

Published data on the rigidity of the most rigid cyclochain aromatic polymers—poly (para-benzamide) (Table 2, polymer 26), poly(para-phenylene tere- phtalamide) (Table 2, polymer 25), and polyami- dohydrazide (Table II, polymer 42)—are contradictory. The discrepancies in the literature are not only those in evaluating the rigidity parameters, which would not be surprising in view of formidable experimental difficulties, but also in calculating the free rotation values [71, 128-130].

As to the experimental parameters $\langle R^2\rangle/M$ and A, values for most rigid polymers—para-aromatic polyamides (Table 2, polymers 25 and 26) and polyamidohydrazide (Table 2, polymer 42)—found following the Bohdanecky model (model II) from measurements in concentrated sulfuric acid, in DMAA + LiCl, and in DMSO + LiCl, are all much lower than free-rotation values calculated by several authors [3, 89]. However, an error is hardly possible since, as is seen from experimental data in Figures 11 and 12, in spite of appreciable scatter, all the values fit a common dependence for a wide range of molecular weights and various solvents (H_2SO_4, DMAA + LiCl, DMSO).

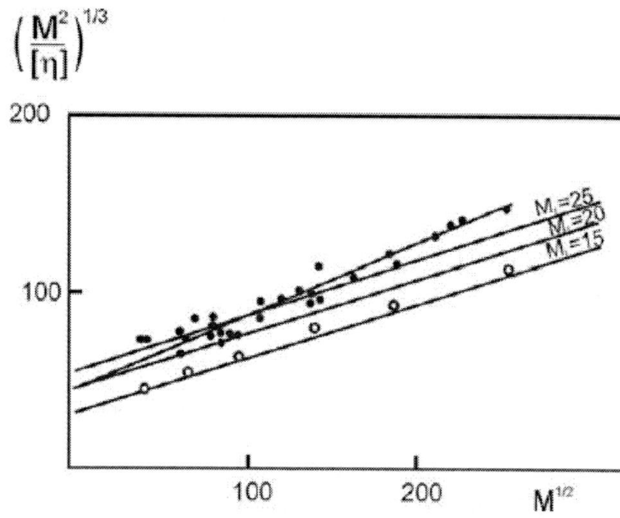

Figure 11. Dependence of $(M^2/[\eta)])^{1/3}$ on $M^{1/2}$ for polymers 25 and 26 in sulphuric for polymers 25 and 26 in sulfuric acid (X: polymers 25 and 26; •: polymer 25 with substituent CN in the *ortho* position [78]; O: polymer 25 in DMAA). Solid lines are drawn using eqs. (10) and (11) with A = 940 Å and d = 5 Å.

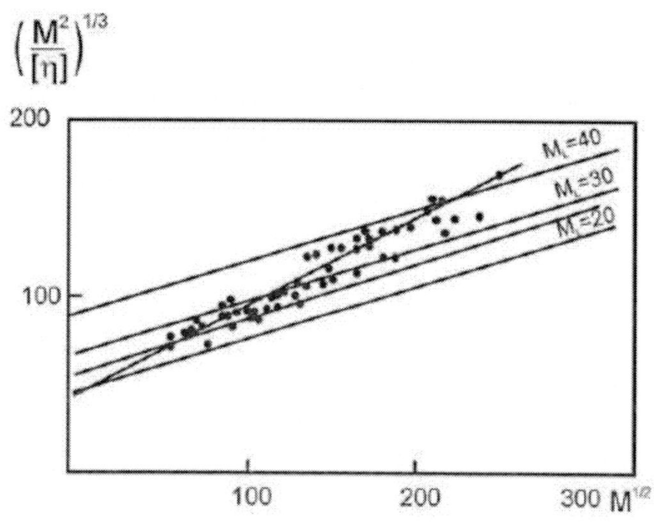

Figure 12. Dependence of $(M^2/[\eta])^{1/3}$ on $M^{1/2}$ for polymer 42 in DMAA + LiCl (•) and DMSO (X). Solid lines are drawn using eqs. (10) and (11) with A = 940 Å and d = 5 Å.

The linear dependence drawn through all these points has a steeper slope than that calculated from Esq. (10) and (11) at $A = 940$ Å and $d = 5$ Å. Nevertheless, the intercept of the experimental line on the *Y* axis coincides with the calculated value for the same parameters. The observed discrepancies in calculated and experimental values of the rigidity parameter can be explained in several ways:

1) By macromolecular aggregation, which leads to discrepancies because of an increase in the measured molecular weight of aggregated particles.
2) By errors in determination of A_{fr} due to as proposed in ref. 135, due to deformations of valence angles and bonds.

It is difficult to judge the importance of these effects since experimental data are scarce, and conclusions drawn by the various authors are sometimes contradictory, even though they are often based on the same experimental results.

At this juncture, it is of interest to mention several peculiarities of *para*-aromatic polyamides. Judging by values of the second virial coefficients obtained by light scattering, concentrated sulphuric acid as well as DMAA + LiCl, $C1S0_3H$, and CH3SO3H are thermodynamically good solvents. Intrinsic viscosities determined in sulphuric acid and DMAA + LiCl are close (within 10-15%) [77, 81, 131]. However, molecular weights measured by the light scattering method in DMAA + LiCl are higher than those measured in sulphuric acid by approximately 30% [77], or according to refs. 82 and 132, three to four times greater.

On the other hand, molecular weights measured in chlorosulphonic acid coincide with those in H_2S0_4 but the intrinsic viscosity is twice as large [77]. Such behaviour is evidently related to a phenomenon that is new in principle and caused by coexisting high rigidity and strong intermolecular interactions. As is well known, local orientation effects are an essential aspect of the structure of homophase solutions of rigid chain polymers. These effects arise at concentrations much lower than the critical one, i.e., before the formation of alyotropic liquid-crystalline phase. In Ref. 133 the anisotropic component of the scattered light of *para*-phenylene terephtalamide in sulphuric acid is appreciable, while the light scattering due to concentration fluctuations is comparatively small, as would be expected for a thermodynamically good solvent.

Introduction of the depolarization correction for the scattered light decreases the measured molecular weight by a factor of 2-3 [80]. According to current ideas on lyotropic liquid-phase separation, [133] this effect must increase with increasing molecular weight. Depolarization decreases sharply at a critical temperature [132]. This temperature depends on the solution concentration and, probably, on the molecular weight. In the same temperature range, a decrease in the molecular weight measured by the light scattering and in the maximum temperature dependence of the second virial coefficient is observed [134]. The nonmonotonic character of the change of the second virial coefficient implies an alteration in the predominant dissolution mechanism.

The temperature dependence of depolarization, molecular weight, and the second virial coefficient in sulphuric acid permits us to suggest that the primarily entropic character of the dissolution of strongly solvated and partly oriented macromolecules is replaced, with increasing temperature, by an enthalpy-driven dissolution of solvated macromolecules.

At this juncture, we believe a further analysis of depolarization of the scattered light and the introduction of adequate corrections will make it possible to elucidate the role of solvation, orientation, and association of macromolecules.

Now we consider in detail a second cause of discrepancies between calculated and experimental values of the rigidity parameter.

For rigid-chain aromatic cyclochain polymers, such as poly(*para*-phenylene terephthalamide) (PPTA) or poly(*para*-benzamide) (PBA), the experimental Kuhn segments A_{exp}, are significantly smaller than those calculated on the assumption of free rotation in the chain A_{fr}.

Previously, it was demonstrated that the rotation about Ph-CO and Ph-NH bonds is not hindered and -CO-NH- bonds are predominantly in the *trans* conformation. In that case, A_{fr} must correspond to the lower margin for the experimental values deduced from hydrodynamic studies of dilute solutions of these polymers.

Tsvetkov [135] suggested an alternative mechanism of chain flexibility in polyamides, which was associated with out-of-plane deflections of amide groups from the planar *trans* structure. These deflections were assumed to be modulated by the Brownian motion of a molecular chain, which caused random deformations of valence angles and bonds of the amide groups.

Let us examine the feasibility of this mechanism. The Monte Carlo method was used to calculate the conformational parameters according to the procedure described in the paragraph I. The procedure was modified by specifying the maximum deviations from valence angles (maximum deformation of valence angles, $\Delta\vartheta_m$). In a computer-aided Monte Carlo simulation of macromolecular chains, for each valence angle, it was determined whether the angle would decreases or increase (both were assumed equiprobable). In the interval of uniformly distributed random numbers from 0 to 1, a random number was generated, which was multiplied by $\Delta\theta_m$ and added, with the corresponding sign, to the valence angle. This was accomplished for all valence angles of the polymer chain.

Table 8 lists the geometric parameters of the two aromatic rigid-chain polyamides. For comparison, the data for two aromatic flexible-chain polymers are also presented.

The Kuhn segments for these polymers in the absence of deformation are listed in Table 9 for different magnitudes of $\Delta\vartheta_m$.

Table 8. Geometric parameters of some polymers

No	Structural Repeat Unit	l_i, Å	θ_1, deg	Rotation
1	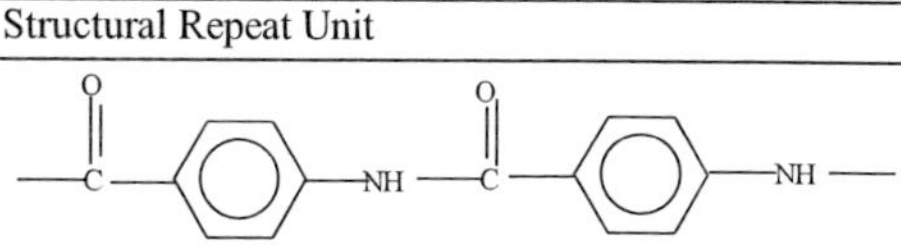	5.73	57.0	Free
		1.32	66.0	Banned, *trans**
2	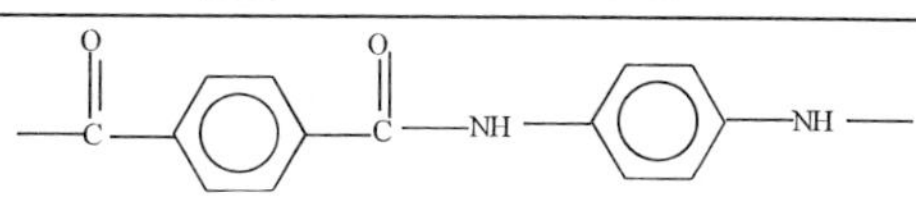	5.75	66.0	Free
		1.32	57.0	Banned, *trans*
		5.71	57.0	Free
		1.32	66.0	Banned, *trans*
3	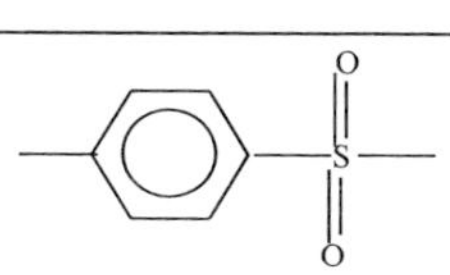	6.33	75.0	Free

Table 8. (Continued)

No	Structural Repeat Unit	l_i, Å	θ_1, deg	Rotation
4	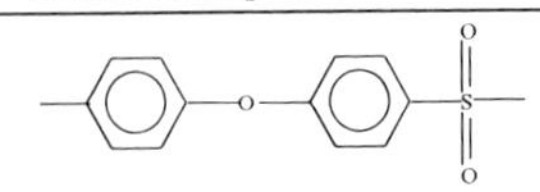	5.92 5.92	54.0 75.0	Free Free

* trans conformation with respect to consider bond.

As can be seen from Table 9, for flexible-chain polymers, the Kuhn segments are almost not affected by the deformation of valence angles as high as ±10°. Rigid-chain polymers are much more sensitive to the variation in valence angles. To reduce the rigidity by two times, the deformation of angles at CO and NH groups must not be less than ±10°. Is this value large or small?

To answer this question, let us consider the mean amplitudes of Brownian vibrations of the atoms in a molecule. Usually, mean-square amplitudes of the vibrations, which are defined as

$$l_{ij} = \left\langle \left(R_{ij} - R^{e}_{ij}\right)^{2} \right\rangle^{1/2},$$

where indexes i and j refer to arbitrary (bonded or not bonded) pairs of atoms, R_{ij} is the internuclear distance at a specified time instance, and R^{e}_{ij} is the corresponding equilibrium state] are considered. Then, quantities entering into this equation are measured in gas electronographic studies or calculated from the microwave, IR, or Raman spectra.

As an example, let us consider the data of electronographic studies of the molecules of anisole [136] and sulphonyl chloride [137, 138].

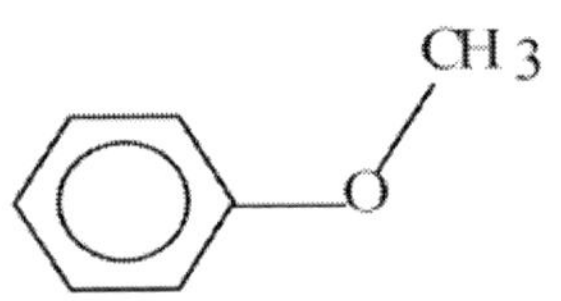

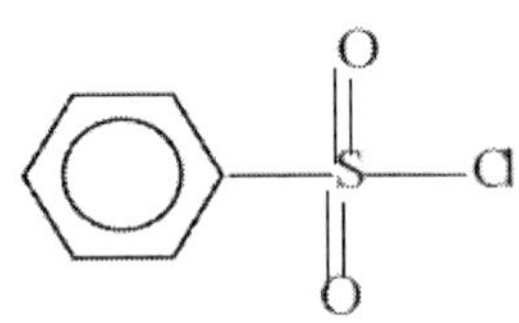

In both compounds, the rotation may occur about the Ph-Y bond (Y stands for the atoms of oxygen or sulphur). Table 10 lists the mean-square amplitudes of vibrations for the valently unbonded atoms in both molecules. The results of the studies of these molecules demonstrate that, in those cases when the rotation is possible, the mean-square amplitudes are rather high and are controlled by the temperature at which the study is conducted. For those bonds that are not affected by the internal rotation, the mean-square amplitudes of the vibrations are not affected by the temperature. Having analyzed the vast data of electronographic studies and the amplitudes of vibrations calculated from spectra, we came to a conclusion that, for unbonded atoms, the mean-square amplitudes of vibrations are about 0.1 - 0.2 Å [139]. It must be kept in mind that the mean-square amplitude of the vibration for an atom in a molecule is an average over all feasible directions

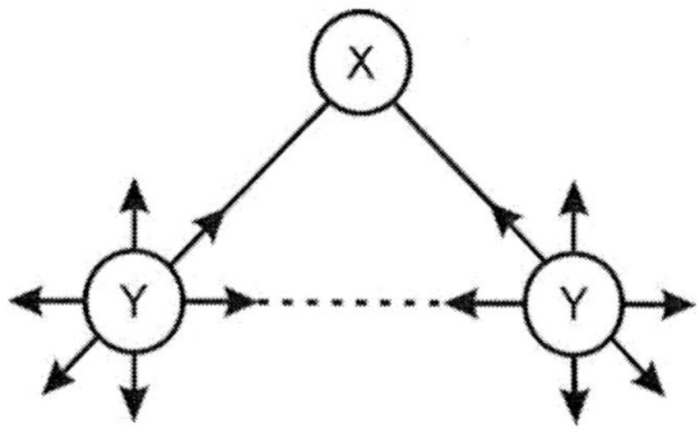

However, assuming that this quantity is associated solely with the Y-Y bond, maximum variation of the valence angle is

Atom O S N
$\Delta\theta_m$ ±4 ±5 ±6

For polyamides, the rotation about the Ph-CO and Ph-NH bonds is possible and, therefore, the mean-square amplitude of the vibrations of the CO--NH bond would be modulated by the torsional vibrations about the Ph-CO and Ph-NH bonds, so that the contribution to the deformation of angles 6 would be even smaller. This suggests that it is hardly reasonable to consider the deformation of valence angles as a factor that can significantly reduce the rigidity of a polymer.

It was previously assumed that the contributions of rotational and deformational components to conformational rigidity complies with the additivity pattern and must be described by the reciprocal additivity scheme

$$\frac{1}{A_{\varphi,\vartheta}} = \frac{1}{A_\varphi} + \frac{1}{A_\vartheta} \qquad (13)$$

where A_φ is the Kuhn segment for the chain with free rotation, and A_θ is the Kuhn segment calculated on the assumption of banned rotation about the bonds but allowed deformation of valence angles. Table 9 lists the Kuhn segments $A_{\Delta\theta}$ for the case of banned rotation about virtual bonds. These data demonstrate that, for a polymer with a single valence angle (polymer 3), banning the rotation about virtual bonds leads to a rigid rod with $A_{\Delta\theta} = \infty$. In this case, relationship (13) seems to be valid, because $A_{\varphi,\theta}$ is not affected by the variation of $\Delta\theta$, whereas the second term in equation (13) is close to zero. When there is more than one valence angle, the valence angles are not equal, the rotation about virtual bonds is banned, and the valence angles are not subject to deformation, the polymer chain twists to a full circle and comes to the initial position in $k = 360/(\theta_1 - \theta_2)$ units. In polymer 1(Table 8), this takes place in 40 structural units; in polymer 2, in 20, and in polymer 4, in 17 units. In all cases, the concept of Kuhn segment is not valid for the resultant cycle. For the polymers examined here, the mean-square end-to-end distance is equal to zero; therefore, the Kuhn segment is also equal to zero. When the deformation of valence angles is allowed, the closed circles are no longer formed, and the Kuhn segment acquires a certain small nonzero value. After substituting the values of the Kuhn segment for polymers 1, 2, and 4 into equation (13), one does not obtain to equality of both parts of the relationship, because the contribution of $1/A_{\Delta\theta\,\theta}$ is significantly greater than that of $1/A_{\varphi}$. As a result, the obtained values of Kuhn segment $A_{\varphi\theta}$ are very small and do not agree with the values deduced in a computer-aided simulation.

Table 9. Kuhn segments of polymer at different maximum deformation angles

$\Delta\theta_m$, deg	Kuhn segment (Å) for polymer*			
	1	2	3	4
With internal rotation				
±0	1000	1022	13.64	16.06
±2	975	979	14.30	16.08
±4	880	889	14.13	16.13
±6	719	785	14.00	16.56
±8	626	623	13.43	16.27
±10	535	526	13.12	16.86
Without internal rotation				
±0	0	0	∞	0
±10	5.13	5.33	1996	1.67
±20	18.11	19.32	546	6.33

*The polymers are numbered as in Table 8.

Table 10. Mean-square amplitudes of vibrations of unbonded atoms

Bond	Molecule of Ph-OCH_3 ($X=C_7$, Y=O)		Molecule of Ph-SO_2Cl X=O, Y=S)
	55^{oC}	250°C	138°C
C_2...X	0.059	0.116	0.123
C_3...X	0.066	0.123	0.128
C_4...X	0.068	0.140	0.108
C_2...Y	0.066	0.055	0.070
C_3...Y	0.068	0.068	0.070
C_4...Y	0.069	0.069	0.068

Thus, the values of the Kuhn segment of aromatic cyclochain polymers with a set of different valence angles cannot be represented as a sum of the rotational and deformational contributions.

With regard to the fact that the magnitude of A_{exp}, is underestimated as compared with A_{fr}, other reasons seem to be more appropriate to explain it. Of these reasons, chain defects are the primary factor. In polyamides, a defect may be associated with phenylene rings in the *meta* position. In addition, Zubkov *et al.* [140, 141] demonstrated that the *trans* conformation of the -CO-NH bond is favourable in the amide bridges. However, under severe conditions of the synthesis or as a result of dissolution of the polymer in sulphuric acid, a certain number of CO-NH bonds may occur in *cis* conformation. Let us examine this suggestion. Using the Monte Carlo procedure described in the paragraph I, the values of the Kuhn segment were calculated for a system in which PBA was treated as a random copolymer composed of units A, with *trans* configuration with respect to the CO-NH bond, and units B, with *cis* conformation of the unit. The probability for unit A to add to unit B was proportional to its fraction in the copolymer. The figure 13 shows the Kuhn segment as a function of the fraction of units B in the copolymer (solid line). As is seen for PBA, even at low content of *cis* conformation (about 1%), the values of A_{fr}. and A_{exp} become closer. A similar calculation, in which units B were identified with phenylene rings in the *meta* position, was carried out for the system with a variable content of units B. The results of this calculation are represented by the dashed line. The agreement between experimental and calculated Kuhn segments is attained at 2.5% content of meta-substituted phenylene rings.

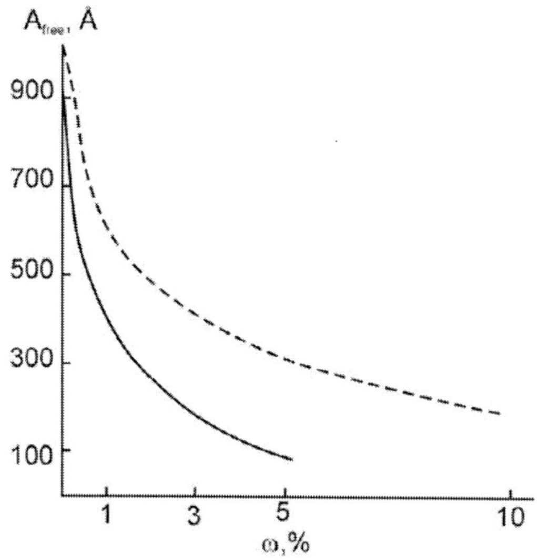

Figure 13. Kuhn segment A_{fr} as a function of the mole fraction ω of units containing phenylene ring in *meta* position (dashed line) or the CO-NH bond in *cis* conformation (solid line) for polybenzamide.

Thus, with polyamides as an example, we demonstrated that, for rigid aromatic cyclochain polymers, the underestimated, as compared with A_{fr}, value of the experimentally determined Kuhn segment is explained by a combined effect of many factors; moreover, the contribution due to the deformation of valence angles is significantly smaller than that associated with structural defects.

V. Calculation of the Free, Occupied and Assume Volumes

Many physical properties of polymers are dependent on chains packing in a glassy state. The knowledge of a free volume in a polymer matrix can effectively explain different behavior of the polymer.

In [142], a new rather simple rolling-over method was developed for calculation of Van der Waals, V_w, occupied, V_{occ}, and accessible V_{acc} volumes. The successful application of this method requires that the geometry of the monomer unit be accurately evaluated. The geometry of the repeating monomer unit was refined by the AM1 quantum-chemical method [14]. The as-calculated coordinates of atoms were used for occupied volume calculations by the rolling-over method [142].

The monomer unit of a polymer was considered a system of overlapping hard spheres with coordinates of centers corresponding to coordinates of atoms and radii equal to the Van der Waals radii of corresponding atoms. The Van der Waals volume of the monomer unit is equal to the volume of a body formed by these overlapping spheres. The rolling-over method includes the calculation of the Van der Waals volume of the monomer unit through the Monte Carlo procedure, and any changes in the unit geometry that arise from a change in the composition of groups contained in this unit are taken into account.

V_f is the non-occupied volume in polymer matrix, and it is given by equation (14):

$$V_f = 1/\rho - (N_A/M_o)V_w, \qquad (14)$$

where ρ is the polymer density, M_o is the molecular weight of the repeating unit, N_A is the number of Avogadro, V_w is Van der Waals volume of the repeating unit. It was called as free volume.

The occupied volume of the monomer unit perceived by a certain gas is determined as a sum of the Van der Waals volume and a volume nearby the monomer unit inaccessible to the gas molecules because of geometry constraints. In what follows, this inaccessible volume will be referred to as the dead volume (fig.14).

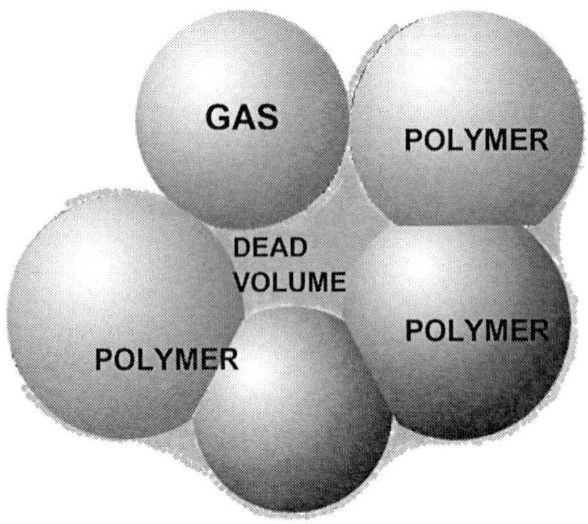

Figure 14. Definition of occupied volume.

In this case, occupied volume may be interpreted as specific characteristic of the gas–polymer system.

The accessible volume for a given gas (V_{acc}) was calculated with equation (14) where the Van der Waals volume was replaced by the occupied volume V_{occ}.

Among the gas molecules considered in this work, only methane has spherical symmetry characterized by a single radius. All other molecules (diatomic ones and triatomic CO_2) are more accurately approximated by ellipsoids. Therefore while scanning the surface of the repeat units, the two radii characterizing these molecules were considered, namely those equal to the length of the major ($2R_{max}$) and minor ($2R_{min}$) axes of the ellipsoids. The size and the shape of the dead volume are determined by only those spheres that touch the surface of the repeat unit. We will call them ''rolling over spheres''. For an accurate determination of the dead volume the number of the spheres should be sufficiently large. Thus, for a single atom from the repeat unit one should introduce about 10^6 virtual gas spheres that can intersect each other.

Table 11. Maximum and minimum radii of spheres and van der Waals volumes of the gas molecules

Gas moleculeR_w, Å		r_{ij},Å	R_{max},Å	R_{min},Å
CO_2 R_{wC}=1.80	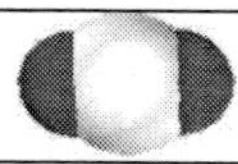	C-O 1.123	2.52	1.8
CH_4	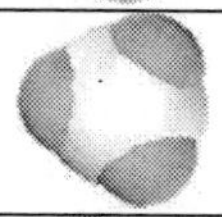	C-H 1.08	2.25	2.25
CO R_{wO}=1.36	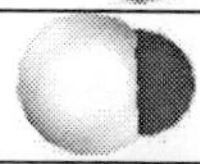	C-O 1.123	2.14	1.8
N_2 R_{wN}=1.57		N-N 1.095	2.07	1.57
O_2		O-O 1.208	1.96	1.36
H_2 R_{wH}=1.17	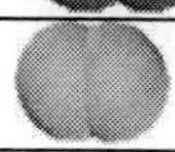	H-H 0.741	1.54	1.17

Previously, it has been shown that for all the polymers, one can note anomalies in the behavior of CO_2 concerning the dependence of diffusion coefficient D versus V_{occ}, if R_{max} values were used in the calculation of V_{occ}, strong deviations are observed for carbon dioxide. These deviations disappear or become much less pronounced if, instead, R_{min} values are used for this molecule [144]. It was shown that the quadrepole CO_2 interacts through its both oxygen atoms with hydrogen atoms of the polymer matrix and is situated relatively perpendicular on the direction of the gas stream [145].

VI. The Effect Conformational Rigidity on Several Physical Properties of These Polymers

As is known from the X-ray and neutron diffraction studies of the behavior of a macromolecular coil in the amorphous state, in the solid state the coil retains the Gaussian type of distribution typical to coils in dilute solutions [146]. This means that variation of the kinetic mobility of a macromolecule with temperature is governed mostly by its capability to conformational transitions, i.e. by the conformational rigidity of the chain. All this allows us to assume that some physical properties of polymers possibly correlate with the conformational rigidity of their chains calculated in the assumption of free rotation.

Such physical properties of polymers as, e.g., the glass transition temperature, friction coefficients, and module of elasticity are governed by the relation between the entropy and enthalpy factors. The Kuhn segment length (which characterizes the conformational rigidity of a polymer chain) contributes to the entropy term of the thermodynamic potential of the system $J = H - TS$ [147]. In the case when there is a correlation between physical properties of a polymer and the conformational rigidity of its chain, the entropy term must increase. If there is no such correlation, then, because of intermolecular interactions, the enthalpy contribution to the thermodynamic potential exceeds the entropy one. A correlation between a physical property of a polymer and the conformational rigidity of its chain shows that contribution of the conformational rigidity to this polymer property is significant.

As it was known, such polymer properties as friction coefficients, elasticity modulus, and yield strength are directly dependent on the glass transition temperature [148]; therefore, we will consider simultaneously the friction coefficients and glass transition temperatures or the elastic properties and glass transition temperatures.

In conclusion to this section, we emphasize that not all the observed correlations between the polymer property and the conformational rigidity of the chain can be directly explained. Nevertheless, the correlation does exist, which makes one ponder on.

To exclude the crystalline effect, we will consider physical properties of polymers merely in the amorphous state.

1. Glass Transition Temperature

The glass transition temperature is a conventional characteristic that determines the thermostability of the polymeric solid. The glass transition temperature is the temperature, at which the deformability of a sample sharply increases on heating. It depends on the heating rate, the character of load, the chemical composition, and the structure of the polymer material. Therefore, we tried to compare the glass transition temperatures of different polymers that were obtained under the same conditions.

Let us first consider the dependences of the glass transition temperature $T_{g.}$ on the conformational rigidity of a polymer chain for several classes of polyheteroarylenes (each of these classes having its specific type of heterocycle) [149,150]. Not all polyheteroarylenes fit the same $A_{fr}(T_g)$ dependence. There is a specific $A_{fr}(T_g)$ dependence within each class of polyheteroarylenes (Table 12).

Table 12. Dependence of the glass transition temperature on A_{fr} for various classes of the polyheteroarylenes

No	Structural Repeat Unit	A_{fr}, Å	T_g, °C
	Polyarylates		
1	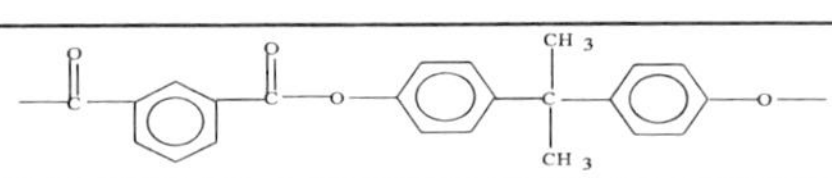	15.0	270
2	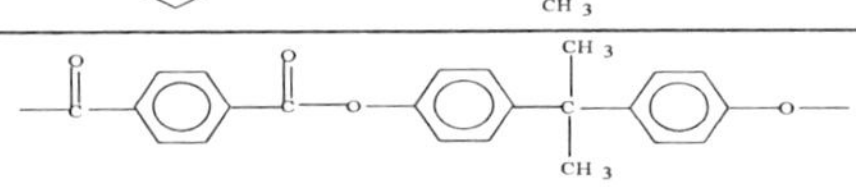	25.7	320
3	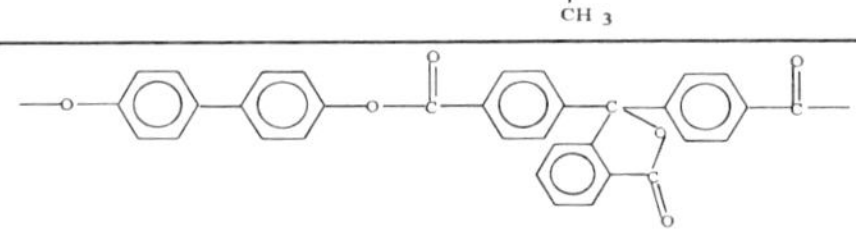	31.7	340
4	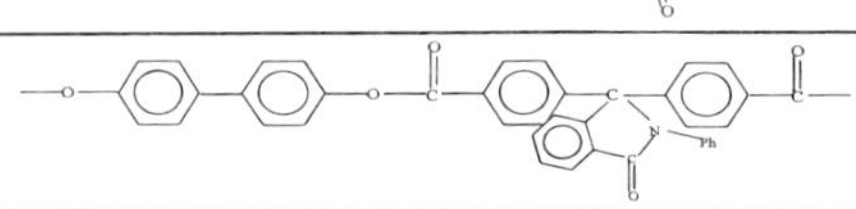	33.7	350

No	Structural Repeat Unit	A_{fr}, Å	Tg.,oC
	Polyimides		
5	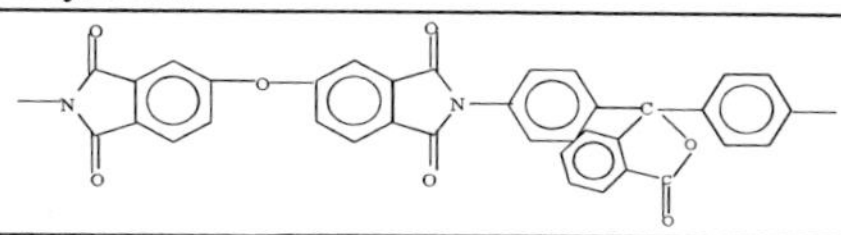	30.3	360
6	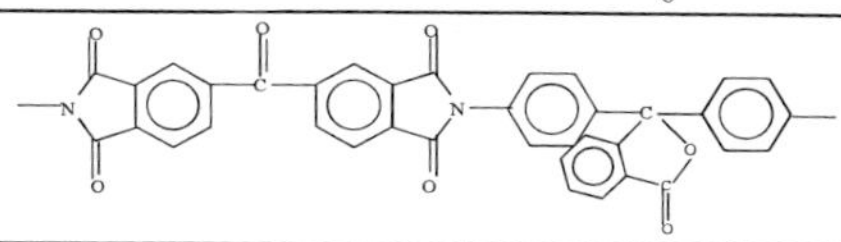	32.8	370
7	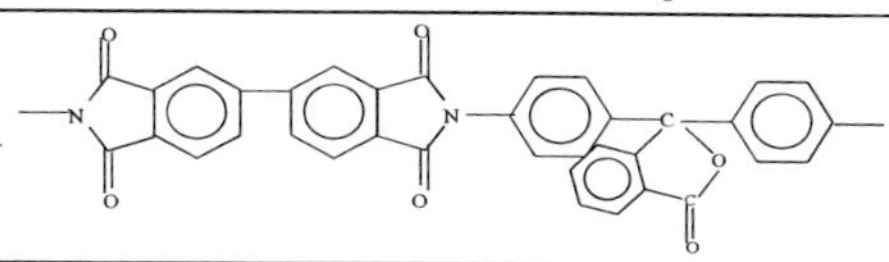	50.8	420
8	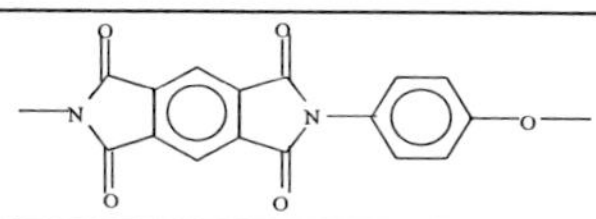	78.1	345
9	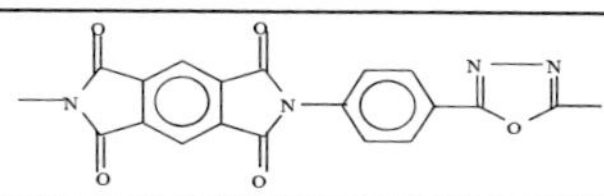	157.5	450
	Polyoxadiazoles		
10	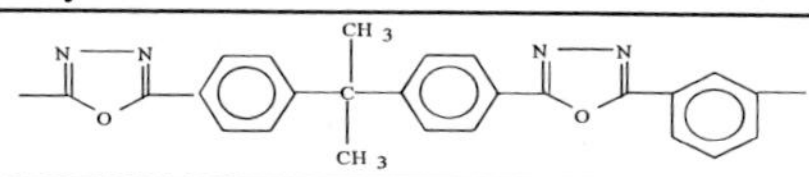	23.3	290
11	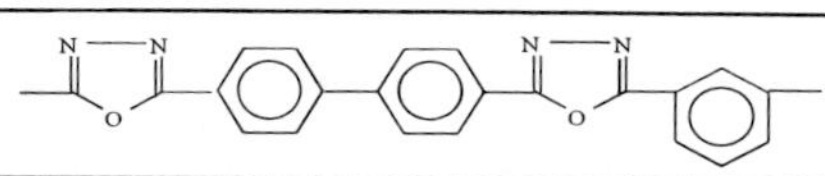	34.6	460
12	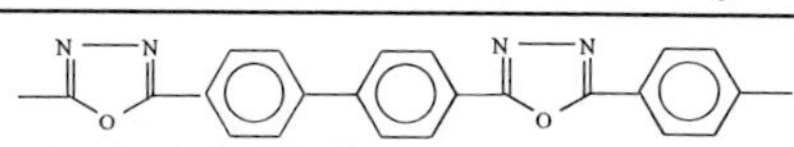	81.1	500
	Polybenzoxazoles		
13	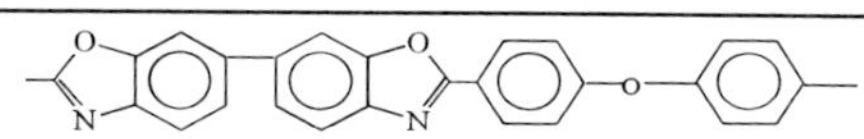	31.1	295
14	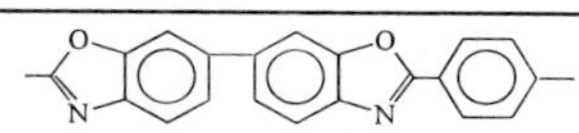	121.6	>460
	Polybenzimidazoles		
15	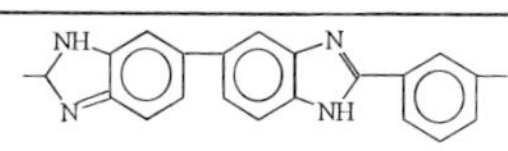	34.1	295
16	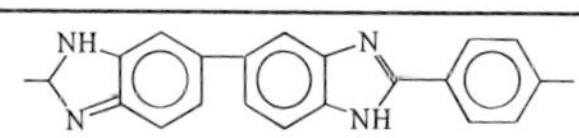	126.0	550
	Polybenzquinazolones		
17	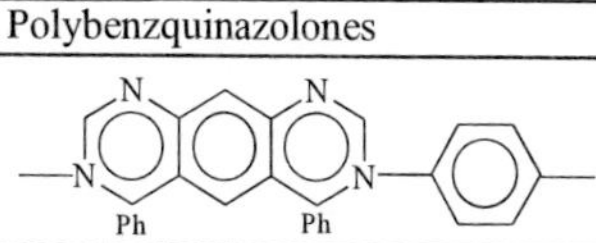	34.8	394
18	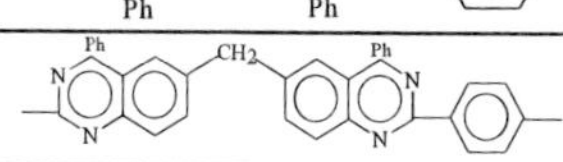	40.4	370

Table 12. (Continued)

No	Structural Repeat Unit	A_{fr}, Å	T_g, °C
19	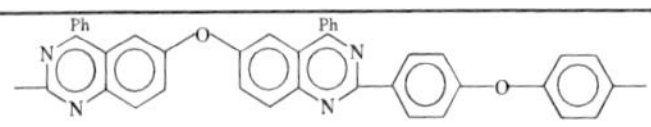	45.4	284
20	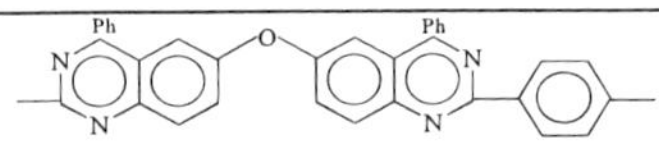	73.8	250

Inside each class, for the polymers containing heterocyclic fragments of different structure (compare the polyimide polymers 5-7 and 8, 9) the $A_{fr}(T_g)$, dependences are also different. For all classes of polyheteroarylenes under consideration the glass transition temperature increases with the increase of the conformational rigidity of the polymer chain. The only exception is polyphenylquinazolones, where the glass transition temperature decreases with the increase in the conformational rigidity of the polymer chain. It might be the case, when the contribution of the enthalpy term (arising from intermolecular interactions) to the thermodynamic potential of the system exceeds that of the entropy term (associated with the conformational rigidity of the polymer chain).

Let us consider the relationship between the conformational rigidity of polyphenylquinoxalines (PPQO) and the glass transition temperatures [70]. As can be seen from Table 13, the glass transition temperature of PPQO increases in accordance with the increasing conformational rigidity of a polymer chain. Such dependence of the glass transition temperature on A_{fr} shows that in this case the chain kinetic flexibility is to an essential degree determined by its conformational flexibility, whereas the intermolecular interaction plays a less important role in this relative change of softening temperatures.

To study the effect of the chemical structure of a polymer chain on its conformational rigidity, we consider PPQO. As can be seen from Table 13, in six PPQO polymers one fragment of the structural unit is similar and the other fragment is different. About half contour length of the whole structural unit is the phenylquinoxaline fragment, which contains four virtual bonds. The remaining contour length is the R fragment. Its geometric structure determines the value of A_{fr} for PPQO under consideration. With increasing number of virtual bonds at the same contour length, the flexibility of fragment R increases, thus increasing the flexibility of the whole polymer chain. For example, a 3,3'-benzophenone fragment (polymer 1) has six virtual bonds, and 4,4'-benzophenone fragment (polymer 3) has two virtual bonds. So, for the polymer containing the first fragment, A_{fr} is considerably smaller than for the polymer containing the second fragment. Polymers containing diphenyl oxide and 4,4'-benzophenone fragments (polymers 2 and 3, respectively) have the same number of virtual bonds with close bond lengths and angles. So, these polymers have close A_{fr}. With increasing virtual bond length in phenyl, diphenyl, and phenylnaphhtnylimide fragments (polymers 4-6). A_{fr} increases from 33 to 71 Å. On the example of PPQO, it is possible to show that the geometric structure of a structural unit (i.e., the number and length of virtual bonds and angles) reflects its chemical structure and, therefore, it is taken into account in calculation of the Kuhn segment length.

As can be seen from figure 15 the dependence of glass transition temperature on Kuhn segment is linear with high coefficient correlation. The glass transition temperature increases with the increase of conformational rigidity.

Table 13. Physical properties of some PPQO depending on the conformational rigidity of polymeric chain [18]

N	Structural Repeat Unit	A^{*}_{fr}, Å	T_g, °C	The friction coefficient of the	
				polymer	Filled systems
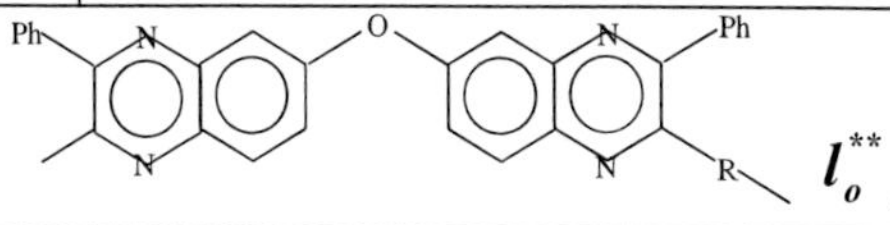 l_o^{**} =13.1, Å					
1	R= 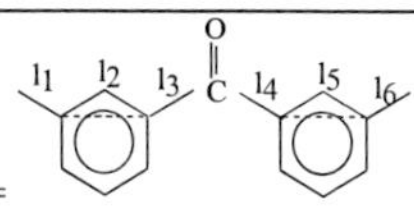 l_o=10.0 Å	17.4	210	0.76	0.28
2	R= 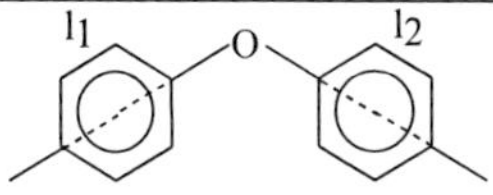 l_o=9.4 Å	30.3	215	0.63	0.18
3	R= 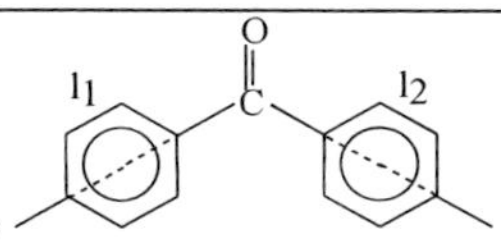 l_o=9.6 Å	32.2	245	0.80	0.32
4	R= 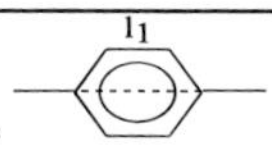 l_o=5.8 Å	35.2	275	0.68	0.23
5	R= 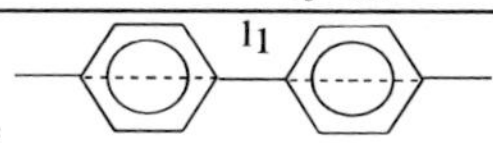 l_o=8.7 Å	48.9	295	0.73	0.26
6	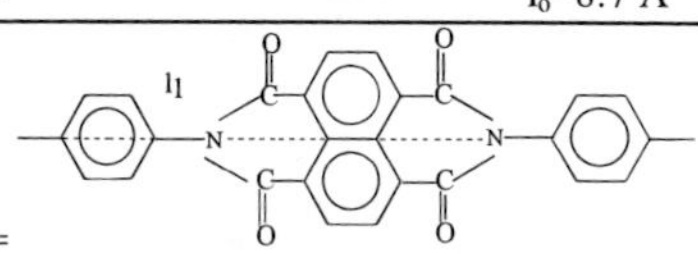 R= l_o=17.1 Å	71.9	375	0.82	0.33

* - A_{fr} corresponds to equal contributions of all isomers (Table I).
** - contour length of fragment R.

Another example of such dependence is the dependence on Figure 16 for polyphenylenes (Table 14). Coefficient of correlation in this case equals 99.66% [151].

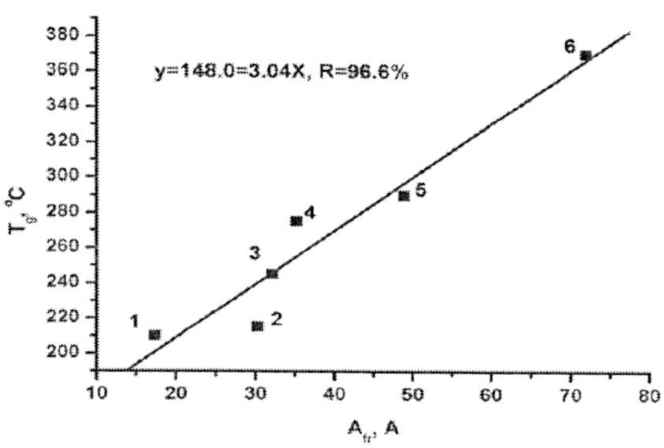

Figure 15. The Dependence of glass transition temperature on Kuhn segment for polymer PPQO (Table13).

Table 14. Conformational parameters and glass transition temperature of polyphenylenes

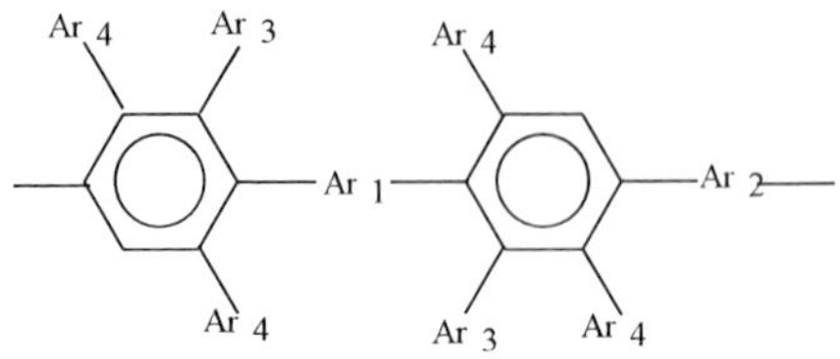

No	Ar_1	Ar_2	Ar_3	Ar_4	l_o.Å	A_{fr}, Å, p	T_g,°C
1	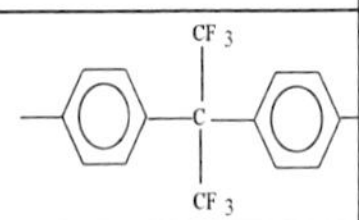	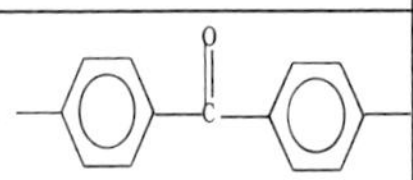			25.33	23.1 12,3	285
2	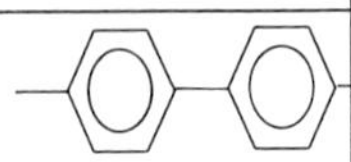	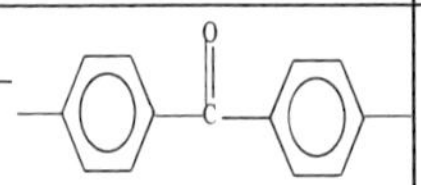			24.89	32.6 16,1	305
3	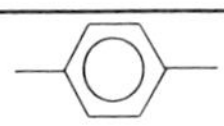	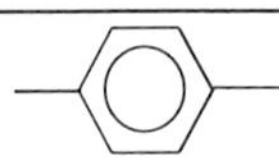	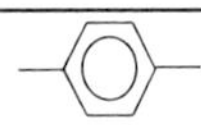	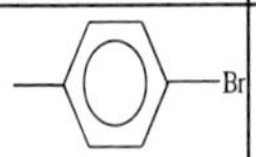	16.00	31.1 22.4	346
4	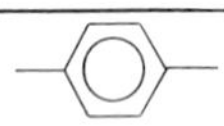	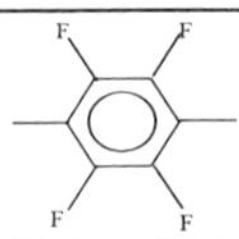	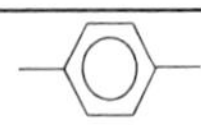	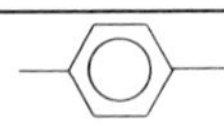	16.00	31.1 24.6	367

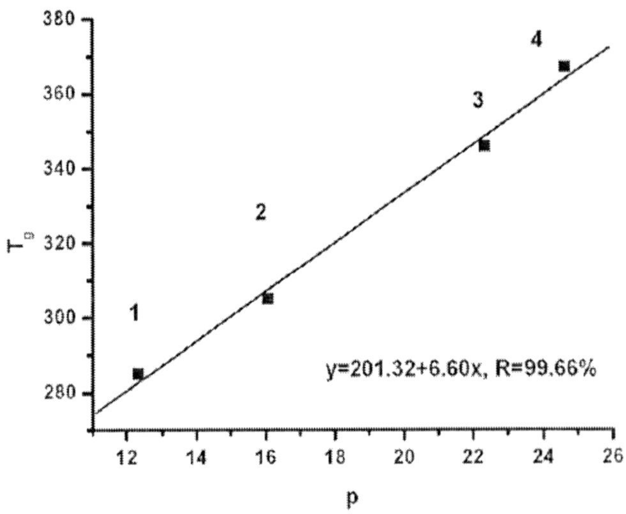

Figure 16. The dependency of glass transition temperature on the number of aromatic rings in the Kuhn segment for polyphenylenes (Table 14).

The same behavior of the dependence $T_g(p)$ is inspect for polymers (1,3,4 oxadiazole-amide-ester)s containing in the main chain besides phenylene rings oxadiazole rings, amide and ester bridges (Table 15, figure 17) [152].

Table 15. Conformational parameters and glass transition temperature of poly(1,3,4 oxadiazole-amide-ester)s

N	Structural Repeat Unit	l_o.Å	A_{fr}, Å	C∞	p	T_g, °C	Elastic Modulus (GPa)
1	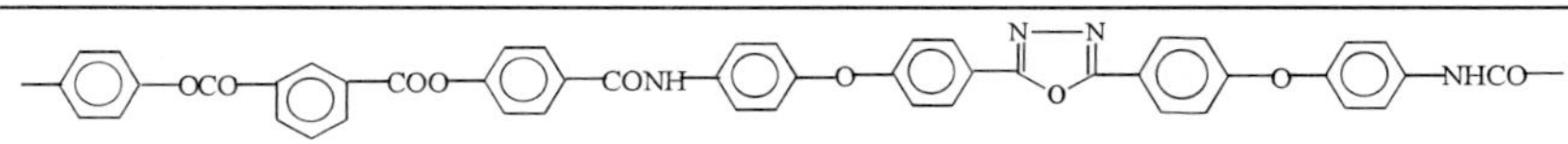	42.85	45.1	1.05	8.42	235	2.83
2	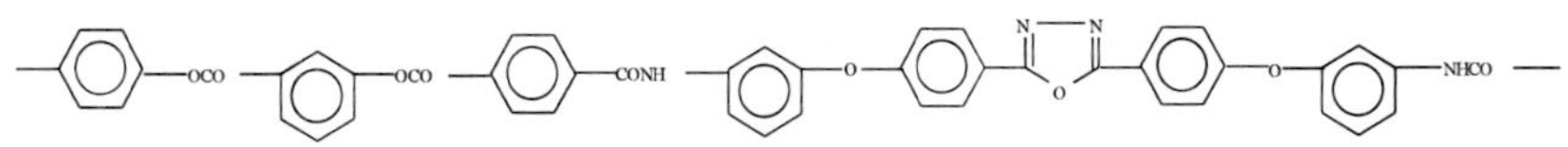	42.84	27.8	0.65	5.19	225	2.22
3	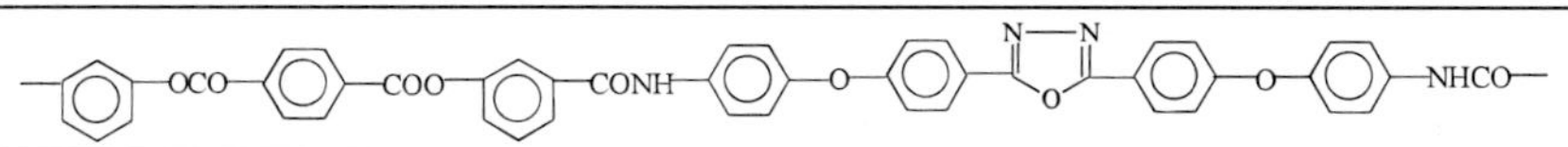	42.58	34.1	0.80	6.40	232	2.51
4	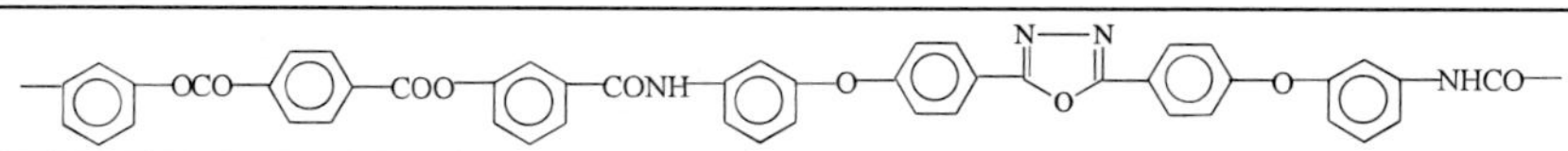	42.58	23.8	0.56	4.47	215	-
5	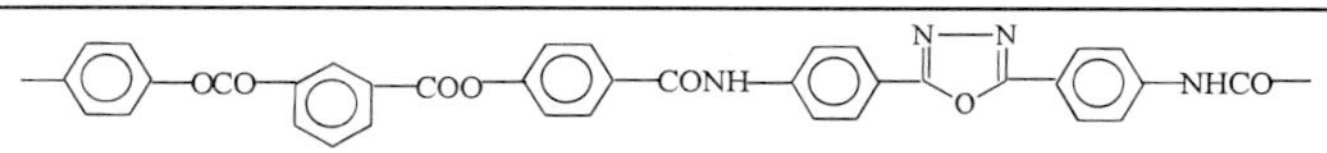	32.98	68.1	2.06	12.39	260	3.98
6	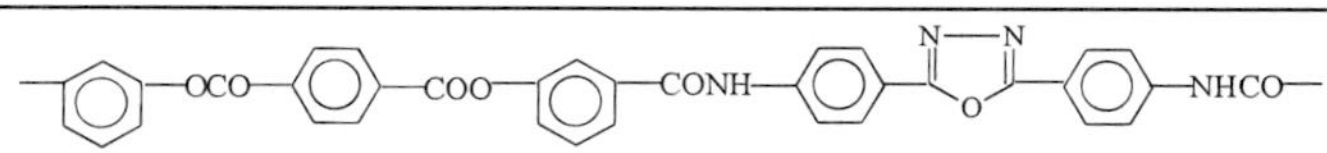	32.69	40.7	1.24	7.46	245	-

Table 16. Physical and conformational properties of polyimidoamides

No	Structural Repeat Unit	Mw g/mol	Mn g/mol	Tg ^0C	IDT ^0C	l_0, Å	A_{fr}, Å	p
1	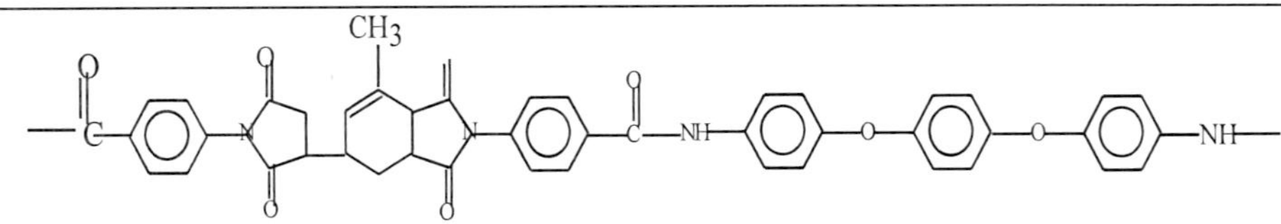	53300	33100	238	350	36.36	5.0	4.80
2	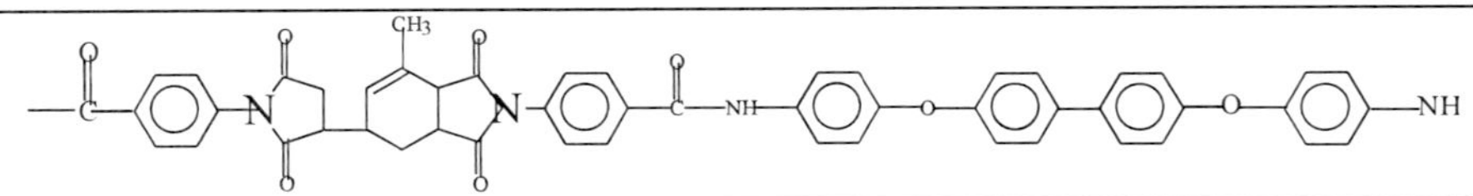	66000	41500	223	370	40.08	27.9	5.57
2a	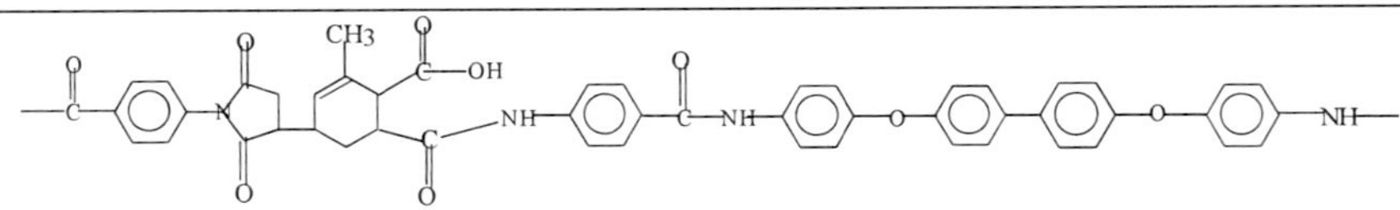	66000	41500	223	370	40.08	23.4	4.67
2b	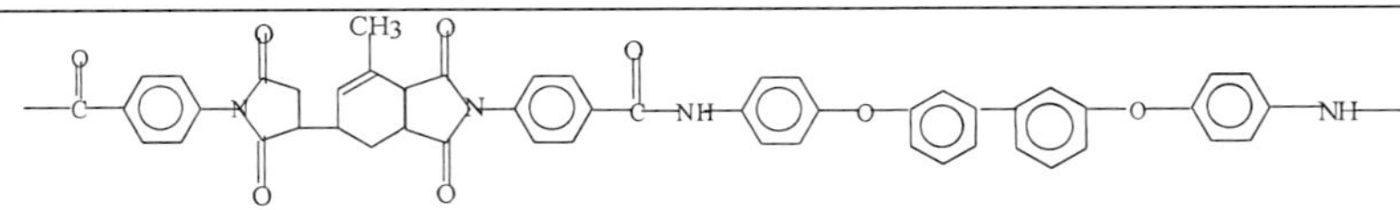	66000	41500	223	370	40.54	21.6	4.26
3	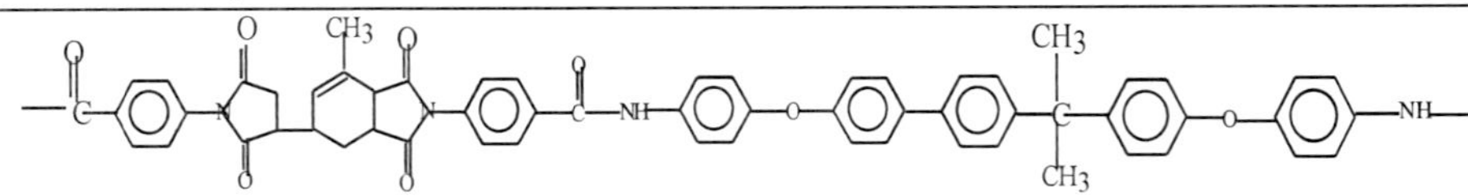	50000	31500	234	370	41.03	25.4	4.78
4	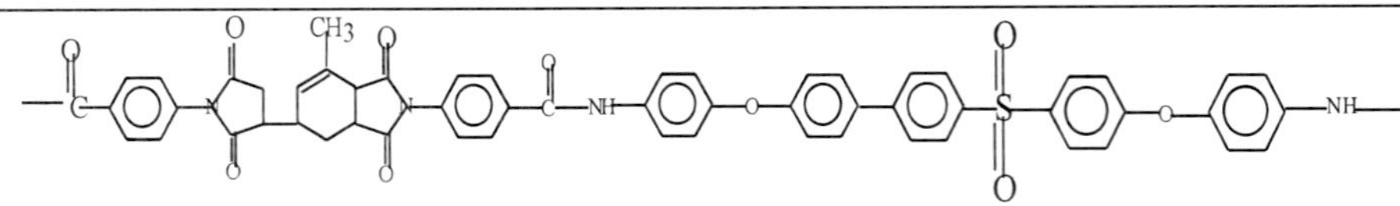	48600	29200	226	380	41.35	22.3	4.32

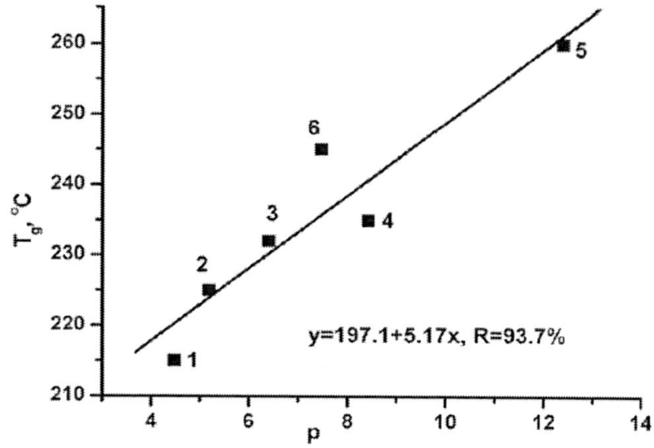

Figure 17. The Dependence of glass transition temperature on the number of aromatic rings in segment Kuhn for poly(1,3,4 oxadiazole-amide-ester)s (Table 15).

The above mentioned examples show the inherence of the linear dependence of glass transition temperature upon the conformation rigidity to all polyheteroarilens while considering the groups of polymers with an identical heterocycle. So one can reveal reasons for declining of glass transition temperatures from a general dependence.

a) The possibility of structure defects determination

The dependencies of glass transition temperatures upon the Kuhn segment are depicted in Figure 18(a) for polymers from the Table 16 [153]. As can be seen the linear dependence is observed for a majority of polymers with high correlation coefficients, and the correlation coefficient with the Kuhn segment is lot of high. The polymer 2 falls out from that dependence.

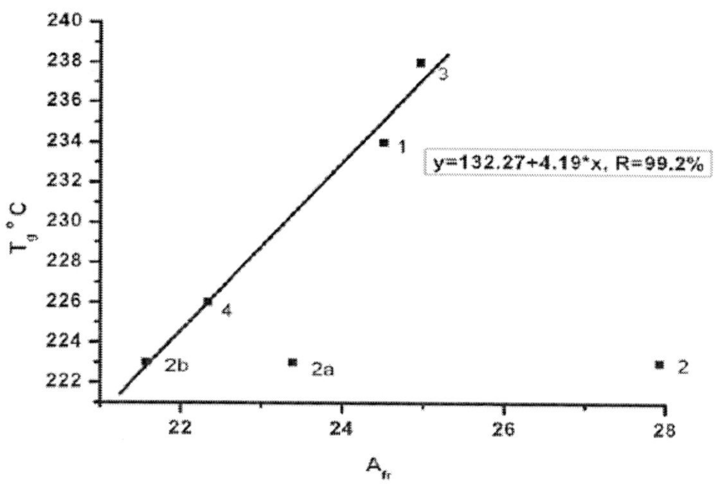

a

Figure 18. (Continued).

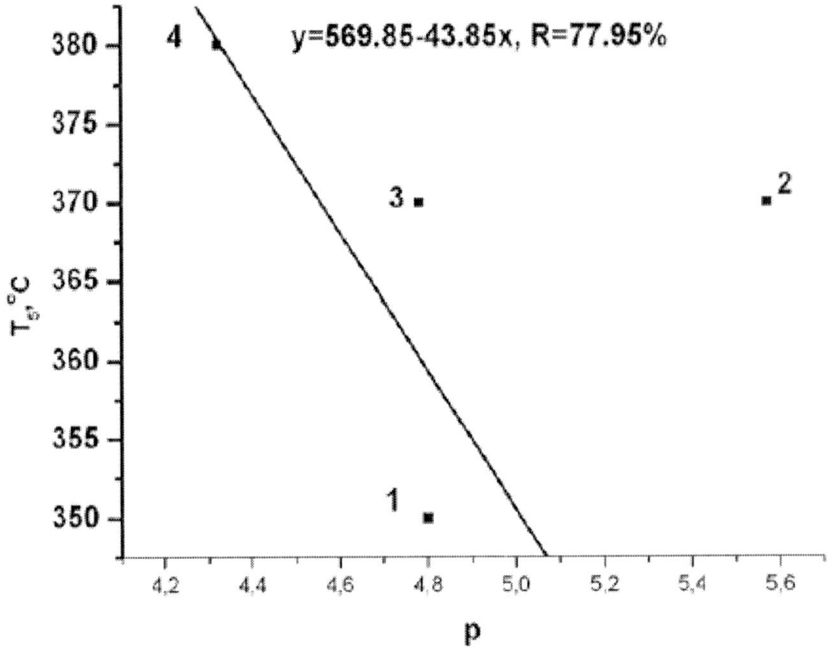

b

Figure 18. The dependence of glass transition temperature on Kuhn segment (a) and number of geterocycles in Kuhn segment (b) for polymers from Table 16.

A similar behavior is observed for a temperature of the 5% weight loss but with the lower correlation coefficient (Figure 18b). Molecular mass in this polymer group are almost the same. One can suppose the appearance of defect structures during the synthesis process: the part of imide cycles are cyclized incompletely or the part of phenilen rings came into a chain in the meta-position. To check these suggestions Kuhn segments for polymers 2a and 2b have been calculated. While a cyclization is not complete the Kuhn segment was decreased but that decrease is not sufficient to satisfy with the dependence. At the same time the Kuhn segment for the polymer 2b with two phenylen rings in meta-position is equal to 21.58 Å. This value is in good accordance with the rest of points in the linear dependence $T(A_{fr})$, with the correlation coefficient of 99.2%. These calculations suggest the probable existence of both defect structures during the synthesis process. Recently the partial cyclization of parabanic rings has been also observed [154] in polyimides containing tiophenic fragment and this fact has been experimentally supported afterwards.

b) the possibility of determinate the glass transition temperature, which it can not measure.

Let as consider the group of polymers poly(1,3,4-oxadiasole-ether)s (Table 17) [155].

For some of the present polymers (Table17) one must take into account the hindrance of rotation. This is connected to the isopropyliden bridge. In this case the side substituents are close to phenyl rings of main chain. Contacts between hydrogen (fluorine) atoms of isopropyliden group and atom of phenyl rings are less the sum of Van der Waals radii of hydrogen atoms - 2.34 Å (2.67). These produce to the hindrance of rotation. It is necessary to calculate by quant chemical method AM1 [14] the dependence of potential function from the rotation angle around the bonds with hindrance.

For three polymers – 1, 2 and 4 the dependence of the glass transition temperature on the number of aromatic rings in Kuhn segment is linear (see Fig. 19). For 5 polymer the glass transition temperature was not measured. The point that corresponds to polymer 3 lies out of line. Since points that correspond to 1, 2 and 4 polymers lie on the same line with the correlation coefficient that equals to 99.92%, it is possible calculate the glass transition temperature for polymer 5. From the dependence of Tg (p), y= 153.87 + 5.236x, the Tg of polymer **5** has been calculated. It is equal to 210 °C.

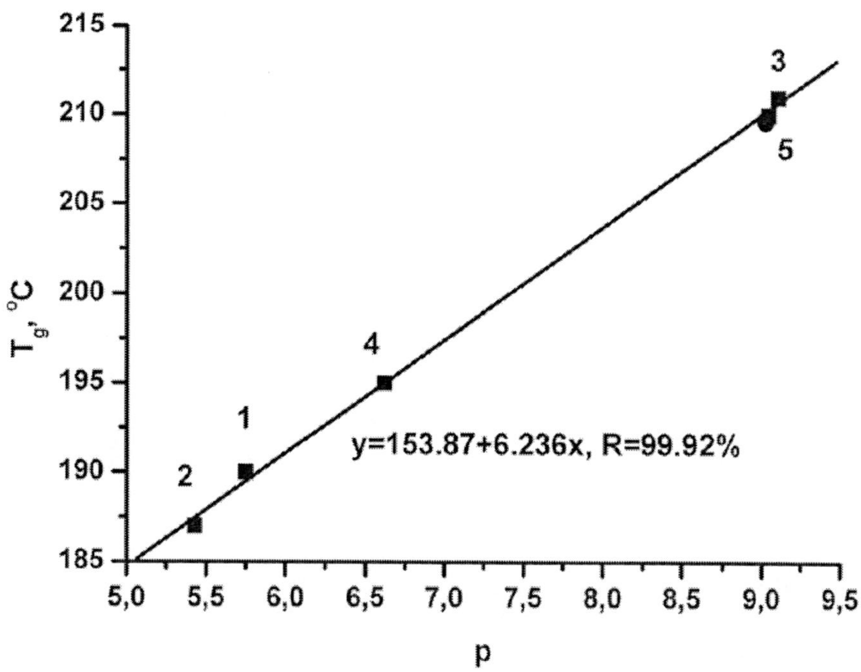

Figure 19. The dependence glass transition temperature on the number aromatic rings in segment Kuhn of poly(1,3,4-oxadiasole-ether)s (Table 17).

c) Identification of polymers with deficiently high molecular mass.

The group of polymers polyesteramides is shown in the Table 18.The samples of polymers 2 and 6 are out of the straight line (fig.20) which could be due to the low molecular weight of these two polymers. In such cases gaussian coil has not formed yet, the packing of the chains is loose, and the conformational transitions take place very easy in glass transition state. Later on the low molecular weight has been confirmed by experiment.

Table 17. Conformational parameters and glass transition temperature of poly(1,3,4-oxadiasole-ether)s

No	Structural Repeat Unit	l_o, Å	A_{fr}, Å	A_{hin}, Å	p	T_g, °C
1	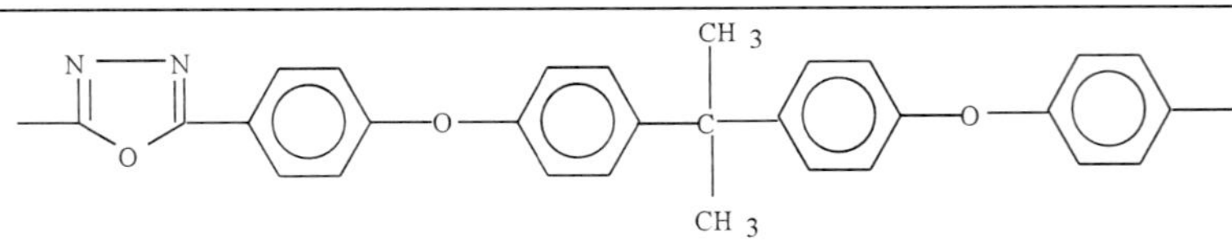	22.16	24.5	25.2	5.53 5.70	190
2	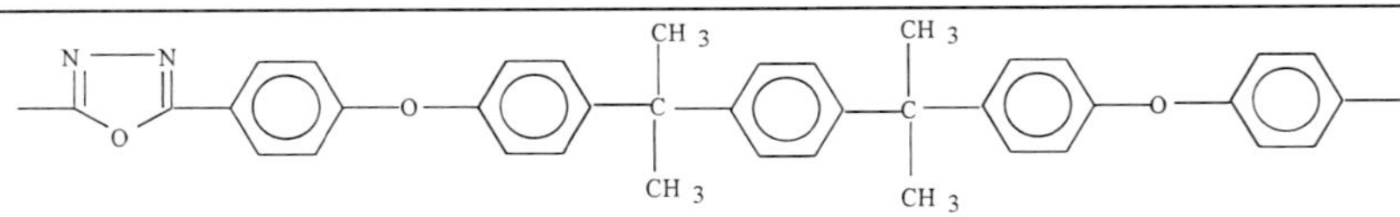	26.95	22.8	24.2	5.07 5.38	187
3	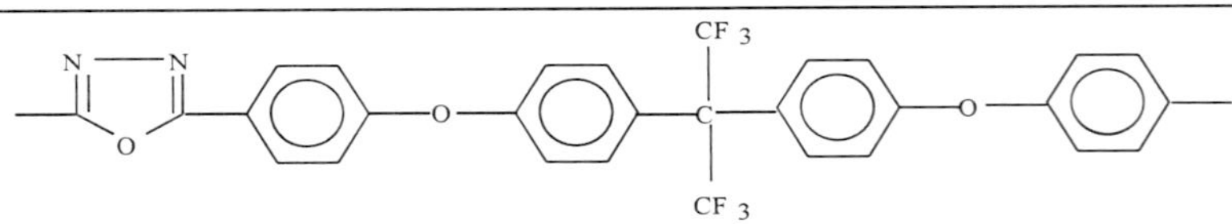	22.16	24.5	40.6	9.16	211
4	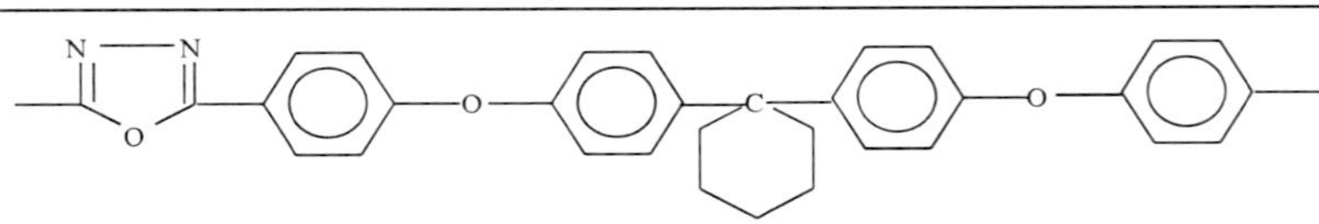	22.08	23.6	24.4	6.42 6.62	195
5	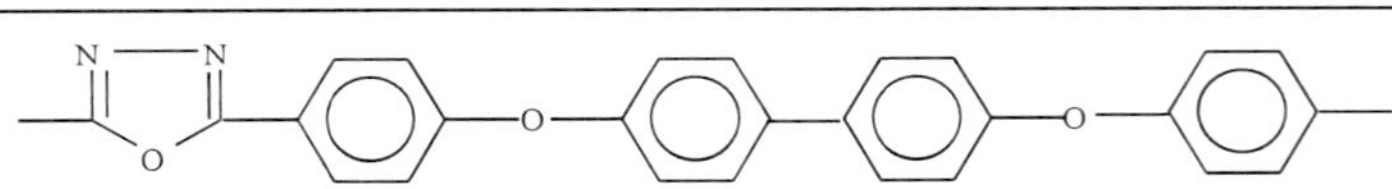	21.34	38.6	-	9.04	ND 210

Table 18. Conformational parameters and glass transition temperature of polyesteramides

N_o	Structural Repeat Unit	l_o, Å	A_{fr}, Å	T_g, °C	V_w, $Å^3$	V_f (cm^3/g)
1	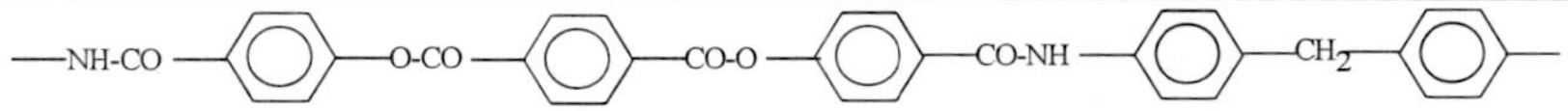	30.83	72.5	218	521.673	0.2416
2	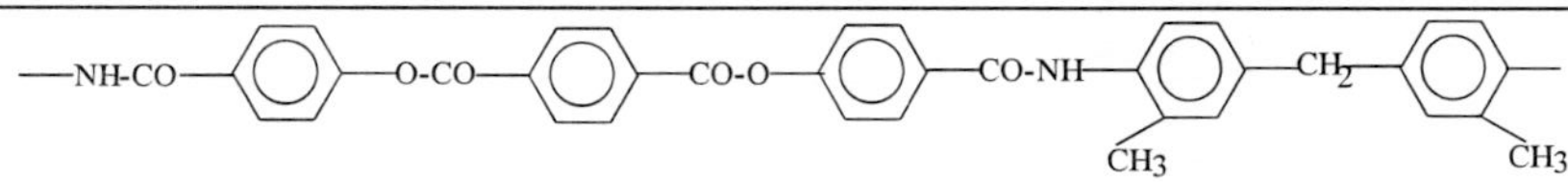	30.81	72.3	190	553.099	0.2385
3	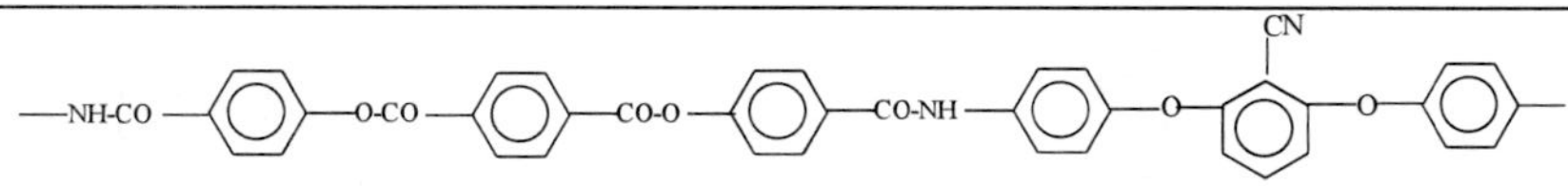	36.06	71.4	221	614.855	0.2358
4	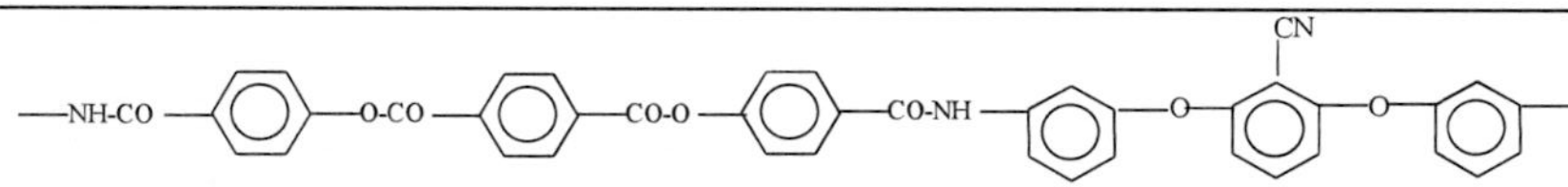	35.47	28.1	205	614.851	0.2428
5	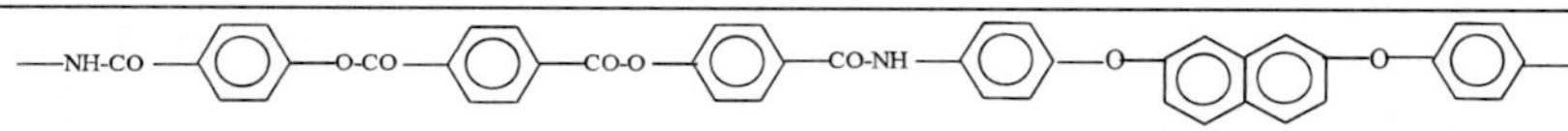	38.45	41.9	210	642.950	0.2538
6	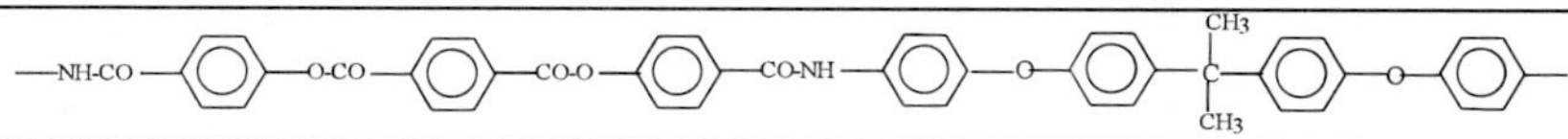	40.61	40.6	190	721.688	0.2673

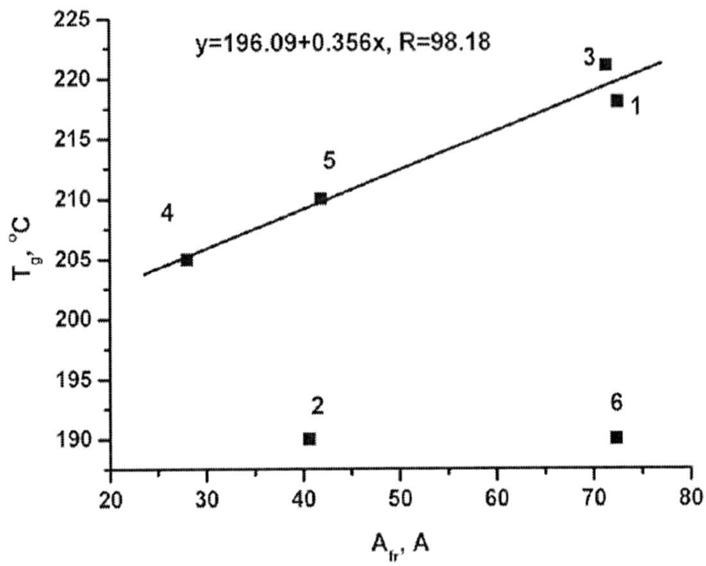

Figure 20.Dependence of glass transition temperature (T_g) on Kuhn segment (A_{fr}) for polyesteramides from Table 18.

d) The correlation of glass transition temperatures with free volume allows determinate the presence of the solvent in polymer matrix.

In the polymers of Table 18 (figure 21) the dependence of glass transition temperature on the free volume is linear for three polymers: 1, 3 and 5. In the case of sample 4 the free volume is significantly smaller than it should be according to the general dependence in this series.

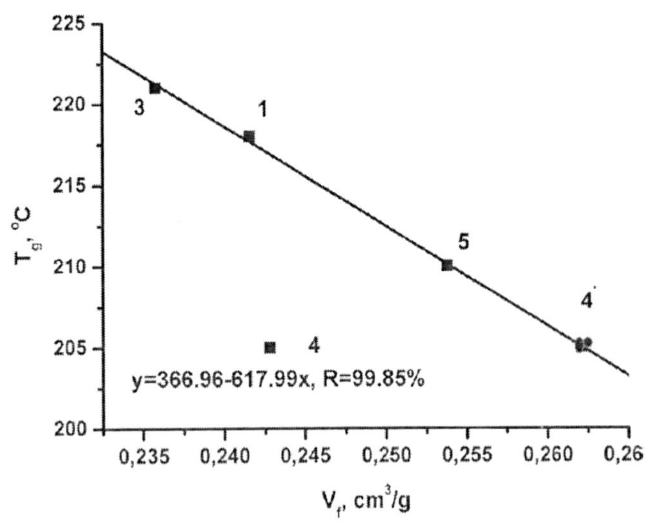

Figure 21. The dependence of the glass transition temperature on free volume for polymers from table 18.

It shows that in case of this later polyheteroarylene, when the film was prepared, part of the solvent remained inside it and did not evaporate by drying until constant weight. This part of the free volume can be calculated. Experimental free volume for polymer 4 is 0.2467 cm3/g. The solvent used to cast the films was N-methyl-pyrrolidinone (NMP). Van der Waals volume (1 gram of molecule N-MP is 0.6157 cm3/g) was calculated. By the equation describing the dependence of glass transition temperature on the free volume (y=366.96-617.99x, R=99.85%) the free volume of polymer 4 was found to be 0.262 cm^3/g. The difference between the experimental value of free volume and the calculated one is 0.0192 cm^3/g, or 7.9 % the free volume of this polymer. That is the part of free volume which is occupied by the solvent and this is why the density of the polymer was higher. The value of the free volume occupied by the solvent corresponds to 2.4 %.

2. Coefficient of Friction

Let us consider the relationship between the conformational rigidity of polyphenylquinoxalines (PPQO) and friction coefficients and thermofrictional properties of plastics on their basis of these polymers (Table13) [70].

A linear dependence is observed for the friction coefficients of PPQO (Table 13); (Fig.22). The presence of frictional filler results in the overall decrease of the friction coefficient, whereas the trend of the increase in the friction coefficient with increasing A_{fr} is retained (Fig.23).

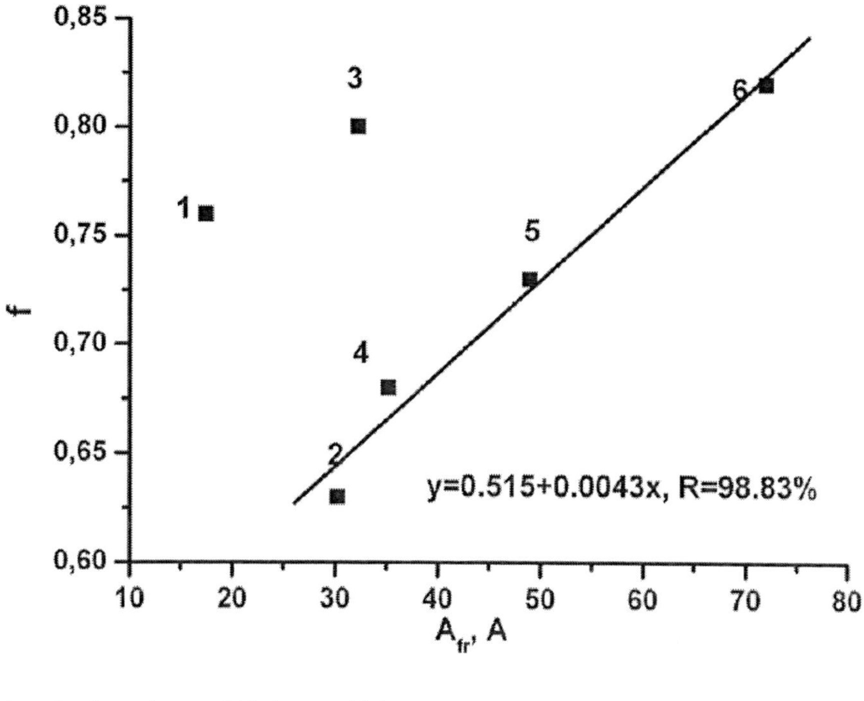

Figure 22. The dependence of friction coefficient on Kuhn segment for polymer table 13.

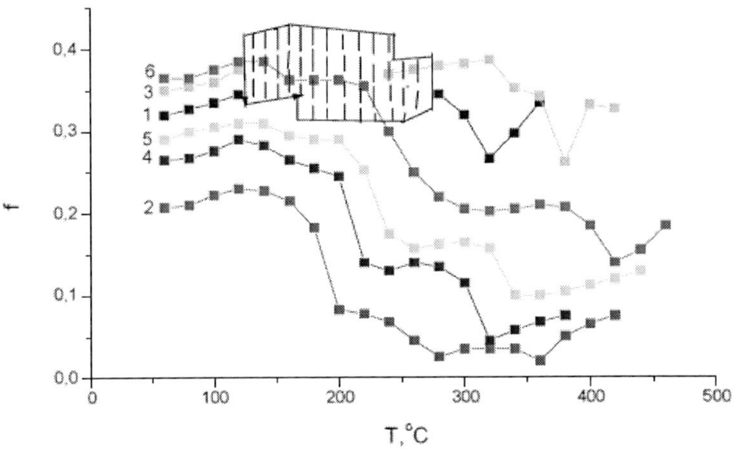

Figure 23. Dependence of friction properties of filled PPQO systems on temperature (the curve numbers correspond to the numbers in the table 13).

For polymers 1 and 3 containing 3,3- and 4,4-benzophenone fragments, respectively, the friction coefficients deviate from dependence. Despite a significant conformational flexibility of these polymers (A_{fr}=17 and 32 Å), they have relatively high values of the friction coefficient (0.76-0.80 for polymers and 0.28-0.32 for filled systems). These specimens were studied after being subjected to friction; particles of transferred metal were observed in their surface layers, whereas there was no metal transfer for other PPQO during the test period (1 hour).

Thermofrictional and relaxation properties were studied in a wide temperature range from 40 to 400°C. It was important to reveal whether the dependence of f on A_{fr} retains or not at high temperatures. As can be seen from Fig. 23, with the only exclusion of systems 1 and 3 all other systems retain the dependence of the friction coefficient on the conformational rigidity in the whole temperature range studied. The friction coefficient increases with A_{fr} increasing. For example, for the system based on polymer 6 (A_{fr} = 71 Å), the friction coefficient is the highest in the whole temperature range and lower thermofrictional curves correspond to smaller A_{fr} (see curves 5, 4, and 2 corresponding to polymers with A_{fr} = 49, 33, and 30 Å). In addition to this, the obtained data (Fig. 23) show that for filled systems based on PPQO the friction coefficient decreases under thermal treating. Probably, this is due to decrease of intermolecular interaction and decrease of the friction coefficient of filler (graphite).

A slower decrease in the friction coefficient with temperature is observed for systems based on polymers 1 and 3. In the region of the glass transition temperature the process of friction is metastable, which is not found for systems based on other polymers. Such a behavior of PPQO 1 and 3 and the systems based on these polymers can be associated with a reactive electron-acceptor carbonyl group in their macromolecular repeat unit, which favors the formation of partly or completely crosslinked structures during the friction process. The crosslinking must result in an increase of the friction coefficient. A high conformational flexibility of macromolecules of these polymers provides a low glass transition temperature and favors the process, which results in an unstable character of the change in the friction coefficient of plastics at the glass transition temperature and above it (from 200 to 270°C).

Thus, as is shown on the example of PPQO, different softening temperatures, friction coefficients, and thermofrictional properties of filled systems based on these polymers are probably related to the conformational rigidity of the backbone, which has different structure of the repeat unit for different PPQO polymers.

3. Elastic Properties

Several aromatic cyclochain polymers are capable of forming fibers and films, which possess high elastic module. Such properties of film- and fiberforming polymers are due to the conformational rigidity of the polymer chain determined by the structure of a macromolecule and alternation of active groups and intermolecular interactions of these groups in polymer solution as well as in solid. The properties of products based on these polymers depend on the same factors.

It is of interest to try to relate the conformational rigidity of a polymer chain (A_{fr}) with such characteristics (measured for films and fibers) as the elasticity modulus (E) and the yield strength (Σ). The experimental data on the orientational stretching of polyamidobenzimidazoles (PABI) are given in [156]. The polymer samples listed in Table 19 have the molecular weight in the range of 6-8 10^4. All the samples are amorphous. The conditions of sample preparation and test were the same as described in [157]. Table 19 summarizes the elastic module of nonoriented isotropic films and fibers immediately after their preparation (E_1) and after thermal treating (E_2).

Table 19. The Kuhn segments and properties of films and fibers of PABI [156]

No	Structural Repeat Unit	A_{fr}, Å	Films		Fibers	
			E_1 GPa	E_2 GPa	E_1 GPa	E_2 GPa
1	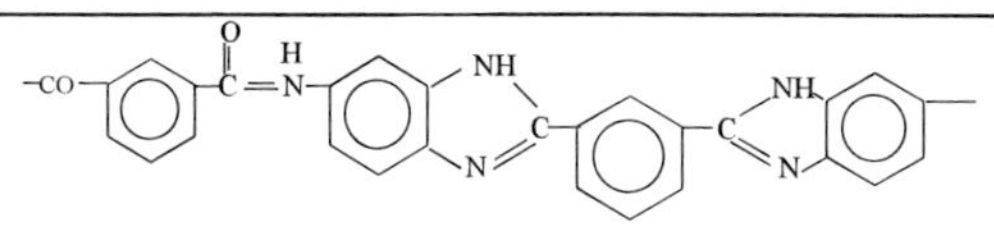	32.8	9.6	10.6	11.0	14.0
2	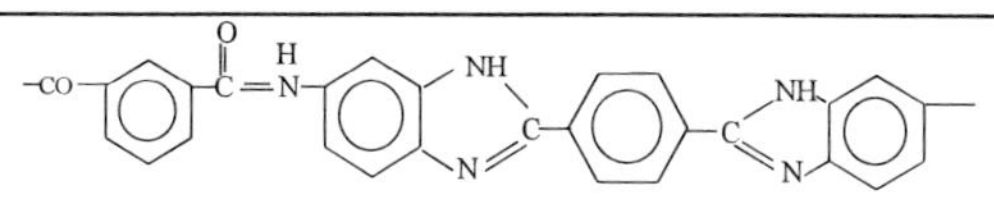	48.8	10.8	12.3	25.1	33.3
3	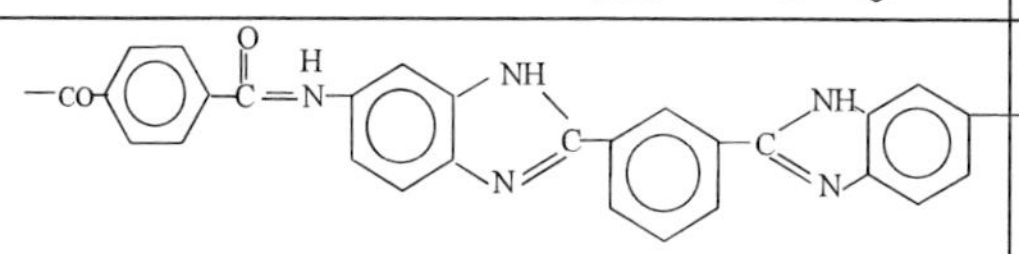	50.9* 64.2	11.3	13.3	30.4	38.2
4	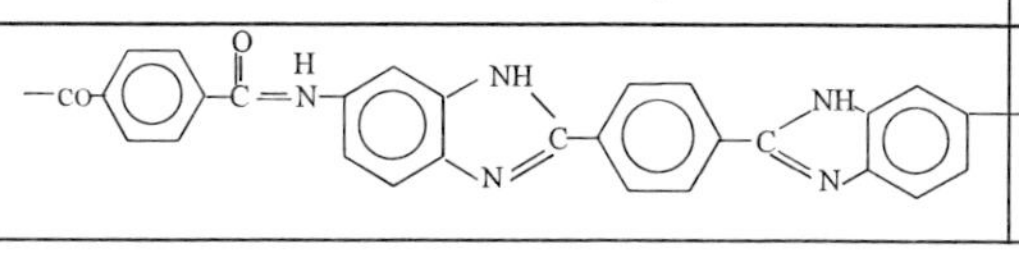	167.8	12.9	15.3	54.3	95.7

The dependences of the elastic module E_1 and E_2 for films and fibers of PABI on the conformational rigidity of the polymer chain ($1/A_{fr}$) are presented in Figs. 24 and 25, respectively. The conformational rigidity of a polymer chain in a non-oriented specimen has little effect on the elastic modulus of the films but considerably affects the properties of fibers. With the conformational rigidity of the polymer chain increasing, the elastic modulus increases by the factor of 4.5 for fibers and merely by a factor of 1.3 for films. After thermal treating, the elastic modulus increases drastically (by the factor of 6.6) for fibers and much less (by the factor of 1.44) for films having an increasing conformational rigidity of the polymer chain.

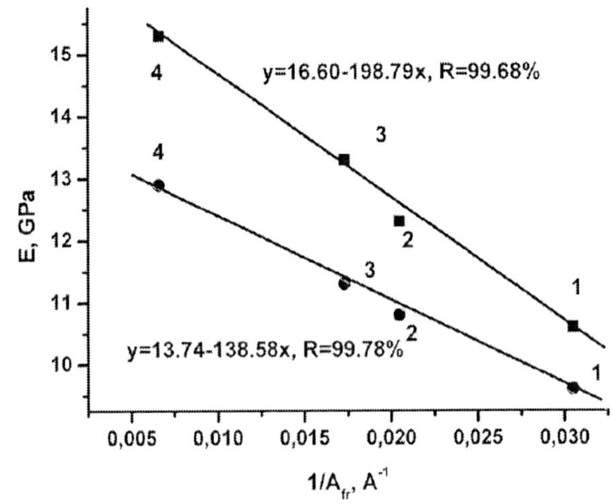

Figure 24. Dependence of the elasticity module of the PABI films (Table 19) on the conformational flexibility of the polymer chain. (Equal probability of the presence of isomers in the chain was assumed, so that A_{fr} was calculated as arithmetic average.)

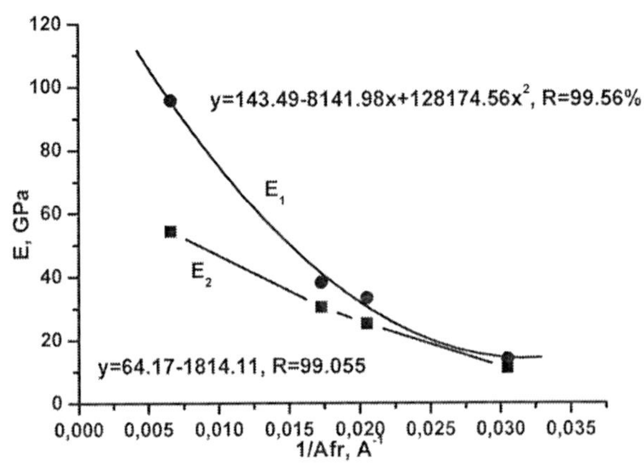

Figure 25. Dependence of the elastic modulus of the PABI fiber (Table 19) on conformational flexibility of the polymer chain.

After thermal treating, the elastic modulus of the most rigid PABI polymer (Table 19, polymer 4) increased approximately by the factor of two. The authors of [156] believe that formation of mesamorphic phase is possible in this case. The phase is obtained in the course of thermal treating of a preoriented (by repeated stretching) polymer.

Under these conditions disintegration of the system of original hydrogen bonds occurs. This gives rise to segment mobility, which results in mutual orientation of macromolecular fragments and the following formation of a highly oriented phase with a new system of hydrogen bonds. Similar dependences of the elastic modulus and yield strength of films on the conformational rigidity of the polymer chain can be observed also for copolymers.

For copolymer [158]

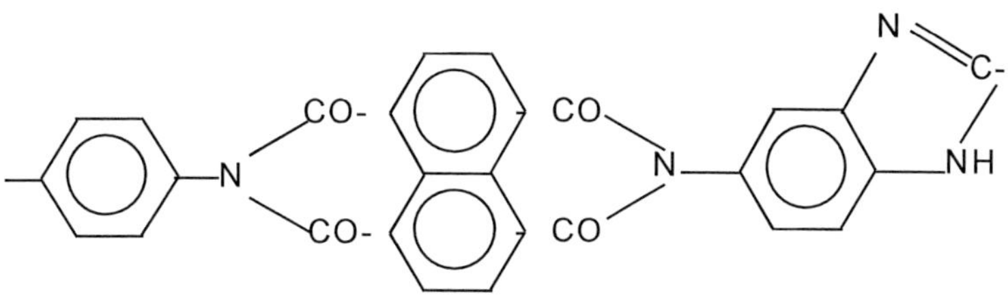

component **A**

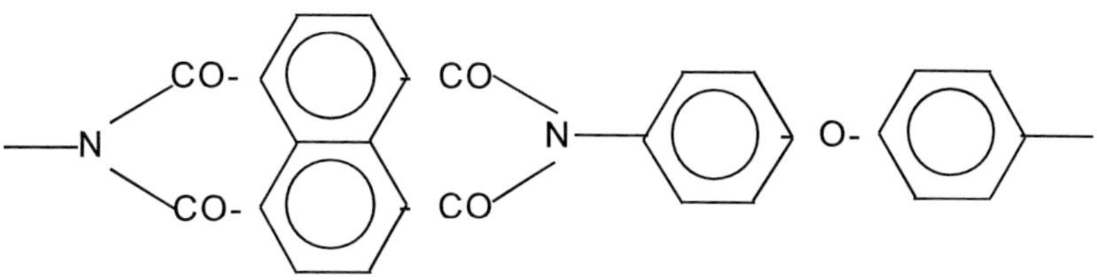

component **B**

the elasticity module and the yield strength (Σ) of films were measured for different ratios of **A** and **B** components.

Table 20. Dependence of the films yield strength and elasticity modulus on the copolymer composition

A:B(%)	Intrinsic η (5% dl/g)	A_{fr},* Å	Σ, (MPa)	E (GPa)
100	6.0	296	300(350)	4.56(5.10)
50:50	6.5	118	185(214)	2.47(2.33)
40:60	4.7	104	169(185)	2.54(2.47)
20:80	8.1	84	177(165)	2.86(2.31)

* calculated for the statistical copolymer [160]; in brackets values of elasticity modulus and yield strength for heated films are given.

As can be seen from Table 20, there is a good correlation between the yield strength of the heated films and the Kuhn segment length at different component ratios. This correlation is somewhat worse for the elasticity modulus. However, one is to take into account that the measurements were carried out for the specimens with different reduced viscosities, i.e. with different molecular weights.

Table 21 summarizes the yield strengths (Σ) values for PPQCO of different structures [159]. In this case correlation between Σ and A_{fr} is observed as well. The only exclusion is polymer 2. In this polymer the bridge between the quinoxaline cycles is very short (one phenylene ring), as compared with bridges in other polymers. It might give rise to steric hindrances, which impede free rotation of the fragments with respect to each other in the macromolecule. This can be explained by the fact that the interatomic distances between isolated fragments of the repeat unit become shorter than the sum of the van der Waals radii. These results in that rotations about several virtual bonds become hindered. Hence, the real value of the Kuhn segment for this polymer is slightly higher than that corresponding to free rotation.

Table 21. Dependence of the yield strength of PPQCO fills on the conformational rigidity

N	Structural Repeat Unit	A_{fr}, Å	T_g,°C	Σ,MPa
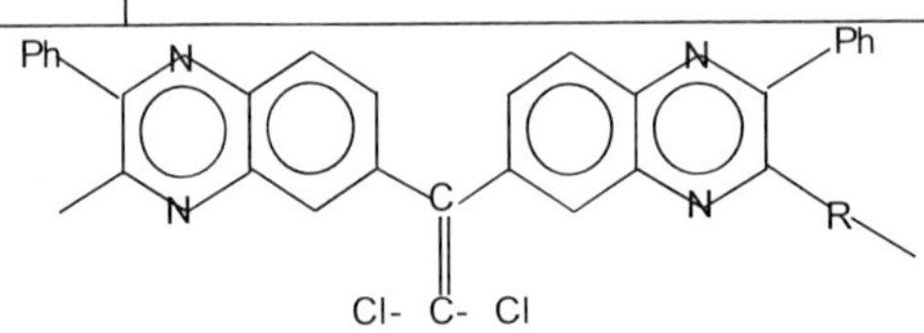				
1	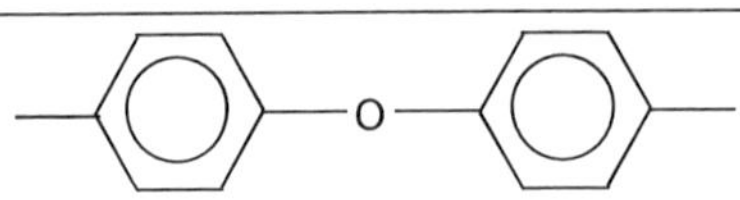 R=	25.5	280	65.5
2	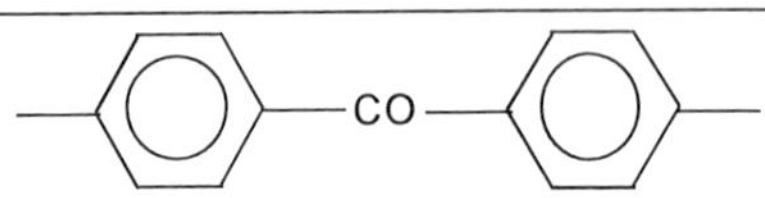 R=	27.0	290	72.04
3	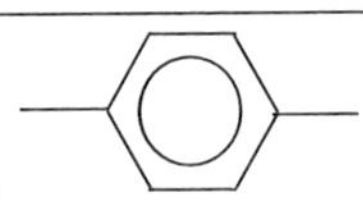 R=	29.7	330	81.4
4	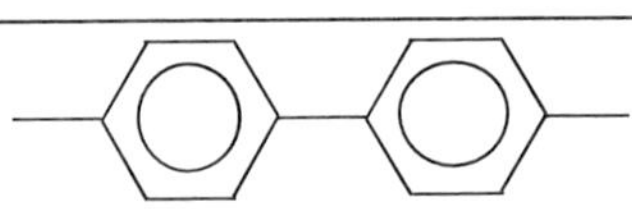 R=	40.4	325	76.0
5	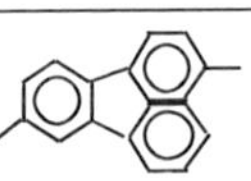 R=	45.1	350	84.0

The behavior of the glass transition temperatures is similar to that of the yield strengths.

Another example of dependence of elastic modulus on Kuhn segment is poly(1,3,4-oxadiasole-ether)s films (Table 15) [152]. In this case the dependence is linear on Kuhn segment (Fig.26).

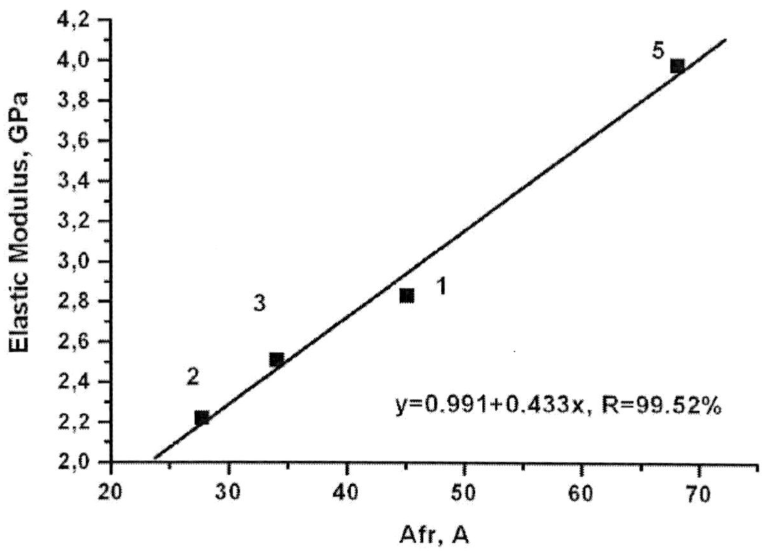

Figure 26. The dependence of elastic modulus on segment Kuhn for polymer of poly(1,3,4-oxadiasole-ether)s films (Table 15).

4. The Relationship between the Fire Resistance and the Kuhn Segment

Studies of ignition and flammabilility processes in polymers and polymer-based composites are acquiring ever-increasing importance for the development of laser technologies, theoretical foundations of synthesis of polymers with reduced ignition, and prediction of service conditions for flame-retardant polymers that would exclude ignition or suppress spreading of fire. The complexity of the process of ignition and combustion of a polymer makes one take into account mutually related chemical and physical transformations both in gas and condensed phases. We have made an attempt to find a correlation between the fire-resistance parameters and conformational rigidity of a polymer chain, which seems to be accidental at first sight.

Table 22 summarizes the fire-resistance parameters of a number of polyimides, such as the coking value (CV), the oxygen index (OI) and the ignition delay time (IDT) [161]. The coking value is the weight of the dry residue in percentage when heated at 900° for 30 minutes. The oxygen index is the fraction of oxygen in the nitrogen-oxygen mixture, at which stable combustion is sustained. The ignition delay time is the time interval (in seconds) between the moment of action of the ignition source and appearance of flame under fixed conditions (ambient temperature, a constant rate of the flux of the nitrogen-oxygen mixture) [162].

As is known, introduction of halogen results in increasing IDT because of the reduction of concentration of flammable substances and a lower rate of their formation [163]. The rate of attaining critical concentration of combustibles and highly volatile decomposition products close to the surface of a polymer specimen depends on its chemical structure. The rate of formation of those products is related to the heating rate of the polymer specimen. The rate of thermal conductivity of a specimen in its turn must depend on numerous factors including mobility of the chain elements of the polymer, i.e., the conformational rigidity of the polymer chain [126].

Table 22. Parameters of the fire resistance of some polyimides

N	Structural Repeat Unit		OI %	CV %	IDT %	A_{fr}, Å	*p*
1	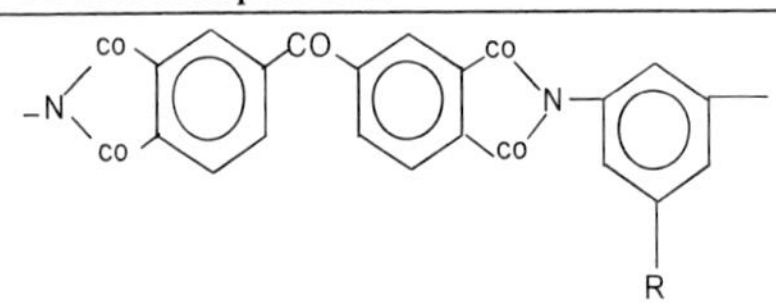	R= Br	50.0	50.0	0.80	43.6[a]	7.60
2		R= H	44.0	42.0	0.30	25.8	4.50
3	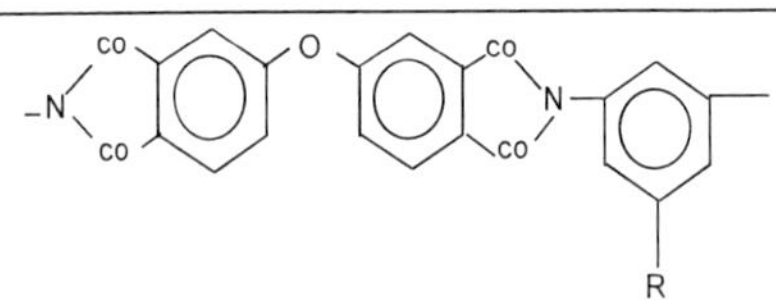	R= Br	51.8	51.0	0.90	46.3[a]	8.16
4		R= H	45.0	43.0	0.35	27.4	4.82
5	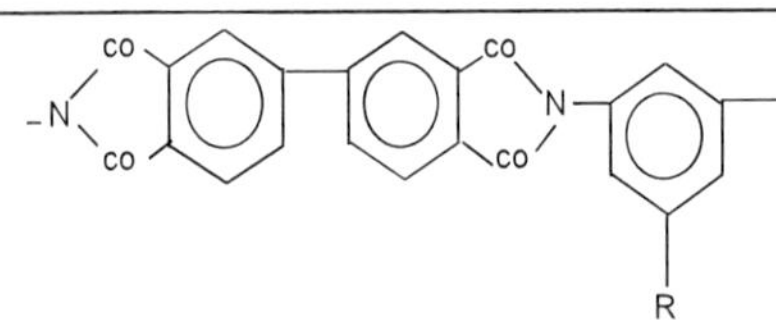	R= Br	59,06[b]	58,07[b]	1,47[b]	62.4[a]	11.8
6		R= H	49.0	47.6	0.69	36.7	6.9
7	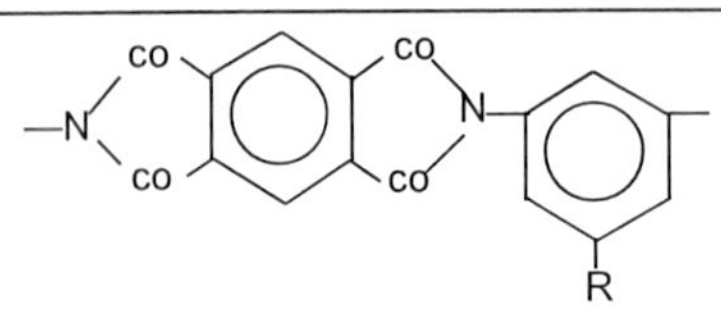	R= Br	57.0	55.0	1.30	63.5[a]	10.75
8		R= H	48.0	47.0	0.53	37.6	6.36

[a] -A_{hind} calculated in the assumption of hindered rotation [127,164];
[b] - found from dependences OI(*p*), CV(*p*) and IDT(*p*).

Table 22 presents the data for the two series of polymers with the structures differing only by the presence of a bromine atom as a substituent. For polymers containing the bromine atom as a substituent in the phenylene rings, the Kuhn segment was calculated as described in [127,164] with consideration of hindered rotation. The map of conformational energy was calculated as a function of the rotation potential energy by the angle φ about the heterocycle-Ph(Br) bond. The calculated energy maps show the values angel φ, at which the distance between the Br atom and other atoms of the structural unit of the polymer become smaller than the sum of Van der Waals radii. The conformers with such rotation angles were not taken into account in construction of the polymer chain. The Kuhn segment corresponding to this case was denoted as A_h. Figures 27(a, b, c) present the dependences of CV, OI, and IDT on A_{fr} (this is A_{fr} for bromine free polymers and A_h for bromine containing polymers Figures 28(a, b, c) present the dependences of CV, OI, and IDT on p

As is known, the values of CV, OI and IDT are governed by the presence of aromatic nuclei in the polymer chain. One can try to find a correlation between these parameters and the number of aromatic nuclei in the polymer repeat unit. However, polymers 1-6 have equal number of aromatic nuclei in the polymer repeat unit, whereas their CV, OI and IDT values differ and change with variation of the conformational rigidity of the polymer chain.

Therefore, we will consider the number of aromatic nuclei not in the polymer repeat unit but in the Kuhn segment.

As can be seen from Table 22 and Figs. 27, 28, the dependences of CV, OI, and IDT on *p* are also linear.

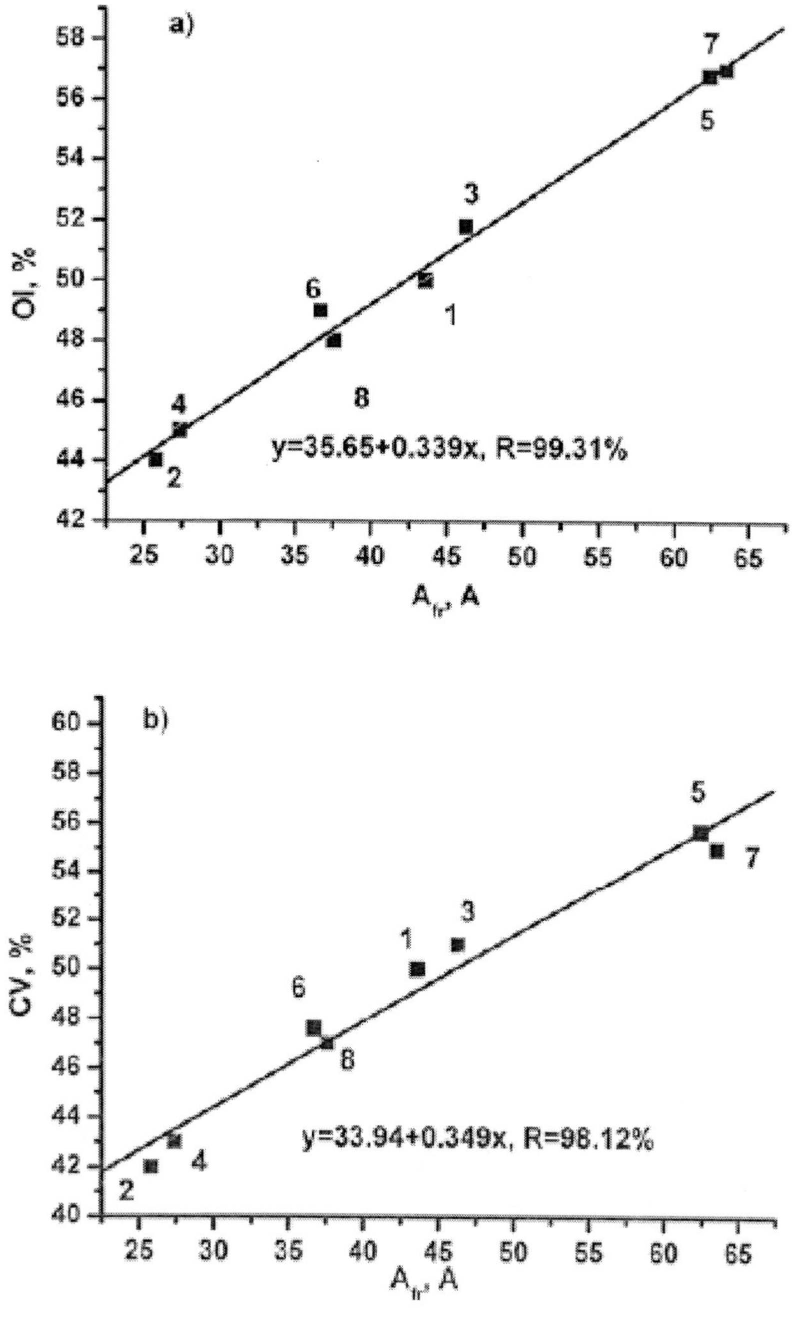

Figure 27. (Continued).

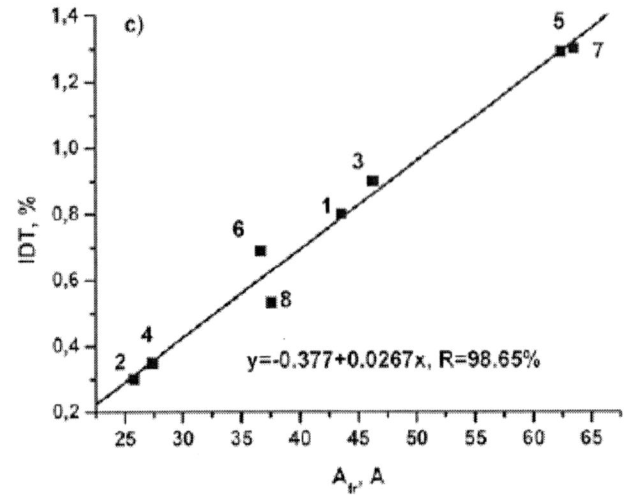

Figure 27. Dependence of OI, CV and IDT on Kuhn segment (polymers of Table 22).

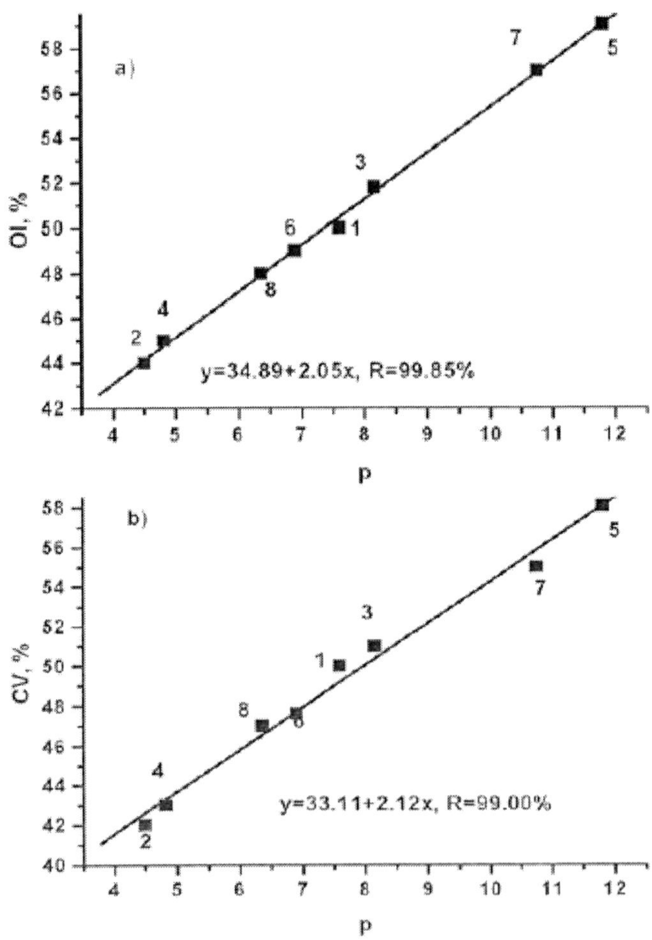

Figure 28. (Continued).

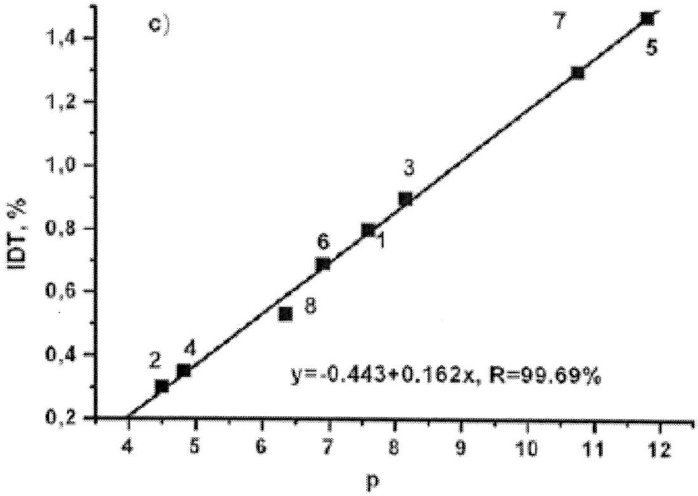

Figure 28. Dependence of OI, CV and IDT on the number of aromatic rings in the Kuhn segment (Table 22).

However, for the dependences on *p* the correlation is better. We explain the correlation of the fire-resistance parameters with the polymer chain rigidity in the following way. The Kuhn segment characterizes mobility of a polymer chain. For chains with smaller Kuhn segment, the number of chain conformations is larger, i.e., the chain mobility is higher and the surface, which is to ignite, softens and melts more readily. So, the period of inflammation becomes shorter. Large Br atoms sharply decrease chain mobility $A_{hin}/A_{fr} = 1.7$, and polymers with the side Br substitute have considerably higher parameters of fire resistance than similar polymers not containing Br atoms. High correlation coefficients for dependencies of CV, OI and IDT on *p* (more than 99%) permit to estimate these values for polymer 5 which have not been determined in experiment (Table 22).

Similar dependences for polynaphtoylenbenzimidazoles (PNBI) containing different bridging groups (Table 23) [165] have been obtained. Two isomeric forms of the naphtoylenbenzimidazole cycles, *cis* and *trans*, are formed in the course of the reaction.

Table 23. Dependences glass transition temperature, T_g, initial decomposition temperature, T_d and OI on conformational rigidity of PNBI

N	Structural Repeat Unit	A_{fr} ,Å	T_g°C	T_d °C	OI%
	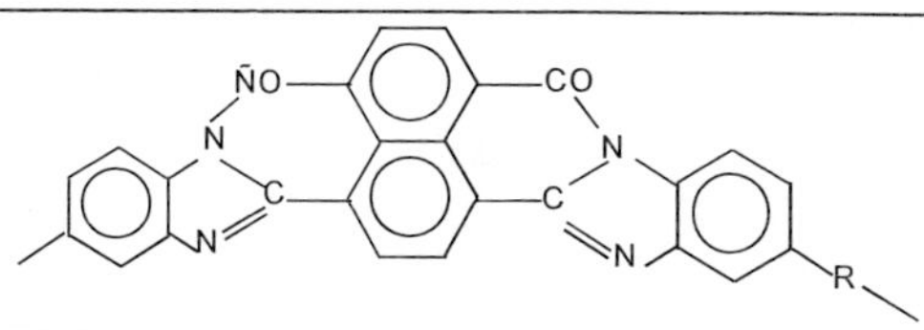				
1	R=-SO_2	27.9	500	420	0.50
2	R=-CO_2-	37.0	500	450	0.50
3	R=-O-	46.3	510	480	0.56
4	R=-	81.2	520	550	0.95

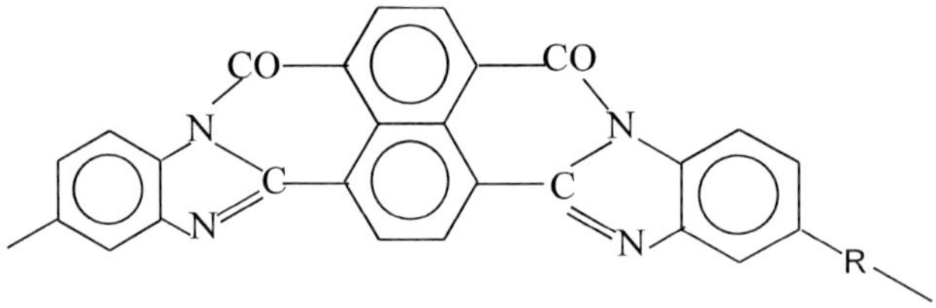

4,4'-*cis*

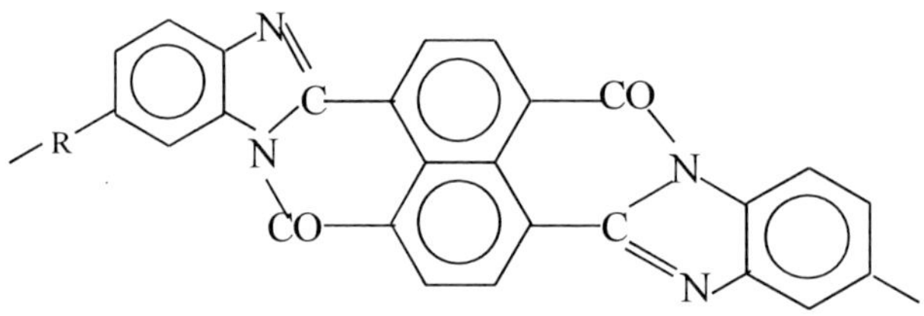

3,3'-*trans*

Each of both forms can enter into the chain in three different ways, so that the polymer will contain six isomers. The Kuhn segment values calculated for polymers containing only a single type of these isomers differ essentially [166, 167]. Therefore, we considered PNBI as statistical copolymers with equiprobable participation of all isomers in the chain [160]. As is seen from Table 23, both fire resistance parameters and glass transition temperatures depend on the Kuhn segment value. This fact suggests the idea that this seeming incredible correlation between the conformational rigidity of the polymer chain and the fire resistance is based on factors associated with changes in the conditions of diffusion of products of thermal destruction of macromolecules to the surface upon thawing the chain mobility.

5. Conformational Rigidity and Transport Properties

Table 24 shows the chemical structure of the repeating unit, glass transition temperature T_g, density ρ [168, 169] and conformational parameters A_{fr} and C_∞ , calculated with the equations (1) and (2), for 12 polyimides.

In this group of polyimides it was investigated the influence of conformational rigidity on the packing of macromolecules in glassy state and, consequently, on distribution of the free volume in polymer matrix. As seen in table 24, the conformational rigidity, that is Kuhn segment, of the studied polyimides varies in a relatively large interval, from 22 Å to 44 Å, and glass transition temperature from 190°C to 410 ° C. Thus, the selected series of polyimides is quite representative.

The dependence of glass transition temperature (T_g) on rigidity (A_{fr}) divides these selected polymers in three groups (figure 29). Each of them behaves in a normal way: with increasing rigidity the glass transition temperature increases [145]. The correlation coefficients in all these three groups of polyimides are high enough, in the range of 97.06–99.99%.

The dependence of glass transition temperature (T_g) on free volume (V_f) is linear and situates all these polymers on one line with the exception of polymer **PAI** (figure 30). The values of V_f are given in table 25. The correlation coefficient is high enough, 96.53%. With increasing free volume of the polymer, the glass transition temperature decreases. Higher the free volume, more conformational transitions can take place in polymer matrix at heating and the highly elastic state appears faster.

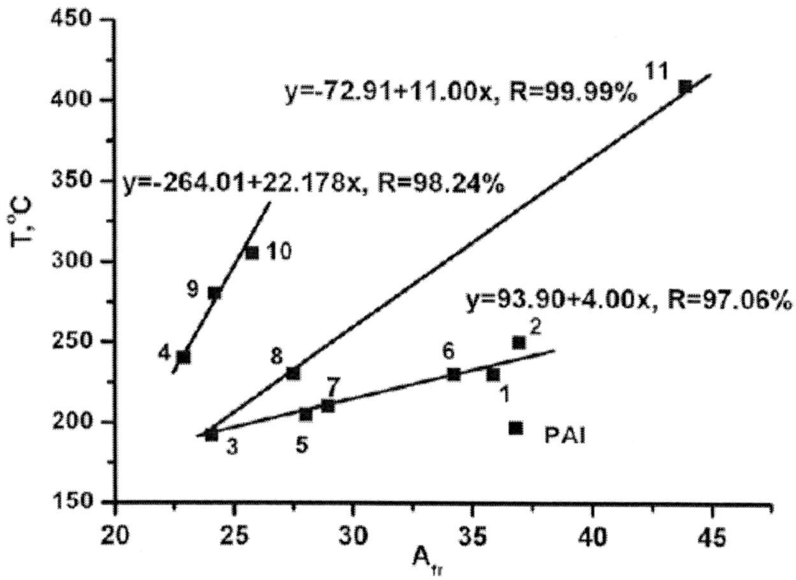

Figure 29. The dependence of glass transition temperature (T_g) on Kuhn segment (A_{fr}), for polyimides from table 24.

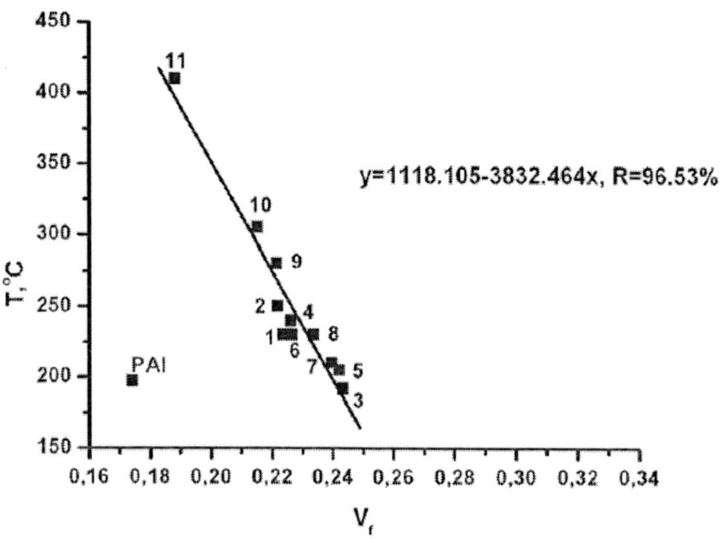

Figure 30. The dependence of glass transition temperature (T_g) on free volume (V_f), for polyimides from table 24.

Table 24. Structural repeating units and properties of the group of polyimides

Polymer	Structural Repeat Unit	T_g, °C	ρ, g/cm³	l_o (Å)	A_{fr}, Å	C_∞
Pi-1	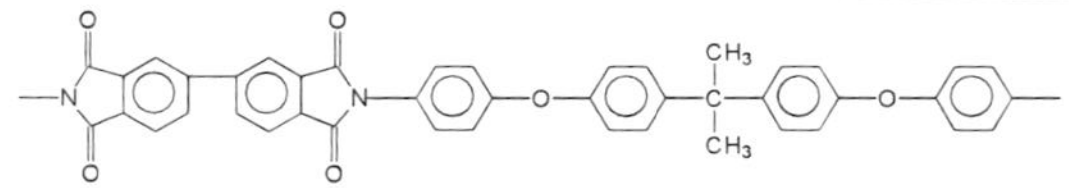	230	1.292	30.97	35.9	1.16
Pi-2	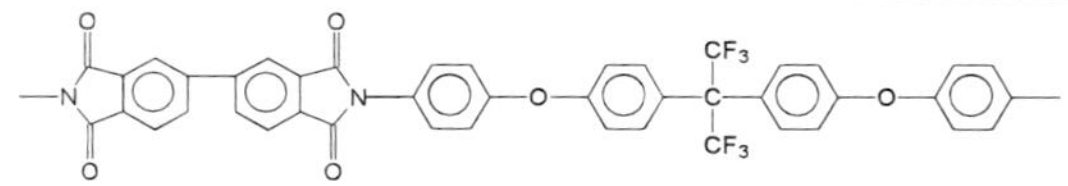	250	1.390	30.97	37.00	1.19
Pi-3	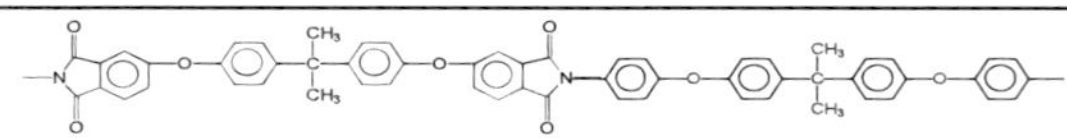	192	1.252	42.05	24.1	0.573
Pi-4	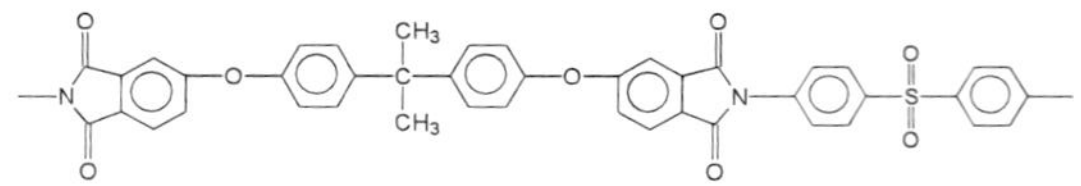	240	1.336	32.63	22.9	0.702
Pi-5	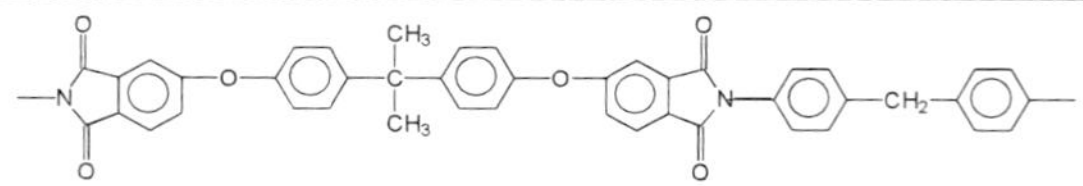	205	1.260	32.24	28.0	0.870
Pi-6, copolymer 1:1	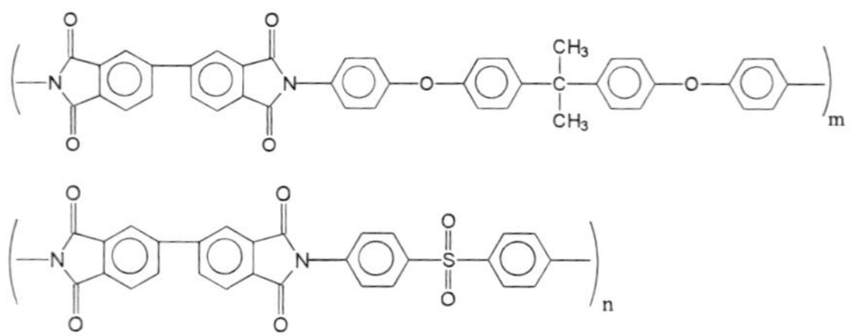	230	1.331	52.51	34.2	0.652

Polymer	Structural Repeat Unit	T_g, °C	ρ, g/cm^3	l_o (Å)	A_{fr}, Å	C_∞
Pi-7 copolymer 1:1	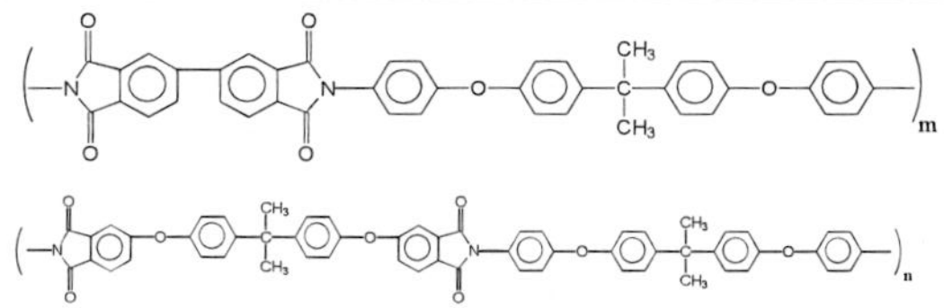	210	1.262	75.25	29.0	0.385
Pi-8	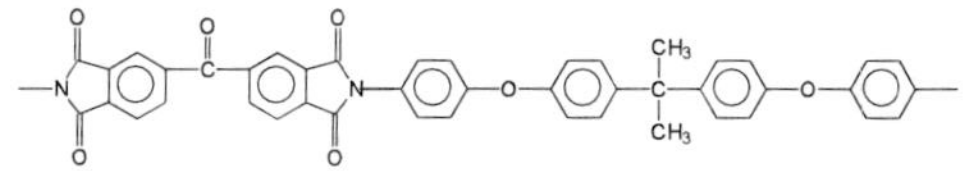	230	1.290	32.05	27.5	0.847
Pi-9	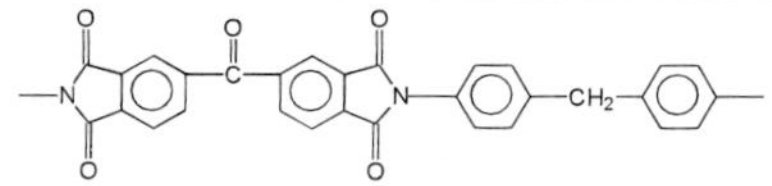	282	1.330	22.41	24.2	1.08
Pi-10	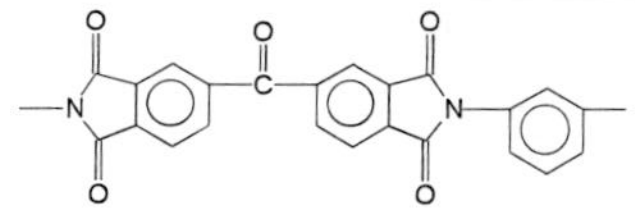	305	1.349	17.22	25.8	1.498
Pi-11	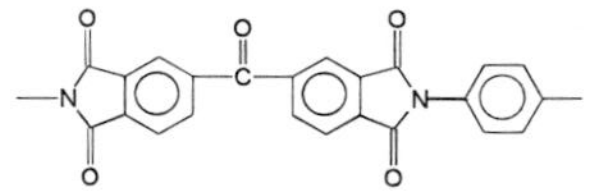	410	1.400	17.67	43.9	2,48
PAI	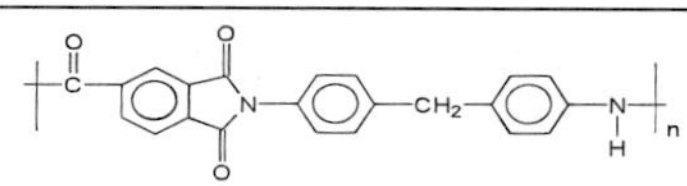	197	1.259	17.40	36.8	2.12

Table 25. Van der Waals (V_w), free (V_f), occupied (V_{occ}) and accessible (V_{acc}) volumes of polyimides from Table 24

Polymer	V_w (Å^3)	V_f (Å^3)	H_2		CO		CO_2		CH_4	
			V_{occ} (Å^3)	$\mathbf{V_{acc}}$ (cm^3/g)	V_{occ} (Å^3)	$\mathbf{V_{acc}}$ (cm^3/g)	V_{occ} (Å^3)	$\mathbf{V_{acc}}$ (cm^3/g)	V_{occ} (Å^3)	$\mathbf{V_{acc}}$ (cm^3/g)
Pi-1	610.900	0.2239	646.051	0.1919	660.039	0.1794	652.406	0.1866	662.307	0.1777
Pi-2	641.686	0.2220	679.993	0.1923	695.243	0.1804	686.717	0.1817	697.961	0.1783
Pi-3	825.166	0.2434	876.832	0.2088	899.047	0.1939	886.774	0.2021	902.953	0.1913
Pi-4	635.673	0.2263	672.665	0.1958	688.284	0.1830	679.516	0.1902	691.053	0.1807
Pi-5	625.359	0.2422	661.567	0.2103	676.362	0.1972	668.369	0.2042	679.144	0.1947
Pi-6	1024.818	0.2268	1087.410	0.1948	1108.728	0.1838	1097.122	0.1898	1112.292	0.1820
Pi-7	1436.066	0.2397	1522.883	0.2060	1559.086	0.1921	1539.180	0.1997	1565.260	0.1897
Pi-8	626.346	0.2339	660.248	0.2046	675.624	0.1987	667.042	0.1987	678.383	0.1889
Pi-9	426.696	0.2216	449.094	0.1937	458.160	0.1825	453.330	0.1885	459.656	0.2402
Pi-10	333.975	0.2153	349.409	0.1910	356.025	0.1806	352.390	0.1863	357.052	0.1790
Pi-11	333.975	0.1883	349.409	0.1640	356.025	0.1520	352.390	0.1593	357.052	0.1520
PAI	325.572	0.2167	342.356	0.1870	349.811	0.1738	345.713	0.1810	351.117	0.1714

At the same time, the free volume of the polymer is determined by its capacity to pack in glassy state and consequently it depends on its conformational rigidity. Figure 31 shows the dependence of free volume (V_f) of the studied polyimides on conformational rigidity (A_{fr}). It can be seen that there are three dependences (three lines), and in each of them enter the same polymers as in figure 29. The correlation coefficients of all three dependences are very high, in the range of 97.53–99.99%.

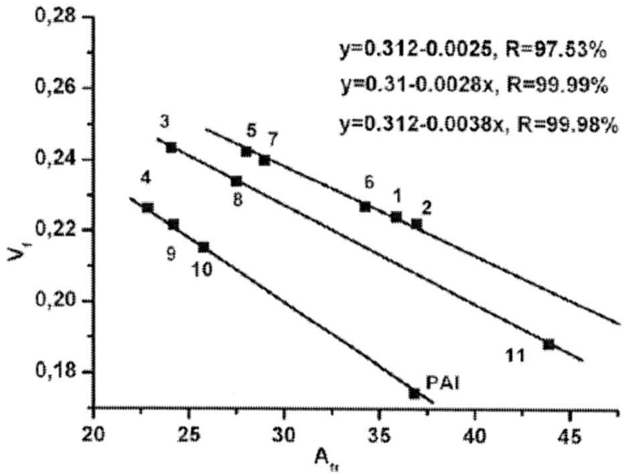

Figure 31. The dependence of free volume (V_f) on Kuhn segment (A_{fr}), for polyimides from table 24.

The influence of the conformational rigidity on the permeability coefficients (***P***) of the membranes made from these polymers was examined. Table 26 shows the permeability coefficients to four gases H_2, CO, CO_2 and CH_4, taken from publications [168, 169]. For these polymers we calculated the Van der Waals volume, free volume, occupied volume and accessible volume (Table 25).

Since the accessible volume (V_{acc}) is considered as the volume of the space inside the polymer which can be occupied by the centers of gas molecules so that the Van der Waals spheres of the gas should not overlap on the Van der Waals spheres of the atoms of polymer chain [170], one can approximate the free volume by using the equation (15):

$$P = A_{exp}(-B/V_{acc}), \tag{15}$$

However, more often is used the so-called fractional accessible volume (***FAV***), without any dimensions, which gives a better concordance with the coefficients of diffusion and of permeability [170], that is given by equation (16):

$$FAV = V_{acc}/V_{sp} = V_{acc} \cdot \rho, \tag{16}$$

where V_{sp} is the specific volume of the polymer, equal to ρ. Thus, the equation (5) can be written as shown in equation (17):

$$P = A\exp(-B/FAV), \tag{17}$$

Figure 32 shows the dependence of the permeability coefficients of the first series of studied polyimides on the fractional accessible volume in the system one polymer - different gases. All these dependences are linear, with high correlation coefficients, being in the range of 99.05–99.91% (Table 27).

Table 26. The permeability coefficients, Barrer [168,169] for polyimides from Table 24

Polymer	H_2	CO	CO_2	CH_4
Pi-1	2.53±0.02	0.0567±0.0015	0.644±0.012	0.0329±0.0007
Pi-2	5.91±0.04	0.145±0.005	1.64±0.02	0.0504±0.0033
Pi-3	3.20±0.03	0.0678±0.0012	0.762±0.011	0.0305±0.0027
Pi-4	3.56±0.01	0.0836±0.0016	0.891±0.039	0.0418±0.0021
Pi-5	5.07±0.03	0.136±0.001	1.62±0.09	0.080±0.007
Pi-6	1.79±0.01	0.0395±0.0016	0.428±0.002	0.0099±0.0002
Pi-7	4.43±0.04	0.250±0.01	1.57±0.04	0.205±0.008
Pi-8	5.59±0.01	0.102±0.004	1.47±0.08	0.089±0.002
Pi-9	3.14±0.04	0.0367±0.0017	0.594±0.035	0.016±0.001
Pi-10	1.25±0.01	0.00397±0.00019	0.0627±0.0029	0.0011±0.0001
Pi-11	1.14±0.02	0.0176±0.0007	0.1971±0.0033	0.00795±0.00028
PAI	1.63±0.01	0.0342±0.0005	0.435±0.004	0.0163±0.0003

Table 27. The slopes (*A* and *B*) and correlation coefficients (*R*) for curves described by equation (17) for the system one polymer–various gases, for polyimides from Table 24

Polymer	A	–B	**R** (%)
Pi-1	24.302	5.918	99.91
Pi-2	27.326	7.076	99.54
Pi-3	22.815	5.821	99.87
Pi-4	24,066	6.142	99.89
Pi-5	24.482	6.281	99.38
Pi-6	31.700	8.121	99.05
Pi-7	16.950	4.234	99.55
Pi-8	23.920	6.111	99.15
Pi-9	32.757	8.294	99.77
Pi-10	43.919	11.311	99.71
Pi-11	27.210	6.235	99.97
PAI	22.832	5.315	99.69

Table 27 presents the parameters of equation (17) and the correlation coefficients. Parameter ***B*** means the slope of the line representing the dependence of permeability coefficients on the fractional accessible volume. This parameter reflects the common selectivity of the respective polymer, which means that it shows how fast one gas penetrates through the polymer membrane compared with another gas. Higher the value of ***B***, better the selectivity of the respective polymer, which means that the respective polymer will better separate one gas from another one. This behavior is determined by the packing of the polymer in glassy state, i.e., that is by the distribution of microcavities in polymer matrix. The packing of polymer is significantly determined by its conformational rigidity. In table 27 it can be seen that polymers **6, 9** and **10** have the highest values of coefficient ***B***, being 8.121, 8.294 and 11.311, respectively.

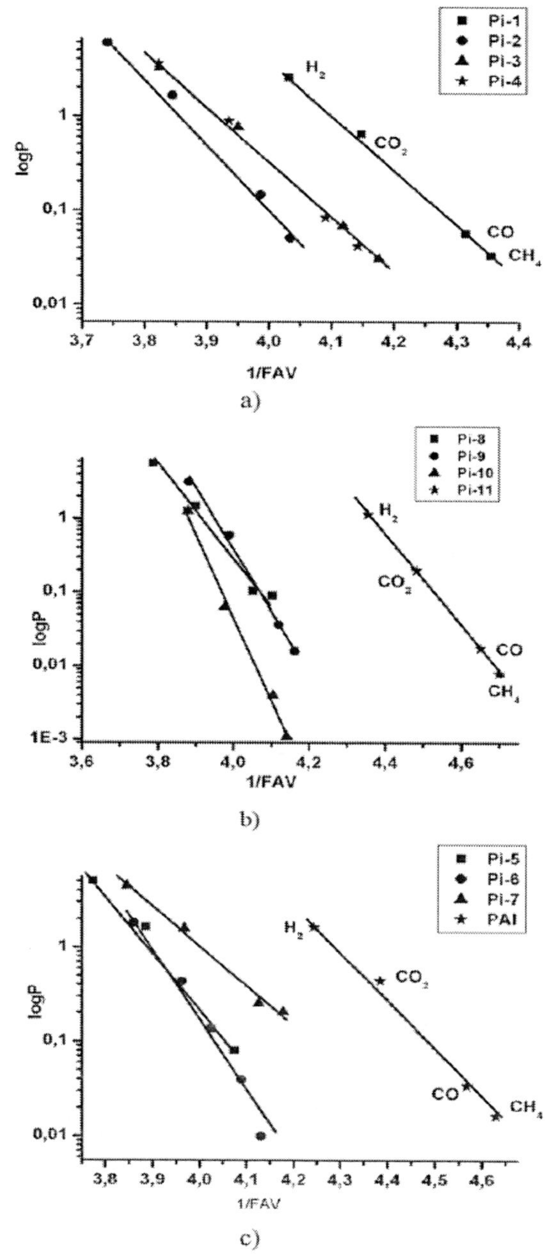

Figure 32. The dependence of permeability coefficients (***P***) on fractional accessible volume (***FAV***), for polyimides from table 24.

Figure 33 presents the dependence of parameter B on characteristic ratio C_∞. It can be seen that the selectivity of the polymer reaches the maximum when the characteristic ratio is 1.5, which means that Kuhn segment contains 1.5 repeating units. At such a value of C_∞ the packing of polymer in glassy state gives presumably a narrow distribution of microcavities and the permeability of one gas is significantly higher than that of the other, which leads to high selectivity. On the other hand, as seen in the dependence of free volume on conformational rigidity, with increasing rigidity the free volume decreases, and in case of polymer **PAI** and **Pi-11** it is the lowest, which means that the packing of macromolecules in these two polymers is more dense.

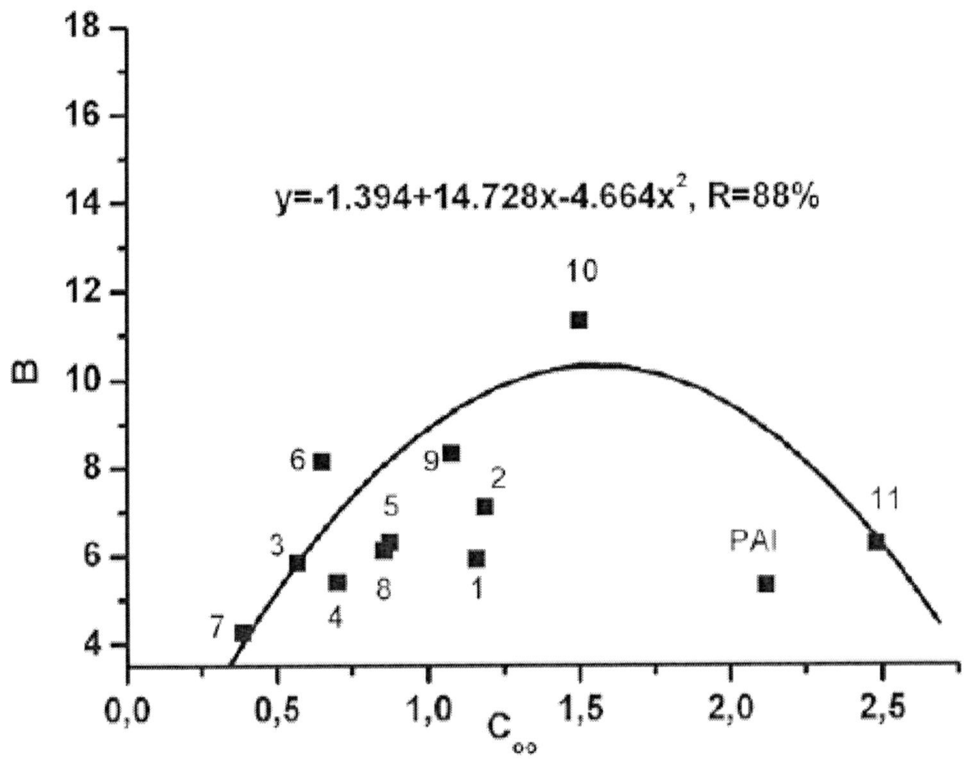

Figure 33. The dependence of slope ***B*** of equation (5) on characteristic ratio C_∞, for polyimides from table 24.

While examining the dependence of permeability coefficients on fractional accessible volume for one gas and different polymers (figure 34), it can be seen that the points corresponding to polymers **6, 9** and **10** fall down visibly from the general dependences, for all four studied gases: CO_2, CO, H_2 and CH_4. It shows that even if the selectivity is high, the permeability of these polymers is low. The geometrical structure of the repeating unit determines the conformational rigidity of the polymer, its packing in glassy state, and subsequently its membrane characteristics: permeability and selectivity.

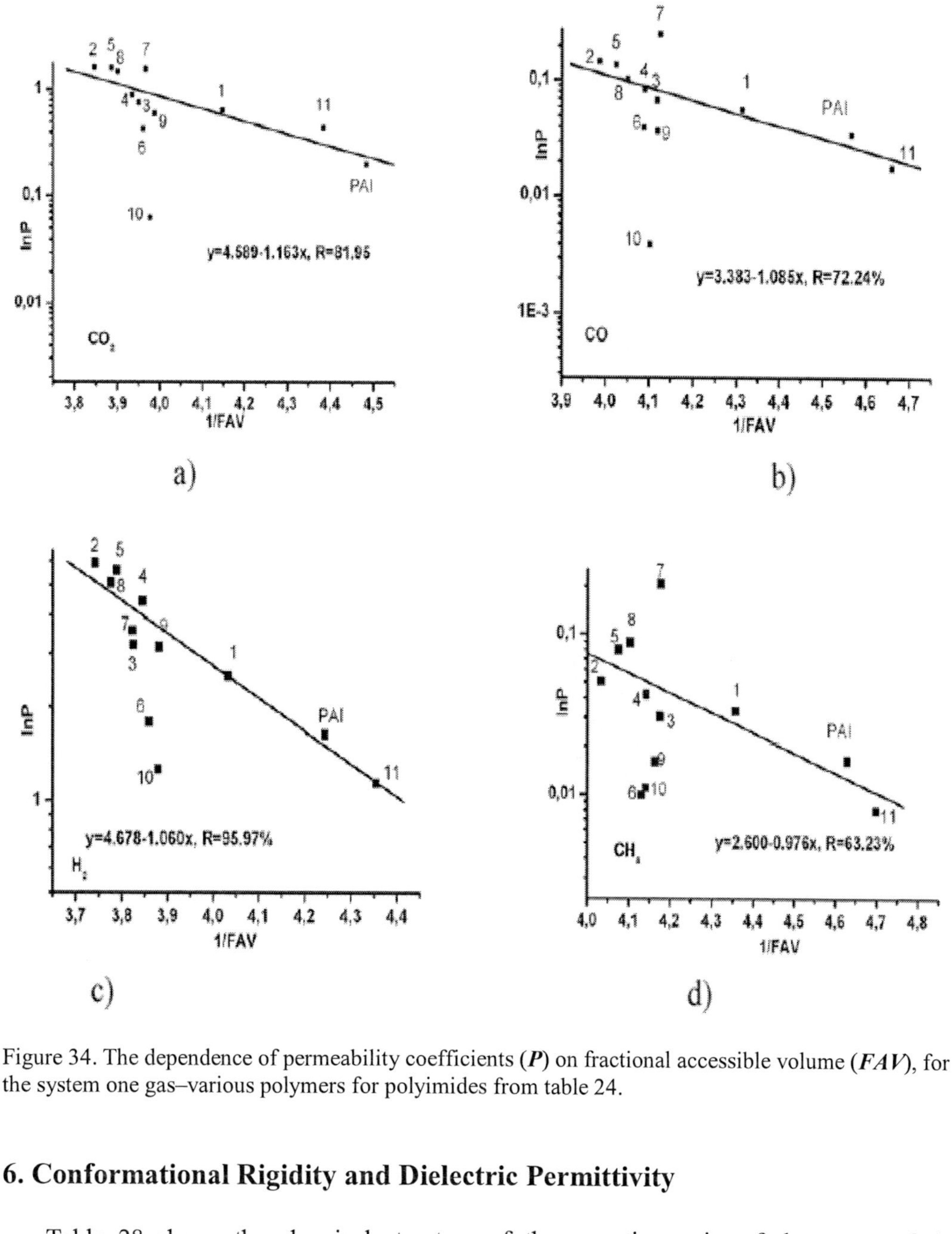

Figure 34. The dependence of permeability coefficients (***P***) on fractional accessible volume (***FAV***), for the system one gas–various polymers for polyimides from table 24.

6. Conformational Rigidity and Dielectric Permittivity

Table 28 shows the chemical structure of the repeating units of the group of the polyimides and some of the physical properties [171]. Figure 35 presents the dependence of glass transition temperature on Kuhn segment. It can be seen that the behavior is as usual: with increasing of Kuhn segment the glass transition temperature increases [4].

Table 28. Structures and properties of the group of polyimides

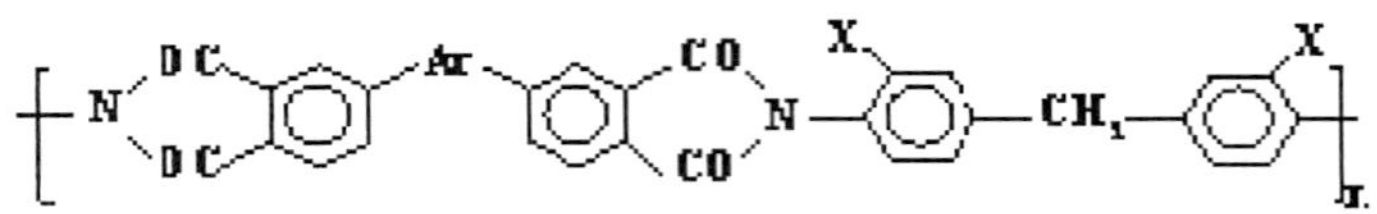

Polymer	Ar	X	ρ, g/cm^3	Td ,^0C	Tg ^0C	l_o ,Å	A_{fr} ,Å	C_∞	V_W, $Å^3$	V_f, cm^3/g	ε_o
1	-O-C_6H_4-$C(CH_3)_2$-C_6H_4-O-	-	1.200	493	200	32.32	25.4	0.79	625.359	0.2915	2.5448
2	-O-C_6H_4-$C(CH_3)_2$-C_6H_4-O-	CH_3	1.169	479	228	32.32	26.2	0.81	656.092	0.3056	2.1099
3	CO	-	1.398	509	-	24.53	28.0	1.14	426.696	0.1889	3.2754
4	CO	CH_3	1.337	498	278	24.53	28.9	1.18	460.613	0.2085	2.8454
5	$C(CF_3)_2$	-	1.546	508	275	22.37	28.6	1.28	488.651	0.1648	2.3099
6	$C(CF_3)_2$	CH_3 CH_3 CH_3	1.488	501	287	22.37	29.5	1.32	522.676	0.1793	1.99

ρ = density of polymer film; Td = the initial temperature of decomposition ; Tg = glass transition temperature;
l_o = contour length of a repeating unit; A_{fr} = Kuhn segment; C_∞ = characteristic ratio; V_w = Van der Waals volume;
V_f = free volume; ε_o = dielectric constant.

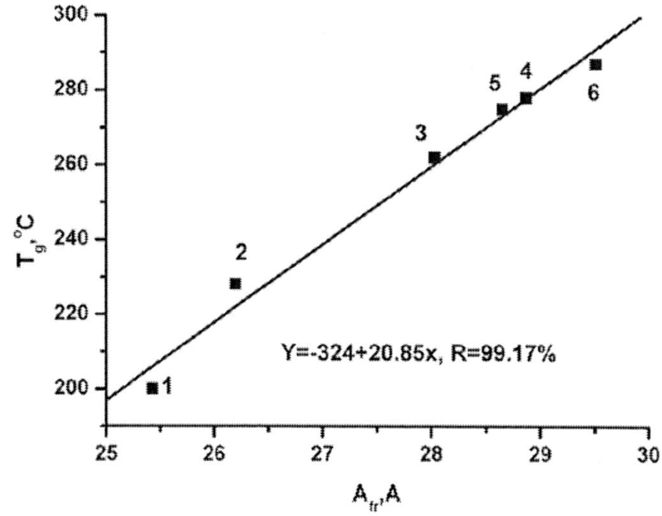

Figure 35. The dependence of glass transition temperature (T_g) on Kuhn segment (A_{fr}), for polyimides from table 28.

Since all the points of this dependence are situated on a straight line, described by equation ***y=324+20.85x*** with a high correlation coefficient R=99.17%, the glass transition temperature of polymer **3** can be calculated because it could not be measured experimentally. Thus, its calculated value is 262°C. The value of glass transition temperature of this polymer is well situated on the line showing the dependence of glass transition temperature on free volume (figure 36). This dependence shows that with the increasing of free volume, the glass transition temperature in general decreases because the higher free volume allows the molecular fragments to change their conformation and therefore an increased mobility in polymer matrix appears which leads to the decrease of glass transition temperature.

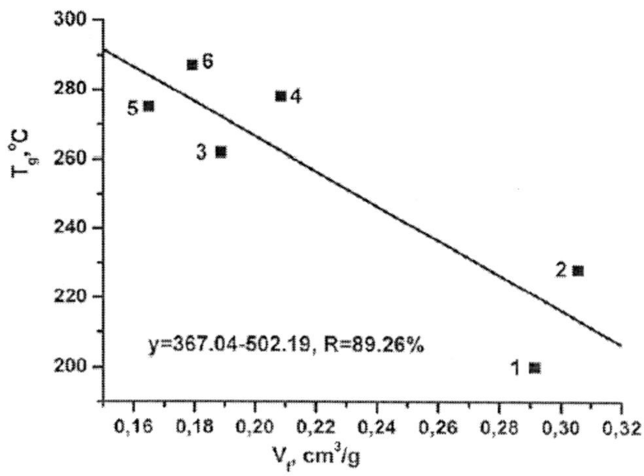

Figure 36. The dependence of glass transition temperature (T_g) on free volume (V_f), for polyimides from table 28.

Figure 37 presents the dependence of free volume on conformational rigidity. This dependence is linear, with a high correlation coefficient, 97.24%. It confirms that with increasing rigidity of polymer matrix, the free volume decreases.

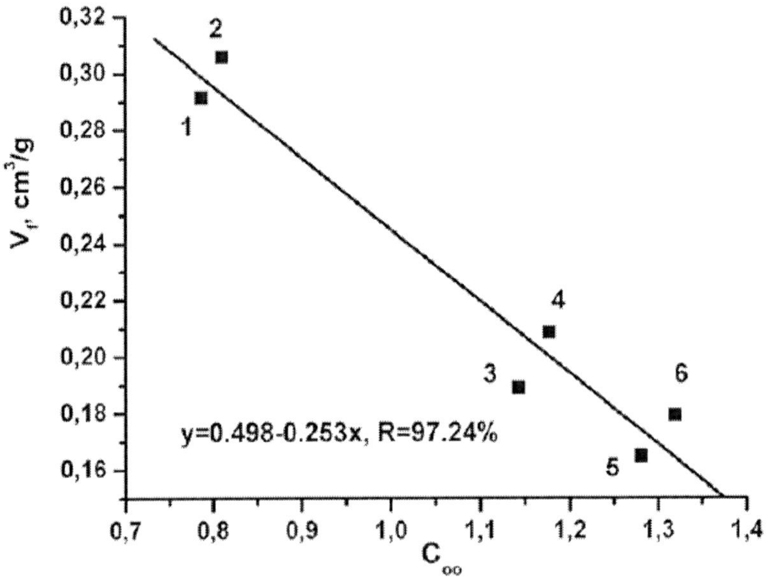

Figure 37. The dependence of free volume (V_f) on characteristic ratio C_{∞}, for polyimides from table 28.

When comparing figures 35-37, a contradiction appears in this series of polymers: with increasing rigidity, the glass transition temperature should increase and the free volume should decrease. However, by going from the polymers without any substituents (**1, 3** and **5**) to the polymers containing side methyl groups (**2, 4** and **6,** respectively) not only the glass transition temperature increases, but also the free volume increases. This can be explained in the following way: the introduction of methyl substituents on phenylene rings in *ortho*-position towards the bond between phenyl and imide rings increases the hindrance of rotation and consequently the rigidity of the polymer increases. The amount of conformational isomers and the mobility of one repeating unit towards the others when heating decreases, and consequently the glass transition temperature increases. On another hand, the methyl substituents do determine a loosing of packing, but the polymer can not benefit of the increased free volume because the rotation hindrance does not allow the formation of all conformations which should occupy that free volume when heating the samples. This is why the glass transition temperature of polymers containing methyl substituents does not decrease, but it increases in each pair of these polymers.

Table 30 shows also the values constant of dielectric constant ε_0 .For each polymer in this series dielectric permeability was measured at six frequencies, from 1 Hz to 100 kHz, and it was approximated at the frequency equal to zero, as shown in figure 38.

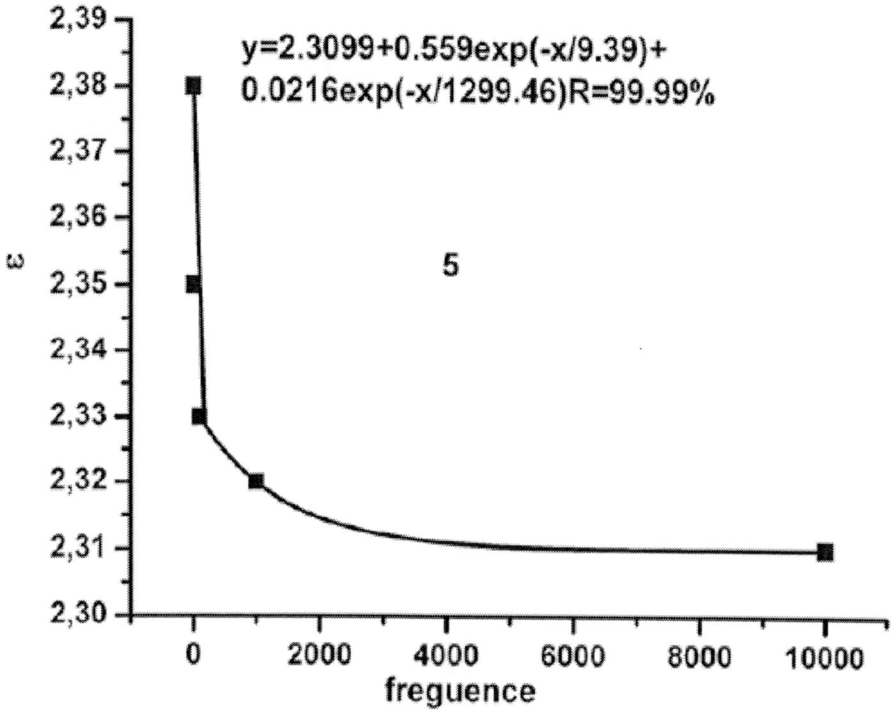

Figure 38. The dependence of dielectric permittivity (**ε**) on frequency, for polymer **5** for polyimides from table 28.

Figure 39 presents the dependence of dielectric constant of the studied polymers on characteristic ratio. It can be seen that the points are also situated in pairs. The values of ε_o of the polymers containing methyl substituents is lower than those of the related polymers which do not contain methyl groups. By introduction of methyl substituents, the packing of the polymer becomes loose. However, the general dependence divides into two branches: polymers **1** and **2** are situated on the up-going part, while polymers **3–6** are situated on the down-going part of this diagram. The presence of hexafluoroisopropylidene bridge between imide rings increases the loosing effect of methyl groups.

It is known that fluorine atom has a highly electronegative effect [172]. Therefore, it is interesting to study the distribution of electron density in the repeating unit of macromolecules. We used the quantum-chemical method AM1 to calculate the charges on atoms which constitute the repeating units of both polymers **5** and **6**. The charge on fluorine in hexafluoroisopropylidene segment was -0.17 e, while the charge on each hydrogen atom of isopropylidene group was only -0.08 e.

The packing of fragments of two macromolecules was examined: one containing hexafluoroisopropylidene and the other containing isopropylidene group. It was supposed that oxygen atoms of ether groups are situated close to each other, at a distance equal to the sum of their Van der Waals radii, as shown in scheme below.

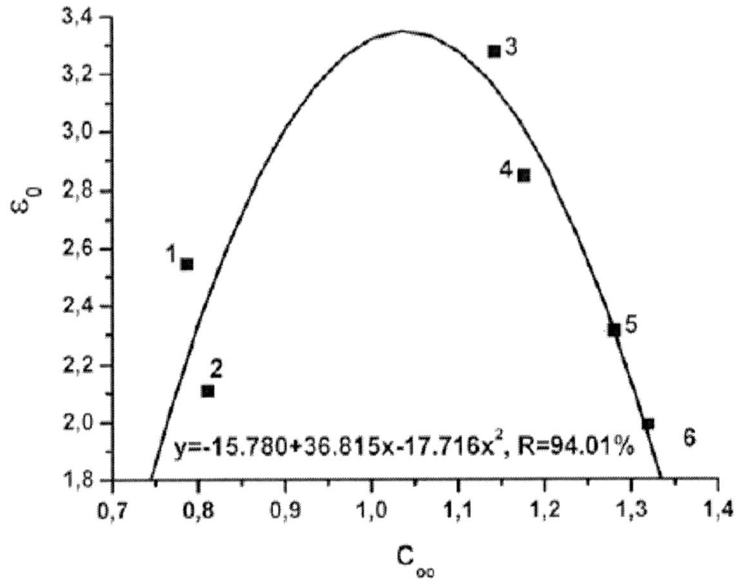

Figure 39. The dependence of dielectric constant (ε_0) on characteristic ratio C_∞ for polyimides from table 28.

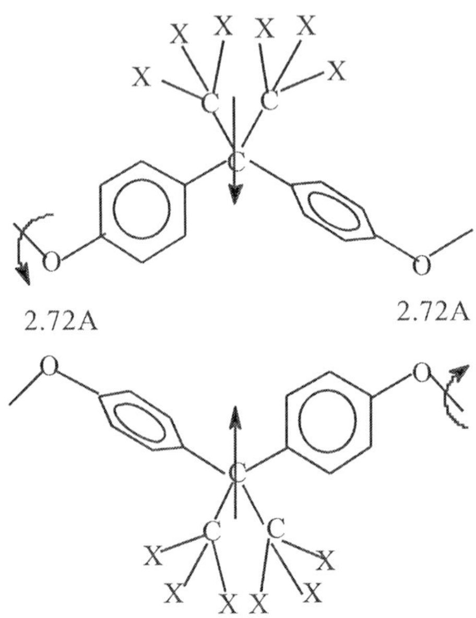

When packing, it is possible that hexafluoroisopropylidene groups of different chains will be positioned as shown in this scheme. This position permits modeling of their mutual interaction. It was presumed that the groups come out of the plane of this drawing, to meet each other and so the energy of repulsion which appears between hydrogen or fluorine atoms during this movement as a function of the distance between these atoms was compared. The energy of repulsion between the three atoms of one group and the three atoms of the other group is:

$$E = \sum_i \frac{q_i q_j}{r_{ij}},$$

where q_1 is the charge on hydrogen or fluorine atom, r is the distance between atoms. The sum is done at the same value of angle at which the two groups come out of the plane. Figure 40 presents the energy of repulsion between groups during their movement towards each other in the process when they come out of the plane.

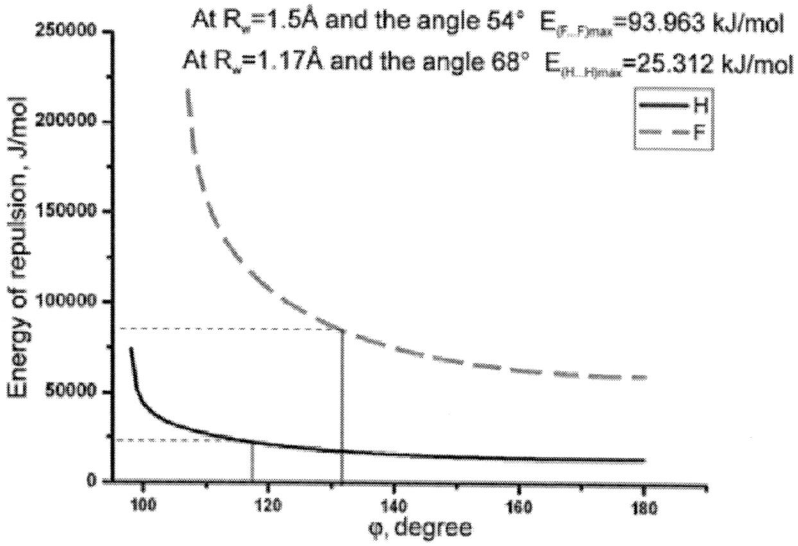

Figure 40. The dependence of repulsion energy on angle φ between hexafluoroisopropylidene groups (red line) and on the angle between isopropylidene groups (black line). Van der Waals radius of fluorine atom is 1.5 Å; Van der Waals radius of hydrogen atom is 1.17 Å.

In both cases, the groups can approach each other at a distance not shorter than the sum of Van der Waals radii of the corresponding atoms. When the distance between hydrogen or fluorine atoms is equal to the sum of Van der Waals radii of the corresponding atoms, the energy of repulsion in case of fluorine atoms is higher than in case of hydrogen atoms. Similar repulsion can appear between hexafluoroisopropylidene groups and oxygen atoms of carbonyl groups in imide rings. It follows that in glassy state a repulsion appears between fluorine atoms of hexafluoroisopropylidene groups which leads to the formation of supplemental microcavities in this polymer.

The increase of free volume of polymers containing methyl substituents leads to the decrease of dielectric constant compared with related polymers which do not contain such substituents.

Other examples of inter bonding of dielectric permittivity and conformational rigidity was shown in papers [173, 174].

The structural repeat units, conformational parameters, glass transition temperature and dielectric permittivity of poly(phenylquinoxaline-imide)s containing dimethylsilane [173] are shown in the Table 29. The dependence of dielectric permittivity ε on conformational parameters C_∞ for polymer Table 29 is presented in figure 41. The measurements of dielectric permittivity were carried out at 20°C and 10 kHz using a Novocontrol system composed of an Alpha frequency respond analyzer and Quattro temperature controller. The samples were prepared in the form of films with a thickness of 20 μm with gold electrodes evaporated in vacuum.

Table 29. The structural repeat unit, conformational parameters, glass transition temperature and dielectric permeability of poly(phenyl quinoxaline-imide)s containing dimethylsilane

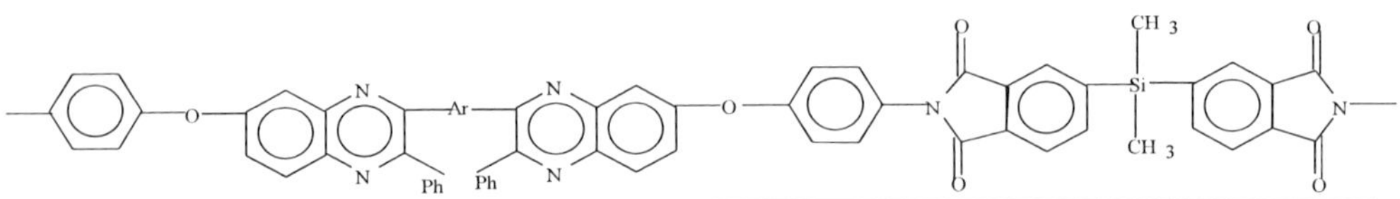

No	Ar	l_o, Å	A_{fr}, Å	C_∞	Tg °C	ε
1	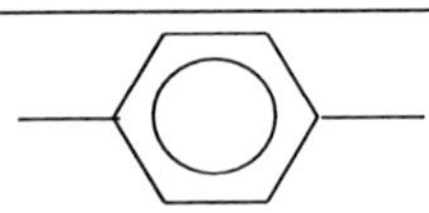	40.55	27.3	0.67	265	3.08
2	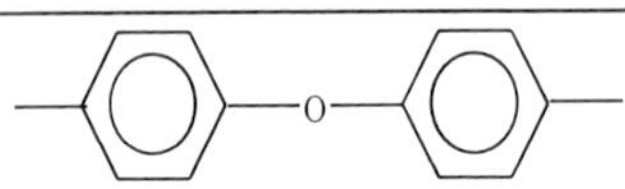	45.27	23.0	0.51	245	2.94
3	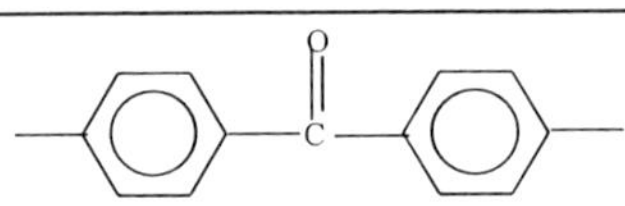	45.44	24.0	0.53	255	2.97

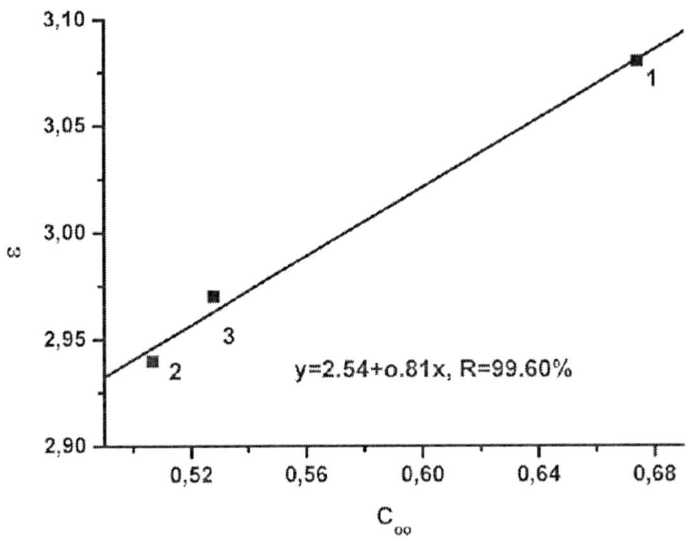

Figure 41. The dependence of dielectric permittivity on characteristic ratio for polymers from table 29.

Dielectric permittivity of polymers increases as chain rigidity increases. Coefficient of correlation is very good value (99.6%). The low values of dielectric permittivity of this polymer can be explained by the presence of bulky dimetilsilane and phenylquinoxaline groups in the main chain, which increased the free volume and thus lowered the polarization by decreasing the number of polarizable groups per volume unit.

The influence of bulk side substitutes on dielectric permittivity is well manifested in behavior of polyamides containing pendant acetobenzamide groups (Table 30) [174]. The dependence of dielectric permittivity on number aromatic rings in segment Kuhn is shown on Figure 42.

The dependence is described by the second order equation. As was shown above, the free volume decreases as the rigidity increases, hence the dielectric permittivity. In this polyamide row this property is high enough in comparing with polyimides in spite of large bulk side substitute in polymers. Introducing the 1,3,4 oxadiasole cycles into the main chain of these polymers results in the increase of its rigidity and slight decrease of dielectric permittivity. Due to this the linear dependence transforms into parabolic one.

Table 30. The structural repeat unit, conformational parameters, glass transition temperature and dielectric permeability of polyamides containing side acetoxybenzamide groups

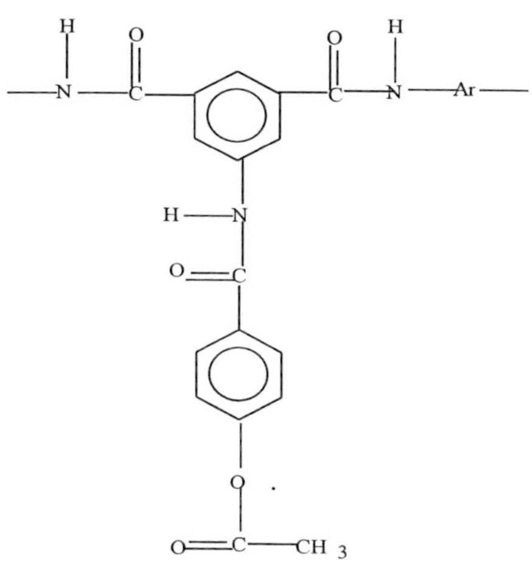

No	Ar	l_o, Å	A_{fr}, Å	A_{hin}, Å	C_∞	Tg ^{0}C	ε
1	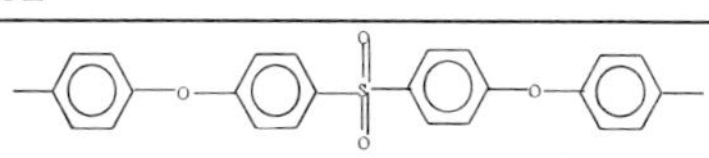	27.54	23.6	-	0.86	243	3.42
2	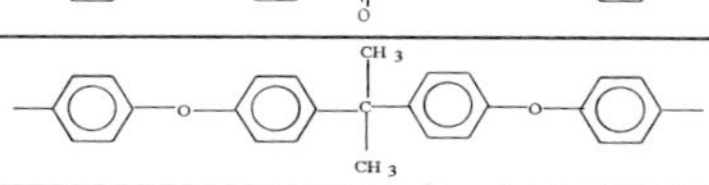	27.25	24.6	25.3	0.93	263	3.53
3	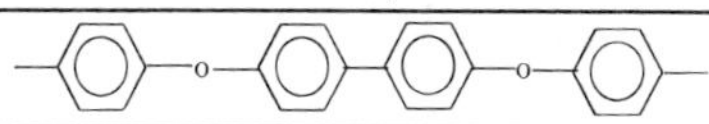	26.41	33.6	-	1.27	290	4.13
4	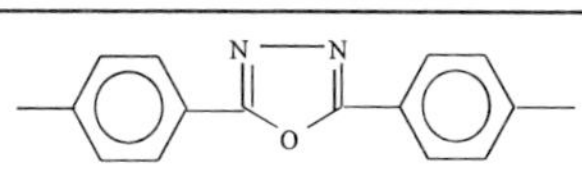	20.25	44.8	-	2.21	280	4.90
5	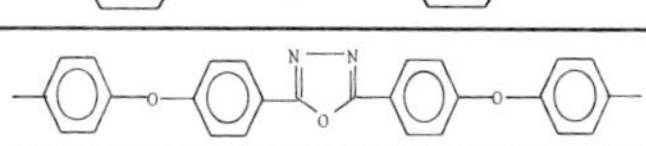	30.05	34.2	-	1.14	270	3.93

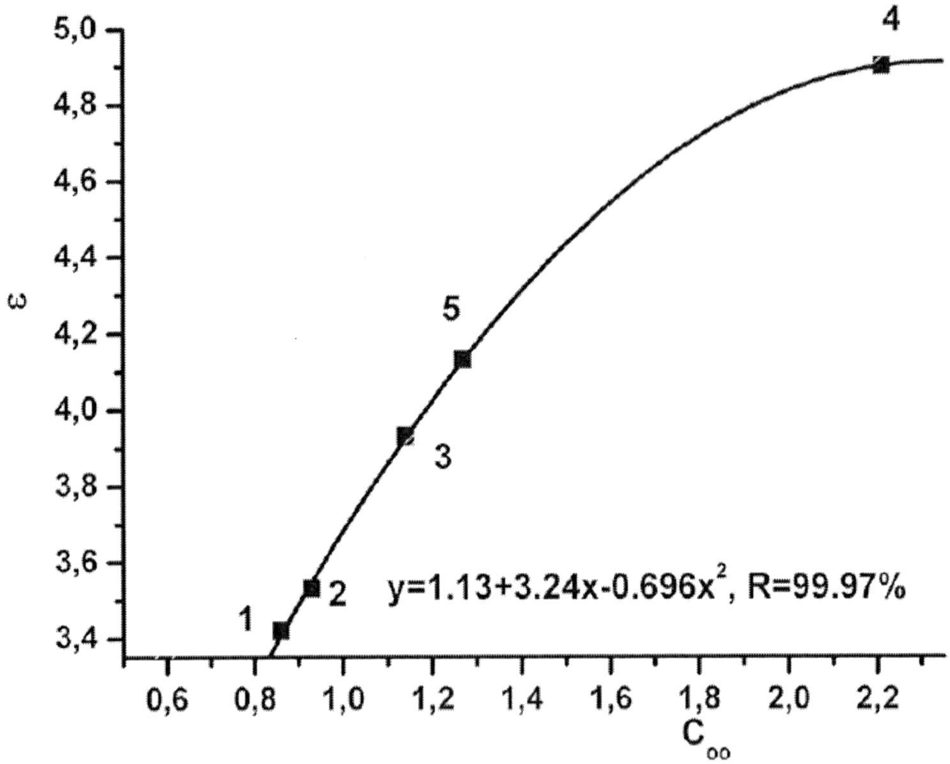

Figure 42 The dependence of the dielectric permittivity on characteristic ratio for polymers from Table 30.

6. Conformational Rigidity and Electrets Properties

Table 31 presents the conformational and physical characteristics [175] of the еще одной группы of studied polyimides. We examine now the glass transition temperature of these polymers. Figure 43 shows the dependence of glass transition temperature on Kuhn segment.

This dependence is linear, with a high correlation coefficient (99.15%) and it shows usual behavior: with increasing of conformational rigidity, the glass transition temperature increases [4]. Two points corresponding to polymers **4** and **8** come out of the general dependence. If we presume that in these two polymers not all the imide rings were closed, then the Kuhn segment was even longer and the points corresponding to these polymers move to the right. In addition, the reason that these points are out of the general dependence can be the lower molecular weight of those polymers as confirmed by publication [175]. In this case the macromolecular coil could not form, the conformational transitions take place easier and consequently the glass transition temperature decreases significantly.

Table 31. Conformational and physical properties of some polyimides

Polymer	Repeating unit	l_o (Å)	A_{fr} (Å)	C_∞	T_g (°C)	F (wt %)	V/V_o^{24h} (%)	ε at 1kHz
1	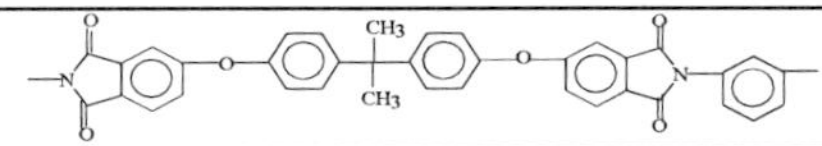	27.13	19.8	0.73	221	0	3	
2	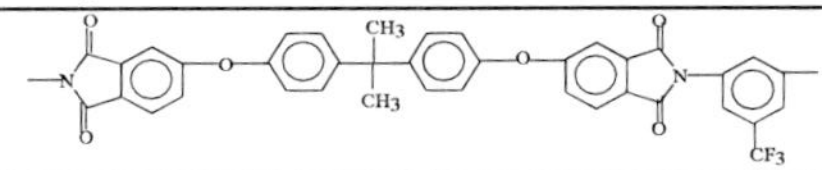	27.13	20.3	0.75	224	8.6	61	3.12
3	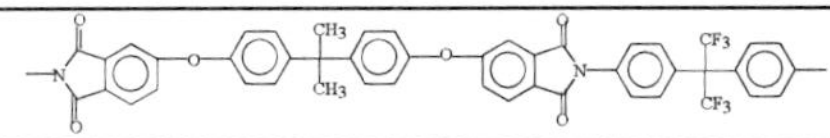	32.44	23.6	0.73	232	13.9	56	3.08
4	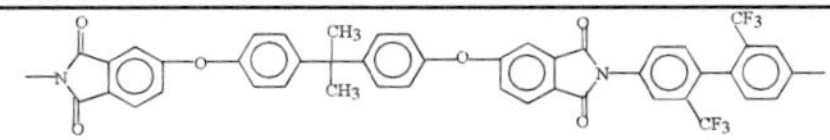	31.86	29.2	0.92	224	14.1	76	3.11
5	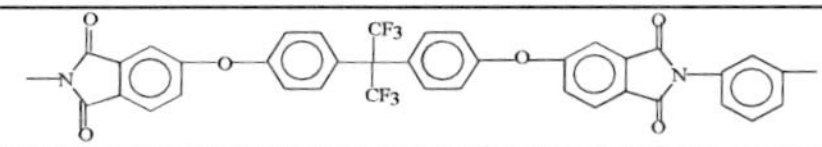	27.13	23.2	0.86	234	16.3	87	
6	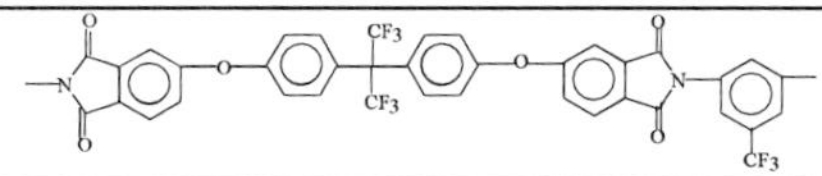	27.13	23.9	0.89	236	22.2	85	3.03
7	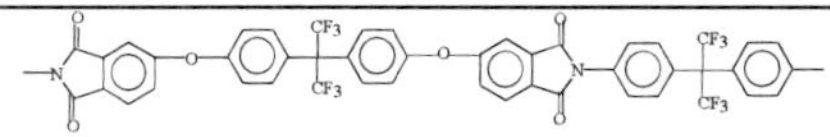	32.40	28.2	0.87	247	24.6	79	3.06
8	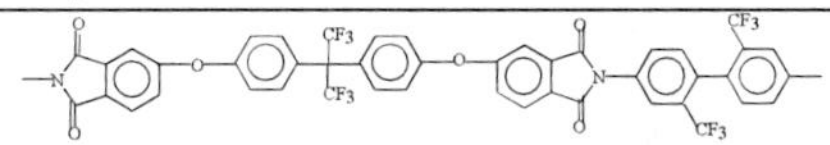	31.85	35.0	1.10	235	25.0	63.	3.09

l_o = contour length of a repeating unit; A_{fr} = Kuhn segment; C_∞ = characteristic ratio; T_g = glass transition temperature; F = fluorine content; V/V_o^{24h} = normalized surface potential ; ε = dielectric permittivity.

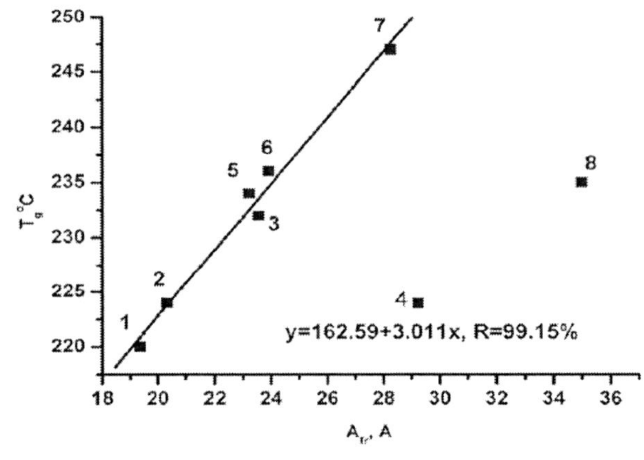

Figure 43. The dependence of glass transition temperature (T_g) on Kuhn segment (A_{fr}), for polyimides from Table 31.

We have studied the correlation between the structure and the electrets properties of these polymers. A dielectric is called ''electret'' if it is capable of maintaining an internal polarization over a long period of time. Depending on the charging technique, this polarization can be due to oriented dipoles, surface charges, or spaceharges. In corona-charged polymer electrets, where real surface charges are generated, the charge storage capability is attributed to the presence of so-called ''traps''. The nature of such traps can be of diverse nature: impurity ions, additives, free volumes, interfaces between different materials or phases, and particular sites in the molecular chain have been all postulated. There are many to produce electrets, in which certain dielectrics maintain the electric charges for a couple of days or tens of years. We examine now the behavior of the normalized surface potential as a function of the mass fraction of fluorine atoms in repeating unit of the polymer. As seen in figure 44, this dependence is described by an equation of second order. It shows that the normalized surface potential is maximum when the mass fraction of fluorine is 17-18, which corresponds to polymers **5** and **6**.

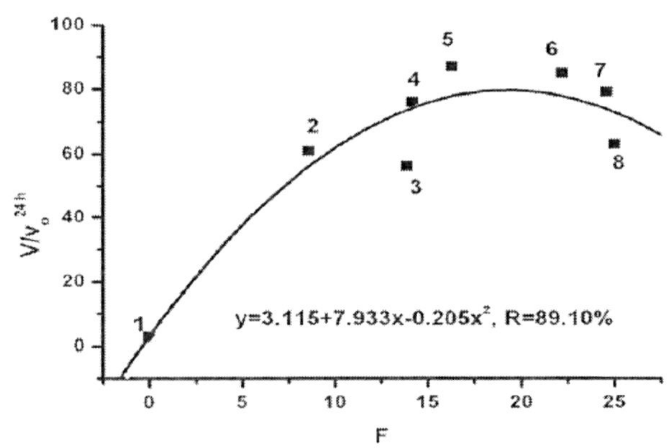

Figure 44. The dependence of remaining normalized surface potentials (V/V_o^{24h}) of corona-charged films on fluorine content (F) in the repeating unit of the polyimides from Table 31.

To see how the conformational rigidity of the polymers influences the normalized surface potential we have drawn figure 45 which presents the dependence of normalized surface potential on characteristic ratio. This dependence is linear, with a high correlation coefficient for all polymers with the exception of polymer **1, 2** and **3**.

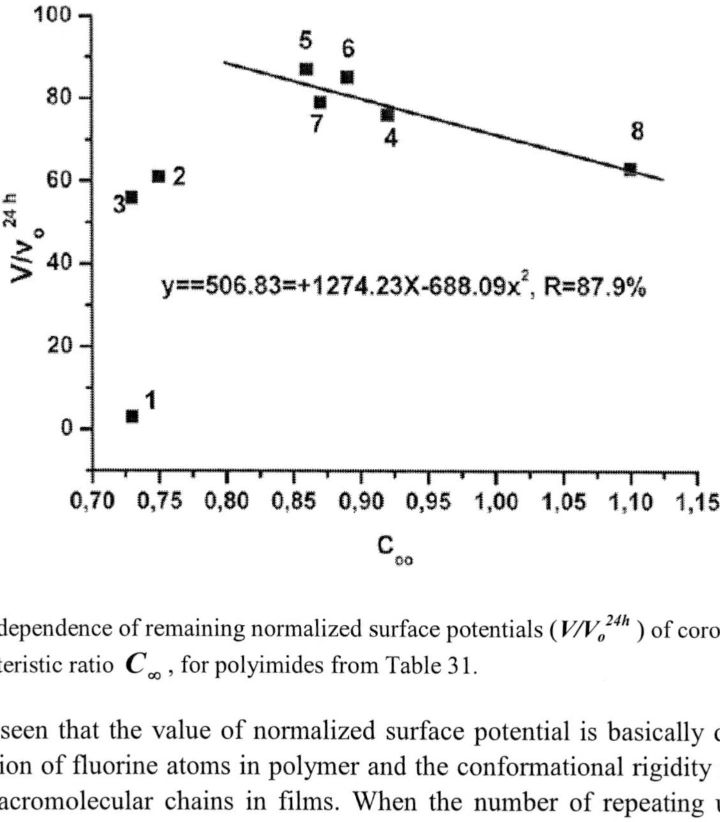

Figure 45. The dependence of remaining normalized surface potentials (V/V_o^{24h}) of corona-charged films on characteristic ratio C_∞, for polyimides from Table 31.

It can be seen that the value of normalized surface potential is basically determined by the mass fraction of fluorine atoms in polymer and the conformational rigidity influences the packing of macromolecular chains in films. When the number of repeating units in Kuhn segment increases to 0.85-0.95, the normalized surface potential is maxim. Then, with further increasing of the number of repeating units in Kuhn segment, it decreases.

The dependence of dielectric permittivity (ε) at a frequency of 1 kHz on Kuhn segment (figure 46) divides these polymers in three groups: the first group of three polymers (**6, 7** and **8**) is situated on a line with correlation coefficient of 99.21%. With increasing of rigidity, the dielectric permittivity increases. This is explained in the following way: with increasing of rigidity, the free volume decreases [176]; the polymer chains pack stronger with increasing rigidity. For polymer **5**, by using the dependences of all these three polymers, it is possible to evaluate the dielectric permittivity: it is equal to 3.029, practically identical with the dielectric permittivity of polymer **6**. In case of these polymers, the values of Kuhn segment and of normalized surface potential are very close. In the second group we have the polymers **3** and **4**. They have almost identical mass fraction of fluorine. Polymer **2** does not enter in any of these groups, because it has a very low mass fraction of fluorine. The different behavior of dielectric permittivity measured at the frequency of 1 kHz in this series of polymers compared to the behavior of dielectric constant in the second series of polymers is related with the fact that in the second series of polymers the measurements were performed at certain frequency levels and they were approximated at zero frequency. Thus, the maximum values of normalized surface potential in the studied polyimides is reached when the mass fraction of fluorine is in the domain of 17-18 and the characteristic ratio is in the range of 0.85-0.95. It means that the contour length of the repeating unit must be somewhat longer than Kuhn segment. The minimum value of dielectric permittivity is reached when the mass fraction of fluorine is 16.3, and with its increase the dielectric permittivity increases. The presence of fluorine in the corresponding dianhydride monomer makes the dielectric permittivity to significantly decreases in these series of studied polyimides. The introduction of fluorine in the diamine monomer leads to the increase of dielectric permittivity.

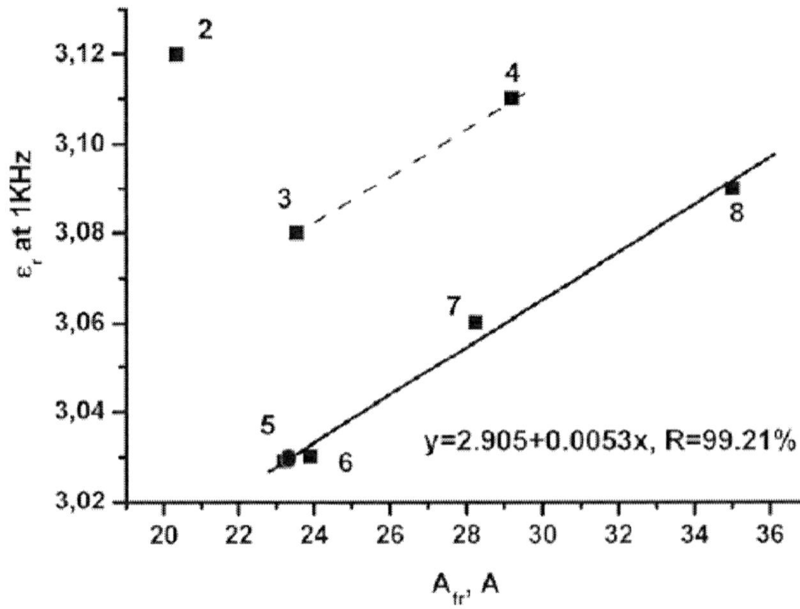

Figure 46. The dependence of dielectric permittivity (ε) on Kuhn segment (A_{fr}), for polyimides from Table 31.

Conclusion

Taking large number of polyheteroarilens as examples it was shown that geometry of a repeat unit determines the conformational rigidity of a polymer macromolecule which in turn determines the packing of polymer chains into a glassy state. The calculation of the Kuhn segment in the framework of the free rotation in absence of bulk side substitutes permits to avoid a tedious experimental measurement of a conformational rigidity.

As it occurs many physical properties are dependent significantly upon the conformation rigidity of a chain. This means that in many cases the behavior of polyheteroarilen macromolecules in glassy state is determined by the entropy contribution in thermodynamic potential of the system. This contribution is exceeding intermolecular interactions, i.e. the entropy contribution.

REFERENCES

[1] A. E. Stepanjan, E. P. Krasnov, N. V. Lukashova, and P. G. Khalatur, *Vysokomol. soed.,* 21 A, 355 (1979).

[2] A. G. Morozov, A. V. Pavlov, A. E. Chernikov, V. A. Isaeva, and V. V. Korshak, *Vysokomol. soed.,* 26A, 1437 (1984).

[3] A. K. Diborova, G. E. Prozorova, A. M. Tshetinin, and S. P. Papkov, *Chimicheskie volokna,* 2, 9 (1973).

[4] A. L. Izjumnikov, E. I. Akhmetjeva, N. N. Barashkov, E. N. Teleshov, A. N. Grimberg, Z. P. Jushina, and A. N. Pravednikov, *Vysokomol. soed.,* 25A, 582 (1983).

[5] A. L. Spatorico, *J. Polym. Sci., Polym. Phys. Ed.,* 12, 159 (1974).

[6] A. Ronova, A. R. Khokhlov, B. V. Shchukin *Polymer Science, Ser.* A, 49(5), 517 (2007)

[7] A. V. Danilova, V. M. Golubev, O. A. Olkhovnikov, P. N. Samoilov, and V. M. Savinov, *Vysokomol. soed.,* 18B, 194 (1976).

[8] A.A. Askadskii. Computational materials science of polymers. Cambridge International Science Publishing (2005).

[9] A.A. Berg, Ju.Monakov, S.N.Salazkin, S.R.Rafikov, V.P.Butov, V.P.Grigorieva, E.S.Filatov and A.E.Egorov, *Vysokomolec soed*., 25A, 2048, (1983)

[10] A.D.Williams and P.J.Flory, *J.Polym.Sci., Polym.Phys. Ed*., 6, 1945 (1968)

[11] A.E.Tonelli, Am.Chem.Soc.Div.Polym.Chem., Polym.Prepr. 13, 1023 (1972)

[12] A.E.Tonelli, *J.Polym.Sci. Polym.Lett.Ed.,* 11, 441 (1973)

[13] A.E.Tonelli, *Macromolecules*, 5, 558 (1972)

[14] A.R.Shultz, A.L.Bridgman, E.M.Hadsel and C.R.McCullough, *J. Polym.Sci., Polym.Phys. Ed.*, 10,273 (1972)

[15] Aharoni, S.M. *J.Macromol.Sci.-Phys.* B13, 173 (1977)

[16] B. Chu, Q. Ying, C. Wu, J. R. Ford, and H. S. Dhadal, *Polymer,* 26, 1401 (1985).

[17] B. Milland and C. Strazielle, *Makromol. Chem.,* 179, 1261 (1978).

[18] Barten'ev, G.M. and Frenkel, S.Ya. in " Physica Polymerov", Khimiya, Sankt.-Peterburg., Ch. 8 (1990). (in Russian)

[19] Brunvol, J. and Hargittai, L, *J. Mol. Struct.,* 30(2), 361 (1976).

[20] C. Hamchiuc, E. Hamchiuc, I.A. Ronova, M. Bruma, *High Perform. Polym*., 9, 177 (1997)

[21] C. Hamciuc, E.Hamciuc, M. Bruma, I. A. Ronova . *High Perfoprmance Polymers* 18 (1), 45 (2006)

[22] C. P. Wong, H. Ohnuma, and G. C. Berry, Am. *Chem. Soc., Div. Polym. Chem., Polym. Prepr.,* 18, 167 (1977)

[23] C.Hamchiuc, E. Hamchiuc, I.Ronova *Bulletin of the Polytechnic Institute of Iasi* 51(1-2), 67 (2006)

[24] Cyvin, S., Molecular Vibrations and Mean Square Amplitudes, Amsterdam: Elsevier, 1968

[25] D.A.Rusov, I.A.Ronova and A.Yu. Alentiev Proceedings 14th Russian Conference on "The structure and Dynamic of molecular system" Yoshkar Ola, 1, 241(2007)

[26] D.W. van Krevelen "Properties of Polymers Corelations with Chemical Structure" Elsvier Publ. Am.,L.,N.-Y. 1972

[27] E. A. Gladkova, S. S. A. Pavlova, L. V. Dubrovina, T. P. Bragina, and S. N. Salazkin, *Vysokomol. soed.,* 26A, 53 (1984).
[28] E. A. Sokolova, I. A Ronova Proceedings 14th Russian Conference on "The structure and Dynamic of molecular system" Yoshkar Ola, 1, 250 (2007)
[29] E. Bianchi, A. Ciferi, A. Tealdi, and W. R. Krigbaum, *J. Polym. Sci., Polym. Phys. Ed.*, 17, 2691 (1979).
[30] E. Bianchi, A: Ciferi, J. Preston, and W. R. Krigbaum, *J. Polym. ScL, Polym. Phys. Ed.,* 19, 863 (1981).
[31] E. D. Molodtsova, G. I. Timofeeva, Ja. S. Vygodsky, S. V. Vinogradova, V. V. Korshak, and S. S. A. Pavlova, *Vysokomol. soed.,* 19A, 346 (1977).
[32] E. D. Molodtsova, G. I. Timofeeva, S. S. A. Pavlova, Ja. S. Vygodsky, S. V. Vinogradova, and V. V. Korshak, *Vysokomol. soed.,* 16A, 2183 (1974).
[33] E. M. Rozhkov, B. V. Schukin, and I. A. Ronova, *Eur. J. Chem.* 1, 402 (2003).
[34] E.E. Said-Galiev, L.N.Nikitin, N.M. Belomoina, and E.S.Krongaus, Abstracts of the "1-st Intenational Conference on Polymeric Materials of low Flammability" Alma-Ata 1990, 76 (in Russian)
[35] Erhard DP, Giesa R, Altstädt V, Schmidt HW *Macromol Chem Phys* 208:1522 (2007)
[36] F. Laupretre and L. Mounerie, *Eur. Polym. J.,* 14, 415 (1978).
[37] G. C. Alfonso, E. Bianchi, A. Ciferri, S. Russo, F. Salaris, and B. Valenti, *Am. Chem. Soc., Div. Polym. Chem., Polym. Prepr.,* 18, 179 (1977).
[38] G. C. Berry, *Disc. Farad. Soc.,* 49, 121 (1970).
[39] G. Cowie and S. Bywater, *Polymer, 6,* 197 (1965).
[40] G. E. Prozorova, A. K. Diborova, A. T. Kalashnik, L. P. Milkova, A. M. Tshetinin, M. M. Iovleva, and S. P. Papkov, *Chim. volokna,* 5, 38 (1976).
[41] G. E. Prozorova, A. V. Pavlov, A. M. Tshetinin, M. M. Iovleva, and S. P. Papkov, *Vysokomol. soed.,* 19B, 750 (1977).
[42] G. E. Prozorova, A. V. Pavlov, M. M. Iovleva, R. V. Antipova, V. D. Kalmykova, and S. P. Papkov, *Vysokomol. soed.,* 19B, 111 (1976).
[43] G. E. Prozorova, M. M. Iovleva, B. N. Smirnova, A. V. Tokarev, L. V. Avrorova, A. V. Volokhina, G. I. Kudriavtsev, and S. P. Papkov, *Vysokomol. soed.,* 23B, 2, 134 (1981).
[44] G. E. Prozorova, M. M. Iovleva, V. N. Smirnova, A. V. Tokareva, L. V. Avrorova, A. V. Volokhina, G. I. Kudriavtsev, and S. P. Papkov, *Vysokomol. soed.,* 23B, 134 (1981).
[45] G. I. Timofeeva, E. D. Molodtsova, I. A. Ronova, and S. S. A. Pavlova, *Vysokomol. soed.,* 20A, 286 (1978).
[46] G. I. Timofeeva, S. S. A. Pavlova, L. A. Lopanova, H. H. Ulrich, G. Reinisch, and E. Leibnits, *Acta Polymer.,* 34, 144 (1983).
[47] G. I. Timofeeva, S. S. A. Pavlova, N. M. Katsoeva, and V. V. Korshak, *Vysokomol. soed.,* 15B, 816 (1973).
[48] G. I. Timofeeva, V. V. Korshak, S. S. A. Pavlova, G. M. Tseitlin, N. S. Zabelnikov, and K. A. Krishtul, *Vysokomol. soed.,* 15B, 428 (1973).
[49] G. I. Timofeeva, V. V. Korshak, S. S. A. Pavlova, Ya. S. Vygodsky, and N. A. Churochkina, *Vysokomol. soed.,* 15A, 2650 (1973).
[50] G. M. Pavlov, N. Ya. Kozlov, S. M. Jakobson, S. V. Usova, and L. S. Efros, *Vysokomol. soed.,* 27A, 30 (1985).
[51] G. M. Pavlov, S. G. Semonin, N. A. Shildjaeva, S. M. Jakobson, L. S. Efros, and S. V. Usova, *Vysokomol. soed.,* 27A, 1627 (1985).

[52] G. V. Tarasova, A. E. Polotsky, T. I. Garmonova, V. S. Galenko, A. N. Cherkasov, M. M. Koton, V. A. Gusinskaja, T. V. Batrakova, and K. A. Ro- mashkova, *Vysokomol. soed.,* 21 A, 541 (1979).

[53] G.Allen, J.McAinish and Strazielle, *Eur.Polym.J.* 5, 319 (1969)

[54] G.C.Berry, H.Nomura and K.G.Mayhan, *J.Polym.Sci., Polym.Phys. Ed*., 5, 1 (1967)

[55] G.SH.Papawa, N.A.Maisuradze and L.V.Dubrovina, *Izvestija Gruzinskoi Aka. Nauk SSR, ser. Chim.*, 1, 152 (1975)

[56] G.Sitaramaiah, *J.Polym.Sci.,* 3A, 2743 (1965)

[57] G.V.Schulz and S.Horbach, *Makromol. Chem.*, 29, 93 (1959)

[58] H. Yamakawa and M. Fujii, *Macromolecules,* 6, 407 (1973)

[59] H. Yamakawa arid M. Fujii, *Macromolecules,* 7, 128 (1974)

[60] Hofmann, L. Fritz, J. Ulbricht, C. Shepers, M. Boehning *Macromol. Theory Simulation* 9:293 (2001).

[61] I. A. Baranovskaja, V. V. Kudriavtsev, N. V. Diakonova, V. P. Sklizkova, and M. M. Koton, *Vysokomol. soed.,* 27A, 604 (1985).

[62] I. A. Ronova, B. V. Shchukin, E. A. Sokolov. Proceedings of XIY Conference High technologies in Russian industry (Materials and fabrication of functional devices for electronics and photonics), Technomash, Moscow, 277-282 (2008),.

[63] I.A. Ronova M. Bruma *Struct Chem* 21:1013 (2010)

[64] I.A. Ronova, and I.I. Ponomarev, *Vysokomol.soed.,* 37A, 123 (1995)

[65] I.A. Ronova, L.N. Nikitin, E.A. Sokolova, I. Bacosca, I. Sava, M. Bruma, *J. Macromol. Sci., Part A: Pure Appl.Chem*. 46, 929 (2009)

[66] I.A.Ronova, A.N.Vasilyuk, C.Gaina, V.Gaina *High Performance Polymers* 14, 195 (2002)

[67] I.A.Ronova, I.A. Nyrkova, , S.A. Pavlova and A.R. Khokhlov, *Polymer Science, Ser. A*, 36, 779 (1994)

[68] I.A.Ronova, I.A.Nyrkova, S.A.Pavlova and A.R.Khokhlov, *Polym.Sci.,*36A, 779 (1994)

[69] I.A.Ronova, Ion Sava Ştefan Chişcă, Maria Brumă, V.A.Medvedeva XVI Proceedings of XVI Conference High technologies in Russian industry (Materials and fabrication of functional devices for electronics and photonics), Technomash, Moscow, 302-304 (2010)

[70] I.A.Ronova, L.V.Dubrovina, A.Yu.Kovalevskii, C.Hamciuc and M. Bruma, *Russian Chem.Bill*, 47, 1248 (1998)

[71] I.I Ponomarev, O.G.Nikol'skii, Ju.A.Volkova, and A.V. Zakharov, *Vysokomol.soed.,* 36A, 1429 (1994)

[72] I.Ronova. *Struct Chem* 21:541 (2010)

[73] I.Sava, I.Ronova and Maria Bruma, *Polymer Journal* (Japan).38(9), 940 (2006)

[74] Ion Sava, Maria Bruma, Inga Ronova. *High. Perfoprmance Polymers* 16, 435 (2004)

[75] J. C. Kirkwood and J. Riseman, *J. Chem. Phys.,* 16, 565 (1948).

[76] J. Hearst and W. H. Stockmayer, *J. Chem. Phys.,* 37, 1425 (1962).

[77] J. Hearst, *J. Chem. Phys.,* 38, 1062 (1963).

[78] J. J. Burke, *J. Macromol. Sci., Chem.,* 7,181 (1973).

[79] J. Kovar, J. Fortenly, and M. Bohdanecky, *Makromol. Chem.,* 178, 23 (1977).

[80] J. R. Schaefgeji, V. S. Foldi, F. M. Logullo, V. H. Good, L. W. Gulrich, and F. L. Killian, *Am. Chem. Soc. Div. Polym. Chem., Polym. Prepr.,* 17, 69 (1976)

[81] J. S. Tan, Am. Chem. Soc., Div. Polym. Chem., Polym. Prepr., 14, 340 (1973).

[82] J. S. Tan, J. Polym. Sci., Polym. Phys. Ed., 12, 175 (1974).
[83] J.M.Barrales-Rienda and D.C.Pepper, *J.Polym.Sci., Polym.Phys. Ed.*, 4,939 (1966)
[84] K. Sakurai, K. Ochi, T. Norisuye, and H. Fujita, *Polym. J.,* 16, 559 (1984).
[85] Korshak V.V. in "Thermostoikie Polymery", "Nauka", Moscow. Ch. 4 (1969). (in Russian)
[86] L. Pouling: *General Cemistry*, Freeman and Co, San Francisco, 1970
[87] L. V. Dubrovina, S. S. A. Pavlova, and T. P. Bragina, *Vysokomol. soed.,* 30B, 995 (1988).
[88] L.N.Nikitin, I.A.Khotina, E.E.Said-Galiev, A.L.Rusanov, I.A. Ronova, and A.M. Andrievskiy, *Polymer Science, Ser.* 37A, 779 (1995)
[89] Lindeman, S.V, Ponomarev, I.I., Ronova, I.A., Raevskiy, N.I. and Struchkov,Yu.T. in "The All Union Conference on Application of Crystallochemical Data in Organic Chemistry", 1991, 39
[90] M. Arpin and C. Strazielle, *C. R. Acad. Sci., Paris,* 280, 1293 (1975).
[91] M. Arpin and C. Strazielle, *Polymer,* 18, 591 (1977).
[92] M. Arpin, C. Strazielle, and H. Benoit, *Polymer, 18,* 262 (1977).
[93] M. Arpin, F. Debeauvais, and C. Strazielle, *Makromol. Chem.,* 177, 585 (1976).
[94] M. Bohdanecky, J. Kovar, and J. Fortenly, *Polymer,* 20, 813 (1979).
[95] M. Bohdanecky, *Macromotecules,* 16, 1483 (1983).
[96] M. G. Vitovskaja, P. N. Lavrenko, E. P. Astapenko, V. Okatova, and V. N. Tsvetkov, *Vysokomol. soed.,* 20A, 320 (1978).
[97] M. G. Vitovskaja, P. N. Lavrenko, O. V. Okatova, E. P. Astapenko, V. B. Novakovsky, S. V. Bushin, and V. N. Tsvetkov, *Eur. Polym. J.,* 18, 583 (1982).
[98] M. L. Wallach, Am. Chem. Soc., Div. Polym. Chem., Polym. Prepr., *6, 53 (1965).*
[99] M. M. Iovleva, G. E. Prozorova, and S. P. Papkov, *Vysokomol. soed.,* 20A, 672 (1978).
[100] M. M. Iovleva, G. E. Prozorova, and V. N. Smirnova,*Vysokomol. soed.*, 25B, 15 (1983).
[101] M. M. Iovleva, G. E. Prozorova, V. N. Smirnova, and S. P. Papkov, *Vysokomol. soed..* 23A, 2092 (1981).
[102] M.J.S. Dewar, E.F. Zoebisch, E.F. Healy, J.J. Stewart *J Am Chem Soc* 107:3903 (1985)
[103] M.L.Wallach, Am.Chem. Soc., Div. Polym. Chem., Polym. Prepr., 6, 860 (1965)
[104] M.L.Wallach, *Makromol. Chem.*, 103, 19 (1965)
[105] N. A. Glukhov, T. L. Garmonova, V. S. Skazka, V. Bushin, M. G. Vitovskaja, and L. M. Tsherbakova, *Vysokomol. soed.,* 17B, 579 (1975).
[106] N. Lavrenko, V. B. Novakovsky, O. I. Mikrjukova, V. N. Kolot, and G. I. Kudriavtsev, *Vysokomol. soed.,* 26A, 785 (1984).
[107] N. V. Pogodina, A. B. Melnikov, I. N. Bogatova, V. N. Tsvetkov, V. V. Korshak, S. V. Vinogradova, L. Rusanov, I. I. Ponomarev, and Yu. N. Mar- galitadze, *Vysokomol. soed.,* 31 A, 73 (1989).
[108] N. V. Pogodina, I. N. Bogatova, and V. N. Tsvetkov, *Vysokomol. soed.,* 27A, 1405 (1985).
[109] N. V. Pogodina, L. V. Starchenko, A. Z. Khrustalev, and V. N. Tsvetkov, *Vysokomol. soed.,* 26A, 2057 (1984).
[110] N. V. Pogodina, L. V. Starchenko, V. N. Tsvetkov, V. V. Korshak, A. L. Rusanov, S. Kh. Fidler, and T. V. Lekae, *Vysokomol. soed.,* 28B, 278 (1986).
[111] N. V. Pogodina, N. P. Evlampiev, V. N. Tsvet,kov, V. V. Korshak, A. L. Rusanov, A. M. Berlin, E. R. Bulycheva, and M. O. Shalikiany, *Vysokomol. soed.,* 31A, 20 (1981).

[112] N. V. Pogodina, N. P. Evlampiev, V. N. Tsvetkov, V. V. Korshak, S. V. Vinogradova, A. L. Rusanov, I. I. Ponomarev, and T. V. Lekae, *Vysokomol. soed.,* 30A, 1198 (1988).
[113] N. V. Pogodina, N. P. Evlampiev, V. V. Korshak, S. V. Vinogradova, A. L. Rusanov, and I. I. Ponomarev, *Dokl. Akad. Nauk SSSR,* 301, 905 (1988).
[114] N. V. Vasilieva, G. E. Prozorova, A. M. Tshetinin, A. K. Diborova, V. G. Kulichikhin, and S. P. Papkov, *Chim. volokna,* 2, 53 (1976).
[115] N.Yu.Kobac, L. V. Dubrovina and S. S. A. Pavlova *Vysokomol. soed.,* 36AB, 425 (1994)
[116] O. V. Kallistov, Ju. E. Svetlov, I. G. Silinskaja, V. P. Kudriavtsev, and M. M. Koton, *Eur. Polym. J.,* 18, 1103 (1982).
[117] O. V. Kallistov, Yu. E. Svetlov, I. G. Silinskaja, T. A. Startseva, V. P. Sklizkova, V. V. Kudriavtsev, and M. M. Koton. *Dokl. Akad. Nauk SSSR,* 244, 396 (1979).
[118] P. Cotts and R. Metzger, Polyimides: Synthesis, Character and Application Vol. 1, 1st Technical Conference, Elleville, NY, November 10-12, 1982, New York, London, 1984, p. 223.
[119] P. J. Flory, *J. Chem. Phys.,* 10, 51 (1942).
[120] P. J. *Flory,* Proc. Royal Soc., Ser. A, *234,* 60 (1956).
[121] P. N. Lavrenko, O. V. Okatova, V. V. Korshak, S. V. Vinogradova, A. L. Rusanov, and I. I. Ponomarev, *Vysokomol. soed.,* 32A, 1209 (1990).
[122] P.J. Flory *Statistical mechanics of chain molecules.* (Interscience, New York 1969)
[123] P.J.Akers, G.Allen and M.J.Bathell, *Polymer*, 9, 575 (1968)
[124] Ponomarev, I.I., Ronova, I.A., Lindeman, S.V, Rusanov, A.L., Vinogradova, S.V. and Struchkov, Yu.T. *Vysokomol.soed., A,* 1992, 34, 123
[125] R.M. Aseeva, and G.E. Zaikov"Gorenie polymernykh materialov" Nauka, M 1981 (in Russian)
[126] Rusanov, A.L., Serkov, V.V., Bulycheva, E.G., et al *Macromol. Chem., Macromol.Symp.* 1993, 189.
[127] S. S. A. Pavlova, G. I. Timofeeva, E. S. Krongauz, A. M. Berlin, and V. V. Korshak, *Vysokomol. soed.,* 18A, 863 (1976).
[128] S. S. A. Pavlova, G. I. Timofeeva, I. A. Ronova, S. A. Krojan, V. V. Korshak, A. P. Travnikova, and E. S. Krongauz, *Vysokomol. soed.,* 22 A, 1698 (1980).
[129] S. S. A. Pavlova, L. V. Dubrovina, I. A. Ronova, and N. Yu. Kobak, *Vysokomol. soed.,* 30A, 1505 (1988).
[130] S. S. A. Pavlova, L. V. Dubrovina, S. V. Vinogradova, S. N. Salazkin, and V. V. Korshak, *Vysokomol. soed.,* 12B, 69 (1970).
[131] S. S. A. Pavlova, N. Ju. Kobak, L. V. Dubrovina, T. P. Bragina, and S. N. Salazkin, *Vysokomol. soed.,* 3OB, 650 (1988).
[132] S. S. A. Pavlova, N. Ju. Kobak, L. V. Dubrovina, T. P. Bragina, and S. N. Salazkin, *Vysokomol. soed.,* 31A, 1680 (1989).
[133] S. S. A. Pavlova, S. V. Vinogradova, G. I. Timofeeva, and D. R. Tur, *Vysokomol. soed.,* 13 A, 2653 (1971).
[134] S. S. A. Pavlova, V. V. Korshak, L. V. Dubrovina, and G. I. Timofeeva, IUPAC International Symposium on Macromolecules, Inter Scientias Inc. Leiden, Vol. II, August 31, 1970, p. 933.
[135] S. S. A. Pavlova, Ya. S. Vygodsky, E. D. Molodtsova, S. V. Vinogradova, G. I. Timofeeva, and V. V. Korshak, *Vysokomol. soed.,* 2IB, 100 (1979).

[136] S. V. Vinogradova, V. A. Vasnev, V. V. Korshak, A. V. Vasiliev, and L. V. Dubrovina, *Vysokomol. soed.,* 13A, 770 (1971).
[137] S.A. Pavlova, G.I. Timofeeva, and I.A. Ronova, *J. Polym. Sci., Polym. Phys. Ed.* 18, 1175 (1980)
[138] S.A. Pavlova, G.I. Timofeeva, I.A. Ronova, and L.A. Pankratova, *J. Polym. Sci., Polym. Phys. Ed.* 18, 1 (1980)
[139] S.A. Pavlova, I.A. Ronova, G.I. Timofeeva, and L.V. Dubrovina, *J. Polym. Sci., Polym. Phys. Ed.* 31, 1725 (1993)
[140] S.M.Jakobson, A.P.Verkhovets, O.G.Akulina, I.M.Kralina, I.S.Gugel, G.M.Pavlov, L.S. Efros and O.S. Karchmarchic K*himicheskie volokna*, .2, 5 (1989) (in Russian)
[141] Seip, H.M. and Seip, R., *Acta Chem. Scand.,* 27(10), 4024 (1973).
[142] T. E. Helminiak, C. L. Benner, and W. E. Gibbs, *Am. Chem. Soc., Div. Polym. Chem., Polym. Prepr.,* 11, 291 (1970).
[143] T. Kojima, R. Yokota, M. Kochi, and H. Kambe, *J. Polym. Sci., Polym. Phys. Ed.,* 18, 1673 (1980).
[144] T.M.Birshtein and A.N.Gorjunov, *Vysokomolec soed*., 21A, 1990, (1979)
[145] T.M.Birshtein *Vysokomolec soed*., 19A, 54, (1977)
[146] The USSR Patent no. 664972, (1979)
[147] Tsvetkov, V.N., *Thestkotsepnye Polimery* (Rigid-Chain Polymers), Leningrad: Nauka. 1986
[148] Uno, T, Mochida, K.. and Manai. K.. *Spectrochim. Acta, Part A,* 24(7), 1705 (1968).
[149] V. B. Novakovsky, P. N. Lavrenko, and V. N. Tsvetkov, *Eur. Polym. J.,* 19, 831 (1983).
[150] V. E. Eskin, I. A. Baranovskaja, M. M. Koton, V. V. Kudrjavtsev, and V. P. Skliskova, *Faserforsch. Tex tiltechn.,* 28, 617 (1977).
[151] V. M. Andreeva, A. A. Tager, and V. I. Konevets, *Vysokomol. soed.,* 23B, 538 (1981).
[152] V. N. Smirnova, G. E. Prozorova, M. M. Iovleva, and S. P. Papkov, *Vysokomol. soed.,* 25B, 527 (1983).
[153] V. N. Smirnova, I. A. Ronova, G. E. Prozorova, N. P. Okromchedlidze, and M. M. Iovleva, *Vysokomol. soed.,* 29B, 710 (1987).
[154] V. N. Tsvetkov, G. I. Kudriavtsev, I. N. Shtennikova, T. V. Peker, E. N. Zakharova, V. D. Kalmykova, and A. V. Volokhina, *Vysokomol. soed.,* 18A, 2212 (1976).
[155] V. N. Tsvetkov, G. I. Kudriavtsev, N. A. Mikhailova, A. V. Volokhina, and V. D. Kalmykova, *Vysokomol. soed.,* 20A, 191 (1978).
[156] V. N. Tsvetkov, I. N. Stennikova, P. N. Lavrenko, G. P. Koblina, and O. V. Okatova, *Acta Polym.,* 31, 434 (1980).
[157] V. N. Tsvetkov, P. N. Lavrenko, G. M. Pavlov, S. V. Bushin, E. P. Astapenko, L. A. Boikov, N. A. Shildjaeva, and B. F. Malichevsky, *Vysokomol. soed.,* 24A, 2343 (1982).
[158] V. N. Tsvetkov, V. B. Novakovsky, I. A. Strelina, V. I. Terenin, A. V. Volokhina, A. I. Koretskaja, and M. P. Babushkina, *Vysokomol. soed.,* 28B, 644 (1986).
[159] V. N. Tsvetkov, V. B. Novakovsky, I. A. Strelina, V. V. Korshak, A. L. Rusanov, S. V. Vinogradova, I.I.Ponomarev, and T. V. Lekae, *Vysokomol. soed.,* 31A, 40 (1989).
[160] V. N. Tsvetkov, V. B. Novakovsky, V. N. Kolot, and G. I. Kudriavtsev, *Vysokomol. soed.,* 26B, 245 (1984).
[161] V. N. Tsvetkov, V. B. Novakovsky, V. V. Korshak, A. L. Rusanov, A. M. Berlin, and S. Kh. Fidler, *Vysokomol. soed.,* 27A, 86 (1985).

[162] V. V. Korshak, I. A. Gribova, A. P. Krasnov, G. V. Mamatsashvilli, S. S. A. Pavlova, G. I. Timofeeva, I. A. Ronova, S. A. Krojan, E. S. Krongauz, A. P. Travnikova, and N. M. Belomoina, *Vysokomol. soed.,* 21 A, 2214 (1979).

[163] V. V. Korshak, S. S. A. Pavlova, L. V. Dubrovina, N. Yu. Kobak, and E. A. Gladkova, *Vysokomol. soed.,* 22A, 1458 (1980).

[164] V. V. Korshak, S. S. A. Pavlova, L. V. Dubrovina, V. A. Vasnev, S. V. Vinogradova, and A. V. Vasiliev, *Izvestija Akad. Nauk SSSR, ser. Chim.,* 6, 1409 (1972).

[165] V. V. Korshak, S. S. A. Pavlova, V. A. Vasnev, and S. A. Vinogradova, *Vysokomol. soed.,* 12A, 1308 (1970).

[166] V.N.Odnoralova, E.A. Vasil'eva-Sokolova, in "Khimiya i Tekhnologiya Vysokomolekuljarnykh Soedineniy", Moscow. 12, 147(1970) (in Russian)

[167] V.V., Korshak, E.S.Krongaus, A.L. Rusanov, et al. *Vysokomol.soed ,* 31A, 80 (1989)

[168] Volkenstein, M.V., «*Konfigurationnaya statistica polimernykh tsepey*», (Izd. Akad. Nauk SSSR 1959).

[169] W. H. Stockmayer and M. Fixman, *J. Polym. Sci.,* lc, 137 (1963).

[170] W. R. Krigbaum and S. Sasaki, *J. Polym. Sci., Polym. Phys. Ed.,* 19, 1339 (1981).

[171] W.R.Moore and M.A.Uddin, Eir. *Polym. J.*, 6, 121 (1970)

[172] Yu. N. Lazareva, M. N. Vidyakin, A. Yu. Alentiev, M. Yu. Yablokova, A. A. Kuznetsov, I. A. Ronova, *Polym. Sci. Ser. A,* 51(10), 1068–1074 (2009).

[173] Yu. N. Lazareva, M. N. Vidyakin, Yu. P. Yampolskii, A. Yu. Alentiev, M. Yu. Yablokova, G. K. Semenov, A. A. Kuznetsov, D. Yu. Likhachev, *Vysokomol. Soedin. Ser.* A, 48(10), 1818 (2006).

[174] Z. M. Farmazjan, S. S. A. Pavlova, and L. V. Dub- rovina, *Vysokomol. soed..* 22A, 1582 (1980).

[175] Zubkov, V.L., Birshtein, T.M., and Milevskaya, I.S., *Vysokomol. Soedin., Ser. A,* 17(11), 1955 (1975).

[176] Zubkov, V.L.. Birshtein, T.M., and Milevskaya, I.S., *Vysokomol. Soedin., Ser. A,* 16(12), 2438 (1974).

In: Polymer Chains
Editors: Régine Penzkofer and Yasue Wu
ISBN: 978-1-62100-422-6

Chapter 2

ALIPHATIC POLYESTER-BASED NANOCOMPOSITES

Jung Seop Lim
Department of Fiber and Polymer Engineering, Hanyang University, College of Engineering, Haengdang-dong, Seongdong-gu, Seoul, South Korea

1. OVERVIEW

Nowadays, aliphatic polyerster-based polymers have attracted a great deal of interest as promising biodegradable materials. Nevertheless, in certain practical applications, these polymers suffer from limitations, such as its thermal stability and insufficient stiffness, low-melt viscosity, and poor mechanical properties. Inorganic nanocomposite, made by adding a small amount of inorganic material within a matrix of polymer, is an effective way of improving these drawbacks and widening its application. Such nanocomposites usually exhibit substantially improved thermal, mechanical, and physical properties in comparison to the corresponding neat polymer. They can combine the favored properties of the inorganic material (e.g., increased strength and modulus, high thermal stability and better gas barrier property) and the organic polymer matrix (e.g., flexibility, low density and processability). Furthermore, they have merits when compared with conventional polymer composites, such as lightweight, transparency, and easier processability. Consequently, organic polymer-based nanocomposites are considered as next-generation materials.

Since the commercialization of nylon-6/clay nanocomposites in the early 1990s, several types of polymer/inorganic nanocomposites have been developed, including those consisting of clay, layerd silicate, talc, MMT, etc. Among them, fumed silica is a useful reinforcing agent for polymer/ inorganic nanocomposites. However, this inorganic material is strongly hydrophilic due to the numerous silanol groups on its surface and exhibits a very high surface energy, which in turn leads to its tendency to aggregate in the composites. In the preparation of these nanocomposites, the dispersion of inorganic particles on a nanometer level and the compatibility between it and the polymer matrix are the most critical factors. To improve the interfacial interaction between inorganic particles and polymer matrix, many studies have attempted to convert the hydrophilic particle surface into a hydrophobic surface in the nanocomposites.

On the other hand, the secondary intermolecular interaction, formed between respective components in polymer based nanocomposites, usually strongly influences several parameters regulating the crystallization behavior; for example, nucleation and growth rate, and characteristics of the individual folded chains, which not only affect the extent of crystallinity and the nature of crystalline morphology but also determine the final optical and mechanical properties of a polymer composite. Furthermore, this relationship is important for assessing the industrial processing conditions of the final product such as spinning, extrusion, compression, and injection molding. Therefore, the connection between intermolecular interaction and the crystallization behavior must be carefully examined. Based on these considerations, in this chapter, we discuss the preparation of various types of aliphatic polyester/silica nanocomposites and quantitatively ascertain not only the relationship of dispersibiliy and physical properties but also effects of intermolecular interaction on their crystallization behavior.

2. Effect of Hydrogen Bonding on the Crystallization Behavior of Poly(3-Hydroxybutyrate-Co-3-hydroxyhexanoate)/Silica Hybrid Composites

2-1. Introduction

(PHB-HHx) is a relatively new member of the polyhydroxyalkanoate (PHA) family, which has gained popularity as a useful biodegradable polymer due to superior thermoplasticity, excellent biocompatibility, and biodegradability [1, 2]. Nevertheless, in certain practical applications, PHB-HHx suffers from limitations, such as its thermal stability and insufficient mechanical stiffness. Hybrids made of PHB-HHx and inorganic materials may circumvent these limitations of the PHB-HHx and to expand its applications.

Fumed silica is a useful reinforcing agent for polymer/inorganic hybrids. However, this inorganic material is strongly hydrophilic due to the numerous silanol groups on its surface and exhibits a very high surface energy, which in turns leads to its tendency to aggregate in the composites. Thus, in order to reduce the surface energy of silica and prevent the subsequent aggregation, fumed silica may be hydrophobically modified. Such hydrophobically modified silica particles are more readily dispersed in the polymer matrix and produce organic/inorganic hybrid composites with improved properties [3, 4].

Until now, studies of biodegradable polymer/inorganic hybrid composites have traditionally focused mainly on the dispersibility of the inorganic materials within the polymer matrix and the resulting physical properties. Relatively little has been reported on the influence of hydrogen bonding on the crystallization behavior, which affects the physical properties and processing conditions of final product. Further, few investigations have examined PHB-HHx/inorganic hybrids.

In this study, PHB-HHx/hydrophobically modified fumed silica hybrid composites were prepared by using simple melt blending and the effects of hydrogen bonding on their crystallization behavior were ascertained. The hybrid composites not only showed great potential for advanced material applications, but also helped to clarify how addition of the inorganic component modifies the crystallization mechanisms.

2-2. Experimental

2-2-1. Materials and Sample Preparation

The bacterial polyester, poly(3-hydroxybutyrate-co-3-hydroxyhexanoate) (PHB-HHx) was supplied by the Procter & Gamble Company (Cincinnati, USA). It contained the 3-hydroxy hexanoate content of about 12 mol% and the weight averaged molecular weight (M_w) was 1,388,000 with the polydispersity of about 2.0. Two types of fumed silica, Aerosil R300 and Aerosil R812 (Degussa Chemical, Düsseldorf, Germany), were used to prepare the hybrid composites. Aerosil R300 has pronounced hydrophilic properties, while Aerosil R812 is hydrophobically modified fumed silica produced by treatment with hexamethyl disilane. It has the specific surface area of 260 ± 30 $m^2\ g^{-1}$, average particle size of 7 nm, and degree of hydrophobicity of about 42%.

PHB-HHx/silica hybrid composites were prepared by the direct melt compounding by using a twin roll counter-rotating mixer (Thermo Haake, Paramus, NJ, USA). The barrel temperature was set to be 165°C and the screw speed 100 rpm. The total mixing time was 2 min. Before the melt compounding in a one-step process, PHB-HHx was ground and mixed with silica. After melt compounding, films were prepared using a hot press at 165°C with the holding pressure of 6,000 psi and the holding time of 5 min. The ratios of PHB-HHx to fumed silica in two hybrids were 100/0, 99/1, 97/3, and 95/5 by weight. The prepared films were quenched in cold water.

2-2-2. Measurements

Fourier transform infrared (FT-IR) spectra were obtained at room temperatures on an IFS 88-IR spectrometer (Bruker AXS GmbH, Karlsruhe, Germany). The scanned wavenumber range was 4000 – 400 cm^{-1}. All spectra were recorded at a resolution of 4 cm^{-1}, and 16 scans were averaged for each sample.

Scanning electron microscopy (SEM) of the PHB-HHx/silica hybrids films was undertaken after fracturing the films under liquid nitrogen and drying at room temperature. The samples were gold coated by ion sputtering and examined by using field emission (FE) SEM (JSM-6700F, JEOL, Tokyo, Japan).

The thermal properties of 6 mg sample of each PHB-HHx/silica hybrid composite in a sealed aluminum sample pan were analyzed using a Perkin Elmer DSC7. All the samples were heated at 5°C min^{-1} from – 40 to 180°C and then maintained at 180°C for 3 min, before cooling to – 40°C at 1, 2, 3, or 5°C min^{-1}. Subsequently, the samples were reheated at 5°C min^{-1}.

The isothermal crystallization of the PHB-HHx/silica hybrids was analyzed by heating each sample from –10 to 200°C at 40°C min^{-1} and then maintaining it at 200°C for 5 min to erase the previous thermal history. The samples were then quenched to their crystallization temperatures at 150°C min^{-1} and the isothermal crystallization exotherm was recorded for 60 min.

The spherulitic morphology of the hybrid samples was investigated by using a Nikon polarized light microscope (Nikon, Tokyo, Japan). Each sample was melted on a hot stage at 190°C for 1 min and then quickly transferred to a Mettler-Toledo FP82HT hot stage and equilibrated at the desired crystallization temperature (Mettler-Toledo, Greifensee, Switzerland). Crystallization occurred over 60 min and the spherulitic morphology of the samples was examined when crystallization was complete.

2-3. Results and Discussion

2-3-1. Intermolecular Hydrogen Bonding

Figure 2-1 shows the IR spectra in the hydroxyl group regions of the hydrophilic and hydrophobic silica. The hydrophilic silica shows a broad band centered at 3435cm^{-1}, reflecting the intramolecular hydrogen-bonded silanol hydroxyl groups [4]. In the hydrophobic silica, the hydroxyl band shifts to a higher wavenumbers and its peak occurs near 3441cm^{-1}. These results indicate that the intramolecular hydrogen bonding in the hydrophobic silica is weaker than that of the hydrophilic silica as a result of a relatively small amount of hydroxyl groups [5].

Figures 2-2(a) and (b) show the IR spectra in the 1800–1680cm^{-1} carbonyl region of the PHB-HHx/hydrophobically modified silica and PHB-HHx/hydrophilic silica hybrid composites, respectively. The amorphous carbonyl peak of pure PHB-HHx occurs at about 1730cm^{-1}. As the silica content of the hybrids increases, the carbonyl peak shifts to lower wavenumbers, suggesting that the intermolecular hydrogen bonding in the PHB-HHx/silica hybrids increases with the added silica content. However, compared with the PHB-HHx/hydrophilic silica hybrids, the carbonyl peaks of the PHB-HHx/hydrophobically modified silica hybrids occur at lower wavenumbers.

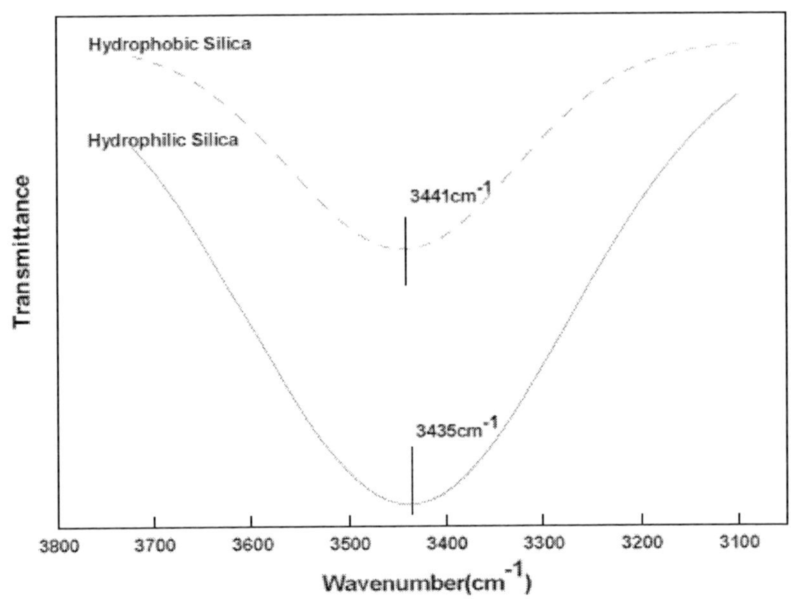

Figure 2-1. IR spectra in the hydroxyl group region of hydrophilic silica and hydrophobic silica.

The maximum extent of the carbonyl peak shift is about 3cm^{-1} for the PHB-HHx/hydrophobic silica (5wt%) composites for which the carbonyl peak lies near 1727cm^{-1}. On the other hand, carbonyl peak of the PHB-HHx/hydrophilic silica (5 wt%) moves down only by about 1cm^{-1}, so that this peak is located at about 1729cm^{-1}. These results demonstrate that hydrophobically modified silica, with a relatively small amount of hydroxyl groups, contains weaker intramolecular hydrogen bonding than hydrophilic silica. Hence, intermolecular hydrogen bonds may be more easily formed with the carbonyl group of the PHB-HHx.

Figure 2-3 shows the IR spectra in the hydroxyl group region of the PHB-HHx/hydrophobically modified silica hybrids. On forming hybrids with silica, the OH band appears at 3418cm^{-1} for PHB-HHx/silica (1wt%) hybrids and gradually shifts to 3407 and 3401cm^{-1}, respectively, as the silica content increases from 3 to 5wt%. This result may be attributed to the fact that with higher silica content, more silica hydroxyl groups can take part in the intermolecular hydrogen bonding with the carbonyl groups of PHB-HHx.

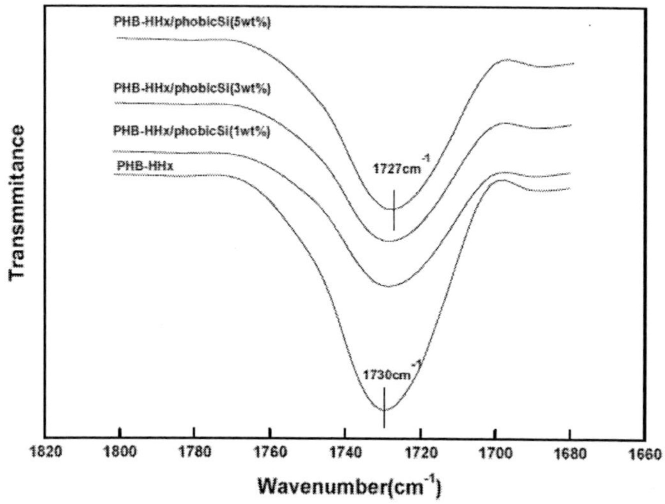

(a)

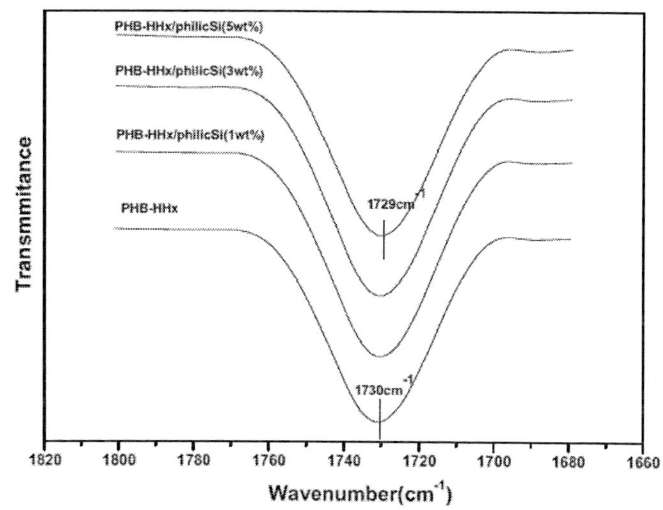

(b)

Figure 2-2. IR spectra of the carbonyl group region of the PHB-HHx/silica hybrid composites; (a) PHB-HHx/hydrophobically modified silica, (b) PHB-HHx/ hydrophilic silica.

To better understand the intermolecular hydrogen bonding interaction between the PHB-HHx and hydrophobically modified silica in the hybrid composites, the vibrational force constant was calculated as shown by our previous work [5] and displayed in Table 2-1. If intermolecular hydrogen bonding occurs between a hydroxyl group of silica and the carbonyl group of PHB-HHx, the hydroxyl bond energy of silica is reduced owing to the interaction between the H atom in the hydroxyl group and the O atom of the carbonyl group. Consequently, the force constant of the hydroxyl group will exhibit lower values. As seen in Table 2-1, the force constant of the PHB-HHx/hydrophobically modified silica decreases gradually as the silica content increases, confirming that the intermolecular hydrogen bond between the hydroxyl group of silica and the carbonyl groups of PHB-HHx is enhanced in the hybrid composites as the silica content increases.

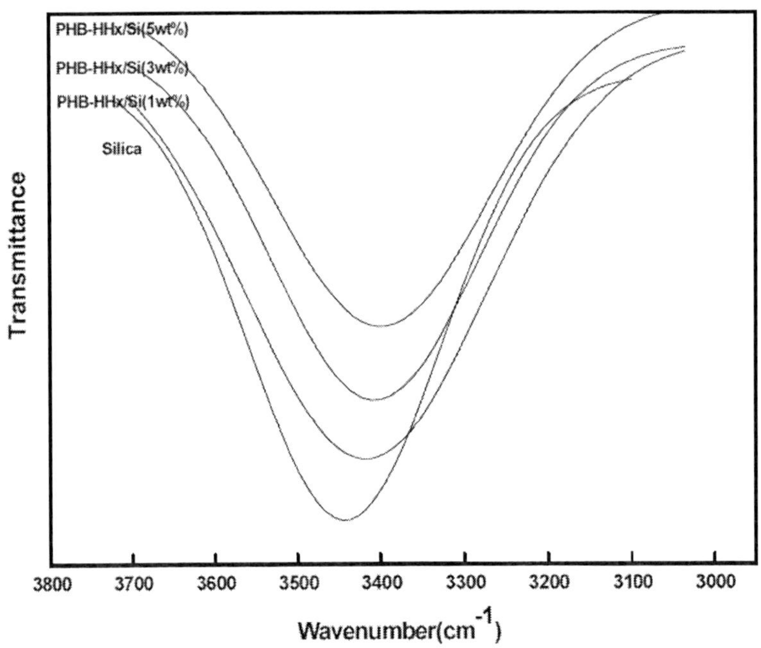

Figure 2-3. IR spectra of the hydroxyl group region of the PHB-HHx/hydrophobically modified silica hybrid composites.

Table 2-1. The wavenumber and force constant of the hydroxyl group for the PHB-HHx/hydrophobically modified silica hybrid composite

Sample	The wavenumber of hydroxyl peak(cm^{-1})	The force constant (dynes/cm)
Hydrophobically modified silica	3441	6.56×10^5
PHB-HHx/Si (1 wt%)	3418	6.47×10^5
PHB-HHx/Si (3 wt%)	3407	6.43×10^5
PHB-HHx/Si (5 wt%)	3401	6.41×10^5

2-3-2. Quantitative Analysis of the Fraction of the Intermolecular Hydrogen Bond

Figure 2-4 shows the IR spectra in the 3100-2700cm^{-1} region of the PHB-HHx/hydrophobically modified silica hybrid composites. The bands in the 2980cm^{-1} and 2880cm^{-1} region can be assigned, respectively, to the CH_3 asymmetric and symmetric stretching peaks, where pure PHB-HHx has a relatively larger CH_3 asymmetric stretching peak than any of the PHB-HHx/hydrophobically modified silica hybrid composites do. Generally, CH_3 asymmetric stretching peak appears when a bonding energy of three CH stretches is not balanced. It is reported that an inter- or intramolecular hydrogen bond can be formed between the O atom of the C=O group and the H atom of one of the C-H bonds of the CH_3 group in the PHB-HHx [6]. Based on this, relatively larger CH_3 asymmetric stretching peak of the PHB-HHx can be explained that the bonding energy in three C-H bonds is not balanced due to hydrogen bond within PHB-HHx/modified silica hybrid composites.

With the increase of the silica content, it can be found that the CH_3 asymmetric stretching peak is gradually decreased. Meanwhile, the intensity of the CH_3 symmetric stretching peak becomes more intense; indicating the bonding energy of three CH stretches within CH_3 groups is same. This can be attributed that the addition of silica breaks down of inter- or intramolecular hydrogen bond between the carbonyl group and the methyl group within the PHB-HHx, leading to the balance of bonding energy in three C-H stretches. Accordingly, more free carbonyl groups of the PHB-HHx contribute to the intermolecular hydrogen bonding with the hydroxyl groups of the silica in the hybrids, resulting in the shift of the carbonyl group in the PHB-HHx as seen in Fig. 2-2. The likely mechanism is shown schematically in Figure 2-5.

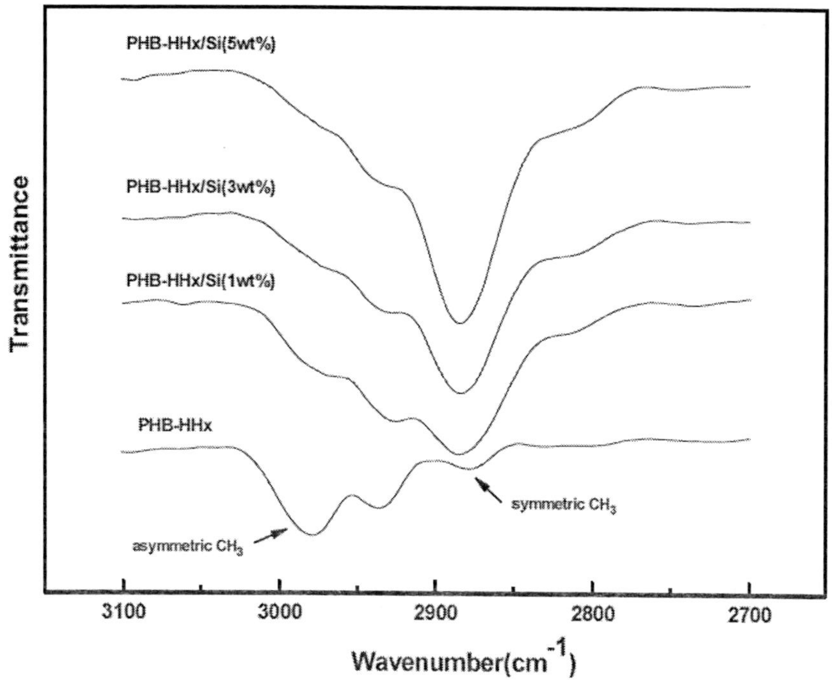

Figure 2-4. IR spectra in the 3100 - 2700cm^{-1} region of the PHB-HHx/hydrophobically silica composites.

The fraction of intermolecular hydrogen bonds formed between carbonyl groups in the PHB-HHx and the hydroxyl groups of silica was estimated from the ratio between the relative peak intensities of the symmetric and asymmetric CH_3 stretching bands. Figure 2-6 shows the variation in the intermolecular hydrogen bond fraction as a function of the silica content in the PHB-HHx/hydrophobically modified silica hybrid composites. The maximum value occurs at 5wt% silica content, indicating that at this silica composition, the intermolecular hydrogen bond is the strongest.

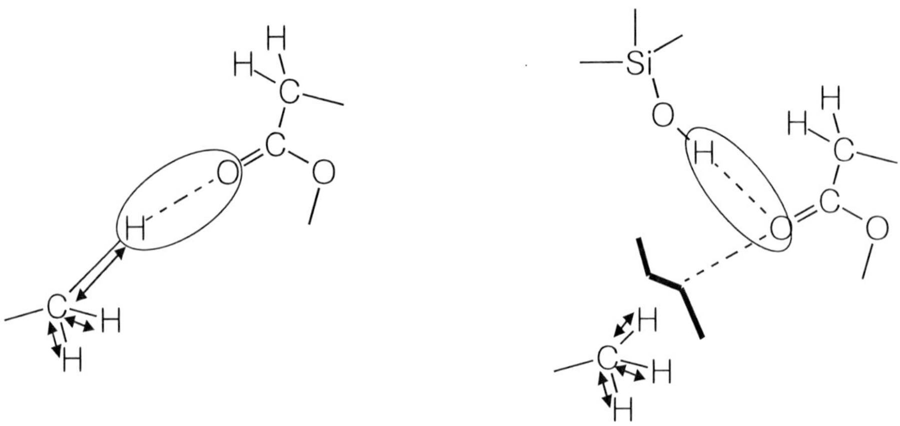

Asymmetric CH_3 band **Symmetric CH_3 band**

PHB-HHx **PHB-HHx/Silica**

Figure 2-5. The mechanism of intermolecular hydrogen bond in the PHB-HHx/hydrophobically modified silica hybrid composites.

As seen in Figure 2-6, the fraction of intermolecular hydrogen bonds increases sharply as the silica content increases from 0 to 1wt%, but increases only moderately at higher silica contents, suggesting that the silica content is not directly proportional to the extent of intermolecular hydrogen bonding. In general, molecules containing hydroxyl groups, such as silica, phenol, and 2-propanol, have intramolecular hydrogen bonds between hydroxyl groups, which may become an important factor determining the extent of intermolecular hydrogen bonding in polymer composites [7]. For example, when a substance possesses a large amount of intramolecular hydrogen bonding within itself, then the presence of this intramolecular hydrogen bond will limit the extent of hydrogen moieties that can associate with the intermolecular hydrogen bond. In our hybrids, consequently, this result is explained as the following; at relatively low silica content, most of the hydroxyl groups participate in creating the intermolecular hydrogen bonding with the carbonyl groups of PHB-HHx. By contrast, at higher silica contents, additional amount of hydroxyl groups not only form intermolecular hydrogen bond with the carbonyl groups of PHB-HHx but also contribute to the intramolecular hydrogen bond. Therefore, the relative increase in intermolecular hydrogen bonding at high silica content may be lower than that observed by the addition of only small amount silica in the PHB-HHx matrix. Figure 2-7 shows this mechanism schematically. This model concurs with the work of Li *et al.*, who showed that in poly(methyl methacrylate) (PMMA)/poly(vinyl phenol) (PVPh) blends, both intra- and intermolecular hydrogen bonds exist at high PVPh compositions, but only intermolecular hydrogen bonding occurs at low PVPh compositions [8].

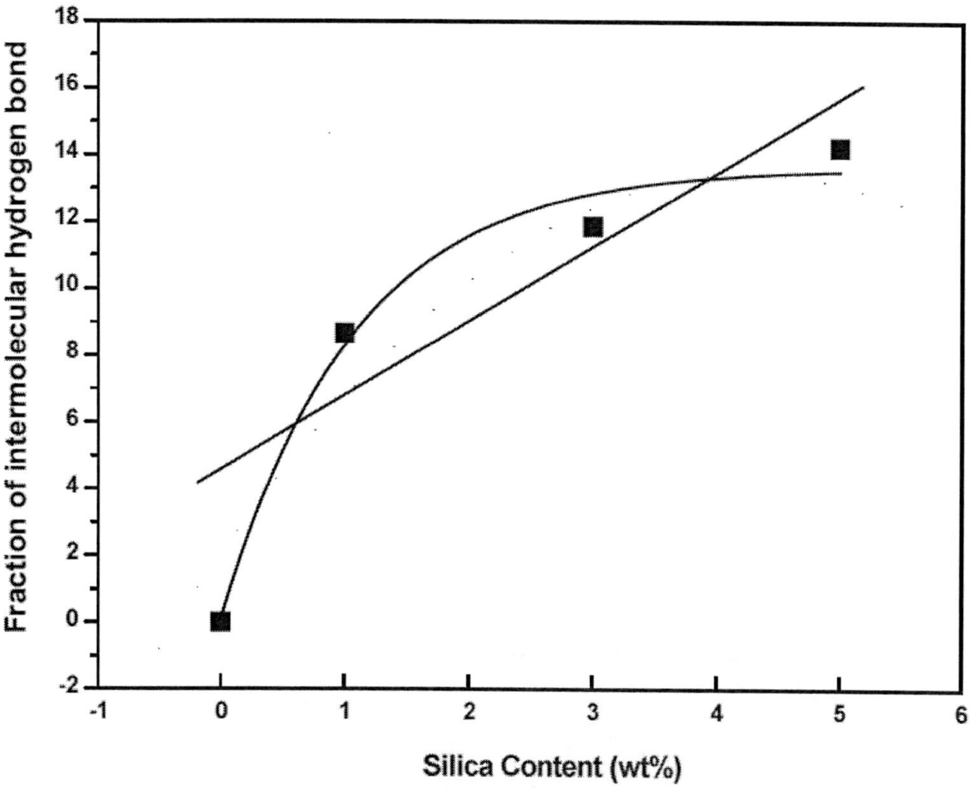

Figure 2-6. The variation of intermolecular hydrogen bond fraction as a function of silica content in the PHB-HHx/hydrophobically modified silica hybrid composites.

Figures 2-8(a)–(c) show the SEM micrographs of the PHB-HHx/hydrophobically modified silica hybrid composites. It is well known that silica readily aggregates with each other due to intramolecular hydrogen bonding [9]. The hybrid composites in Figure 2-8 show various sizes of aggregates for given silica content. For a silica content of 1wt%, the silica particles disperse effectively in the PHB-HHx matrix and exhibits only aggregates with size of less than 100 nm. However, at silica content of 5wt%, the number of large aggregates increases markedly. Some of the aggregates even exceed 250 nm. This result supports the view that the higher the silica content is, the larger the silica aggregates becomes due to the strong intramolecular hydrogen bonding that takes place between the hydroxyl groups in silica.

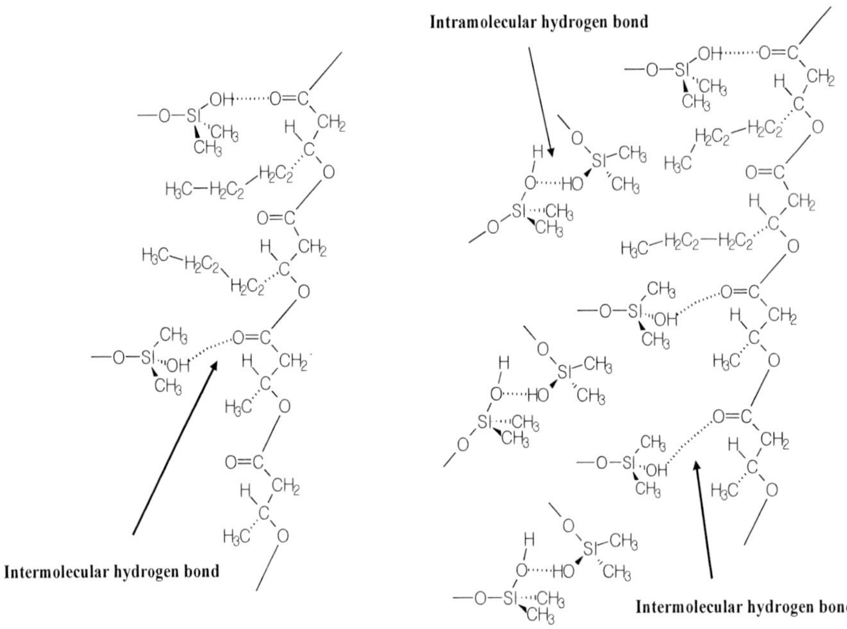

(a) (b)

Figure 2-7. Scheme of hydrogen bonding interaction in the PHB-HHx/hydrophobically modified silica hybrids: (a) low silica composition, exhibiting intermolecular hydrogen bond, (b) high silica composition, showing intra and intermolecular hydrogen bond.

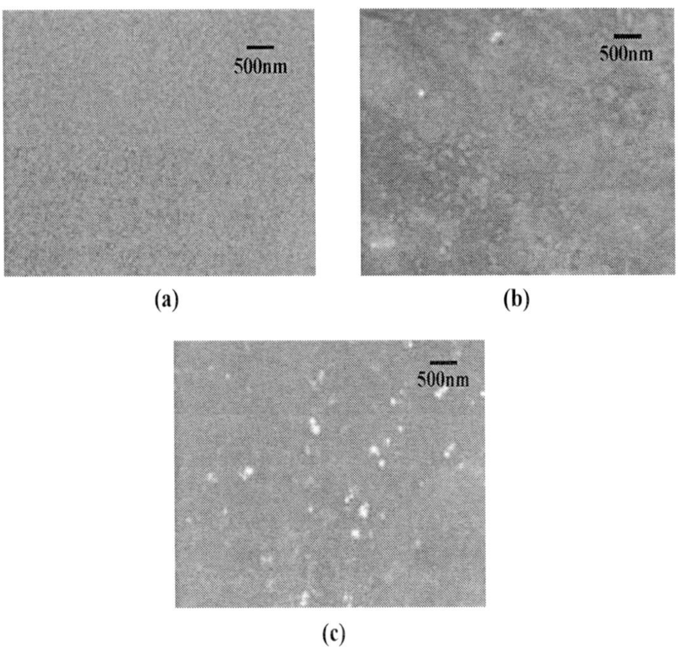

Figure 2-8. SEM images of the PHB-HHx/hydrophobically modified silica hybrid composites (scale bar: 500 nm); (a) PHB-HHx, (b) PHB-HHx/Si (1 wt%), (c) PHB-HHx/Si(5 wt%).

2-3-3. Non-Isothermal Crystallization Behavior

Figures 2-9(a) and (b), respectively, show the DSC cooling curves for the PHB-HHx/hydrophobically modified silica hybrid obtained at cooling rates of 1 and 3°C min^{-1}. All the samples show only one exothermic peak, pointing to the crystallization of a single phase. As the cooling rate increases, the maximum crystallization temperature shifts to a lower temperature for all samples. Generally, when a specimen is cooled quickly, a greater degree of supercooling is required to initiate the crystallization, presumably as the motion of the PHB-HHx molecule is unable to follow the cooling rate. On the other hand, for a given cooling rate, with the increase of silica content, the crystallization temperature of the hybrid composites shifts to a lower temperature with a gradual widening in the range.

By using the observed T_c values, the fastest crystallization time (t_{max}) for various cooling rates was determined. The results are shown in Figure 2-10, where t_{max} is defined as the time from the onset of crystallization to the appearance of the crystallization peak temperature. t_{max} can be expressed as a function of temperature (T) and the cooling rate (R) as [10]:

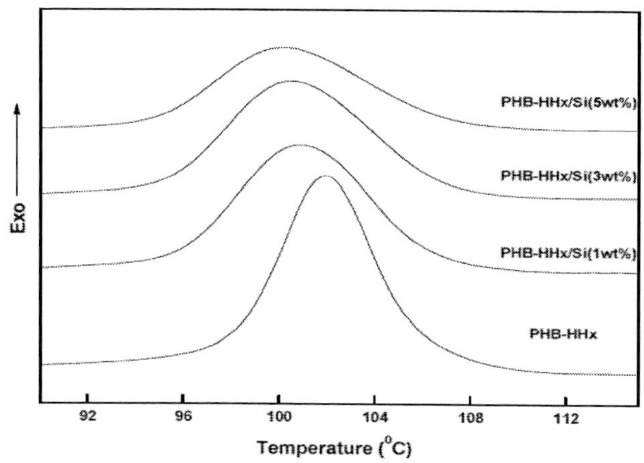

(a)

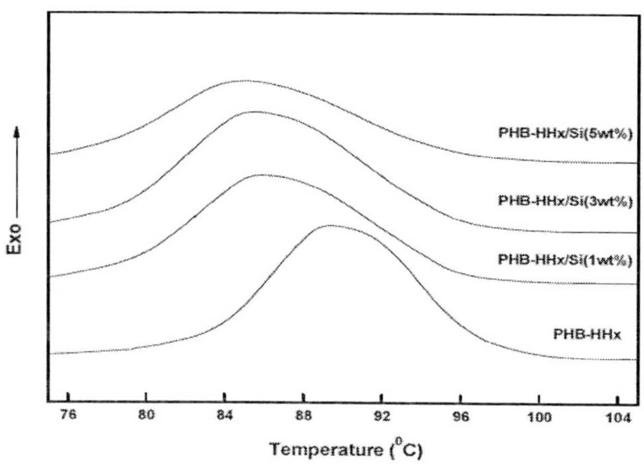

(b)

Figure 2-9. DSC cooling curves for the PHB-HHx/hydrophobically modified silica hybrids; (a) 1°C/min cooling rate, (b) 3°C/min cooling rate.

$$t_{max} = \frac{T_o - T_c}{R} \quad (1)$$

where T_0 is the onset temperature of crystallization. For a given cooling rate, the t_{max} of the hybrid composites increases with the silica content, a result consistent with the intensity of the intermolecular hydrogen bonding between PHB-HHx and silica. Kou *et al.* investigated the effect of hydrogen bond strength on the crystallization behavior of PCL blends, and they concluded that the crystallization rates of blends were reduced because of the strong intermolecular interaction between blend components [11]. On that basis, it could also be suggested that the addition of silica hinders the formation of PHB-HHx nuclei due to the existence of relatively strong intermolecular hydrogen bonding, thereby requiring more time to produce PHB-HHx crystals.

Crystallization rate coefficient (CRC) may be used to characterize the non-isothermal crystallization rate of the PHB-HHx/hydrophobically modified silica hybrid composites in detail [12, 13]. CRC can be defined as;

$$CRC = \frac{\Delta R}{\Delta T_c} \quad (2)$$

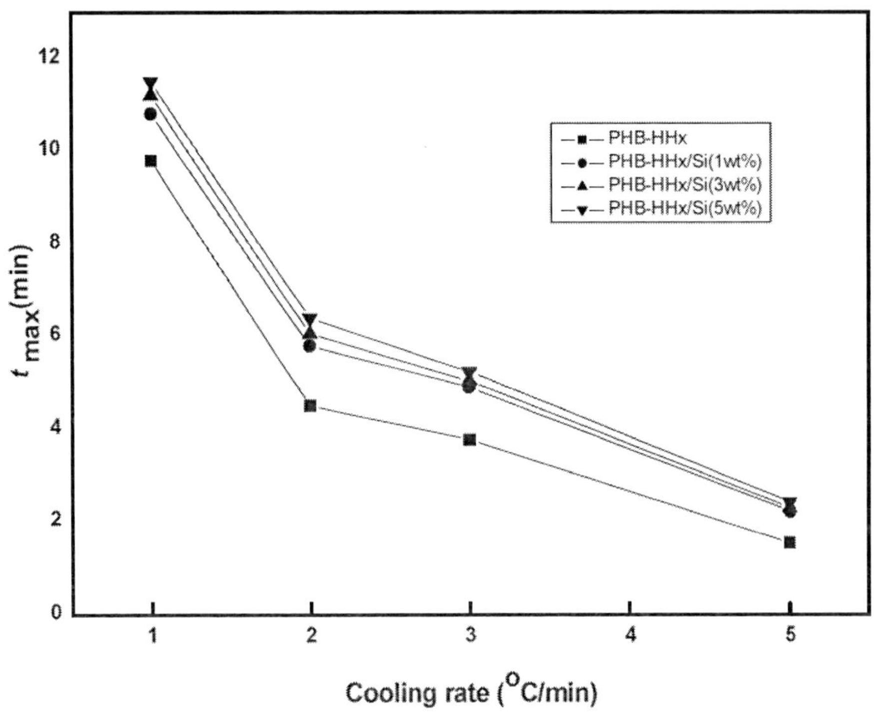

Figure 2-10. The values of fastest crystallization (t_{max}) of the PHB-HHx/hydrophobically modified silica hybrids at various cooling rates.

where R is the cooling rate and T_c is crystallization temperature. It can be calculated from the slope of a plot of cooling rate *vs*. crystallization temperature. The larger the slope is, the faster the crystallization rate is. Figure 2-11 shows the cooling rate *vs*. crystallization temperature for the PHB-HHx/hydrophobically modified silica hybrid composites, where the slope of the regression line can be used to determine the CRC of the hybrid composites (R^2 = 0.975). As seen in the Figure 2-11, the slopes of the PHB-HHx/silica hybrids decreased as the silica content increased. Figure 2-12 shows the variation in CRC as a function of the intermolecular hydrogen bond fraction in the PHB-HHx/hydrophobically modified silica hybrid composites and demonstrates that CRC gradually decreases as the extent of intermolecular hydrogen bonding increases. Clearly, intermolecular hydrogen bonding between the PHB-HHx and silica inhibits the mobility of the PHB-HHx chains and causes the non-isothermal crystallization rate of the PHB-HHx/silica hybrids to slow. The relationship between the extent of intermolecular hydrogen bonding and CRC can be fitted by an empirical second order equation. This result may be explained in such a way that, as the silica content increases, the increasing ratio of intermolecular hydrogen bondingbecomes relatively lower as already seen in the FT-IR results. Thus the relationship between the extent of intermolecular hydrogen bonding and the non-isothermal crystallization rate cannot be described by a simple linear equation.

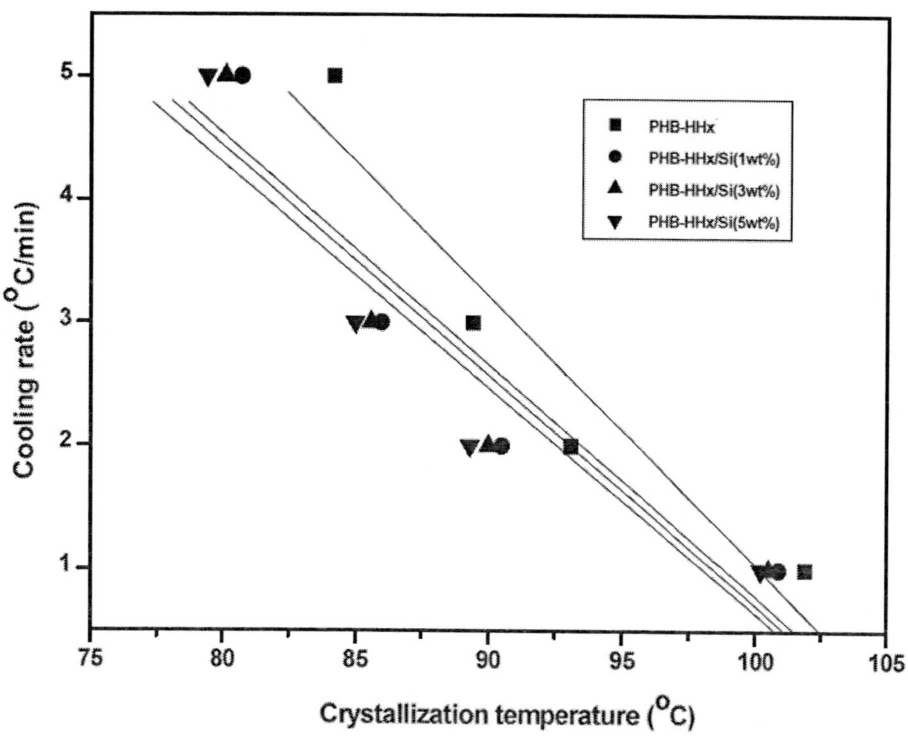

Figure 2-11. Cooling rate *versus* crystallization temperature for the PHB-HHx/hydrophobically modified silica hybrids composites.

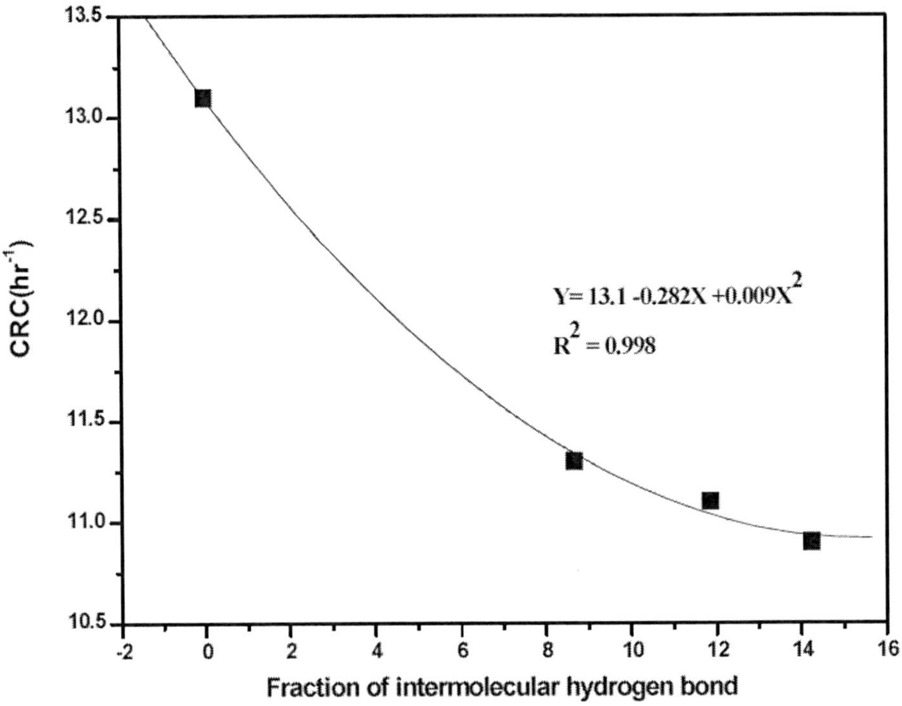

Figure 2-12. The variation of CRC as a function of intermolecular hydrogen bond fraction in the PHB-HHx/hydrophobically modified silica hybrid composites.

2-3-4. Isothermal Crystallization Kinetics

Figure 2-13 exhibits the development of relative crystallinity with time for the PHB-HHx/hydrophobically modified silica hybrid composites at the isothermal crystallization temperature of 70°C, where the relative crystallinity, X_t, was obtained from the ratio of the area of the exotherm up to time t, divided by the total isothermal crystallization exotherm. All isotherms exhibit a sigmoidal dependence on time. On the whole, the isotherms shift gradually toward longer crystallization times with increasing silica content. The induction time also increases with the silica content, proving that the isothermal crystallization rate becomes slower with increasing silica content in the hybrid composites.

Figures 2-14(a) – (c) show the spherulitic morphology of pure PHB-HHx and PHB-HHx/hydrophobically modified silica hybrid composites crystallized at 70°C after melting. The pure PHB-HHx spherulites show the familiar Maltese cross birefringent pattern and concentric extinction bands. For the hybrid composites, the band spacing increases with the silica content. Generally, the banded structure of spherulites is believed to be due to the presence of twisted lamellae resulting from stress build up during crystallization and it probably occurs within disordered fold surfaces of the polymer crystal [14, 15]. It has been reported that the band spacing usually increases as the crystal growth rate decreases and the crystallization temperature increases [16]. Consequently, this result strongly supports the hypothesis that silica reduces the isothermal crystallization rate of the PHB-HHx in their hybrids.

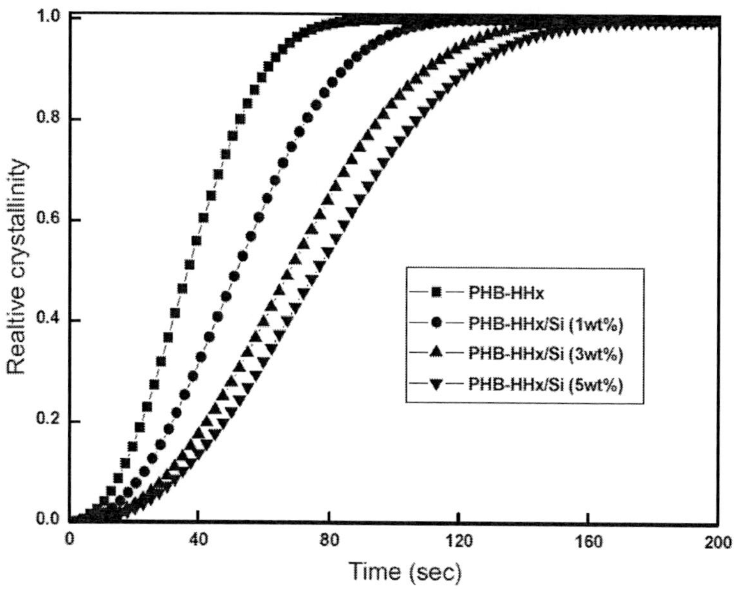

Figure 2-13. The relative crystallinity *versus* time at the 70 °C Isothermal crystallization temperature for the PHB-HHx /hydrophobically modified silica hybrid composites.

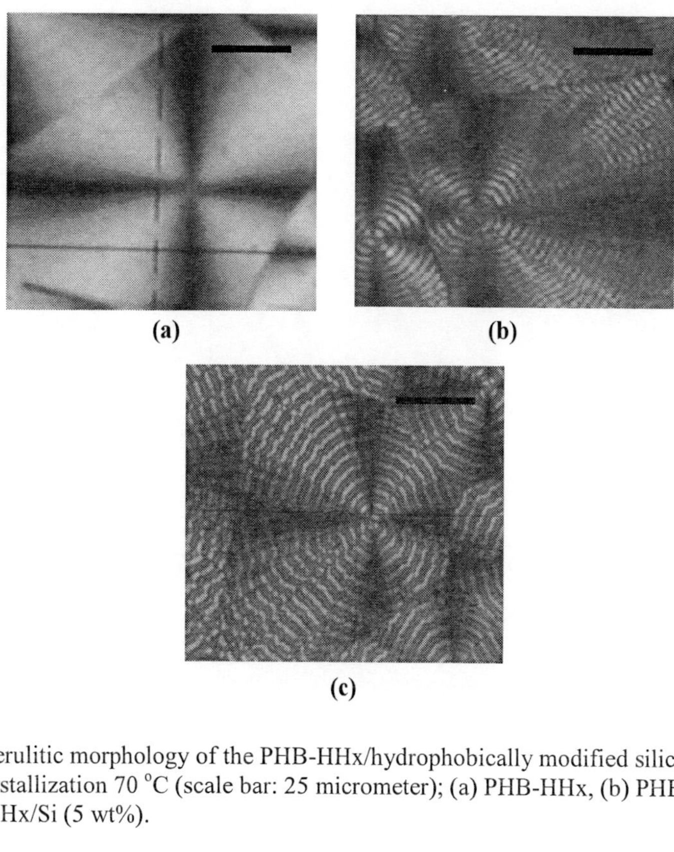

Figure 2-14. Spherulitic morphology of the PHB-HHx/hydrophobically modified silica hybrid composites at crystallization 70 °C (scale bar: 25 micrometer); (a) PHB-HHx, (b) PHB-HHx/Si (1 wt%), (c) PHB-HHx/Si (5 wt%).

The isothermal crystallization rates of the PHB-HHx/hydrophobically modified silica hybrid composites were investigated in detail by measuring the half-time of crystallization, $t_{1/2}$. Their values are listed in Table 2-2. For a given crystallization temperature, the $t_{1/2}$ for a hybrid composite is slower than that of pure PHB-HHx. It also becomes slower with the increase of silica content. Hence, as the silica content increases, the segmental mobility of the PHB-HHx molecules is restricted due to a much greater degree of intermolecular hydrogen bonding and the presence of silica themselves, which results in slowing of the isothermal crystallization rates in the hybrid composites.

Figure 2-15 shows the Avrami plot of PHB-HHx/hydrophobically modified silica hybrid composites at $T_c = 70$°C. The Avrami equation is defined as [17, 18]:

$$log[-ln(1 - X_t)] = logK + nlogt \quad (3)$$

the Avrami exponent, n, is correlated to the nucleation mechanism and the geometry of crystal growth. The overall crystallization rate, $\tau_{0.5}$, is defined as the reciprocal of the half-time of crystallization, and it is commonly used to describe the crystallization rate. The values of the rate parameters n and $\tau_{0.5}$ obtained from the Avrami equation are listed in Table 2-2.

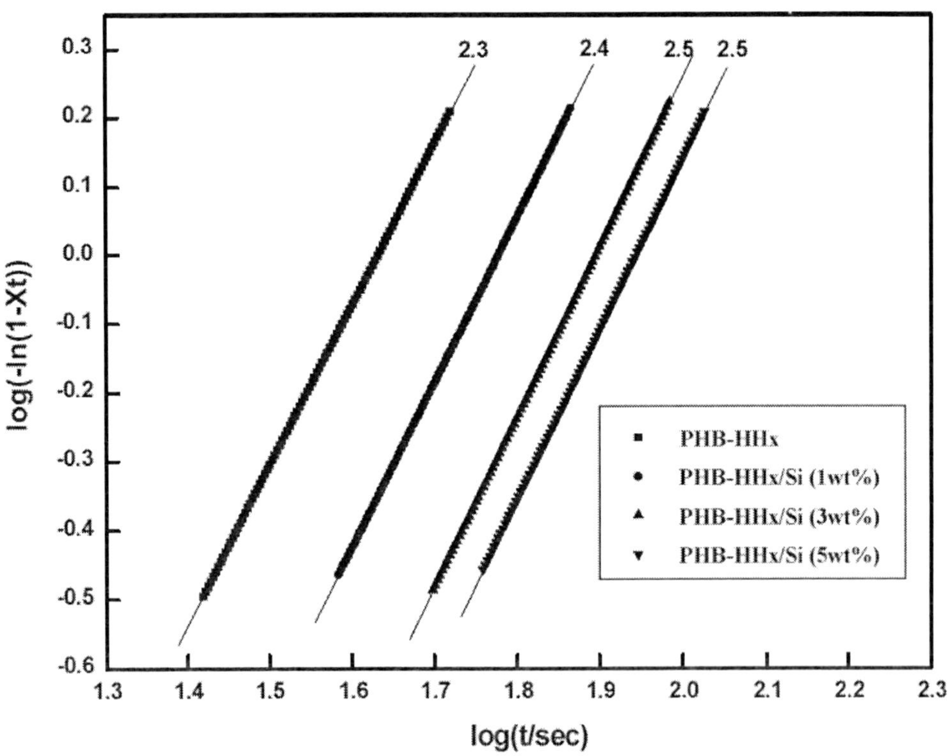

Figure 2-15. The Avrami plot of PHB-HHx/hydrophobically modified silica hybrid composites at T_c = 70 °C.

Table 2-2. The rate parameters obtained from the Avrami equation for the PHB-HHx/hydrophobically modified silica hybrid composites

Sample		65°C			70°C	
	n	$t_{1/2}$ (sec)	$\tau_{0.5}\times10^3$ (sec^{-1})	n	$t_{1/2}$ (sec)	$\tau_{0.5}\times10^3$ (sec^{-1})
PHB-HHx	2.3	39.1	25.6	2.3	36.7	27.2
PHB-HHx/Si(1 wt%)	2.4	58.2	17.2	2.4	51.8	19.3
PHB-HHx/Si(3 wt%)	2.4	86.6	11.5	2.5	68	14.7
PHB-HHx/Si(5 wt%)	2.5	87.4	11.4	2,5	76.1	13.1
Sample		75°C			85°C	
	n	$t_{1/2}$ (sec)	$\tau_{0.5}\times10^3$ (sec^{-1})	n	$t_{1/2}$ (sec)	$\tau_{0.5}\times10^3$ (sec^{-1})
PHB-HHx	2.4	57.2	17.5	2.5	82.7	12.1
PHB-HHx/Si(1 wt%)	2.5	58.3	17.2	2.6	87.7	11.4
PHB-HHx/Si(3 wt%)	2.5	94.5	10.6	2.4	101.9	9.8
PHB-HHx/Si(5 wt%)	2.4	102.8	9.7	2.5	127.6	7.8

Regardless of the silica content and isothermal crystallization temperature, n is essentially constant. This result indicates that the silica has little effect on either the nucleation mechanism or the geometry of crystal growth of the PHB-HHx in their hybrid composites. On the other hand, for a given crystallization temperature, the $\tau_{0.5}$ value of the PHB-HHx/hydrophobically modified silica hybrids decreases gradually as the silica content increases, supporting the notion that the overall crystallization rates of the PHB-HHx/hydrophobically modified silica hybrids are retarded by the addition of silica.

The activation energy of crystallization, ΔE, was determined by using the Arrhenius equation for the PHB-HHx/hydrophobically modified silica hybrid composites [19]. Figure 2-16 shows the plots of $(1/n)lnK$ *versus* $1/T_c$ for the PHB-HHx/silica hybrid composites. The slope of the regression line in Figure 2-16 can be used to determine the ΔE of the hybrid composites. The ΔE of neat PHB-HHx was – 48.6 kJ mol^{-1}. It can be seen that the higher the silica content, the larger the absolute value of ΔE is for the PHB-HHx/hydrophobically modified silica hybrid composites. For the PHB-HHx/silica (5wt%) hybrids, ΔE was –54.5 kJ mol^{-1}. Figure 3-17 shows the correlation of absolute activation energy as a function of the intermolecular hydrogen bond fraction in the PHB-HHx/hydrophobically modified silica hybrid composites. The greater the extent of intermolecular hydrogen bonding, the higher the activation energy, confirming that more activation energy is required for crystallization in the PHB- HHx/silica hybrid composites, most likely because the intermolecular hydrogen bonding between the PHB-HHx and silica retards the overall crystallization rate of PHB-HHx in their hybrid composites. Their variation also corresponds to the empirical second-order equation, as found for the non-isothermal crystallization rates. This result again supports the point that the relationship between the extent of intermolecular hydrogen bonding and the crystallization rate cannot have a linear equation, since the added silica content is not directly proportional to the extent of intermolecular hydrogen bonding.

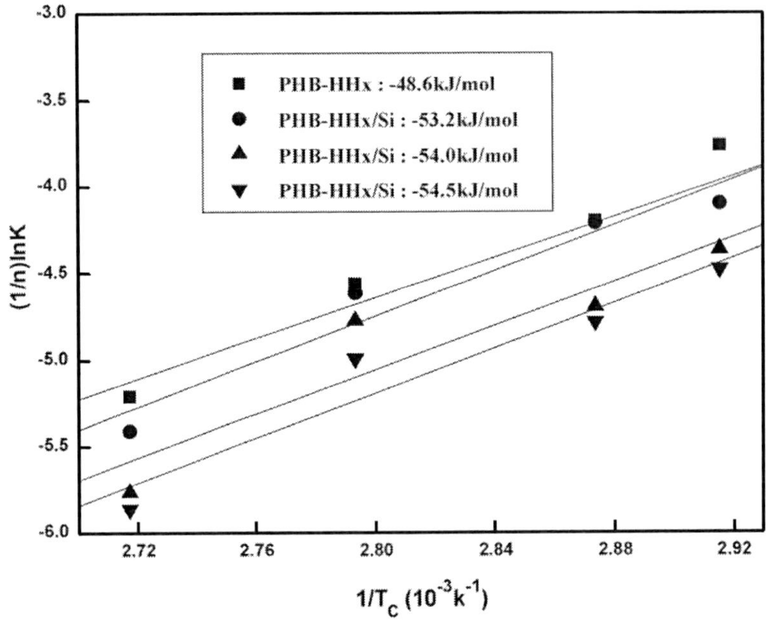

Figure 2-16. The plots *(1/n)lnK versus 1/*T_c for the PHB-HHx/hydrophobically modified silica hybrid composites.

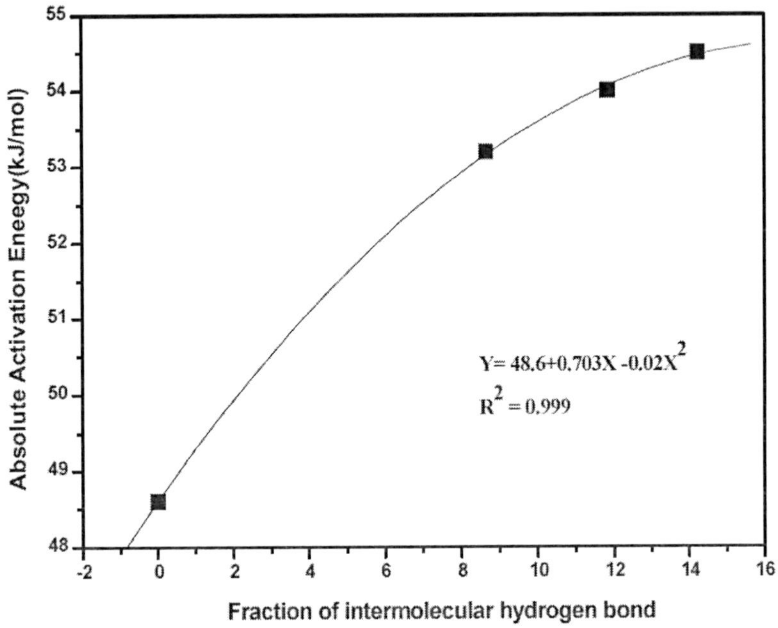

Figure 2-17. The absolute activation energy as a function of intermolecular hydrogen bond fraction in the PHB-HHx/hydrophobically modified silica hybrid composites.

2-4. Summary

1) Compared with the PHB-HHx/hydrophilic silica hybrids, the PHB-HHx/ hydrophobically silica hybrids had stronger intermolecular hydrogen bonds. As the silica content increased, the intermolecular hydrogen bonding between the PHB-HHx and hydrophobically modified silica became stronger. However, the extent of intermolecular hydrogen bonding was not directly proportional to the amount of added silica.
2) The silica particles reduced the non-isothermal and isothermal crystallization rates of PHB-HHx in the hybrids, and they were ordered by the strength of intermolecular hydrogen bonding between PHB-HHx and silica. The values obtained for the activation energy of crystallization supported that silica retarded the overall crystallization rate of PHB-HHx as a result of the formation of intermolecular hydrogen bonds between PHB-HHx and silica in the hybrid composites.
3) Finally, it was observed that the relationship between the extent of intermolecular hydrogen bond and crystallization rate was described by the empirical second order equation.

References

[1] B. Fei, C. Chen, H. Wu, S. Peng, X. Wang, L. Dong, and J. H. Xin, *Polymer*, 45, 6275 (2004).
[2] D. Li and J. Brisson, *Polymer*, 39, 793 (1998).
[3] H. Barthel, M. Heinemann, M. Stintz, and B. Wessely, *Part. Syst.*
[4] H. D. Keith and F. J. Padden, *Polymer*, 25, 28 (1984).
[5] H. Sato, R. Murakami, A. Padermshoke, F. Hirose, K. Senda, I. Noda, and Y. Ozaki, *Macromolecules*, 37, 7203 (2004).
[6] J. P. Penning and R. S. J. Manley, *Macromolecules*, 29, 84 (1996).
[7] J. S. Lim, I. Noda, and S. S., Im, *J. Polym. Sci., Part B: Polym. Phys.*, 44, 2852 (2006).
[8] L. L. Madison and G. W. Huisman, *Microbiol. Mol. Biol. ReV.*, 63, 21 (1993).
[9] L. Zhehui, M. Philippe, and J. Robert, *Polymer*, 38, 5149 (1997).
[10] M. Avrami, *J. Chem. Phys.*, 7, 1193 (1939).
[11] M. Avrami, *J. Chem. Phys.*, 9, 177 (1941).
[12] M. Lemoigne, *Bull. Soc. Chim. Biol.*, 8, 770 (1926).
[13] P. Cebe and S. D. Hong, *Polymer*, 27, 183 (1986).
[14] S. C. Chung, W. G. Hahm, S. S. Im, and S. G. Oh, *Macromol. Res.*, 10, 221 (2002).
[15] S. Viswanathan, M. D. Dadmun, *Macromolecules*, 35, 5049 (2002).
[16] S. W. Kuo, S. C. Chan, and F. C. Chang, *Macromolecules*, 36, 6653 (2003).
[17] T. P. Khanna, *Polym. Eng. Sci.*, 30, 1615 (1990).
[18] *Technical Bulletin,* vol 6, no 11, Degussa Corporation, Akron, OH, (1989).
[19] X. Kong, X. Yang, G. Li, X. Zaho, E. Zhou, and D. Ma, *Eur. Polym. J.*, 37, 1855 (2001).

3. In Situ Polymerized Poly(Butylene Succinate)/Silica Nanocomposites: Physical Properties and Biodegradation

3-1. Introduction

Nowadays, PBS-based polymers have attracted a great deal of interest as promising biodegradable materials. Among the manufacturing methods available for PBS-based polymers, copolymerization and blending have proven to be effective with regard to improvement of the mechanical properties and biodegradability [1-6]. However, these methods often give rise to problems affecting the physicochemical properties of the polymer. Incorporation of a secondary component above 20 mol% reduces the crystallinity of the parent polymer and significantly reduces the melting point due to imperfect lateral packing and/or isomorphism, and adversely affects the temperature range over which they can be used [7-9].

Another manufacturing method for biodegradable PBS-based polymers is as a nanocomposite with an inorganic material, such as silica or clay, acting as a structural reinforcement. However, the inorganic components tend to aggregate when processed into a nanocomposite due to strong hydrophilic interactions. Therefore, a number of studies of biodegradable polymer/inorganic composites have focused mainly on the dispersibility of inorganic materials within the polymer matrix and the resulting physical properties [10-12].

Recently, using small-angle X-ray scattering (SAXS) and transmission electron microscopy (TEM), Kim et al. [13] investigated the thermal stability of a PBS/organoclay composite and demonstrated that a chemical reaction between the epoxy groups of the organoclay and the end groups of PBS improved the thermal stability of the composite. Lim et al. [14] manufactured a poly (3-hydroxybutyrate-co-3-hydroxyhexanoate) (PHBeHHx) /silica composite using simple melt compounding and found that the formation of intermolecular hydrogen bonds between PHB-HHx and silica slowed the rate of crystallization.

Despite these studies, little is known about the physical properties and biodegradation of in situ polymerized PBS/silica composite. The present study was performed to investigate the chemical structure of PBS/silica composites prepared by in situ polymerization. Melt rheological properties and biodegradability were evaluated during biodegradation in compost.

3-2. Experimental

3-2-1. Materials and Polymerization

Succinic acid, 1,4-butanediol, titanium tetrabutoxide, and fumed silica (Aerosil R300) were obtained from Aldrich (St. Louis, MO) and used without further purification. PBS/silica composites were prepared in the presence of fumed silica (-5.5 wt% of total monomer) by in situ two-step polymerization consisting of direct esterification and polycondensation. Slurry composed of succinic acid, 1,4-butanediol, and fumed silica was prepared in a reactor equipped with an overhead stirrer and a temperature controller. A flask containing succinic acid and 1,4-butanediol (1:1.2 by mole ratio) was heated to 190 ^{o}C in an oil bath with mechanical stirring at 300 rpm. The fumed silica was then added under a nitrogen atmosphere. The resulting monomer mixture was melted and stirred at 190 °C for 2 h, evolving water condensate. The reaction temperature was then raised to 240 °C over a period of 30 min, with gradual application of reduced pressure to remove low molecular weight condensates, and maintained at 240 °C for 3-5 h depending on the feed composition. The resulting poly(butylene succinate)/silica composites are denoted as follows: e.g., PBS/silica-1.8 represents poly(butylene succinate) containing 1.8 wt% silica.

3-2-2. Solid-State ^{29}Si NMR Spectroscopy and Polymer Characterization

Solid-state ^{29}Si NMR spectra were recorded using a 200 MHz spectrometer (Unity Inova 200; Varian, Palo Alto, CA). These spectra yield quantitative information about Si atom coordination or substitution around reacted $R'_nSi(OR)_{4-n}$ units, where n = 1, 2, or 3. For a tetrafunctional silicate with SiO_4 tetrahedral subgroups, five Q peaks (Q_0, Q_1, Q_2, Q_3, Q_4) are present in the NMR spectra. Inherent viscosity was determined using a Cannon-Ubbelohde microviscometer at a concentration of 0.3 g dL^{-1} in chloroform at 25±0.1 °C. Samples for DSC, DMA, WAXD, and tensile testing were prepared on a compress mold machine. The temperature was set 30-50 °C above T_m. After melting, samples were quenched in ice water. Before use, all samples were dried in a vacuum oven at 40 °C for 48 h. The thermal properties of PBS and PBS/silica nanocomposites were analyzed using a Perkin Elmer DSC-7 differential scanning calorimeter. Upon melting, the samples were heated to 160 °C at a rate of 10 °C/min. The apex of the melting peak was taken as the melting temperature. The loss modulus was measured as a function of temperature by dynamic mechanical analysis (TA-DMA 2980) at a frequency of 1 Hz under a nitrogen atmosphere. Samples were quenched to -100 °C and then heated gradually to 130 °C at a rate of 2 °C/min. The peak position of loss modulus was taken as the glass transition temperature.

3-2-3. Melt Rheological Analysis

The rheological properties of the nanocomposites in the molten state were measured at 150 °C using a dynamic oscillatory rheometer (ARES; Rheometric Scientific Inc., Piscataway, NJ) equipped with parallel plates with a diameter of 20 mm and a gap of 1.0 mm. The frequency ranged from 0.1 to 400 rad/s. Rheological specimens were dried for 24 h in a vacuum oven at 60 °C prior to measurements.

3-2-4. Stress-Strain Measurements

Mechanical properties were measured on a TA 4465 instrument with a 1 kN load cell and a grip distance of 26 mm at a cross-head speed of 10 mm/min. Polymer films were prepared on a dumbbell-shaped film with the following dimensions: 3.5 mm (width), 14 mm (gage length), and 0.3 mm (thickness).

3-2-5. Wide Angle X-Ray Diffraction (WAXD) and Transmission Electron Microscopy (TEM)

Micro-structural changes in the nanocomposite materials were investigated by wide angle X-ray diffraction (WAXD) using an X-ray diffractometer equipped with a nickel-filtered Cu Ka source (40 kV, 30 mA, Model DMAX 2000; Rigaku Denki, Tokyo, Japan). Spectra were obtained at room temperature at a scan speed of 10°/min over a range of 5-40°. The interfacial surface morphology was examined by transmission electron microscopy (TEM) using a JEOL 2000EX(II) (JEOL Ltd., Tokyo, Japan). Samples for TEM analysis consisted of ultra- thin sections ranging from 40 to 60 nm in thickness prepared with a diamond knife at a temperature of -40 °C on a ReicherteJung Ultracut microtome (Leica, Wetzlar, Germany).

3-2-6. Preparation of the Mature Compost

Animal fodder (Samyant Oil & Feed Corporation, Incheon, Korea) was used as the compost substrate. The composition of fiber, fat, and protein in this mixture was similar to human food garbage. Composting was performed at a maximum temperature of 58 °C during the thermophilic phase with the moisture content maintained at 65%. The composting process lasted for 50-60 days.

3-3. Evaluation of Biodegradability in the Mature Compost

All specimens for biodegradation testing were cryogenically ground into a powder (particle size: ca. 80 mm) and dried to a constant weight. Biodegradation of the nanocomposites was conducted in the laboratory-scale compost described above according to ASTM D5209-92 (1992) and ASTM D5338-92 (1992). Airflow was maintained at 40 mL/min. A mixture of mature compost (wet weight: 200 g) and nanocomposite material (5% on a dry basis) was incubated at 58 °C. Carbon dioxide produced from the compost was absorbed by a solution of 0.4 N potassium hydroxide and 2 N barium chloride and quantified by titrating the solution with 0.2 N HCl.

3-4. Results and Discussion

Fig. 3-1 shows representative results of solid-state ^{29}Si NMR of PBS/silica-3.5, providing information regarding Si atom coordination. The spectral data related to Q peaks were as follows: Q_0: -72 to -82 ppm, Q_1: -82 to -89 ppm, Q_2: -92 to-96 ppm, Q_3: -100 to --104 ppm and Q_4: > -110 ppm [15-18]. The shoulder at -106.8 ppm indicates covalent bonding between Si and the PBS polymer backbone chain.

Additional proof for this covalent bonding state was provided by the FTIR spectrum shown in Fig. 3-2, at which pure PBS exhibited the characteristic carbonyl bands at 1723 cm^{-}1. Interestingly, as for PBS/silica-1.8, a small shoulder-like band may be assigned around 1717 cm^{-1}. Furthermore, the intensity of this band became developing in proportion to the added silica content. On the whole, if specific intermolecular interaction is formed between the carbonyl group of polyester and the functional group of other materials, some modification near this carbonyl group can be detected. For example, the wavenumbers of carbonyl group may be shifted [14, 19], or new carbonyl groups near the major carbonyl group can be created [20, 21]. Based on this, this new peak at 1717 cm^{-}1 can be considered to be a result of the interaction between the silanol group and carbonyl groups and the strength of this intermolecular bonding became more powerful with the increases of silica contents. A similar result was reported in the work of Lim et al., who showed that the carbonyl group of PBS molecules interacted with the –NHCO- groups of urethane between silica particles and PBS molecules, giving rise to the formation of intermolecular interaction in PBS/isocyanate-modified silica nanocomposites [22].

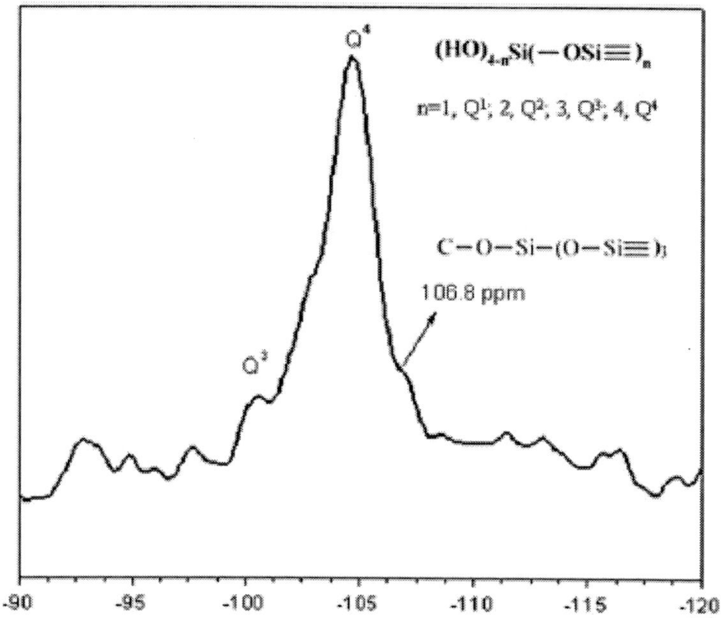

Figure 3-1. Solid-state ^{29}Si NMR spectra for PBS/silica-3.5.

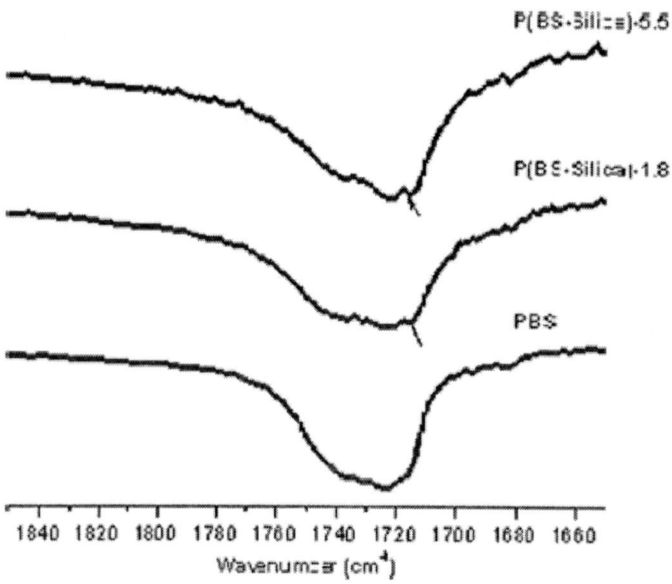

Figure 3-2. FTIR spectra for PBS/silica nanocomposites.

Table 3-1 summarizes the physical characteristics of PBS and PBS/silica nanocomposite materials. The inherent viscosity of PBS/silica nanocomposites ranges between 0.71 and 1.28 dL g^{-1}. Consistent increases in η_{inh} were observed with the incorporation of -3.5 wt% silica. Conversely, a silica content of 5.5 wt% or higher markedly decreased η_{inh}. These results may reflect increased reactivity of polyester in the presence of fumed silica during the polycondensation process [23, 24]. Consequently, these results suggested that 3.5 wt% of silica leads to an increase in the degree of polymerization in the PBS/silica nanocomposite by acting as a catalyst during the polymerization process. PBS/silica-1.8 and 3.5 showed slight increases in glass transition temperature with increasing silica content without significant changes in the melting temperature. The crystallinity of the parent PBS was 52.5%, while the values for nanocomposites were significantly lower. In the case of PBS/silica-3.5, the crystallinity was 36.2%; a decrease of about 38.7% from that of the parent PBS. WAXD patterns of all specimens exhibited distinct crystalline reflections of the (020), (021), (110), and (111) planes (2 θ = 19, 21.5, 22.5, and 28°, respectively) (data not shown). As silica content increased, WAXD curves broadened without any further significant changes.

Table 3-1. Polymer characterization of PBS and PBS/silica nanocomposites

Sample code	Silica content (wt%)[a]	η_{inh}[b] (dL g^{-1})	T_g[c] (°C)	T_m[d] (°C)	ΔH_f (J/g)	X_c[e] (%)
PBS	0	1.18	−10.9	115.4	80.6	52.5
PBS/silica-1.8	1.8	1.20	−9.1	114.6	70.3	35.4
PBS/silica-3.5	3.5	1.28	−9.2	114.6	70.4	36.2
PBS/silica-5.5	5.5	0.71	–	115.4	82.3	–

T_g and X_c of PBS/silica-5.5 could not be measured because compress molded film was extremely brittle.

[a] Weight ratio of silica to total monomer in feed.

[b] At 25 °C, 0.3 g dL^{-1} in chloroform solution.

[c] Measured by DMA, 5 °C/min.

[d] Measured by DSC, 10 °C/min.

[e] Calculated by deconvolution of the WAXD patterns using curve-fitting profram.

The loss tangent (tan δ) curves for parent PBS and PBS/silica nanocomposites prepared via in situ polymerization in the molten state are shown in Fig. 3-3(a). Tan δ is generally measured as an indicator of the solid-like elastic or liquid-like viscous properties of a system. The tan d δ curves of PBS/silica-1.8 and PBS/silica-3.5 showed broader peaks with higher intensity than the parent PBS due to significant phase segregation in the nanocomposite system. This behavior is detailed in the Colee-Cole plots shown in Fig. 3-3(b). While the slope of the parent PBS was approximately 2.0 over all frequencies, the slopes of PBS/silica-1.8 and 3.5 were 0.97 and 0.73, respectively. Another conspicuous observation was that inflection points appear on the Cole-Cole plots in the vicinity of G"= 2390 Pa. Based on the report by Han et al. [3] indicating that a slope approaching 2.0 corresponds to an isotropic and homogeneous system, the PBS/silica nanocomposites were thought to behave as heterogeneous systems below critical shear force. This may be attributed to the existence of a network structure originating from interactions between silica particles, or from interactions between silica particles and polymer chains. With increasing shear force, the network structure is broken and thus the PBS/silica nanocomposite material appears to approach a homogeneous state. In addition, the viscosity curve in Fig. 3-3(c) shows that the PBS/silica nanocomposites exhibited higher viscosity than the parent PBS polymer at all frequencies examined, probably due to the filler effect of silica particles on the PBS matrix. In contrast, the melt viscosity of PBS/silica nanocomposites decreased slightly with increasing frequency, while the parent PBS maintained a nearly constant melt viscosity irrespective of increasing frequency. In our previous study [23] of the melt rheological properties of PET/silica nanocomposites, significant collapse of the network structure between polymer chains and silica was found to occur above critical shear force. Cho and Paul [25] also reported that a decrease in melt viscosity in nylon-6 nanocomposites was related to either the slip between the polymer matrix and the filler material or degradation of the polymer matrix due to high shear heating. However, the rate of decrease in the melt viscosity for PBS/silica composites was not comparable with that of PET/silica composites. This implies that PBS/silica nanocomposites have the potential for enhanced processability in a melt-processing manufacturing system.

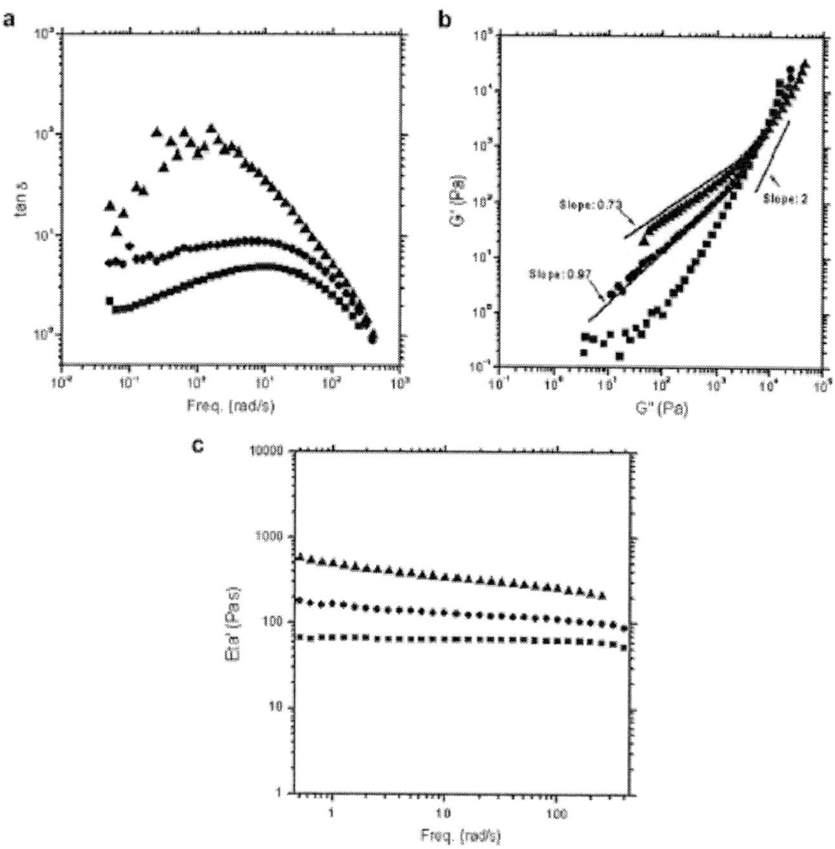

Figure 3-3. Variation of (a) tan d, (b) ColeeCole plot, and (c) viscosity of PBS/silica nanocomposites as a function of frequency at 150 °C. PBS (■); PBS/silica-1.8 (●); PBS/silica-3.5 (▲:).

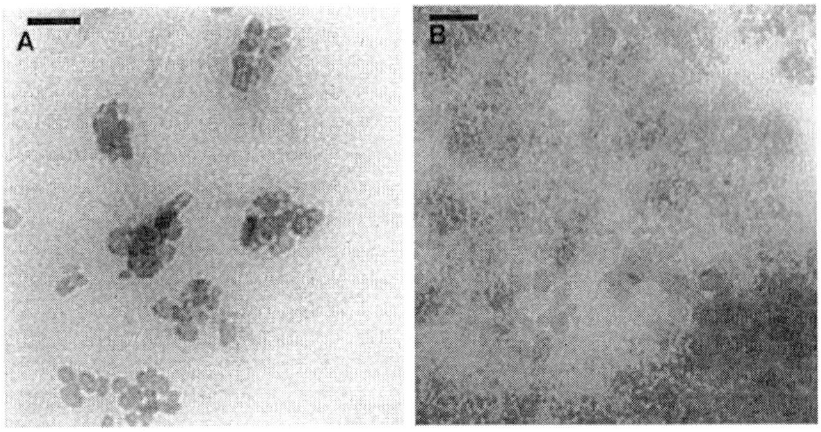

Figure 3-4. TEM images of the interfacial area for PBS/silica-3.5 (a) and PBS/silica-5.5 (b). (scale bar: 242 nm).

TEM images (Fig. 3-4(a) and (b)) of the fractured surfaces of specimens revealed the aggregation of silica particles dispersed in the PBS matrix. For PBS/silica-3.5, most of the agglomerates were comprised of 10-15 silica particles. In composites formed using fumed silica, significant aggregation of silica particles in the polymer matrix was usually observed below 2 wt% silica due to the high hydrophilicity of fumed silica [23]. Nevertheless, these data indicate that (1) the PBS/silica-3.5 composite approximates the optimum silica content and (2) that the in situ polymerization method yields an acceptable distribution of silica throughout the nanocomposite.

Table 3-2. Tensile properties of PBS and PBS/silica nanocomposites

Sample code	Stress at break (MPa)	Elongation at break (%)	Modulus (MPa)
PBS	26.3(6.9)	96.0(7.3)	331.5(48.5)
PBS/silica-1.8	31.4(4.0)	410.3(27.4)	303.4(14.3)
PBS/silica-3.5	38.6(4.6)	514.5(75.6)	309.5(7.0)
PBS/silica-5.5	–	–	–

Figures in parentheses are standard deviations. Tensile properties of PBS/silica-5.5 could not be measured because compress molded film was extremely brittle.

The tensile properties of PBS/silica nanocomposites are shown in Table 3-2. PBS/silica-3.5 exhibited markedly improved tensile properties as compared to parent PBS. For PBS/silica nanocomposites containing more than 3.5 wt% silica, the tensile properties could not be measured as the compressed, molded film was extremely brittle. In the case of PBS/silica-3.5, the stress at break and the strain at break were approximately 39 MPa and 515%, respectively. In general, tensile properties are known to depend largely on the molecular weight of the polymer. However, the combination of WAXD patterns and TEM images obtained in the present study suggest that low crystallinity and silica distribution morphology contribute to significant improvement of tensile properties.

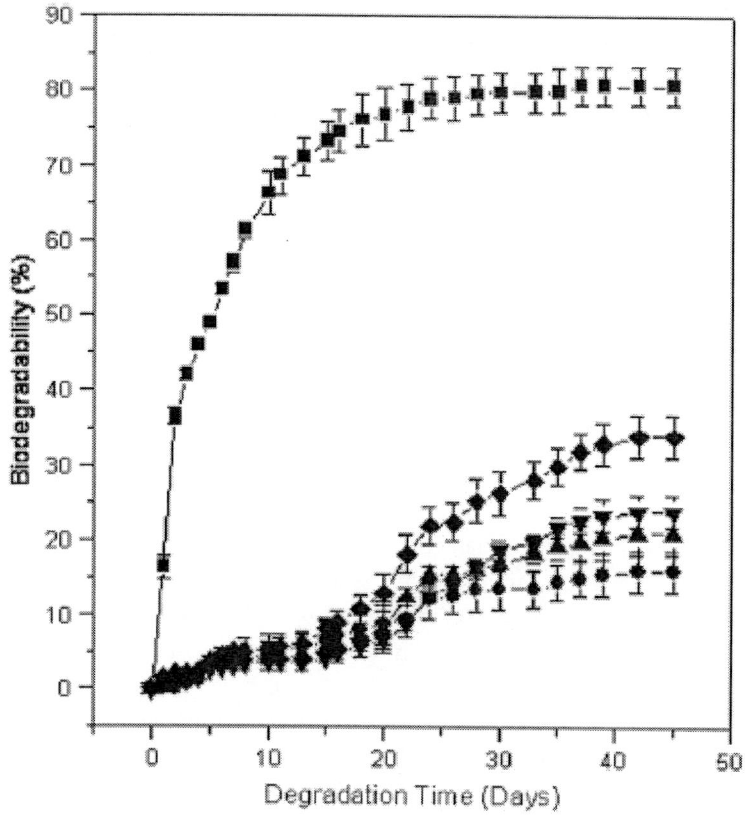

Figure 3-5. Biodegradation of PBS/silica nanocomposites in animal fodder compost.

Cellulose (■); PBS (●); PBS/silica-1.8 (▲); PBS/silica-3.5 (▼); PBS/silica-5.5 (♦).

Fig. 3-5 shows the biodegradability of PBS and PBS/silica nanocomposites. The general expression describing biodegradability is shown in Eq. according to ISO 14853.

$$\text{Biodegradation}\% = \frac{\left[\sum(CO_2)_T - \sum(CO_2)_B\right]}{ThCO_2} \times 100$$

Here, $\Sigma(CO_2)_T$ is the amount of CO_2 evolved from the test material between the start of the test and time t, and $\Sigma(CO_2)_B$ is the amount of CO_2 evolved from the blank test bottle between the start of the test and time t. $ThCO_2$ is the theoretical amount of CO_2 evolved from the test material assuming that all available carbon is transformed into CO_2. As shown in Fig. 3-5, the profile of biodegradation, calculated from the accumulated amount of CO_2 evolved, was sigmoidal.

In general, low molecular weight materials composed of aliphatic polyesters are subject to biodegradation because of the susceptibility of the aliphatic ester linkage to abiotic hydrolysis [26]. The relative rate of biodegradation observed for the PBS/silica nanocomposites was faster than that of the parent PBS polymer, indicating dependence on silica content. PBS/silica-5.5 nanocomposite films were difficult to prepare for mechanical testing due to their low molecular weight, which should lead to rapid degradation. After 45 days in compost, the rate for biodegradation of PBS/silica-3.5 was approximately 50% faster than that of the parent PBS, which may be attributed to the relatively low crystallinity of the PBS/silica nanocomposites. However, despite the similar degree of crystallinity in both PBS/silica-1.8 and PBS/silica-3.5, the latter showed greater biodegradability. These observations imply that increased amounts

of silica result in preferential microbial interactions, leading to the improved biodegradability. Han et al. [27] reported that for aliphatic copolyesters, an increase in hydrophilicity due to free amino groups improved the interactions between the polymer and the degradation enzyme, thereby accelerating degradation. An analogous mechanism may be involved here in which increased hydrophilicity due to the hydroxyl groups on the fumed silica surface may enhance susceptibility to microbial attack.

3-5. Summary

PBS/silica nanocomposites were successfully synthesized in the presence of fumed silica by an in situ two-step polymerization method. Solid-state ^{29}Si NMR and FTIR data provided evidence for covalent silanol-bound carbonyl groups. Melt rheological data showed an increased melt viscosity even under high shear force, suggesting that both adequate distribution of silica and the reduced crystallinity are critical factors affecting the physical properties of PBS/silica nanocomposite materials. PBS/silica nanocomposites containing 3.5 wt% of silica had high tensile strength (38.6 MPa) and elongation (515%). The biodegradation rate of the PBS/silica materials was markedly accelerated with increasing silica content. PBS/silica nanocomposites containing 3.5 wt% silica may represent one of the most useful biodegradable materials in terms of acceptable physical properties and enhanced biodegradation rate

References

[1] Abe H, Doi Y, Hori Y, Hagiwara T. Physical properties and enzymatic degradability of copolymers of (R)-3-hydroxybutyric acid and (S,S)-lactide. *Polymer.* 1998;39(1):59-67.

[2] Alam TM, Assink RA, Loy DA. Hydrolysis and esterification in Organically modified alkoxysilanes: a Si-29 NMR investigation of methyltrimethoxysilane. *Chem. Mater.* 1996;8(9):2366-74.

[3] Babonneau F, Thorne K, Mackenzie JD. Dimethyldiethoxysilane/tetraethoxysilane copolymers: precursors for the SieCeO system. *Chem. Mater.* 1989;1(5):554-8.

[4] Chen Y, Zhou S, Yang H, Wu L. Structure and properties of polyurethane/nanosilica composites. *J. Appl. Polym. Sci.* 2005;95(5): 1032-9.

[5] Cho JW, Paul DR. Nylon 6 nanocomposites by melt compounding. *Polymer.* 2001;42(3):1083-94.

[6] Chung SC, Hahm WG, Im SS, Oh SG. Poly(ethylene terephthalate) (PET) nanocomposites filled with fumed silicas by melt compounding. *Macromol. Res.* 2002;10(4):221-9.

[7] Hahm WG, Myung HS, Im SS. Preparation and properties of in situ polymerized poly(ethylene terephthalate)/fumed silica nanocomposites. *Macromol. Res.* 2004;12(1):85-93.

[8] Han SI, Im SS, Kim DK. Dynamic mechanical and melt rheological properties of sulfonated poly(butylene succinate) ionomers. *Polymer.* 2003;44(23):7165-73.

[9] Han SI, Kang SW, Kim BS, Im SS. A novel polymeric ionomer as a potential biomaterial: crystallization behavior, degradation, and in-vitro cellular interactions. *Adv. Funct. Mater.* 2005;15(3):367-74.

[10] Han SI, Kim BS, Kang SW, Shirai H, Im SS. Cellular interactions and degradation of aliphatic poly(ester amide)s derived from glycine and/or 4-amino butyric acid. *Biomaterials.* 2003;24(20):3453-62.

[11] Han SI, Lee WD, Kim DK, Im SS. Morphology and crystallization behavior of poly(butylene succinate)-based ionomers (PBSi). *Macromol. Rapid. Commun.* 2004;25(6):753-8.

[12] Han SI, Yoo Y, Kim DK, Im SS. Biodegradable aliphatic polyester ionomers. *Macromol. Biosci.* 2004;4(3):199-207.

[13] Kim HS, Chen GX, Jin HJ, Yoon JS. In situ copolymerization of butylene succinate with twice functionalized organoclay: thermal stability. *Colloids Surf. A* 2008;313e314:56-9.

[14] Lim JS, Hong SM, Kim DK, Im SS. Effect of isocyanate-modified fumed silica on the properties of poly(butylene succinate) nanocomposites. *J. Appl. Polym. Sci.* 2008; 107(6): 3598-608.

[15] Lim JS, Noda I, Im SS. Effect of hydrogen bonding on the crystallization behavior of poly 3-hydroxybutyrate-co-3 hydroxyhexanoate)/silica hybrid composites. *Polymer.* 2007; 48 (9):2745-54.

[16] Lim JS, Noda I, Im SS. Miscibility and crystallization behavior of poly(3-hydroxybutyrate-co-3-hydroxyhexanoate) and methoxy poly(ethylene glycol) blends. *J. Polym. Sci. Part B Polym. Phys.* 2006;44(19): 2852-63.

[17] Park JW, Kim DK, Im SS. Crystallization behaviour of poly(butylene succinate) copolymers. *Polym. Int.* 2002;51(3):239-44.

[18] Peeters MPJ, Wakelkamp WJJ, Kentgens APM. A Si-29 solid-state magic-angle-spinning nuclear-magnetic-resonance study of Teos-based hybrid materials. *J. Non-Cryst. Solids.* 1995;189(1-2):77-89.

[19] Ray SS, Maiti P, Okamoto M, Yamada K, Ueda K. New polylactide/layered silicate nanocomposites. 1. Preparation, characterization, and properties. *Macromolecules.* 2002;35(8):3104-10.

[20] Takiyama E, Fujimaki T. Characteristics of biodegradable aliphatic polymere bionolle. *Plastics.* 1992;43:87.

[21] Tian D, Blacher S, Dubois P, Jerome R. Biodegradable and biocompatible inorganiceorganic hybrid materials e 2. Dynamic mechanical properties, structure and morphology. *Polymer.* 1998;39(4):855-64.

[22] Tian D, Dubois P, Jerome R. A new poly(epsilon-caprolactone) containing hybrid ceramer prepared by the solegel process. *Polymer.* 1996; 37(17):3983-7.

[23] Woo EM, Chiang CP. Glass transition behavior and miscibility in blends of poly(vinyl p-phenol) with two homologous aliphatic polyesters. *Polymer.* 2004;45(25):8415-24.
[24] Yang HS, Yoon JS, Kim MN. Dependence of biodegradability of plastics in compost on the shape of specimens. *Polym. Degrad. Stab.* 2005;87(1): 131-5.
[25] Yoo YT, Ko MS, Han SI, Kim TY, Im SS, Kim DK. Degradation and physical properties of aliphatic copolyesters derived from mixed diols. *Polym. J.* 1998;30(7):538-45.
[26] Yoo YT, Lee BJ, Han SI, Im SS, Kim DK. Physical properties and biodegradation of poly(butylene adipate) ionomers. *Polym. Degrad. Stab.* 2003;79(2):257-64.
[27] Young SK, Jarrett WL, Mauritz KA. Nafion (R)/ORMOSIL Nanocomposites via polymer-in situ solegel reactions. 1. Probe of ORMOSIL phase nanostructures by Si-29 solid-state NMR spectroscopy. *Polymer.* 2002;43(8):2311-20.

4. Effect of Isocyanate-Modified Fumed Silica on the Properties of Poly(Butylene Succinate) Nanocomposites

4-1. Introduction

Poly(butylene succinate) (PBS) is a biodegradable aliphatic polyester synthesized by the polycondensation of 1,4-butanediol with succinic acid. [1–4] PBS has several interesting properties, including biodegradability, melt processability, and chemical resistance, which gives rise to wide potential applications. [5, 6] However, other properties of PBS pose shortcomings, such as insufficient stiffness, low-melt strength and viscosity, and poor mechanical and thermal properties.

The nanocomposite is an effective way of improving these drawbacks and widening its application. Furthermore, they have merits when compared with conventional polymer composites, such as lightweight, transparency, and easier processability. [7, 8] In the preparation of these nanocomposites, the dispersion of inorganic particles on a nanometer level and the compatibility between it and the polymer matrix are the most critical factors. [9] However, it is difficult to disperse nanoparticles in polymeric matrix homogeneously because of the strong tendency of nanoparticles to agglomerate, which is caused by the poor interaction between the hydrophilic nanoparticles and hydrophobic polymer matrix. To improve the interfacial interaction between inorganic particles and polymer matrix, many studies have attempted to convert the hydrophilic particle surface into a hydrophobic surface in the nanocomposites. [10–14]

Among them, the surface of nanoparticles grafted with polymer exhibits more hydrophobic property due to the nature of the grafting polymers, which may improve the compatibility between the polymer matrix and inorganic nanoparticles. In addition, the entanglement between the grafting polymers is known to enhance the interfacial interaction substantially. [12, 15]

In this study, by grafting method, we sought to improve the dispersibility of silica and the interfacial interaction between silica particles and the PBS matrix in a PBS nanocomposite. This novel PBS nanocomposites may be anticipated that the compatibility between the grafted

PBS and silica can be enhanced markedly, leading to the enhancement of physicochemical properties for PBS. The thermal and mechanical properties and dynamic rheological behavior of the nanocomposite were investigated to confirm the effects of grafting PBS molecules on the nanocomposite.

4-2. Experimental

4-2-1. Materials

Hydrophilic silica (Aerosil1 300) was obtained from Degussa (Du¨ sseldorf, Germany). It has a specific surface area of 300 6±30 m^2 g^{-1} and an average primary particle size of 7 nm. 1,4-Butanediol (99%), succinic acid (99%), titanium (IV) isobutoxide (97%), HDI, and dibutyltin dilaurate were purchased from Aldrich Chemical (Milwaukee, WI) and used without further purification. Hydrophilic silica was used after drying for over 24 h in a vacuum oven.

4-2-2. Modification of the Silica Particles

Five grams of silica were suspended with stirring in 100 mL of toluene in a 500-mL round flask. An excess of HDI (5 mL) was then added to the dispersion. Several drops of dibutyltin dilaurate were introduced to the mixture as the catalyst. The mixture was stirred at 30°C under a nitrogen atmosphere for 24 h. The modified silica was obtained as slurry after centrifugation at 6000 rpm for 20 min. The slurry was washed with tetrahydrofuran (THF), and then redispersed and recentrifuged several times to remove the unreacted HDI. The solvent in the slurry was removed completely under vacuum. The resulting powder was dried in a vacuum oven at 40°C for over 24 h before characterization.

4-2-3. Preparation of PBS Nanocomposites

The PBS nanocomposites containing unmodified and modified silica (0.5, 2.0 wt %) was prepared using in situ polymerization, applying conventional two-step polymerization. Modified silica was first dispersed in 1.2 mol 1,4-butanediol, using an ultrasonic homogenizer for 30 min. Then 1,4-butanediol slurry was heated to 90°C with stirring under a nitrogen atmosphere for 1 h. Afterward, the slurry was mixed with 1- mol succinic acid and 0.03 wt % titanium (VI) butoxide, and the temperature was increased to 190°C slowly with mechanical stirring. Finally, esterification was carried out for 3.5 h continuously removing the water formed. The polycondensation reaction was carried out at 235°C under a vacuum below 0.1 Torr until a sudden rise in torque occurred (2–3.5 h). The reaction scheme for the PBS nanocomposite containing silica-g-PBS is briefly shown in Scheme 1.

4-2-4. Separation of Silica Powder from the Nanocomposites

For the quantitative determination for the extent of the covalently attached isocyanate groups per silica particle and the degree of grafting for PBS molecules, the silica powder grafted with PBS molecules was separated from the nanocomposites by using centrifugation method. First, the prepared nanocomposites (5g) were diluted with the 50 mL chloroform solution. Afterward, the dilute solution was separated by centrifugation machine (MICRO 22R) with speed of 6000 rpm. The obtained silica powder was redispersed and recentrifuged more than five times to remove free PBS molecules, which cannot be reacted with the silica particles. Finally, the obtained samples were dried in a vacuum at around 308C for 24 h before the detailed analysis.

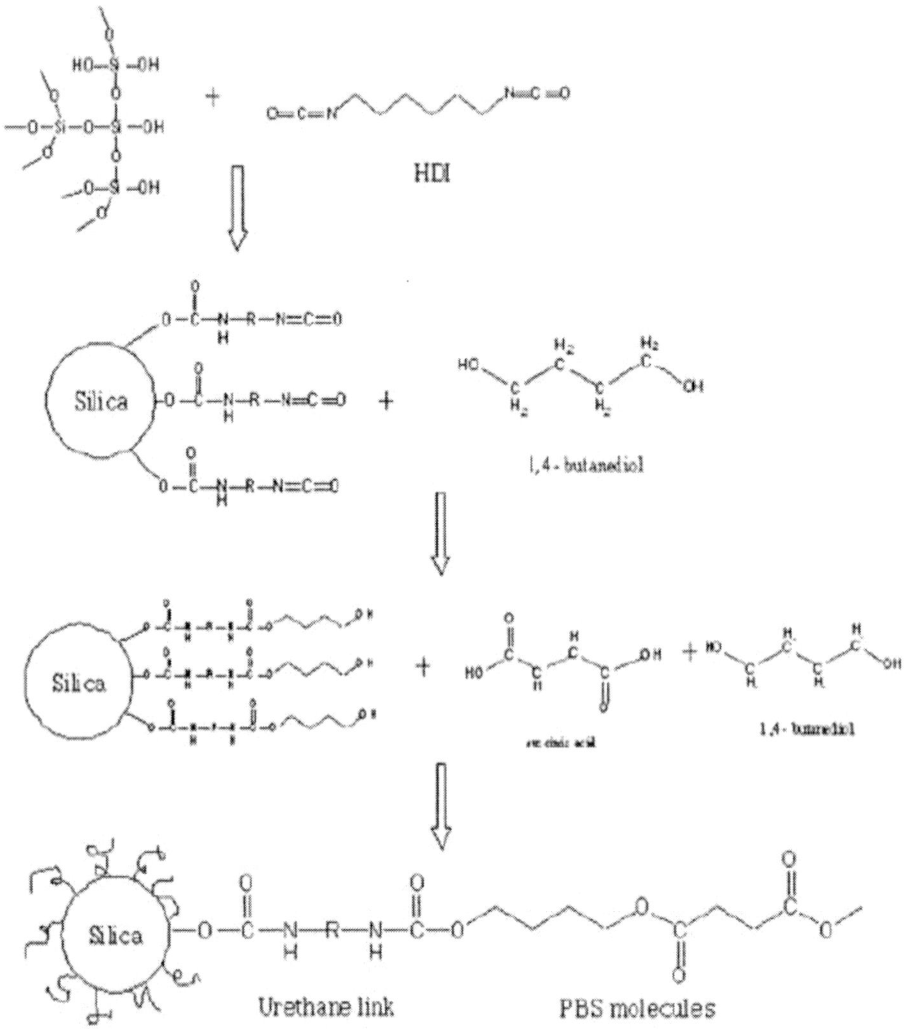

Scheme 1. Reaction scheme for PBS nanocomposites containing silica-g-PBS.

4-2-5. Characterization of the Modified Silica Particles and PBS Nanocomposites

To verify the chemical structure of the modified silica particles and nanocomposites, Fourier transform infrared (FTIR) analysis was conducted using a Nicolet 760 Magna-IR spectrometer (Nicolet-Magna, Madison, WI). Thermogravimetric analysis (TGA) was used to determine the amount of the grafting polymer on the surface of the silica particles and to investigate the thermal stability of the nanocomposites. TGA was performed under an air purge at a heating rate of 5°C/ min $^{-1}$ using a Pyris 6 thermal analyzer (PerkinElmer, Wellesley, MA). The inherent viscosity of the pure PBS and PBS/ silica nanocomposites were determined by using a Cannon–Ubbelohde microviscometer (Cannon Instrument, State College, PA). The measurement was done on the material from which silica was removed. For example, 2.00 g of the nanocomposite sample were dissolved in 100-mL chloroform with a concentration of 0.2 g/dL. After cooling to room temperature, the mixture was centrifuged at 5000 rpm for 20 min and then supernatant was carefully separated and was filtered by PTFE membrane filter. The inherent viscosity of samples was calculated from:

$$\eta_{rel} = \frac{t}{t_0} = \frac{\rho}{\rho_0} \qquad \eta_{inh} = (\ln \eta_{rel})/C$$

where t_0, q_0: flow time and viscosity of the solvent; t, q: flow time and viscosity of the solvent; C: concentration of the solution. The observation of the dispersibility of silica particles on the PBS matrix in the nanocomposites was carried out using transmission electron microscope (TEM; model TACNAI200; FEI, Hillsboro, OR). Ultra-thin films of the nanocomposites were prepared by solution casting. A carbon-coated mica sheet was dipped into a 0.1 wt % solution of nanocomposites and dried at room temperature. Then, the ultra-thin film supported by carbon layer was detached from the mica sheet onto a water surface, and picked up on a copper grid. The specimen on the grid was shadowed with platinum (Pt) to increase the image contrast and calibrate electron diffraction patterns. The shadow angle was $\tan^{-1}(1/3)$. The thermal properties of the nanocomposites were measured using a TA instrument. To remove the previous thermal history, all the samples were first heated at 150°C for 5 min and then a second scan of the heating and cooling run was performed at a rate of 10°C /min. To determine the mechanical properties of PBS/ silica nanocomposites, films were prepared using a hot press at 150°C with the holding pressure of 6000 psi and the holding time of 5 min. The fabricated films were quenched in cold water and dried at 40°Cfor the overnight. Afterward, dried films were cut in the shape of dog-bone with 6 cm X 1 cm. Mechanical tests were conducted with a tensile testing machine (Instron 4465; Instron, Norwood, MA) at a crosshead speed of 20 mm min^{-1} at room temperature. The rheological properties of the nanocomposite melts were measured using an advanced rheometric expansion system (ARES; Rheometric Scientific, Piscataway, NJ) with parallel plate geometry. The frequency was varied from 0.05 to 500 rad s^{-1} with a 12.5mm diameter parallel plate with a 1.0mm gap at 150°C.

4-3. Results and Discussion

4-3-1. Characterization of the Modified Silica Particles

Figure 4-1 shows the FTIR spectra of the (a) original silica particles, (b) silica-HDI obtained after grafting HDI to the silica surface, and (c) silica g-PBS obtained after reacting the pendant HDI on the silica surface with PBS molecules. In the original silica particles, strong and broad peaks at 3470cm^{-1}, reflecting the intramolecular hydrogen-bonded silanol hydroxyl groups and near 1100 cm^{-1}, which may attribute to the Si-O vibration, are assigned, respectively. [16] As for silica-HDI, C=O stretching peak at 1632 cm^{-1} and an amide N-H vibration peak at 1570 cm^{-1} are observed, which corresponds to the urethane bond. Besides, a peak, N-H stretching, is assigned at 3347 cm^{-1} [17] and a new peak attributable to isocyanate groups, is found at 2290 cm^{-1}. [18] These results suggest that HDI was successfully introduced on the silica particles.

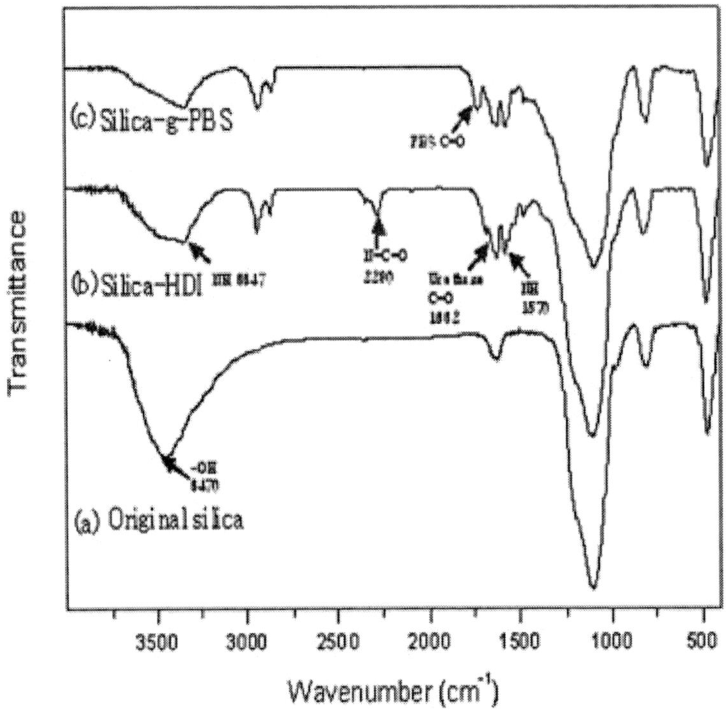

Figure 4-1. FTIR spectra of silica particles.

Compared with the hydroxyl group of original silica, the hydroxyl band of silica-HDI shifts to a lower wavenumber and its peak occurs near 3436 cm-1 with a reduced intensity. These results support that the intramolecular hydrogen bonding in original silica is weaker, as a result of an incorporation of HDI group within the hydroxyl group in the silica particles. However, unique peaks, which are related to the crosslinking of silica particles or intraparticle reaction, cannot be observed in the IR spectra. On the other hand, in the case of silica-g-PBS, the peak of the NCO group disappeared, meanwhile the C=O stretching peak of PBS molecules appeared at 1726 cm^{-1} in the spectra. This indicates that the -NCO group of silica-HDI reacted with the –OH group of the PBS molecules during in situ polymerization, producing the complete silica-g-PBS.

Figure 4-2 shows the TGA curves of the original silica particles, silica-HDI, and silica-g-PBS. For the original silica particles, about 3% weight loss was observed as a result of the separation of physically bound water. For silica-HDI, the approximately 16.5% weight loss occurred because of the thermal decomposition of the HDI moieties. The extent of the covalently attached isocyanate groups (NCO group nm^{-2}) was determined from the TGA data. Since 1.2 g silica-HDI was used to prepare the nanocomposite, 0.198 g HDI was obtained from the weight loss of the silica-HDI. The mole value of the HDI can be obtained from the amount of HDI divided by its molecular weight, and this value was about 1.18 mmol g^{-1}. Then, the number of NCO groups per nm^2 of the silica surface was calculated using the equation as follows:

$$\text{NCO group per nm}^2 = \frac{n_{\mathrm{HDI}} \times N_A}{A \times 10^{18}}$$

where nHDI is the number of moles of HDI on the silica surface, N_A is Avogadro's number, and A is the specific surface area of silica particles. The number of isocyanate groups on the silica surface was about 2.0 NCO nm^{-2}.

For the silica-g-PBS, an approximately 53.8% weight loss is expected from the thermal decomposition of the grafted PBS molecules. The amount of grafted PBS molecules can be calculated from this weight loss. The extent of grafting was confirmed as the percentage of grafting by the following equation: [19]

$$\text{Percentage of grafting (\%)} = \frac{\text{Grafted PBS (g)}}{\text{Used silica particles (g)}} \times 100.$$

Here, the silica particles used and grafted PBS molecules can be determined from the TGA curve of silica-g-PBS. The percentage of grafted PBS molecules is about 124%. As a result, the IR spectra and TGA results strongly suggest that an isocyanate group was introduced on the surface of the silica particles by the reaction of HDI with silanol groups on the silica surface, and the modified silica particles also chemically bonded with the PBS molecules after in situ polymerization.

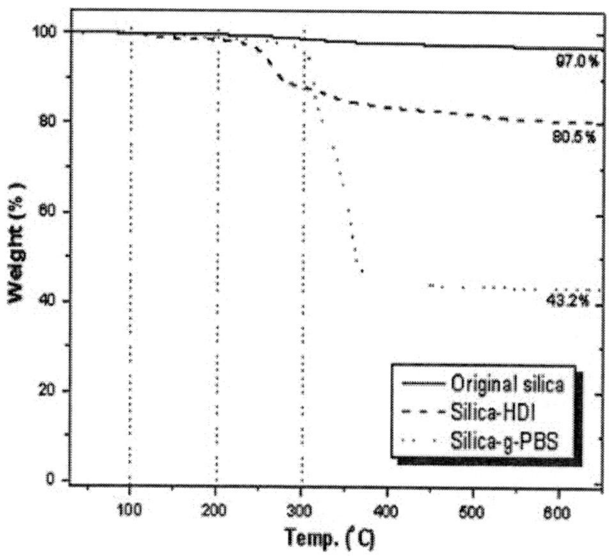

Figure 4-2. TGA curves of silica particles.

4-4. Characterization of the PBS Nanocomposites

4-4-1. Reaction Time and Intrinsic Viscosity

To identify the influence of grafting on in situ polymerization for the nanocomposites, the intrinsic viscosity (IV) was determined. Table 4-1 shows the sample code, reaction time, and IV data for all the PBS nanocomposites. In the sample codes, S and G denote PBS nanocomposites containing original silica particles and silica-g-PBS, respectively. The number following S or G is the fraction of silica in the PBS nanocomposites by weight. As shown in Table 4-1, in the case of S05 and S20, the value of IV was similar to that of neat PBS regardless of the silica content. On the other hand, the IV of the G nanocomposites increased with the silica content and this value was definitely higher than neat PBS, indicating that relative to the unmodified silica, the modified silica may influence strongly the degrees of polymerization of PBS. This phenomenon may be explained that the grafting PBS molecules on the silica particles operates as an active site, which can improve the degree of polymerization, generating the enhancement of the molecular weight of PBS.

Table 4-1. Sample Code, Reaction Time, and IV Value

	Reaction time (h)		
Sample code	Esterification	Polycondensation	[η]
Neat PBS	3.5	2.5	0.759
S05	3.5	3	0.794
S20	3.5	3.5	0.764
G05	3.5	3.5	0.808
G20	3.5	3.5	1.018

4-4-2. FTIR Spectra

We checked the IR spectra of neat PBS, PBS/ unmodified silica, and PBS/modified silica nanocomposites, respectively. Except the characteristic bands of PBS (ester bond, 1264 cm^{-1}; O-H, 3600– 3400 cm^{-1}; C-H, 2941 cm^{-1}; C=O, 1728 cm^{-1}; C-O, 1140 cm^{-1}), [20] there were no new peaks which may be influenced by the addition of silica, indicating that PBS/silica nanocomposites were successfully fabricated without crosslinking or intraparticle reaction of silica (data not shown).

To verify the existence of specific interaction between PBS and silica particles, the FTIR spectrum of neat PBS, S20, and G20 was normalized using the unchanged peaks, and then the S20 and G20 spectra were subtracted from the background spectrum of neat PBS; these results are shown in Figure 4-3. The S20-neat PBS had no characteristic peak, indicating that there is no interaction or chemical bonds created between the original silica particles and the PBS molecules. Interestingly, in the case of the G20-neat PBS, additional peak can be assigned at about 1710 cm^{-1} when compared with neat PBS. [18] On the whole, if specific hydrogen bond is formed between the carbonyl group of polyester and the functional group of other materials, some modification near this carbonyl group can be detected. For example, the wavenumbers of carbonyl group may be shifted [21, 22] or new carbonyl groups near the major carbonyl group can be created. [23, 24] Based on this, this new peak at 1710 cm^{-1} can be thought to be a powerful evidence that the carbonyl group of PBS molecules interacts with the –NHCO- groups of urethane between silica particles and PBS molecules, giving rise to the formation of hydrogen bonds.

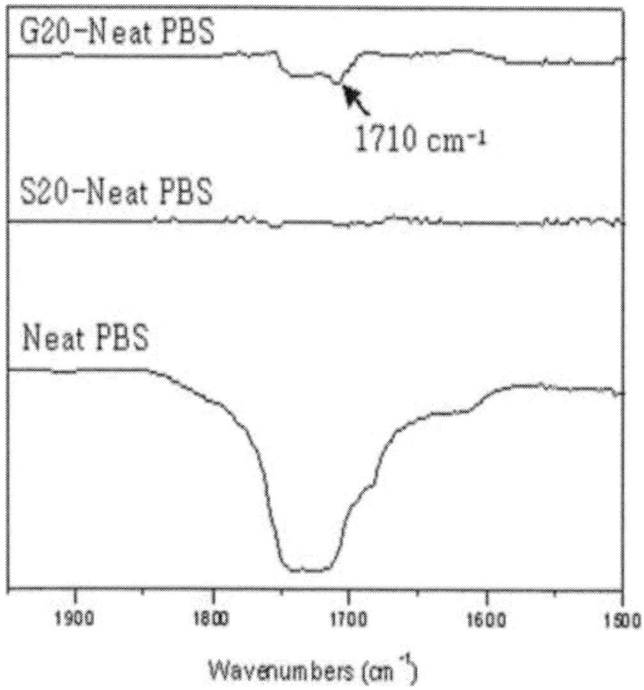

Figure 4-3. FTIR spectra of PBS nanocomposites: G20-neat PBS, S20-neat PBS, and neat PBS.

4-4-3. Dispersibility

The dispersibility of silica particles on the PBS matrix in the nanocomposites was investigated using TEM; the result is shown in Figure 4-4. The silica particles in the G nanocomposites are dispersed more homogeneously than the original silica particles in the S nanocomposites. In addition, the dispersed phases in the G nanocomposite are clearly smaller than those of the original silica particles in the S nanocomposites. These results indicate that the surface energy difference between silica-g-PBS and the PBS matrix is relatively low than original silica and PBS. That is, this can be explained that the surface of the silica-g-PBS is chemically similar to that of the PBS matrix. However, the surface of the original silica particles is quite different from that of PBS molecules. Based on this it can be clearly confirmed that when compared with the original silica particles, silica-g-PBS has much less tendency to aggregate by itself, leading to enhanced adhesion with the PBS matrix.

4-4-4. Thermal Stability

To examine the thermal stability for the nanocomposites, TGA analysis was conducted and the results are summarized in Table 4-2. The initial thermal degradation temperature indicating a 2% weight loss (TD^2) was 273°C for the S and G nanocomposites, which was 9°C higher than that of neat PBS. This confirms that at the beginning of thermal decomposition, the silica particles obstruct the internal diffusion of heat because they act as insulators with a high thermal capacity. Consequently, the nanocomposites are more stable than neat PBS at the initial thermal degradation, independent of the modification of the silica particles.

On the other hand, comparing the thermal degradation temperature of 20 wt % (TD^{20}), neat PBS, and S20 have similar values, while G20 has a value more than 20°C higher, suggesting that the modification of silica particles enhances the thermal stability of the nanocomposites. It may be likely that the dispersion of the silica particles in the PBS matrix is the main factor producing this behavior. That is, the highly dispersed silica-g-PBS particles in G20 act as more effective insulators than the aggregated silica particles in S20, leading to the enhanced thermal stability. Furthermore, this result is in good agreement to TEM micrographs in Figure 4-4.

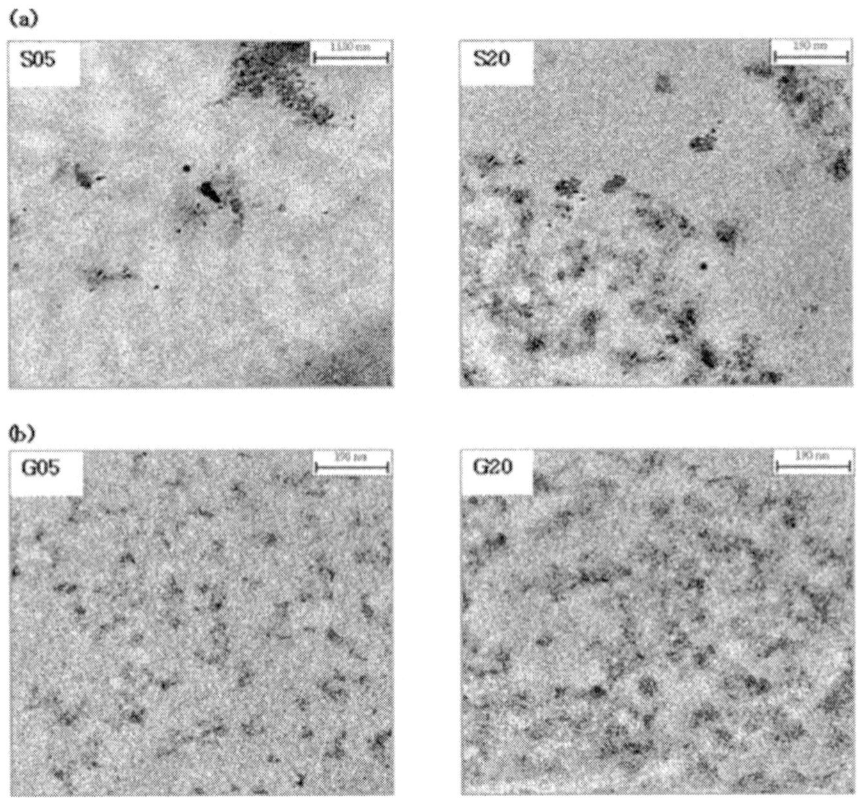

Figure 4-4. TEM micrographs of PBS nanocomposites: (a) S nanocomposites and (b) G nanocomposites.

Table 4-2. Thermal Stability and Thermal Properties of PBS Nanocomposites

Sample	T_c (°C)	T_m (°C)	$\Delta H_{f,m}$ (J/g)	ΔT (°C)	T_D^{2} (°C)	T_D^{20} (°C)
Neat PBS	69.0	115.0	71	46.0	264	318
S05	72.3	115.0	69	42.7		
S20	76.1	115.0	65	38.9	273	319
G05	69.0	115.0	68	46.0		
G20	75.7	115.0	64	39.3	273	340

4-4-5. Thermal Properties

To elucidate the thermal behavior of the neat PBS and the nanocomposites, DSC analyses were performed. The DSC thermograms for the S and G nanocomposites during the cooling and second heating scans are shown in Figure 4-5, and the results are summarized in Table 4-2. Neat PBS exhibited the single cooling exothermic peak (T_c) at 69.0°C. As seen in the Figure 4-5 and Table 4-2, as the silica contents increased, T_c for both S nanocomposites (original silica) and G nanocomposites (silica-g-PBS) shifted to the higher temperature, and their degree of supercooling exhibited lower value than that of neat PBS. This observation can be interpreted that both the original silica and silica-g-PBS particles can act as nucleating agents during the crystallization process and accelerate the crystallization rates in the PBS matrix.

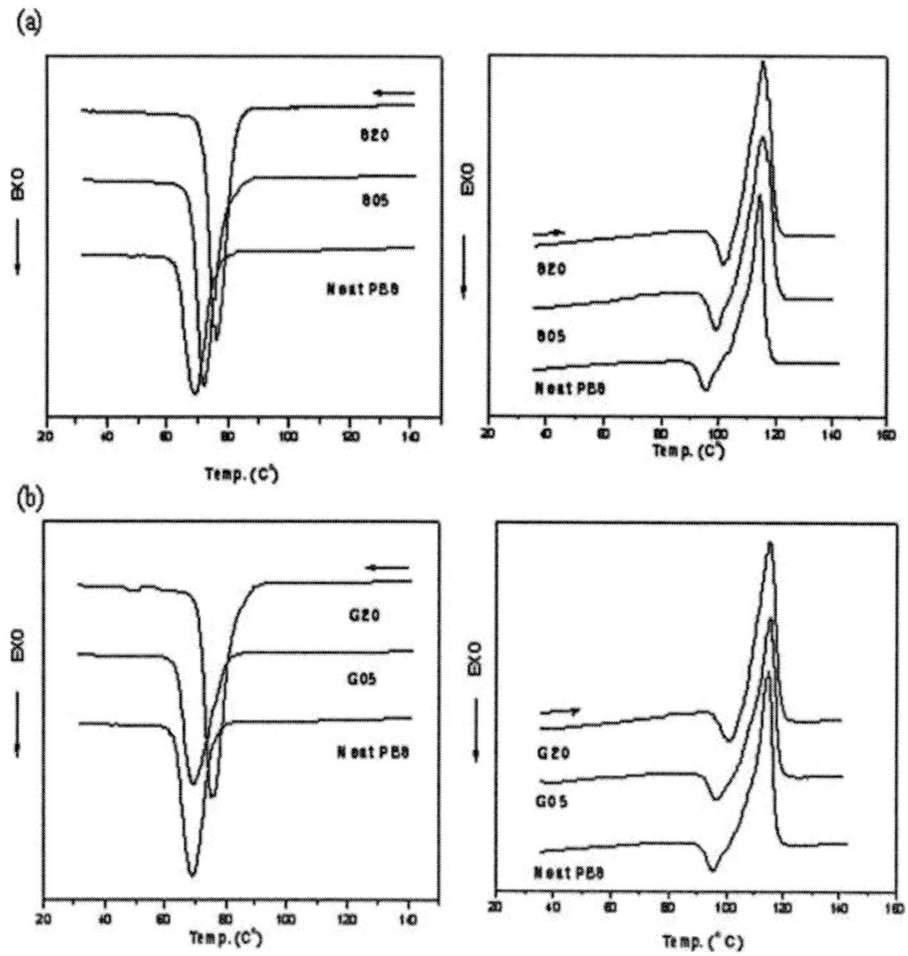

Figure 4-5. DSC cooling and second heating run of (a) S nanocomposites and (b) G nanocomposites.

On the other hand, relative to S nanocomposites, ΔT for the G nanocomposites showed somewhat higher values over same silica contents, confirming that the degree of the nucleating effect of silica-g-PBS is weaker than that of the original silica particles in the PBS matrix. This result may be attributable to the improved interfacial adhesion between the modified silica particles and PBS molecules, which induce a relatively lower nucleating effect. As for the second heating DSC curve, the melting endothermic peaks (T_m) of all of the samples were the same regardless of the modification or silica particle or its content. In summary, it can be concluded that the silica particles strongly affect the crystallization rate of PBS. That is, they contribute favorably to the faster crystallization of PBS molecules, but may not influence with the main PBS crystals.

4-4-6. Mechanical Properties and Morphology of Fractured Surfaces

The mechanical properties of PBS nanocomposites as a function of the contents and modification of the silica particles are shown in Figure 4-6. The tensile strength of the S and G nanocomposites increased with the silica content. In contrast, at the same silica content, the G nanocomposites filled with silica-g- PBS had a much greater tensile strength than the S nanocomposites, indicating that the entanglement of the grafted PBS segment with PBS molecules in the matrix occurred and this induces to the enhancement of the interfacial adhesion between the silica particles and PBS matrix.

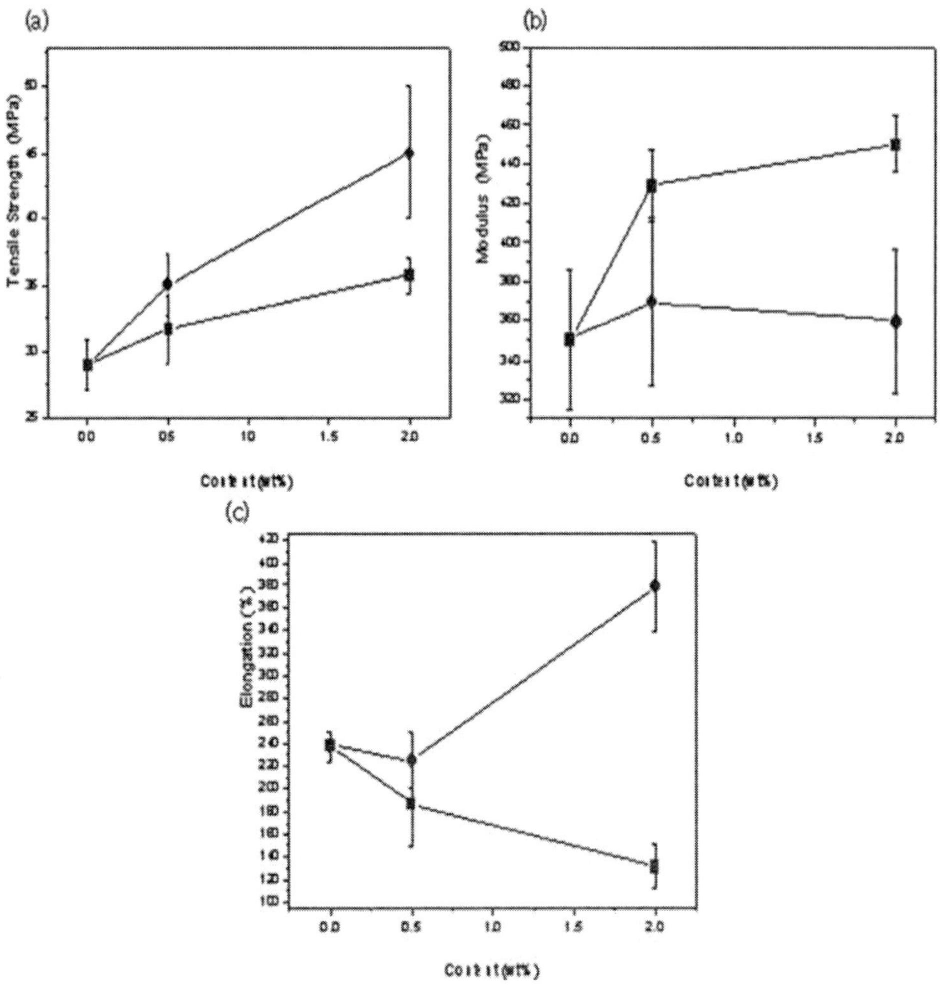

Figure 4-6. Mechanical properties of PBS nanocomposites: (a) tensile strength, (b) modulus, and (c) elongation at breaking.

Figure 4-6(b) shows the Young's modulus of the S and G nanocomposites. The Young's modulus for the S nanocomposites increased with the silica content. The high Young's modulus of the S nanocomposites may be explained that the addition of silica particles increases the stiffness. However, for G nanocomposites, a different phenomenon is observed. The Young's modulus of the G nanocomposites was increased slightly at a silica content of 0.5 wt % and remained unchanged at a silica content of 2.0 wt %. Rong et al. reported similar results for polypropylene composites filled with grafted nanoinorganic particles. [12] They suggested that the compliant layer at the interface of the grafted polymer tends to mask the stiffness of the filler particles. Therefore, the modulus of the nanocomposites containing original silica increases with the silica content, while the grafted polymer greatly decreases the stiffening effect of silica.

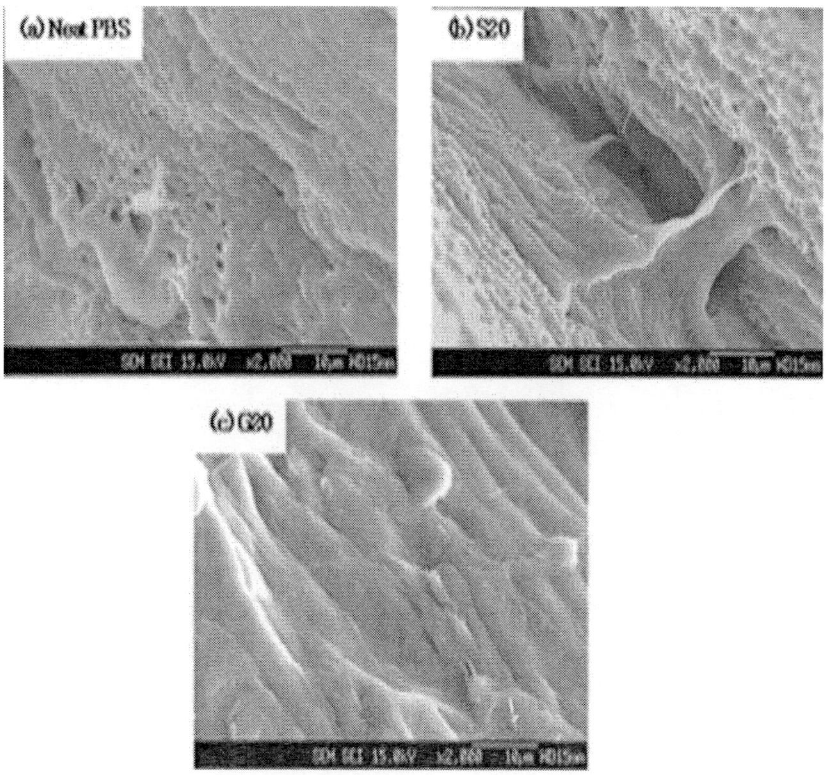

Figure 4-7. SEM fractographs of the tensile fractured surface.

A remarkable difference in the nanocomposites was observed in the elongation at breaking, as shown in Figure 4-6(c). The elongation at breaking of the S nanocomposites decreased with the silica content, which may be due to the aggregated silica particles, as seen in Figure 4-4. [25] Conversely, the elongation at breaking of the G nanocomposites increased with the silica content. This result can be interpreted as follows: first, the interfacial adhesion was enhanced by the entanglement between the grafted PBS molecules and the PBS matrix. Second, the modified silica particles were well dispersed in the PBS matrix. Accordingly, G nanocomposites have a more compact structure, leading to the resistance of the crack propagation.

Scanning electron microscopy was used to observe the fractured surfaces of specimens. As Figure 4-7(b) shows, a coarser appearance and void formation at the interface are observed at the fractured surface of S20. In contrast, a relatively smooth surface and no void formation are observed at the fractured surface of G20 in Figure 4-7(c), supporting the results of the mechanical tests.

The elongated neck area was fractured vertically in the direction of stretching after immersion in liquid nitrogen. Comparing S20 [Fig. 4-8(a)] with G20 [Fig. 4-8(b)], numerous cracks were frequently observed in the surface of S20, indicating poor adhesion between the aggregated silica particles and PBS matrix. In contrast, the surface of G20 showed no cracks and a compactly deformed structure. This clearly suggests that the silica particles grafted with PBS molecules were dispersed homogeneously and the adhesion between the silica particles grafted with PBS molecules and PBS matrix was improved.

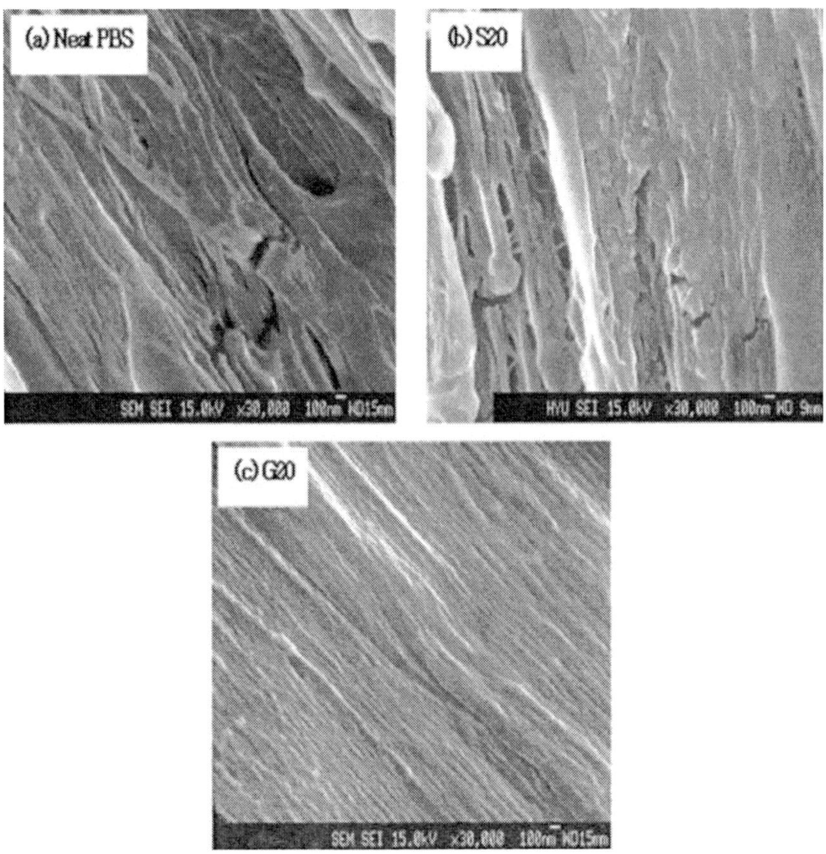

Figure 4-8. SEM fractographs of the horizontally fractured surface along the tensile direction.

4-4-7. Rheological Properties

Figure 4-9 shows the dynamic viscosity of the nanocomposites. S20 has a higher viscosity than neat PBS at a low frequency. This indicates that a network structure forms as a result of particle–particle interactions. As the shear force increases, all the samples exhibit the shear thinning behavior by the breakdown of the network structure. G20 has a higher viscosity than neat PBS over the entire frequency range. This may result from entanglement between the grafted PBS molecules and the PBS matrix molecules at the interface, which substantially enhances the interfacial interaction.

The storage modulus (G') and loss modulus (G") of the nanocomposites are displayed in Figure 4-10. The G' and G" of S20 are higher than neat PBS at low frequency. This confirms that the network structure because of the particle–particle interaction confers confers more elasticity than in neat PBS, and that the dissipation of energy also increases at low frequency. As the shear rate increased, the G' and G" of S20 showed similar value to that of neat PBS, reflecting the collapse of the network structure because of the dominant shear force. The G' and G" of G20 are also higher than those of neat PBS over the entire frequency range. This suggests that grafting PBS molecules on the silica particles leads to strong entanglement with PBS molecules in the matrix.

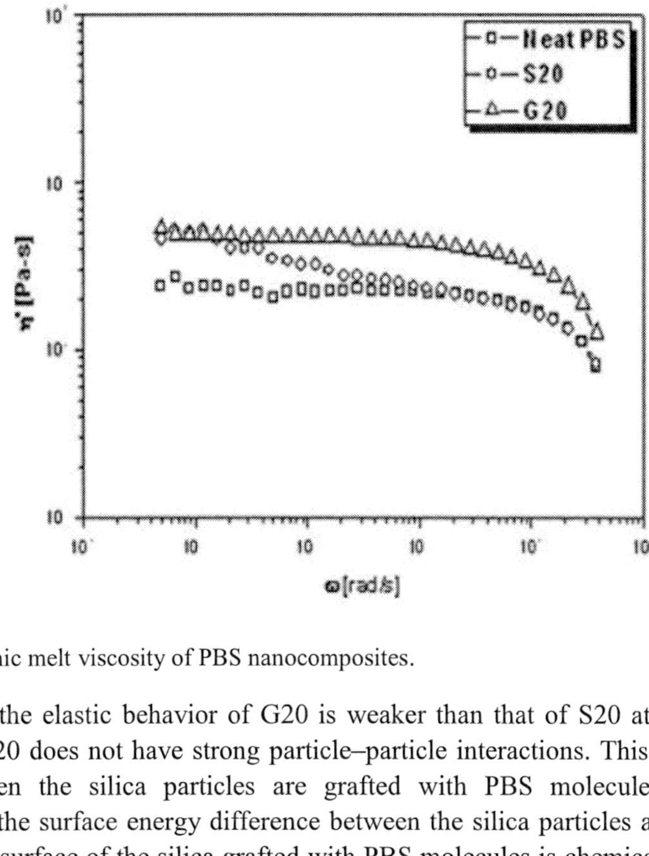

Figure 4-9. Dynamic melt viscosity of PBS nanocomposites.

In contrast, the elastic behavior of G20 is weaker than that of S20 at low frequencies, implying that G20 does not have strong particle–particle interactions. This can be explained as follows: when the silica particles are grafted with PBS molecules during in situ polymerization, the surface energy difference between the silica particles and PBS matrix is low because the surface of the silica grafted with PBS molecules is chemically similar to that of the PBS molecules in the matrix. Therefore, the particle–particle interactions between silica are relatively weakened. As a result, the probability that can be formed as network structure becomes lower.

The Cole–Cole plot of the nanocomposites is shown in Figure 4-11. Neat PBS, G20, and S20 give similar master curves with slopes of 1.75, 1.64, and1.110, respectively. If the polymer melt is isotropic and homogeneous, the curve slope on the Cole–Cole plot is 2.0, irrespective of temperature. [26] Neat PBS gives a master curve with a slope of 1.75, suggesting that little interaction occurs in the molecular structure. S20 gives curves with very shallow slopes, indicating that the system of nanocomposites is heterogeneous and much energy is dissipated due to the presence of the network structures. G20 has a slightly shallower slope of 1.64 when compared with neat PBS. This is in evidence that silica-g-PBS has good interfacial interactions because the grafted PBS molecules mask the surface of the hydrophilic silica particles. Over the loss modulus value about 10^4, however, the slope of curves for all nanocomposites increased and approach to neat PBS curves. This result reflects that some network structure owing to particle–particle interaction (S20) and particle–matrix interaction (G20) was broken down by shear force, and becomes gradually homogeneous system.

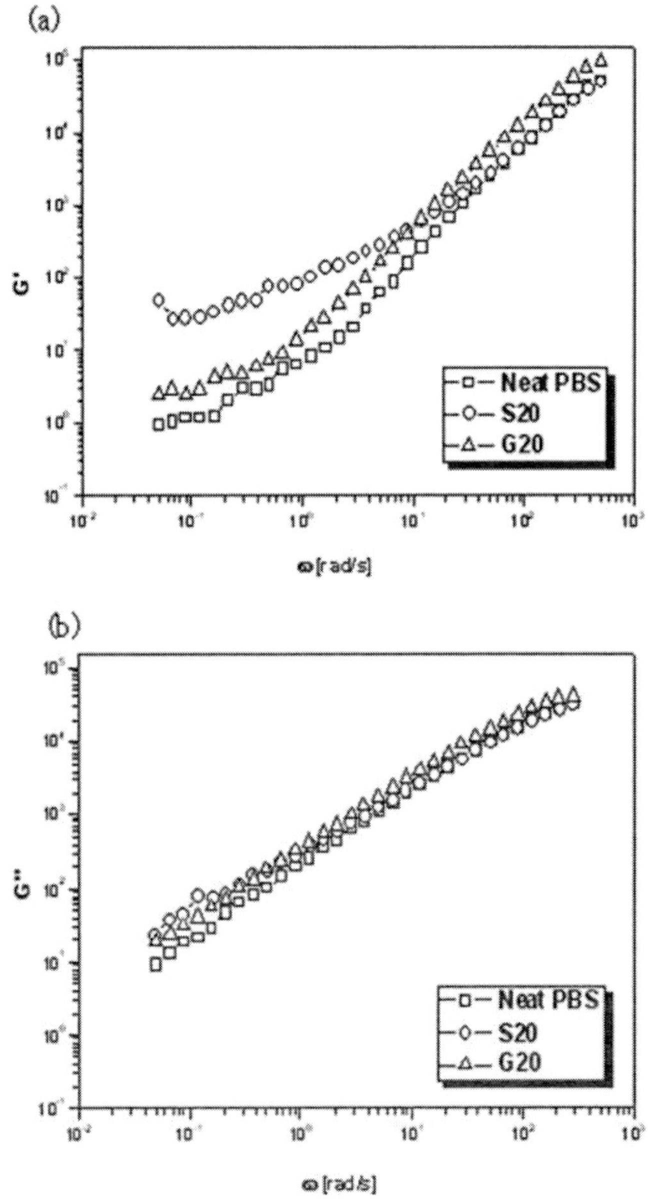

Figure 4-10. (a) Storage modulus (G0) and (b) loss modulus (G00) of PBS nanocomposites.

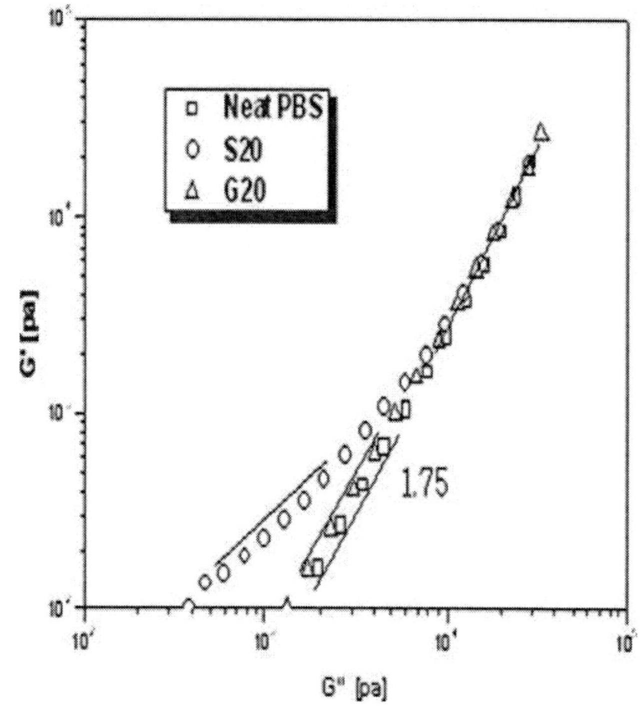

Figure 4-11. Cole–Cole plot for PBS nanocomposites.

4-5. Summary

Functionalized silica particles were synthesized using HDI, and PBS nanocomposites filled with silica-g-PBS were prepared successfully using in situ polymerization. Comparing the differences in the thermal, mechanical, and rheological properties of the S and G nanocomposites, the effect of grafting PBS molecules on the silica particles was investigated. Silica-g-PBS was well dispersed in the PBS matrix and provided good compatibility with PBS molecules, because the surface properties of the silica grafted with PBS molecules were chemically similar to those of the PBS molecules in matrix. Based on the rheology study, the S nanocomposites appeared to have a network structure due to particle–particle interactions, while the G nanocomposites showed strong adhesion between the grafted PBS molecules and PBS molecules in the matrix. Consequently, grafting PBS molecules on silica particles resulted in good dispersion and compatibility with PBS molecules in the matrix, leading to the enhanced thermal and mechanical properties when compared with neat PBS and PBS/original silica nanocomposite.

References

[1] Ahmed, S.; Jones, F. R. *J. Mater Sci.* 1990, 25, 4933.

[2] Bhari, K.; Mitomo, H.; Enjoji, T.; Yoshii, F.; Makuuchi, K. *Polym. Degrad. Stab.* 1998, 62, 551.

[3] Chen, Y.; Zhou, S.; Yang, H.; Wu, L. *J. Appl. Polym. Sci.* 2005, 95, 1032.
[4] Doi, Y.; Kasuya, K.; Abe, H.; Koyama, N.; Ishiwatari, S.; Takagi, K.; Yoshida, Y. *Polym. Degrad. Stab.* 1996, 51, 281.
[5] FTIR Data Bank.
[6] Fujimaki, T. *Polym. Degrad. Stab.* 1998, 59, 209.
[7] Giannelis, E. P. *Adv. Mater.* 1996, 8, 29.
[8] Hahm, W. G.; Oh, S. G.; Im, S. S. *Macromol. Res.* 2002, 10, 221.
[9] Han, C. D.; Kim, J.; Kim, J. K. *Macromolecules*. 1989, 22, 383.
[10] Hernandez, G.; Rodriguez, R. *J. Non-Cryst. Solids*. 1999, 246, 209.
[11] John, V.; Emily, B. Introduction to Spectroscopy, Harcourt; 2001, p 51.
[12] Lee, W. D.; Im, S. S. *Polymer.* 2006, 46, 1364.
[13] Lim, J. S.; Noda, I.; Im, S. S. *J. Polym. Sci. Part. B Polym. Phys*. 2006, 44, 2852.
[14] Lim, J. S.; Noda, I.; Im, S. S. *Polymer.* 2007, 48, 2745.
[15] Novak, B. M. *Adv. Mater.* 1993, 5, 422.
[16] Okada, M. *Prog. Polym. Sci.* 2002, 27, 87.
[17] Rong, M. Z.; Zhang, M. Q.; Zheng, Y. X.; Zeng, H. M.; Walter, R.; Friedrich, K. *Polymer.* 2001, 42, 167.
[18] Sinha Ray, S.; Bousmina, M. *Polymer.* 2005, 46, 12430.
[19] Sinha Ray, S.; Okamoto, K.; Okamoto, M. *Macromolecules.* 2003, 36, 2355.
[20] Tsubokawa, N.; Kogure, A.; Sone, Y. J. Polym. *Sci. Polym. Chem.* 1985, 42, 509.
[21] Uesaka, T.; Nakane, K.; Maeda, S.; Ogihara, T.; Ogata, N. *Polymer.* 2000, 41, 8449.
[22] Woo, E. M.; Chiang, C. P. *Polymer.* 2004, 45, 8415.
[23] Xu, N.; Zhou, W.; Shi, W. *Polym. Adv. Technol.* 2004, 15, 654.
[24] Yoo, E. S.; Im, S. S. *Macromolecules.* 1995, 28, 2460.
[25] Zhang, M. Q.; Rong, M. Z.; Friedrich, K. Handbook of Organic Inorganic Hybrid Materials and Nanocomposites; American Scientific Publishers: California; Vol. 2, 2003, p 113.
[26] Zhang, M. Q.; Rong, M. Z.; Zhang, H. B.; Friedrich, K. *Polym. Eng. Sci.* 2003, 43, 490.

In: Polymer Chains
Editors: Régine Penzkofer and Yasue Wu
ISBN: 978-1-62100-422-6

Chapter 3

BIOPLASTIC-BASED BLENDS

Jung Seop Lim
Department of Fiber and Polymer Engineering, Hanyang University, College of Engineering, Haengdang-dong, Seongdong-gu, Seoul, South Korea

1. OVERVIEW

1.1. Poly(3-Hydroxybutyrate-*Co*-3-Hydroxyhexanoate) (PHB–HHx)

With numerous environmental problems caused by petrochemical plastics waste, poly(hydroxyalkanoates) (PHAs), which can be produced from renewable carbon resources, has attracted a great deal of attention from scientific community and industry as these microbially synthesized polyesters can be completely biodegraded to CO_2 and H_2O under aerobic and anaerobic conditions.

Poly(3-hydroxybutyrate) (PHB), discovered by Lemoigne in 1925, is the most representative member of the PHA family. PHB possesses a melting point close to that of polypropylene (PP), forms a better barrier to oxygen than poly(ethylene terephthalate) (PET) and PP, and has mechanical properties resembling those of polystyrene (PS) and PP. However, it is rigid and stiff because of the high crystallinity and larger spherulites. Furthermore, PHB is also thermally unstable during the conventional melt processing due to the narrow processing temperature window. To overcome these drawbacks of PHB, several PHB copolymers have been biosynthesized in a variety of microorganisms. Table 1-1 shows the various PHB copolymer structures.

Recently, Kaneka Corp. (Osaka, Japan) synthesized poly(3-hydroxybutyrate-*co*-3-hydroxyhexanoate) (PHB–HHx) for the first time from oils and fats using *Aeromonas* sp. OL338 and *Aeromonas* sp. FA440. This new co-polyester has characteristics similar to those of conventional polyethylene and polyester in terms of, for example, its elasticity, toughness, dyeability, and chemical digestibility. In addition, PHB–HHx has been reported to not only possess improved thermoplasticity and mechanical properties over the other PHAs , but also to have excellent compatibility with other biodegradable polymers.

Table 1-1. Microbially synthesized of PHA copolymers containing (R)-3HB as a constituent

Bacterial strain	Carbon substrate	Random copolymer
Ralstonia eutropha	Propionic acid	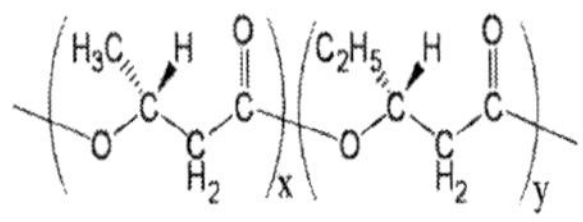 (R)-3HB (R)-3HV
Ralstonia eutropha	Pentanoic acid 3-Hydroxypropionic acid	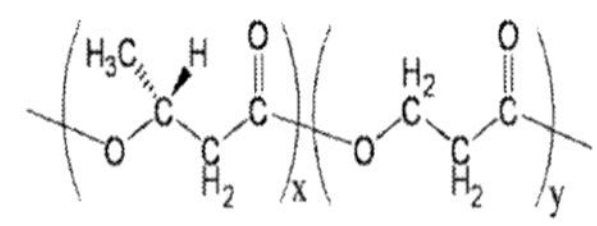 (R)-3HB 3HP
Alcaligenes latus Aeromonas cavie	1,5-Pentanediol Plant oils	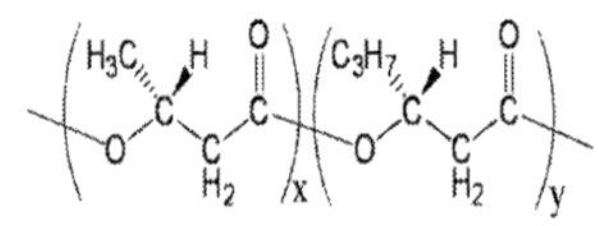 (R)-3HB (R)-3HHx
Pseudomonas sp. Ralstonia eutropha	Sugar 4-Hydroxybutyric acid	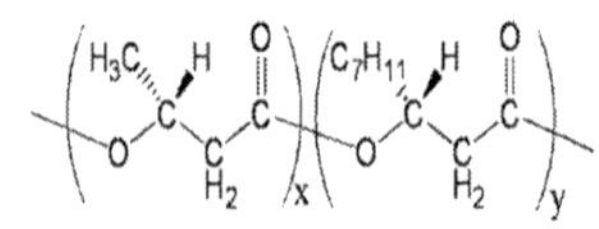 (R)-3HB (R)-3HD
Alcaligenes latus Comamonas Acidovorans	γ-Butyrolactone 1,4-Butanediol 1,6Hexanediol	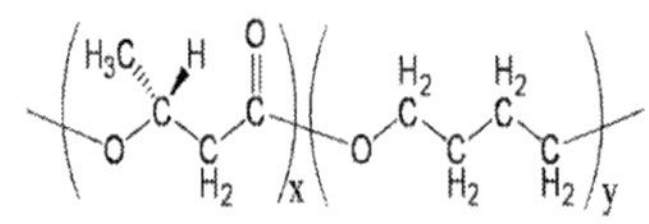 (R)-3HB 4HB

PHB–HHx is a random copolymer composed of 3-hydroxybutyrate and a minute amount of 3-hydroxyhexanoate (3HHx) units with side groups substituted with a longer propyl chain. This PHB-based copolymer shows a wide range of physical properties depending on the comonomer composition. Doi *et al.* found that the X-ray crystallinity of PHB–HHx decreased from 60 to 26%, and the tensile strength of its films decreased from 43 to 20 MPa with the increase in the molar fraction of 3HHx from 0 to 17 mol %. In contrast, the elongation to break increased from 6 to 850%. As seen in Figure 1-1, they also confirmed that the melting temperature of PHB–HHx decreased steeply with the increase in the molar fraction of 3HHx. Based on the relationship between lamellar thickness and melting temperature, randomly distributed second monomer units (3HHx unit) in copolymers were concluded to act as defects in the PHB crystal and excluded from the crystalline lamellae.

Sato *et al.* found that the wide angle X-ray diffraction (WAXD) pattern of the PHB–HHx copolymer at room temperature showed an orthorhombic system ($\alpha = \beta = \gamma = 90°$) with a = 5.76 Å, b = 13.20 Å, c = 5.96 Å (fiber repeat), which is identical to that of the PHB crystal system. However, they suggested that temperature-dependent variations in the lattice parameters *a* and *b* of PHB–HHx were quite different from those of PHB. That is, only the lattice parameter *a* increased with temperature, while the lattice parameter *b* changed very little in the case of the crystalline PHB–HHx. They attributed this to the intermolecular and intramolecular interactions between the C=O group and the CH_3 group, which decreased along the *a* axis of crystalline PHB–HHx with temperature. Structure model of PHB-HHx copolymer is displayed in Figure 1-2, briefly.

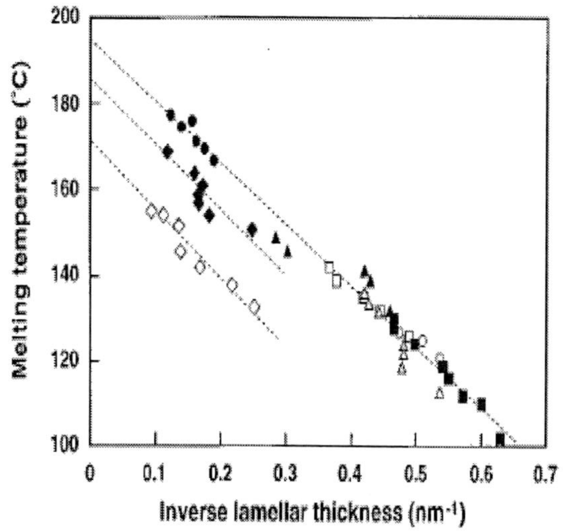

Figure 1-1. Variation in melting temperature (T_m) as a function of the inverse lamellar thickness ($1/l_c$) for melt-crystallized polyester films: (●) P[(R)-3HB]; (◆) P[(R)-3HB-co-6 mol % (R)-3HV]; (◇) P[(R)-3HB-co-16 mol % (R)-3HV]; (O) P[(R)-3HBco-8 mol % (R)-3HH]; (▲) P[(R)-3HB-co-8 mol % 4HB]; (Δ)P[(R)-3HB-co-10 mol % 4HB]; (■) P[(R)-3HB-co-10 mol % 6HH]; (□) P[(R)-3HB-co-5 mol % LA].

Some researchers aimed at exploring the structure and thermal behavior of PHB–HHx by means of infrared (IR) spectroscopy. Padermshoke *et al.* reported two-dimensional (2-D) IR correlation spectroscopy study of PHB and PHB–HHx and suggested that melting of the crystalline structure does not simultaneously induce the formation of the completely amorphous structure. The phase transition process of PHB and PHB–HHx takes place through an intermediate state. They confirmed in the 2-D IR study that the vibrational frequencies of the C=O, C–O–C, and C–H stretching bands due to the crystalline components of PHB–HHx were almost identical to those of PHB. These observations indicate that the helical structure of PHB–HHx is very similar to that of PHB.

In addition, the temperature-dependent IR spectral variations were analyzed for the C-H stretching, C=O stretching, CH_3 deformation, and C–O–C stretching vibration regions, and bands characteristic of crystalline and amorphous parts of PHB-HHx were identified in each region. Based on the anomalous frequencies of the CH_3 asymmetric stretching bands of the PHB–HHx, an inter- or intramolecular interaction (C-H----O hydrogen bond) was found to exist between the C=O group in one helical structure and the CH_3 group in the other helical structure in PHB–HHx .

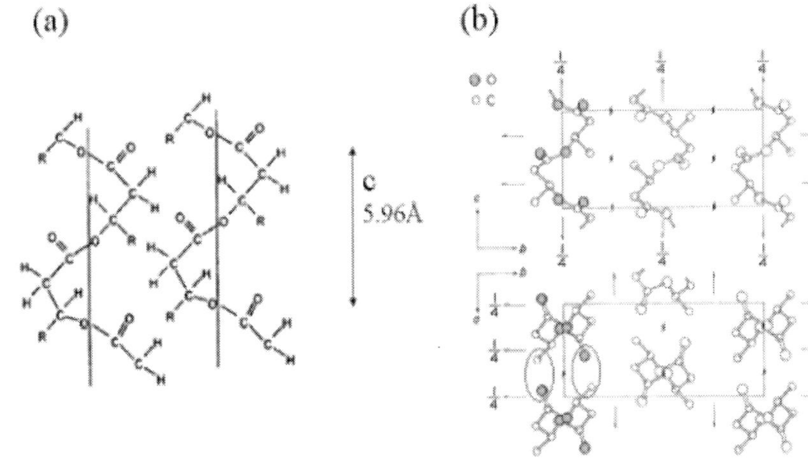

Figure 1-2. (a) Proposed helical structure model of PHB-HHx copolymer, (b) Reported crystalline structure of PHB. The $P2_12_12_1$ orthorhombic system with the lattice parameters of a = 5.76 Å, b= 13.20 Å, c = 5.96 Å (fiber axis).

Recently, the effect of aging on the mechanical properties was investigated for PHB–HHx. Until now, PHB–HHx containing longer side chain 3HHx comonomers that do not co-crystallize was thought to exhibit minimal secondary crystallization and hence no significant aging effect. However, PHB–HHx copolymers with the molar fraction of 3HHx ranging from 2.5 to 8.1mol % were reported to exhibit the typical signs of polymer aging. As a result, the elongation decreased and the strength increased after 11 days of room temperature aging. More recently, the embrittlement of PHB–HHx with 3HHx fractions ranging from 5 to 18mol % was studied, and the mechanism behind this phenomenon was investigated. The high initial crystallinity and the narrower interlamellar amorphous region were found to be the key factors that made the secondary crystallization leading to drastic mechanical effects on the copolymer with a 3HHx fraction lower than 10 mol %. In contrast, because of the lower initial crystallinity and the thicker interlamellar amorphous region, the secondary crystallization has less effect on the mechanical properties of the PHB–HHx copolymers with 3HHx fractions above 10mol %.

1.2. Influence of Intermolecular Interaction on the Miscibility and Crystallization Behavior of the PHB–HHx-Based Composites

As described earlier, PHB–HHx may find uses in a variety of polymer applications because the physicochemical properties can be controlled by changing the comonomer composition to produce forms ranging from fibers to films as well as molded plastic products. These products are expected to be used not only for biodegradable polymers but also for medical materials because of their biocompatibility.

Nevertheless, in certain practical applications, PHB–HHx suffers from limitations such as its thermal instability and insufficient mechanical stiffness, low crystallization rate, hydrophobicity, and high cost of production. In addition, the incorporation of only a small amount of comonomer is not sufficient to overcome the major drawbacks, such as brittleness and a narrow processing window. Consequently, research will have to be directed toward further modifying PHB–HHx. Two modifications are proposed to improve on the shortcomings of PHB–HHx and to obtain useful new materials based on PHB–HHx. One is copolymerization of PHB–HHx and the other is to prepare composites that consist of PHB–HHx. However, copolymerization is inadequate for practical use, since PHB–HHx itself is biosynthesized co-polyester, and the process is relatively complex and expensive. Therefore, we focus on the composites made of PHB–HHx and other bioplastics to circumvent the limitations of PHB–HHx and thus expand its applications.

Generally, the miscibility behavior must be accurately determined prior to the design of new polymeric blend with enhanced properties. Most polymers, however, are immiscible due to the low entropy of mixing . To enhance the formation of a one-phase miscible system in polymer blends, forming favorable specific intermolecular interactions between the two components of the blend is essential. Strong, specific intermolecular interactions bring about an obvious favorable change in enthalpy and also tend to produce strong effects to improve the miscibility. Thus, the extent of intermolecular interactions that can occur between two polymer components is an important parameter in determining the thermodynamics of the blend system. The most commonly observed interactions are of the acid–base type, i.e., hydrogen bonding, dipole–dipole, and charge-transfer interactions. Among the many parameters that impact the extent of intermolecular interaction between two polymers, the intramolecular/intermolecular interaction ratio is very important. For example, if one polymer possesses strong intramolecular interaction with itself, then this intramolecular interaction will inhibit the formation of specific intermolecular interactions between the blend components because this may limit the amount of interaction moieties that are available to participate in an intermolecular interaction. Therefore, in order to design and create an efficient polymer blend, the relationship between controllable parameters influencing the extent of intermolecular hydrogen bonding and the blend miscibility must be more clearly understood. Based on this consideration, we selected polyethylene glycol (PEG) and methoxy polyethylene glycol (MPEG), which possess different amounts of hydroxyl groups, to blend with PHB–HHx and investigated the effect of the intra/intermolecular interaction ratio on the miscibility of the two series of blends. Additionally, the secondary intermolecular interaction, formed between respective components in polymer composites, usually strongly influences several parameters regulating the crystallization behavior; for example, nucleation and growth rate, and characteristics of the individual folded chains, which not only affect the extent of crystallinity and the nature of crystalline morphology but also determine the final optical and mechanical properties of a polymer composite. Furthermore, this relationship is important for assessing the industrial processing conditions of the final product such as spinning, extrusion, compression, and injection molding. Therefore, the connection between intermolecular interaction and the crystallization behavior must be carefully examined. Based on these considerations, we also discuss the preparation of PHB–HHx/PEG blends quantitatively ascertain the effects of hydrogen bonding on their crystallization behavior.

On the other hand, in an attempt to upgrade the properties of PHB–HHx, we first introduce novel PBS-based ionomer as an efficient blend component. Ionomers appear to be excellent candidates for improving the physicochemical properties of the blend samples. During the blending of ionomer with other polymers, the ionic groups within the ionomer often substantially increase the miscibility compared to similar polymers without ionic groups. Such enhanced miscibility seems to occur where the ionic group within the ionomer interacts with other polymers that contain polar groups, such as carbonyl, amine, or amide.

Therefore, ionomers and their interactions in blends with other polymers have been widely studied. Linares *et al.* showed that blends of polyvinylidine fluoride (PVDF) and sulfonated polystyrene had improved compatibility and mechanical stability relative to control blends containing unmodified polystyrene. Lu and Weiss attributed the miscibility in blends of polyamide-6 and manganese sulfonated polystyrene to interpolymer interactions. However, although a understanding of the origin of the specific interactions that can be changed with different ionic and polar groups is critical, most reports have defined this interaction in ambiguous terms, for example, dipole–dipole, ion–dipole, ion pair–ion pair, and acid–base interpolymer interactions.Little systematic research has been published concerning the specific nature of those interactions and its impact on the miscibility and crystallization kinetics. Based on this consideration, we discuss the fabrication of PHB–HHx/biodegradable poly(butylene succinate)-based ionomers (PBSi) blends and investigate the effects of intermolecular interactions on their miscibility and crystallization behavior.

2. The Miscibility and Crystallization Behavior of Poly(3-Hydroxybutyrate-Co-3-Hydroxyhexanoate) and Methoxy Poly(Ethylene Glycol) Blends

2-1. Introduction

Because poly(ethylene glycol) (PEG) has advantages of biocompatibility, biodegradability, and hydrophilicity [1-7], biodegradable polyesters and PEG blends have been examined extensively. Avella *et al.* reported the miscibility of PHB/PEG blends [8-10]. Nijenhuis *et al.* stated that PLLA/PEG blends were miscible in the melt state [11].

However, few studies have dealt with the blends of methoxy poly(ethylene glycol) (MPEG), one of the PEG family, and other biodegradable polymers compared with PEG. Notably, no reports have investigated PHB-HHx/MPEG. In this study, PHB-HHx and MPEG, two semicrystalline biodegradable polymers, were blended using simple melt blending and miscibility was investigated in comparison of the PHB-HHx/PEG blends. In addition, the crystallization behavior in tandem with the miscibility was analyzed for the PHB-HHx/MPEG blends.

2-2. Experimental

2-2-1. Materials and Blend Preparation

Poly(3-hydroxybutyrate-co-3-hydroxy hexanoate) (PHB-HHx) was supplied by P&G Co. (Cincinnati, OH), with a 3-hydroxy hexanoate content of 11.9 mol% and M_w = 1,388,000. Its polydispersity was 2.0. Poly(ethylene glycol) (PEG) and methoxy poly(ethylene glycol) (MPEG) with M_n = 2,000 were purchased form Aldrich Chemical (Milwaukee, WI). Figure 2-1 shows the chemical structures of PEG and MPEG.

PHB-HHx/PEG and PHB-HHx/MPEG blends were prepared using melt blending with a twin roll counter-rotating mixer (Thermo Haake; Thermo Electron Corp., Waltham, MA). The blending conditions were as follows: barrel temperature 160°C, screw speed 100 rpm, and total mixing time 100 sec. Before blending in a one-step process, the polymers were ground and vacuum-dried. The blend compositions were 100/0, 97/3, 95/5, and 93/7 ratios of PHB-HHx/PEG and PHB-HHx/MPEG by weight.

2-2-2. Measurements

A Perkin–Elmer DSC7 (Boston, MA) was used to determine the glass transition temperature, T_g, of the blends. A sealed aluminum sample pan containing 5-10 mg of the blend samples was used in all of the experiments. Samples were heated from -70 to 180°C at a rate of 5°C /min, and maintained at 180°C for 3 minutes before rapid quenching to -70°C and then reheating to 180°C at the same heating rate. To analyze the isothermal crystallization of the PHB-HHx/MPEG blend samples, the samples were heated at a rate of 40°C/min from -10 to 200°C and maintained at 200°C for 5 minutes to remove any thermal history. Then, the samples were quenched rapidly to appropriate crystallization temperatures. Subsequently, the isothermal crystallization exotherm was recorded for 60 minutes.

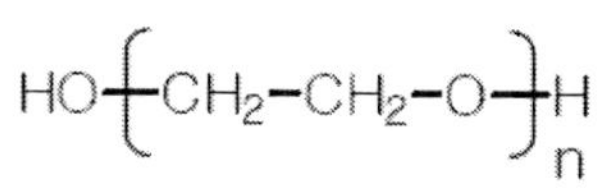

PEG

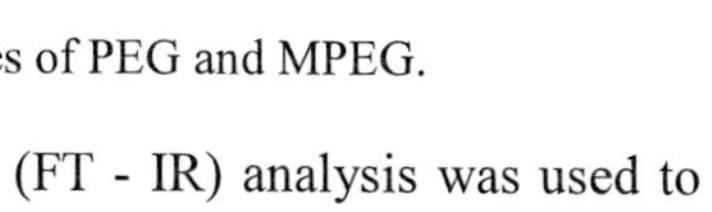

MPEG

Figure 2-1. The chemical structures of PEG and MPEG.

Fourier transform infrared (FT - IR) analysis was used to confirm the interaction in the two series of blends with a Nicolet 760 MAGNA-IR spectrometer (Nicolet Instrument Corp., Madison, WI).

The spherulitic morphology of the blend samples was investigated using a Nikon polarized light microscope (Nikon Corp., Tokyo, Japan). The samples were melted on a hot stage at 190°C for 1 minute, and then were quickly transferred to a Mettler Toledo FP82HT hot stage equilibrated at the desired crystallization temperature (Mettler - Toledo, Inc., Greifensee, Switzerland). They were crystallized for 60 minutes. After crystallization was complete, the spherulitic morphology of the blend samples was monitored.

2-3. Results and Discussion

2-3-1. Miscibility

2-3-1-1. Glass Transition Temperature

Figure 2-2 (a) and (b) show the second heating DSC curves for the PHB-HHx/PEG and PHB-HHx/MPEG blends, respectively. All the blends show only a single T_g between the T_gs of the pure components and their values shift to lower temperatures as the PEG and MPEG content increase. These results indicate that PHB-HHx forms a fully miscible blend over a wide range of PEG and MPEG compositions. However, the T_g of the PHB-HHx/PEG blends was higher than the T_g of the PHB-HHx/MPEG blends of same composition rate, implying that the degree of miscibility in the two series of blends differed. This means that the miscibility of the polymer blends depends on the interactions between the two components during mixing [13]. Sometimes, the strength of interactions in the polymer blends can be obtained from the Gordon - Taylor equation 2-(1) [14, 15],

$$T_g = \frac{W_1 T_{g1} + W_2 T_{g2}}{W_1 + K W_2} \tag{2-1}$$

where W_i is the weight fraction of component i, and T_{gi} denotes the glass transition temperature of the component i. The parameter K, an arbitrary fitting constant, represents the ratio of the thermal expansion cofficient differences above and below T_g for the component. This parameter is also related to the strength of the interaction between the blend components. The greater K is, the stronger the interaction is between components in the blends [13].

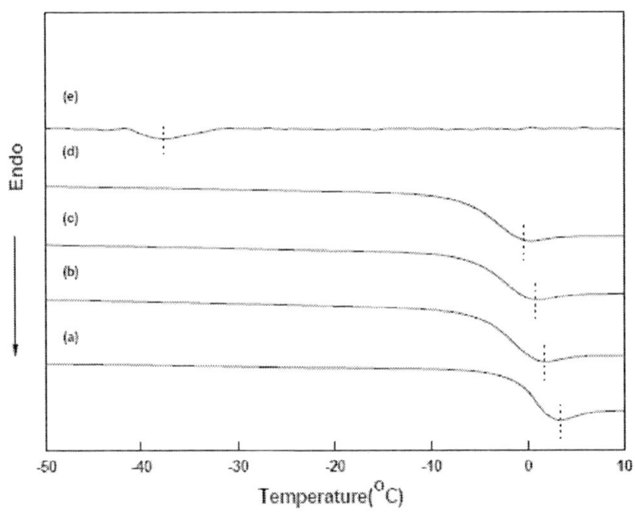

Figure 2-2(a). The second heating DSC curves of PHB-HHx/PEG blends: (a) PHB-HHx, (b) PHB-HHx/PEG (3wt%) blend, (c) PHB-HHx/PEG (5wt%) blend, (d) PHB-HHx/PEG (7wt%) blend, and (e) PEG.

In Figure 2-3, the T_gs of the two series of blends fit those calculated using the Gordon-Taylor equation. However, the K of the PHB-HHx/MPEG blends was higher than that of the PHB-HHx/PEG blends, with K values of 3.18 (PHB-HHx/MPEG) and 1.26 (PHB-HHx/PEG). This indicates that MPEG forms a relatively strong interaction with PHB-HHx and hence PHB-HHx/MPEG blends have a much higher degree of miscibility than PHB-HHx/PEG blends. Further quantitative evaluation of the interaction will be discussed in the following section.

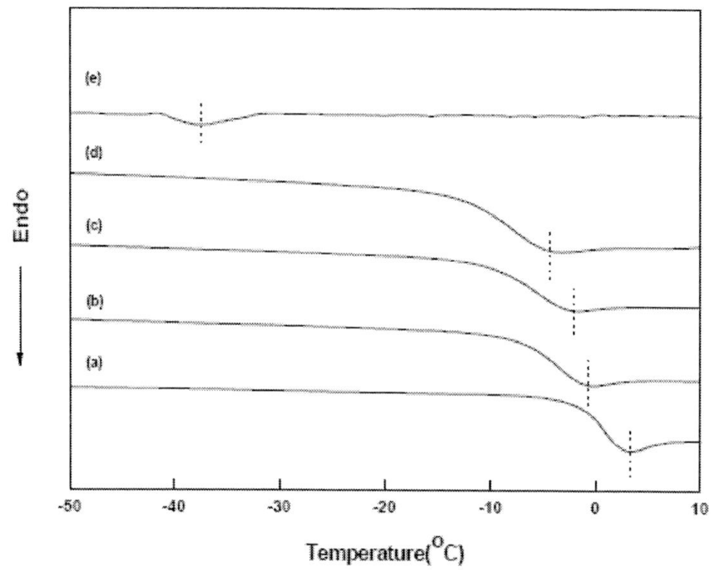

Figure 2-2(b). The second heating DSC curves of PHB-HHx/MPEG blends: (a) PHB-HHx, (b) PHB-HHx/MPEG (3wt%) blend, (c) PHB HHx/MPEG (5wt%) blend, (d) PHB-HHx/MPEG (7wt%) blend, and (e) MPEG.

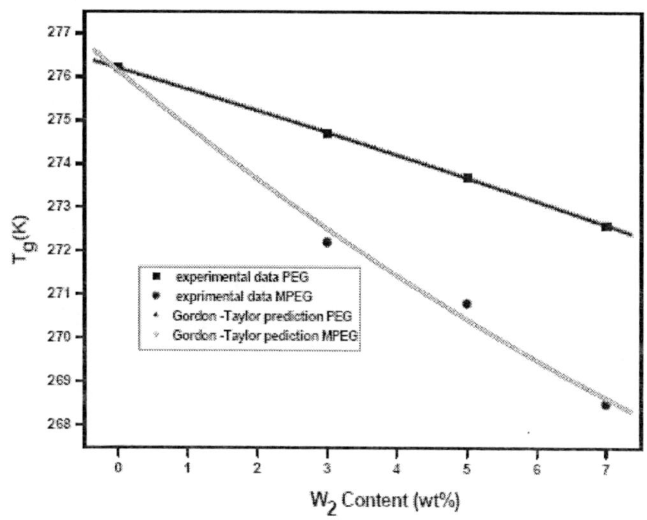

Figure 2-3. The glass transition behavior of PHB-HHx/PEG and PHB-HHx/MPEG blends.

2-3-1-2. Hydrogen Bonding

Figure 2-4 shows IR spectra in the hydroxyl group region of PEG and MPEG. The PEG family involves intramolecular hydrogen bonding between the C-O-C group and the hydroxyl group. In Fig.2-4, PEG has a broad OH band centered on 3400 cm $^{-1}$. Compared to PEG, the OH band of MPEG shifts to a higher wavenumber and the peak intensity is relatively weak, confirming that the intramolecular hydrogen bonding in MPEG is weaker than that in PEG. Rozenberg *et al.* investigated the effect of the hydroxyl end group on the hydrogen bonding in the PEG family and reported that the intramoelcular hydrogen bonding within PEG is gradually increased in proportion to the amount of its hydroxyl group [15].

Figure 2-5 and Figure 2-6 show IR spectra in the 1800 - 1680 cm $^{-1}$ region (carbonyl peak) of the PHB-HHx/PEG and PHB-HHx/MPEG blends. The carbonyl group of PHB-HHx may form intermolecular hydrogen bonds with the hydroxyl groups of PEG or MPEG in their blends. In FT – IR spectroscopy, the frequency of a functional group, such as carbonyl, ether or amide, is usually determined by the force constant and mass of the atoms, which is written as:

$$\overline{V} = \frac{1}{2\pi C}\sqrt{\frac{F}{\mu}} \tag{2-2}$$

where F is the force constant, which is created between inter-bonded atoms in the functional group, and c is the velocity of light (3×10^{10}cm/s). μ is defined as the reduced mass, which can be calculated as $\mu = M_1M_2/(M_1 + M_2)$ (M = the atomic weight). If there is intermolecular hydrogen bonding between the hydroxyl group of PEG or MPEG and the carbonyl group of PHB-HHx, the carbonyl bond energy of PHB-HHx may be reduced due to the interaction between the hydrogen in the hydroxyl group and the oxygen of the carbonyl group. Consequently, F of the carbonyl group will weaken and the peak might shift to a lower frequency [16, 17]. Figure 2-7 outlines this mechanism. As seen in Fig. 2-5 and Fig. 2-6, the carbonyl peak of the blends shifts to lower wavenumbers gradually as the PEG and MPEG contents are increased. This indicates that with increasing PEG and MPEG content, the intermolecular hydrogen bonding in the PHB-HHx/PEG and PHB-HHx/MPEG blends is gradually increased.

On the other hand, the carbonyl peaks of PHB-HHx/PEG blends appear at higher wavenumbers than those of the PHB-HHx/MPEG blends. For PHB-HHx/PEG, MPEG 7wt% blends, the wavenumbers of the carbonyl peak appear at about 1727 and 1725 cm^{-1}, indicating that the intermolecular hydrogen bonding in the PHB-HHx/MPEG blends is stronger than that of the PHB-HHx/PEG blends. Accordingly, this may be ascribed that MPEG, with relatively little hydroxyl group, forms weaker intramoelcular hydrogen bond than PEG and hence carbonyl group of PHB-HHx can participate more easily in intermolecular hydrogen bonding with the hydroxyl group of MPEG. F (dynes/cm) was also calculated by equation 2-(2) and the values are shown in Table 2-1, where the F of the PHB-HHx/MPEG blends is smaller than that of the PHB-HHx/PEG blends. This also suggests that stronger intermolecular hydrogen bonding exists for the carbonyl group of PHB-HHx and the hydroxyl group of MPEG.

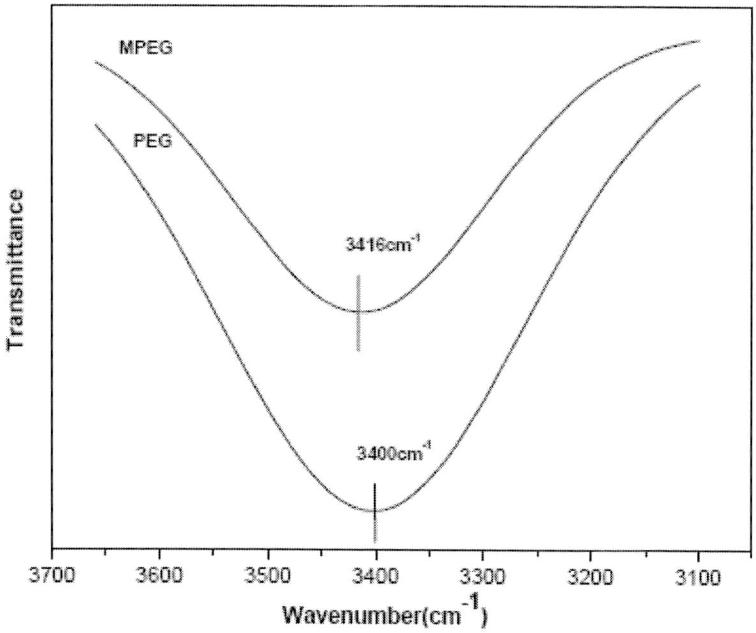

Figure 2-4. FT - IR spectra in the 3050 - 3700 cm $^{-1}$ region of PEG and MPEG.

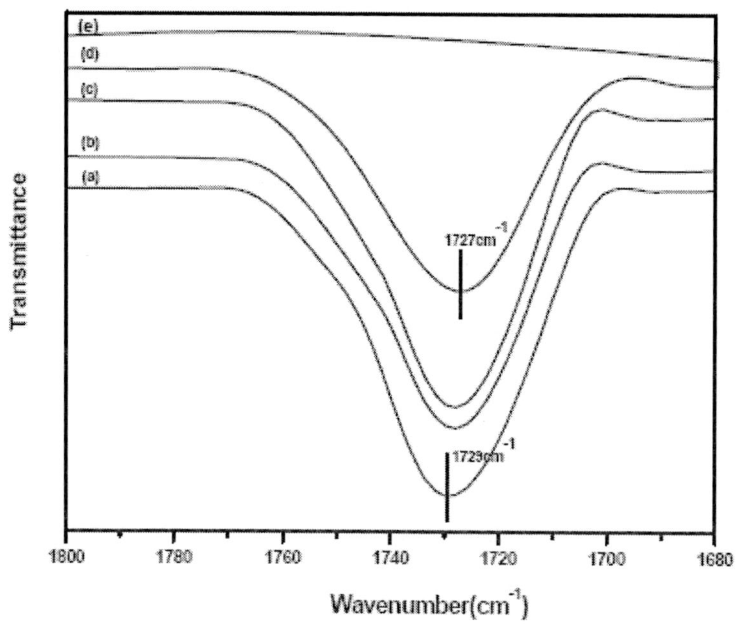

Figure 2-5. FT - IR spectra in the 1800 - 1680 cm $^{-1}$ region of PHB-HHx/PEG blends: (a) PHB-HHx, (b) PHB-HHx/PEG (3wt%) blend, (c) PHB-HHx/PEG (5wt%)blend, (d) PHB-HHx/PEG (7wt%)blend, and (e)PEG.

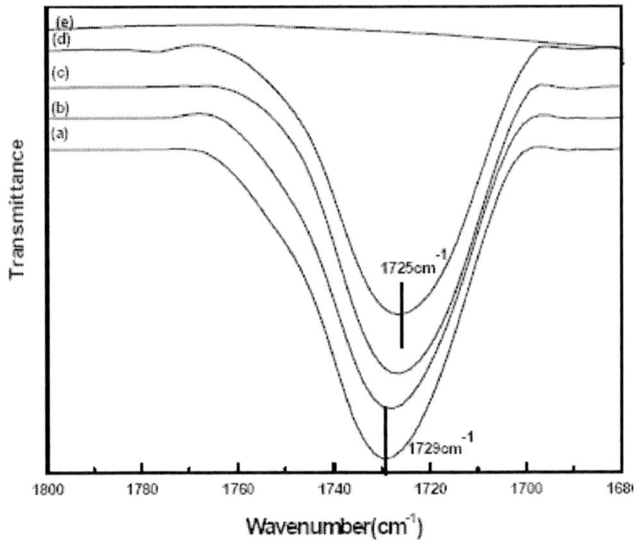

Figure 2-6. FT - IR spectra in the 1800 - 1680cm $^{-1}$ region of PHB-HHx/MPEG blends: (a) PHB-HHx, (b) PHB-HHx/MPEG (3wt%) blend, (c) PHB-HHx/MPEG (5wt%)blend, (d) PHB-HHx/MPEG (7wt%)blend, and (e) MPEG.

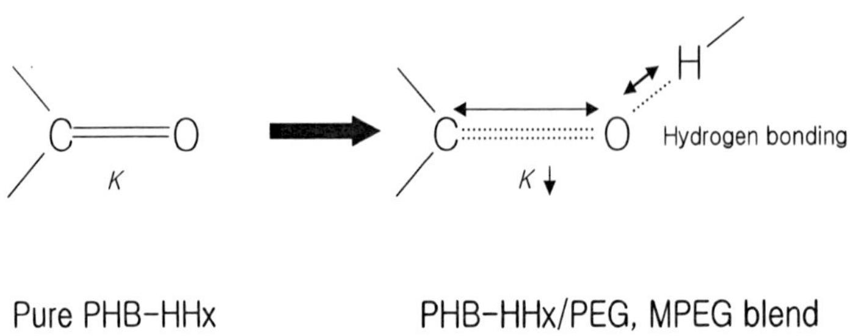

Figure 2-7. The change in the force constant in pure PHB-HHx and PHB-HHx/PEG, MPEG blends.

Table 2-1. The wavenumber and force constant of the carbonyl group in the PHB-HHx/PEG and PHB-HHx/MPEG blends

Sample	The wavenumber of carbonyl peak(cm^{-1})	The force constant (dynes/cm)
PHB-HHx	1729.5	25,699
PHB-HHx/MPEG (3wt%)	1727.7	25,645
PHB-HHx/MPEG (5wt%)	1726.4	25,607
PHB-HHx/MPEG (7wt%)	1725.5	25,580
PHB-HHx/PEG (3wt%)	1728.1	25,657
PHB-HHx/PEG (5wt%)	1727.7	25,645
PHB-HHx/PEG (7wt%)	1727.2	25,630

[1 dyne = 1.020×10^{-3} g· cm/sec^2].

2-3-1-3. Interaction Parameter

The miscibility in the PHB-HHx/MPEG blends was investigated further by calculating the interaction parameter. From a thermodynamic perspective, the change in the Gibbs free energy of mixing (ΔG_m) on mixing two components must be negative to form a miscible polymer blend. The most popular and simplest theoretical model for the change in ΔG_m is the Flory - Huggins theory [18, 19]:

$$\frac{\Delta G_m}{RT} = \frac{\Phi_A}{M_A} ln\Phi_A + \frac{\Phi_B}{M_B} ln\Phi_B + \chi_{AB}\Phi_A\Phi_B \qquad (2\text{-}3)$$

where M_A and M_B are the degree of polymerization of polymer A and B, respectively, and Φ_A and Φ_B are the volume fractions of the respective polymer. In this work, the equilibrium melting temperature (T_m^0) depression was used to determine χ_{AB} due to its simplicity.

PHB-HHx and PHB-HHx/MPEG blends were crystallized isothermally at the desired temperature to obtain the T_m^0. Figure 2-8 shows the DSC curves of pure PHB-HHx and PHB-HHx/MPEG blends crystallized isothermally at 70°C. Double melting peaks were found for PHB-HHx and all the blends. These peaks can be explained by the melting and recrystallization process of PHB-HHx [20]. The lower melting peak is attributed to the melting of crystals formed by recrystallization during heating. As seen in Fig. 2-8, the lower melting peak decreases as the MPEG content increases, while corresponds to the melting of the original crystals formed at a given T_c, while the higher melting peak the higher melting peak is not changed. Therefore, the lower melting peak was introduced to apply the Hoffman - Weeks equation [21].

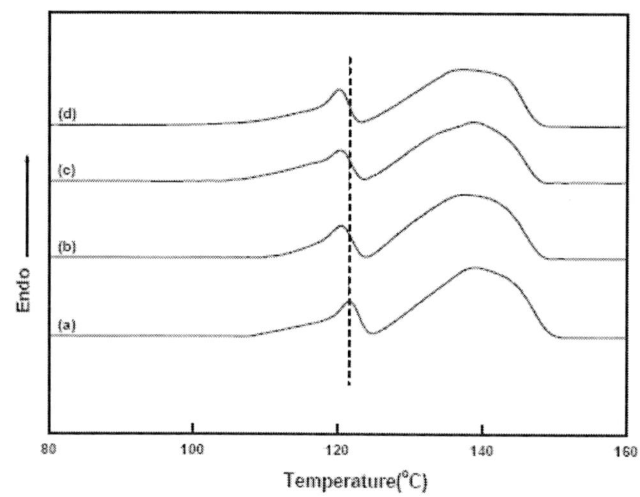

Figure 2-8. DSC curves of isothermally crystallized PHB-HHx/MPEG blends at 70°C: (a) PHB-HHx, (7wt%) blend.

Figure 2-9 shows the Hoffman - Weeks plot of PHB-HHx/MPEG blends using the isothermal crystallization results. The T_m^0 of pure PHB-HHx is observed at about 141.6°C, while the T_m^0 of the blends decreases gradually as the MPEG content increases. The maximum extent of this T_m^0 depression is about 3.5°C for the PHB-HHx/MPEG 7wt% blends, where T_m^0 is approximately 138.2°C.

The data in Fig. 2-9 were analyzed using the Nishi - Wang equation based on the Flory - Huggins theory [22], in which the equilibrium melting temperature depression of the crystallizable component can be written as:

$$\frac{1}{T^0_m(blend} - \frac{1}{T^0_m(pure)} = -\frac{RV}{\Delta H_f^0 V_B}\left(\chi_{AB}\,\phi_B^2\right) \tag{2-4}$$

Using equation 2-(4), a straight line with a negative slope of - 7.16 × 10 $^{-4}$ is obtained by plotting the term on the ordinate in Fig. 2-10 versus Φ_B for the PHB-HHx/MPEG blends, which is equal to χ_{AB}. This demonstrates that PHB-HHx and MPEG are miscible thermodynamically. However, the agreement between the experimental data and the calculated value based on the Nishi - Wang equation was not very good.

The Flory - Huggins theory is based on the assumption that two polymers mix randomly in their blends [23, 24]. Therefore, this assumption is inadequate for the presence of an interaction between the components of blends, such as hydrogen bonding, because the specific interactions might limit the mobility of the polymer chain and hence the chain configuration might be orderly in the mixture [25, 26]. Consequently, the PHB-HHx/MPEG blends are not simply random mixtures without any interactions, but that the PHB-HHx and MPEG interact with each other in their blends via hydrogen bonding.

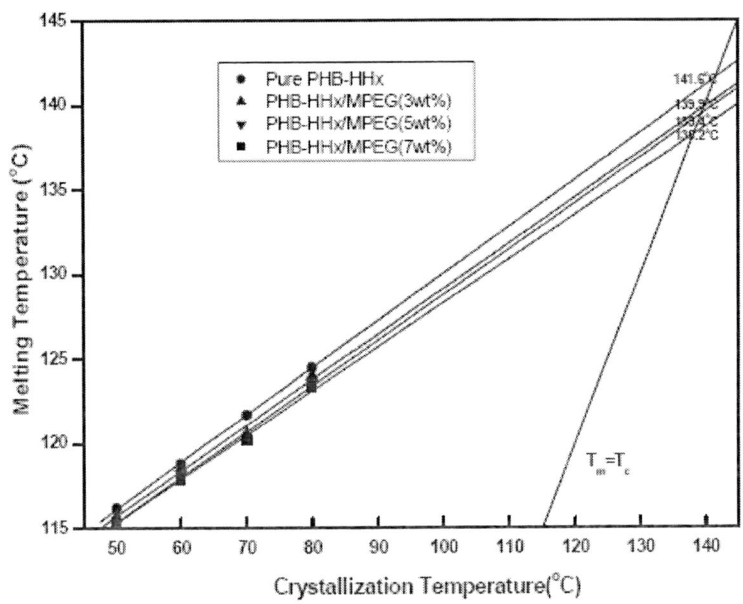

Figure 2-9. Hoffman - Weeks plots for PHB-HHx/MPEG blends: (●) PHB-HHx, (▲)PHB-HHx/MPEG (3wt%) blend, (▼) PHB-HHx/MPEG (5wt%) blend, (■) PHB-HHx/MPEG (7wt%) blend.

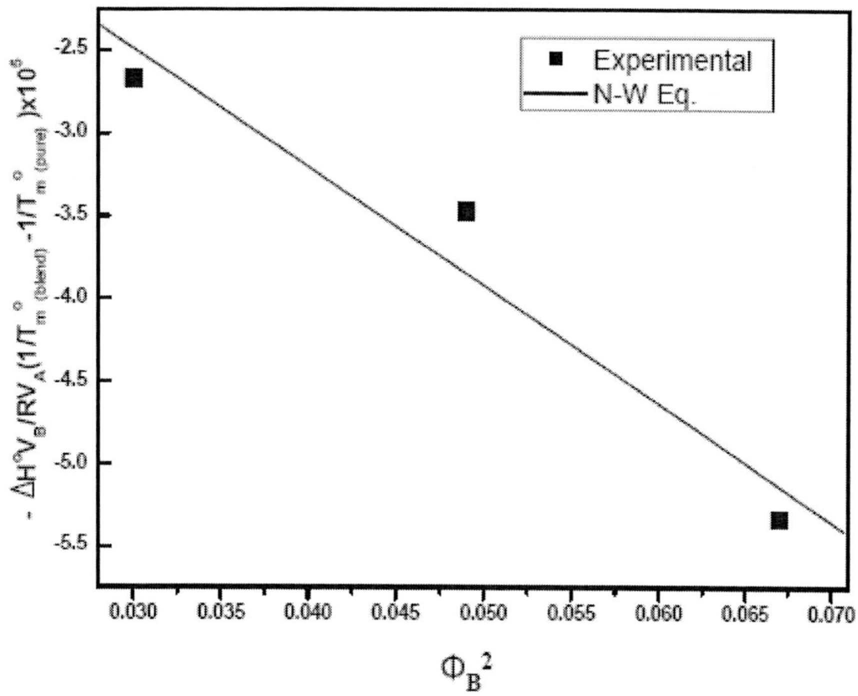

Figure 2-10. Plot according to the Nishi - Wang equation for PHB-HHx/MPEG blends.

2-3-2. Crystallization Behavior

2-3-2-1. Spherulitic Morphology

Figure 2-11 shows the spherulitic morphology of pure PHB-HHx and PHB-HHx/MPEG blends crystallized at 70°C after melting, in which only PHB-HHx crystallizes because the T_c (69°C) of PHB-HHx is higher than the T_m (52°C) of MPEG. Pure PHB-HHx spherulites show the familiar Maltese cross birefringent pattern. For the blends, the band spacing increases gradually with the MPEG content, confirming that MPEG plays a role as a noncrystallizable component in PHB-HHx/MPEG blends above its T_m. It has been reported that the band spacing increases with the increase in the noncrystallizable components [27, 28]. Regardless of the MPEG content, the boundary of PHB-HHx spherulites does not look open and the Maltese cross is not disrupted. This implies that interspherulitic separation does not occur in the PHB-HHx/MPEG blends [29].

Figure 2-12 shows the spherulitic morphology of pure PHB-HHx, pure MPEG, and their blends crystallized at 40°C from the melt, where MPEG crystal may form an insoluble solid particle in the PHB-HHx/MPEG blends, because 40°C is the optimum crystallization temperature of MPEG resulting in the fastest crystallization of MPEG. Pure PHB-HHx and MPEG spherulites show the familiar Maltese cross birefringent pattern, as shown in Fig. 2-12(a) and (d). For the blends, the spherulites are smaller than those of pure PHB-HHx, indicating that MPEG acts as a nucleating agent in their blends below its T_m.

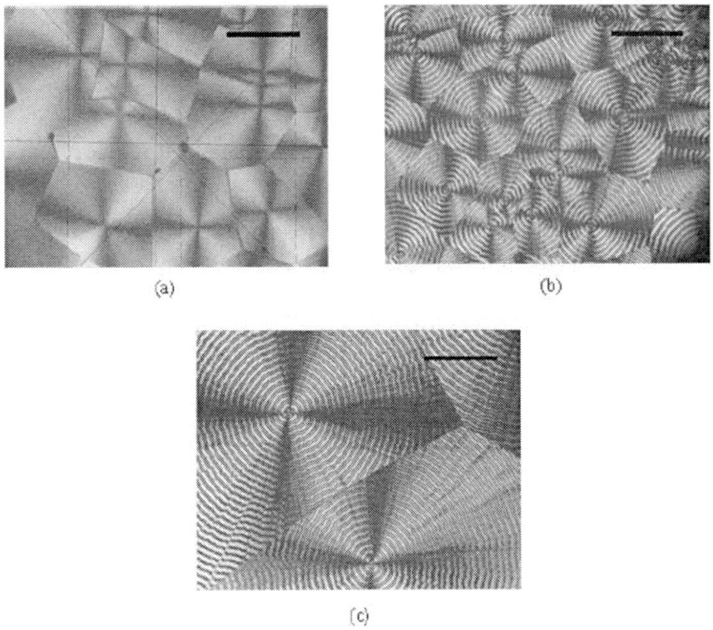

Figure 2-11. Spheruite morphology of PHB-HHx and PHB-HHx/MPEG blends at crystallization 70°C (scale bar: 100 micrometer); (a) PHB-HHx, (b) PHB-HHx/MPEG (3wt%) blend, and (c) PHB-HHx/MPEG (7wt%) blend.

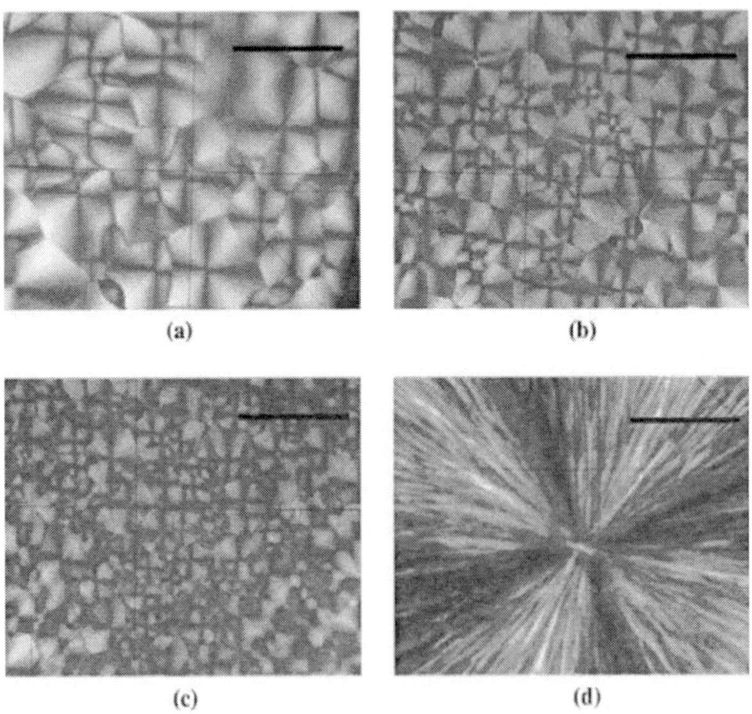

Figure 2-12. Spherulite morphology of the PHB-HHx, PHB-HHx/MPEG blends and MPEG at crystallization 40°C (scale bar: 100 micrometer); (a) PHB-HHx, (b) PHB-HHx/MPEG (3wt%) blend, (c) PHB-HHx/MPEG (7wt%) blend, and (d) MPEG.

To investigate the spherulitic morphology for the blends at 40°C in more detail, Fig. 2-12 was magnified four times (PHB-HHx, PHB-HHx/MPEG 3wt%) and 7.5 times (PHB-HHx/MPEG 7wt%), respectively and the results are shown at Figure 2-13, where the MPEG spherulitic morphology cannot be observed in the blends. In the crystalline/crystalline polymer blends, the spherulitic morphology of the second component may depend on the amount of second component added in the preexisting crystal matrix of the first component [28]. Thus, this result may be attributable to the tiny MPEG content in the blends. In addition, independent of the MPEG content, the disruption of spherulites and distortion of the Maltese cross are not observed in the PHB-HHx/MPEG blends, proving that there is no interspherulitic separation between MPEG and PHB-HHx.

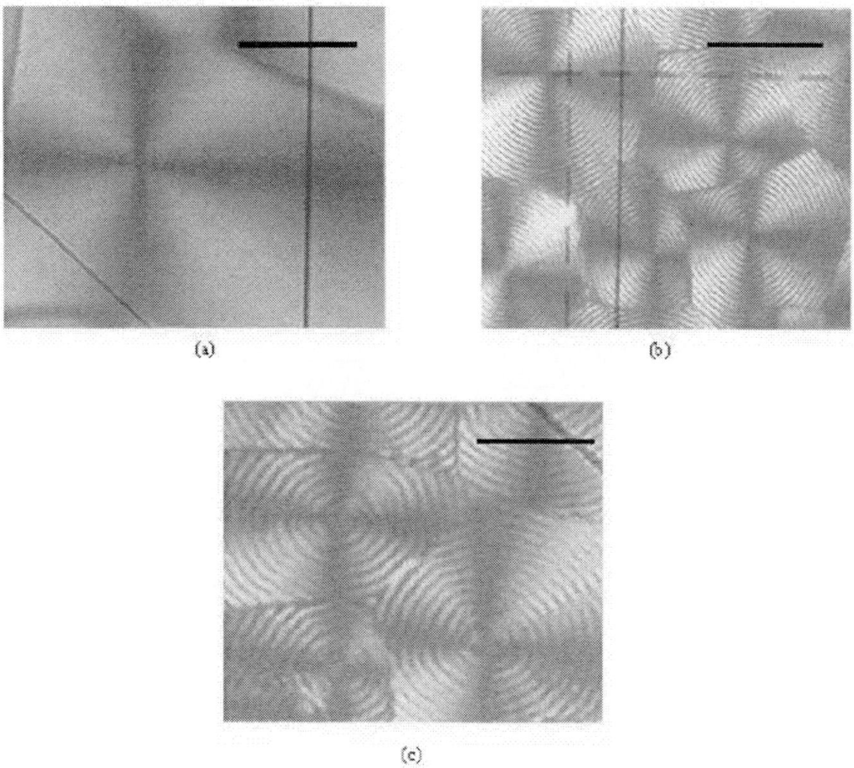

Figure 2-13. Magnified spherulite morphology of the PHB-HHx, PHB-HHx/MPEGblends and MPEG at crystallization 40°C; (a) PHB-HHx (scale bar: 25 micrometer), (b) PHB-HHx/MPEG (3wt%) blend (scale bar: 25 micrometer) and (c) PHB-HHx/MPEG (7wt%) blend (scale bar: 13 micrometer).

2-3-2-2. Isothermal Crystallization Behavior

Figures 2-14(a) and (b) plot the relative crystallinity versus time for pure PHB-HHx and PHB-HHx/MPEG blends at 70°C and at 40°C isothermal crystallization temperature, where relative crystallinity, X_t, was obtained from the ratio of the area of the exotherm up to time t divided by the total isothermal crystallization exotherm. At 70° C, isotherms show a gradual shift toward a longer crystallization time with the increase in the MPEG content in the blends. In contrast, at 40°C the isotherms of the blends are shifted slightly toward shorter crystallization times with the increasing MPEG content.

The crystallization rates of PHB-HHx/MPEG blends were investigated further by measuring the half time of crystallization, $t_{½}$, and the values are listed in Table 2-2. At T_c = 70°C, the $t_{½}$s of the blends are slower than that of pure PHB-HHx as the MPEG content increases. This proves that melting of MPEG plays a role as a noncrystallizable diluent above its T_m and retards the crystallization rate of PHB-HHx in the blends. There are two causes for this observation: the dilution effect reduces the formation of critical nuclei in front of the growing spherulites and the decrease in undercooling due to the T_m depression [30, 31].

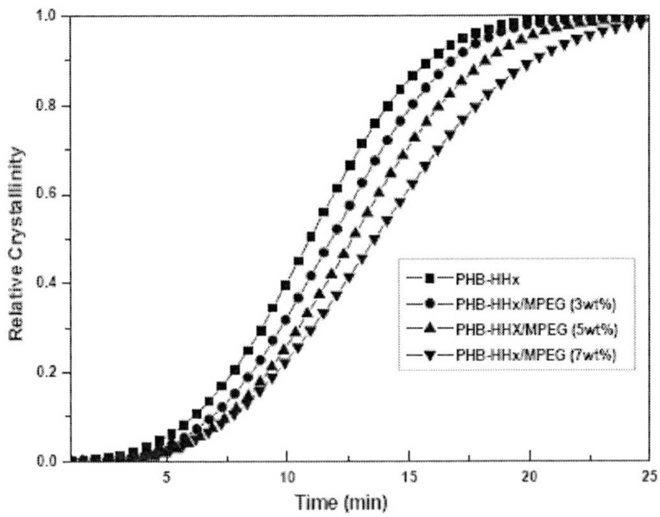

Figure 2-14(a). Relative crystallinity curves for the PHB-HHx/MPEG blends at T_c= 70°C: (■) PHB-HHx, (●) PHB-HHx/MPEG (3wt%) blend, (▲) PHB-HHx/MPEG (5wt%) blend, and (▼) PHB-HHx/MPEG (7wt%) blend.

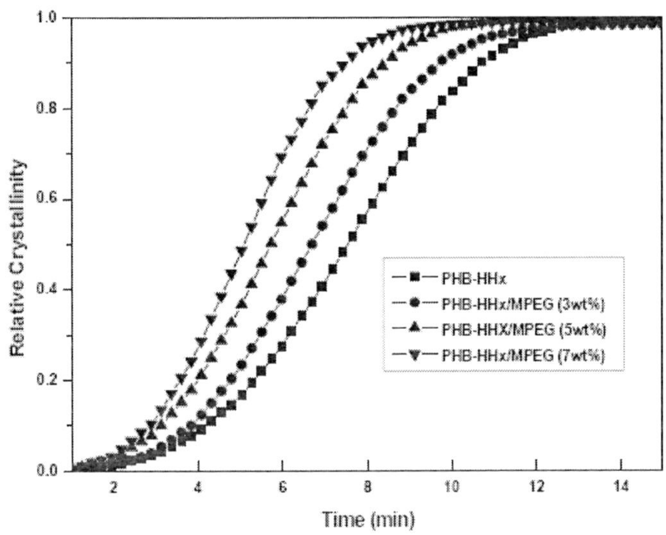

Figure 2-14(b). Relative crystallinity curves for the PHB-HHx/MPEG blends at T_c=40°C: (■) PHB-HHx, (●) PHB-HHx/MPEG (3wt%) blend, (▲) PHB-HHx/MPEG (5wt%) blend, and (▼) PHB-HHx/MPEG (7wt%) blend.

In contrast, at $T_c = 40^{o}C$, the $t_{½}s$ of the blends are faster than that of pure PHB-HHx with the increase in the MPEG content, confirming that the insoluble MPEG particles play a role of a nucleation agent below its T_m, enhancing the crystallization rate of the PHB-HHx in the blends.

Figures 2-15(a) and (b) show the Avrami plots of pure PHB-HHx and PHB-HHx/MPEG blends at $T_c = 70^{o}C$ and $T_c = 40^{o}C$. In addition, the values of n and $\tau_{0.5}$ are shown in Table 2. The Avrami equation is defined as [32, 33].

$$log\,[-\,ln(1 - X_t)] = log\,K + n log t \tag{2-5}$$

Different n values indicate one-, two-, and three-dimensional forms of growth (rod, laminate, or sphere, respectively) [34]. In contrast, the overall crystallization rate, $\tau_{0.5}$, is defined as the reciprocal of the half time of crystallization, and it is usually used to describe the crystallization rate.

Regardless of the blend composition and isothermal crystallization temperature, n is essentially constant, with values of about 3.2 ± 0.2. This implies that MPEG cannot change the three-dimensional spherulitic growth mechanism of the PHB-HHx crystal in the blends. In other words, at $T_c = 70^{o}C$, the $\tau_{0.5}$ values of the blends decrease gradually with increasing MPEG content, while at $T_c = 40^{o}C$, the $\tau_{0.5}$ of the blends increases slightly with the MPEG content. This means that although MPEG does not affect the growth mechanism of PHB-HHx crystal, the MPEG phase, *i.e.*, liquid (melting state) or solid (insoluble state), influences the crystallization rate of the blends strongly.

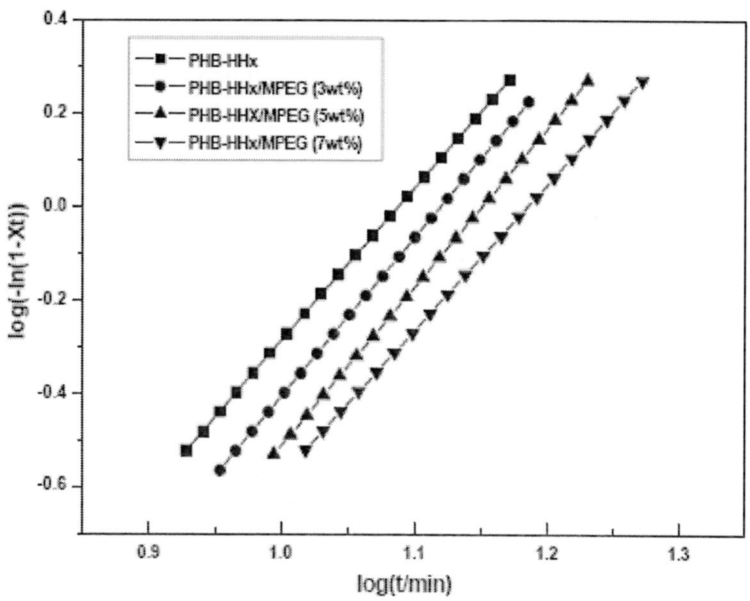

Figure 2-15(a). Avrami plots for PHB-HHx and PHB-HHx/MPEG blends at $T_c = 70\,^{o}C$: (■)PHB-HHx, (●) PHB-HHx/MPEG (3wt%) blend, (▲) PHB-HHx/MPEG (5wt%) blend, and (▼) PHB-HHx/MPEG (7wt%) blend.

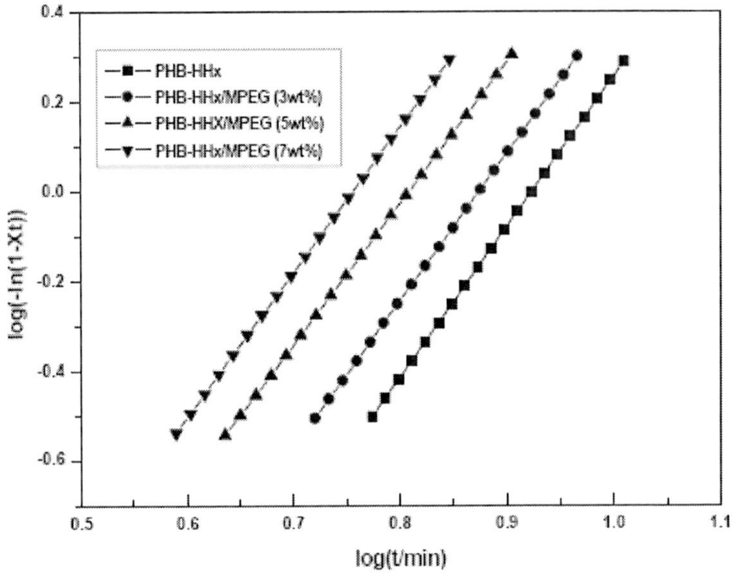

Figure 2-15(b). Avrami plots for PHB-HHx and PHB-HHx/MPEG blends at $T_c = 40°C$:(■)PHB-HHx, (●) PHB-HHx/MPEG (3wt%) blend, (▲) PHB-HHx/MPEG (5wt%) blend, and (▼) PHB-HHx/MPEG (7wt%) blend.

Table 2-2. The $t_{½}$, Avrami exponent and overall crystallization of PHB-HHx and PHB-HHx/MPEG blends at $T_c = 70°C$ and at $T_c = 40°C$

Sample	Crystallization at 40 °C			Crystallization at 70 °C		
	n	$t_{1/2}$(min)	$\tau_{0.5}$ (min $^{-1}$)	n	$t_{1/2}$(min)	$\tau_{0.5}$ (min $^{-1}$)
PHB-HHx	3.3	7.52	0.133	3.3	10.94	0.091
PHB-HHx/ MPEG(3wt%)	3.4	6.68	0.15	3.4	11.83	0.085
PHB-HHx/ MPEG(5wt%)	3.1	5.72	0.175	3.2	12.66	0.079
PHB-HHx/ MPEG(7wt%)	3.1	5.08	0.197	3.2	13.59	0.074

2-4. Summary

1) The Gordon - Taylor equation and FT - IR results strongly suggested the existence of relatively strong hydrogen bonding between PHB-HHx and MPEG; hence, the PHB-HHx/MPEG blends were more miscible than the PHB-HHx/PEG blends. The negative interaction parameter value supported the postulate that PHB-HHx and MPEG form miscible blends with hydrogen bonding.
2) The spherulitic morphology of miscible PHB-HHx/MPEG blends was investigated at two crystallization temperatures (70°C, 40°C). At 70°C, the band spacing of PHB-HHx spherulites increased gradually with the addition of MPEG. At 40°C, the spherulite size of the blends was smaller than that of pure PHB-HHx. However, the

characteristic MPEG morphology could not be observed in the PHB-HHx/MPEG blends.

3) During crystallization at a temperature higher than its T_m, MPEG acted as a noncrystalline diluent that reduced the crystallization rate of the PHB-HHx component in the blends. Conversely, MPEG acted as a nucleating agent that enhanced the crystallization rate of the PHB-HHx component during crystallization at temperatures below the T_m of MPEG.

References

[1] A. Faroane, S. Magazu, G. Maisano, P. Migliardo, E. Tettamanti, and V. Villari, *J. Chem. Phys.*, 110, 1801 (1999).
[2] A. J. Nijenhuis, E. Colstee, D. W. Grijpma, and A. J. Pennings, *Polymer*, 37, 5849 (1996).
[3] C.He, J.Sun, C.Deng, T.Zaho, M.Deng, X.Chen, and X.Jing, *Biomacromolecules*, 5, 2042 (2004).
[4] D.L.Pavia, G.M.Lampman, and G.S.Kriz, *Introduction to Spectroscopy Analysis*; Freedom Academy: USA, 1998.
[5] F. Ravenelle, H. R. Marchessault, *Biomacromolecules*, 3, 1057 (2002).
[6] H. Abe, Y. Doi, H. Aoki, and T. Akehata, *Macromolecules,* 31, 179 (1998).
[7] J. C. Lee, K. Nakajima, T. Ikehara, and T. Nishi, *J. Polym. Sci. Part B: Polym. Phys.*, 35, 2645 (1997).
[8] J. D. Hoffman and J. J. Week, *J. Res. Natl. Bur. Stand.*, 66, 13 (1962).
[9] J. P. Penning and R. S. J. Manley, *Macromolecules*, 29, 84 (1996).
[10] J. W. You, H. J. Chiu, and T. M. Dong, *Polymer*, 44, 4355 (2003).
[11] J. Yi, A. T. Wee, and S. H. Goh, *Macromolecules*, 34, 4662 (2001).
[12] J.M.Harris, S.Zalipsky, *Poly(ethylene glycol);* Chemistry and Biological Application, American Chemical Society, Washington, DC, 1997.
[13] L. Elbert and J. A. Hubbell, *Annu. Rev. Mater. Sci.*, 26, 365 (1996).
[14] L. H. Sperling, *Introduction to Physical Polymer Science*; Wiley, New York, (1986).
[15] M. Avella and E. Martuscelli, *Polymer*, 29, 1731 (1988).
[16] M. Avella, E. Martuscelli, and M. Raimo, *Polymer*, 34, 3234 (1993).
[17] M. Avella, P. Greco, and E. Martuscelli, *Polymer*, 32, 1647 (1991).
[18] M. Avrami, *J. Chem. Phys.*, 7, 1193 (1939).
[19] M. Avrami, *J. Chem. Phys.*, 9, 177 (1941).
[20] M. Gordon and J. S. Taylor, *J. Appl. Chem.*, 2, 493 (1952).
[21] M. L. Huggins, *J. Am. Chem. Soc.*, 64, 1712 (1942).
[22] M. Rozenberg, A. Loewenschuss, and Y. Marcus, *Spectrochiminica Acta Part A*, 54, 1819 (1998).
[23] P. J. Flory, *J. Am. Chem. Soc.*, 87, 1833 (1965).
[24] P. J. Flory, *J. Chem. Phys.*, 10, 51 (1942).
[25] P. J. Flory, *Macromolecules*, 11, 1138 (1978).
[26] P. J. Flory, *Principles of Polymer Chemistry*; Cornell Univ. Press: Ithaca, NY, 1953.
[27] P. X. Xing, L. A. Dong, and Z. Feng, *Macromolecules*, 31, 6898 (1998).
[28] P. Xing, L. Dong, Y. An, and Z. Feng, *Macromolecules*, 30, 2726 (1997).

[29] T. Nishi and T. T. Wang, *Macromolecules*, 8, 909 (1975).
[30] U. W. Gedde, *Polymer Physics*; Chapman & Hall, London, UK, Chapter 4
[31] V. Sriram and D. D. Mark, *Macromolecules,* 35, 5049 (2002).
[32] X. Zhang and G. Steve, *Macromolecules*, 35, 4017 (2002).
[33] Z. Qiu, S. Fujinami, M. Komura, K. Nakajima, T. Ikehara, and T. Nishi, *Polymer*, 45, 4355 (2004).
[34] Z. Shaobing, D. Xianmo, and Y. Hua, *Biomaterials*, 24, 3563 (2003).

3. Effects of Metal Ion-Carbonyl Interaction on Miscibility and Crystallization Kinetic of Poly(3-Hydroxybutyrate-Co-3-Hydroxyhexanoate)/Lightly Ionized PBS

3-1. Introduction

Generally speaking, compared to the non-ionomer/other polymers blend system, ionomer/polymer blends exhibit superior miscibility, consequently resulting in an enhancement in their overall physicochemical properties [1]. This phenomenon may be explained that the ionic group within the ionomer interacts with other polymers that contain polar groups, such as carbonyl, amine, or amide [2]. Therefore, it is necessary to understand the origin of the specific interactions that can be changed with different ionic and polar groups. In addition, this interaction usually alters the crystallization behavior, morphology, and microstructure of the blend system [3-5]. Most importantly, the relationship between the interaction and crystallization behavior must be inspected, as it directly impacts the optimization of the physical properties and processing conditions of the final product.

Compared to other polymer/ionomer blends, little study has been reported for the biodegradable polymer/ionomer blends, such as PHB–HHx and PBS ionomers. No systematic research has been published concerning the origin of the interaction and its impact on the miscibility and crystallization kinetics. In this study, PHB–HHx/PBS ionomer blends were prepared and the effect of intermolecular interaction on their miscibility and crystallization behavior is ascertained. These new PHB–HHx/PBS ionomer blends may provide useful biodegradable composites, since this approach not only reduces crystallinity but also improves mechanical stiffness of PHB-HHx, consequently making the processing of PHB–HHx more straightforward. In addition, this study also gives clarification of the interactions involved in the enhanced miscibility and regulation of the crystallization behavior of the polymer/ionomer blends.

3-2. Experimental

3-2-1. Materials

The bacterial co-polyester, poly(3-hydroxybutyrate-co-3-hydroxyhexanoate) (PHB–HHx) was supplied by the Procter & Gamble Company (Cincinnati, OH, USA). The 3-hydroxyhexanoate content was about 4mol%, the weight-averaged molecular weight (M_w)

and the polydispersity were 827,000 g mol^{-1} and about 2.4. Succinic acid (SA), 1,4-butanediol (BD), and titanium tetrabutoxide were used without further purification. Pure poly(butylene succinate) PBS and PBS ionomer with SDMF content of 1.0 and 3.0mol% were synthesized as in our previous study [6].

3-2-2. Blend Preparations

PHB–HHx/PBS and PHB–HHx/PBS ionomer blends were prepared by direct melt compounding using a twin roll counter-rotating mixer (Thermo Haake, Paramus, NJ, USA). The barrel temperature was set at 170°C and the screw speed at 100 rpm. Total mixing time was 2 min. Before the one-step melt compounding process was undertaken, the two polymers were ground and vacuum-dried. After melt compounding, films were prepared by using a hot press at 170°C with a holding pressure of 6000 psi and holding time of 5 min. The films were quenched in cold water. The blend composition ratios were 0.5, 1.0, and 2.0 wt% for pure PBS or the two PBS ionomers' weights. When more than 2.0wt% PBS ionomer was blended with the PHB-HHx, it was observed that the molecular weight of the blend sample was reduced remarkably compared to the original pure PHB-HHx. Therefore, we focused our effort on the blend of PHB-HHx with only a small amount of PBS ionomers and investigated its miscibility and crystallization behavior.

3-2-3. Measurements

^{23}Na solid – state NMR spectra were acquired at 400M Hz by using a Super conduction FT-NMR spectrometer (Bruker, Avance DSX-400) operating at 105.805MHz. The frequency axis was set using an external NaCl reference. The rotating speed of analysis was 4000sec^{-1}. A pulse delay of 5 sec was applied to obtain fully relaxed spectra.

Fourier transform infrared (FT-IR) spectra were obtained at room temperature on an IFS 88-IR spectrometer (Bruker AXS GmbH, Karlsruhe, Germany). All the samples, were prepared by using a hot press at 170°C with a holding pressure of 6000 psi and holding time of 5 min. The wavenumber range scanned was 4000–400 cm^{-1}. All spectra were recorded at a resolution of 4 cm^{-1}, with 16 scans for each sample.

A Perkin–Elmer DSC7 (Boston, MA, USA) was used to determine the cold crystallization temperature of the blend samples. A sealed aluminum sample pan containing 5 mg of the blend was used in all experiments. The blend samples were heated at 5°C min^{-1} from -20 to 200°C and then maintained for 3 min before cooling to –20°C at a rate of 5°C min^{-1}. The isothermal crystallization of blends was analyzed by heating from -20 to 200°C at 40°C min^{-1} and then holding at 200°C for 3 min to remove any previous thermal history. The samples were then quenched to appropriate crystallization temperatures at 150°C min^{-1}, and the isothermal crystallization peaks were recorded for 60 min. After isothermal crystallization analysis, the fully crystallized samples were quenched to –20°C and a second heating run carried out to 200°C at a rate of 10°Cmin^{-1}.

The dynamic mechanical properties were determined using a dynamic mechanical analyzer (DMA 2980; TA Instruments, New Castle, DE, USA) set in a tension mode at a frequency of 10 Hz under a nitrogen atmosphere. Samples (15 × 5 × 0.5 mm) were cooled to –60°C and then heated gradually to 130°C at a rate of 2°C min^{-1}.

3-3. Results and Discussion

3-3-1. Metal Ion–Carbonyl Interaction

Figure 3-1 shows the carbonyl region of the FT-IR spectra for pure PHB–HHx, PHB-HHx blends with PBS (2wt%), blend with PBS3i (2wt%), and pure PBS3i, respectively. The spectral feature of pure PHB–HHx can be divided into two major peaks at 1740 cm^{-1} and 1723 cm^{-1}, corresponding, respectively, to amorphous and crystalline carbonyl groups [7, 8]. In the case of PBS3i, the characteristic peak of its carbonyl can be assigned at about 1717cm^{-1}. Unlike PHB–HHx/PBS, a new shoulder peak is observed at about 1733 cm^{-1} for PHB–HHx/PBS3i. This result implies that the local environment of the carbonyl group of PHB-HHx may be changed upon blending with the PBS ionomer. The carbonyl group is the most likely site of any interactions with the ionomer, since this functional group can be described as a resonance structure with an interacting site, the carbonyl oxygen. If a specific attractive interaction takes places between PBS ionomer and PHB-HHx, the most likely candidate may be the ion-carbonyl interaction between the carbonyl group of PHB-HHx and sodium counterion of PBS ionomer. The resonance equilibrium of the carbonyl group breaks down when the carbonyl oxygen interacts with the sodium ion. Consequently, the appearance of the shoulder peak may be interpreted as the possible existence of an intermolecular interaction between the sodium ion of the PBS ionomer and the carbonyl group in PHB–HHx.

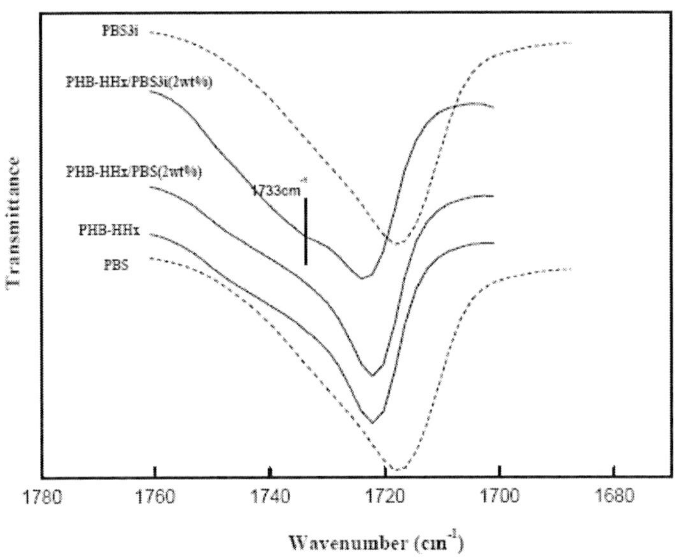

Figure 3-1. Carbonyl region of the FT-IR spectrum for PHB–HHx, its blends with the same amount of PBS and PBS3i (2wt%) and PBS3i ionomer.

We also checked the solid-state ^{23}Na-NMR spectra of pure SDMF with the ionic group and the PHB-HHx/PBS3i (2 wt%) blends (data not shown). In the case of SDMF, an intense peak can be detected at about -37 ppm as a result of the sodium sulfonate ion groups. As for blend sample, this sodium ionic peak was moved downfield (about 13ppm). This phenomenon can be explained such that the electric field gradients surrounding the sodium ion may be changed due to the interaction between the sodium ion and the carbonyl group, consequently generating a complex ionic environment. It has been reported that the metal ion in the ionomer interacts with the carbonyl oxygen of the polyamides in polyamide/sulfonated polystyrene ionomer blends [1, 9]. Based on this background, the intermolecular interaction in the PHB–HHx/PBSi system may most likely be due to a sodium metal ion–carbonyl interaction.

For calculating the fraction of the metal ion-carbonyl interaction peak, carbonyl IR peak was deconvoluted by curve fitting software. Firstly, all the blend samples were normalized, and then spectra were divided by three peaks (amorphous, metal ion-carbonyl, crystalline). Finally, the fraction of metal ion-carbonyl interaction peak was obtained by the following equation;

$$\frac{\text{Area of metal ion-carbonyl interaction carbonyl peak}}{\text{Area of amorphous carbonyl peak + Area of crystalline carbonyl peak+ Area of metal ion-carbonyl peak}}$$

The existence of specific interaction can be confirmed by the appearance of new carbonyl peak upon blending. As the strength of the interaction is increased, the intensity of this carbonyl peak also increases. The relationship between the development of interacted carbonyl peak and the strength of interaction has been discussed [1, 10-12]. The calculated data are summarized in Table 3-1, showing the increase in the ionomer content resulted in the increased fraction of the metal ion-carbonyl interaction peak. This result confirms that as the ionomer content increases, a greater proportion of sodium ion in the PBS ionomer participates in the interaction with the carbonyl groups in PHB–HHx. For the same blend composition, the ionomer with a higher ionic group concentration leads to a larger sodium metal ion–carbonyl interaction. The result demonstrates that a relatively large amount of sodium sulfonate ionic group within PBS3i can take part easily in the metal ion–carbonyl interaction with the carbonyl groups of PHB-HHx. Hence, this interaction may be more intense than PBS1i.

We investigated the degree of ion-carbonyl interaction for the PHB-HHx/PBS and PHB-HHx/PBSi by using the Gordon-Taylor equation [13];

$$T_g = \frac{W_1 T_{g1} + kW_2 T_{g2}}{W_1 + kW_2} \qquad (5\text{-}1)$$

where W_i and T_{gi} are the weight fraction and the glass transition temperature, respectively, of the blend component *i*. The parameter *k* represents the ratio between the thermal expansion coefficient differences above and below T_g for the component. It can be used as a qualitative measure of the strength of the intermolecular interaction between the blend components, i.e., a greater value of *k* indicates a stronger interaction. Using the Gordon–Taylor equation, the *k* value of the three series of blends were calculated; the results are shown in Figure 3-2. The *k* values of the blends were 5.94 (PHB–HHx/PBS3i), 1.97 (PHB–HHx/PBS1i), and 0.88 (PHB–HHx/PBS). This finding strongly suggests that stronger ion-carbonyl interaction between PHB-HHx and PBS ionomer exists with the increase of the ionic group concentration.

Table 3-1. Fraction of sodium metal ion–carboxyl interaction in the PHB–HHx/PBS ionomer blends

Sample / wt%	PHB-HHx/PBS1i	PHB-HHx/PBS3i
0.5	0.08	0.12
1.0	0.10	0.15
2.0	0.14	0.18

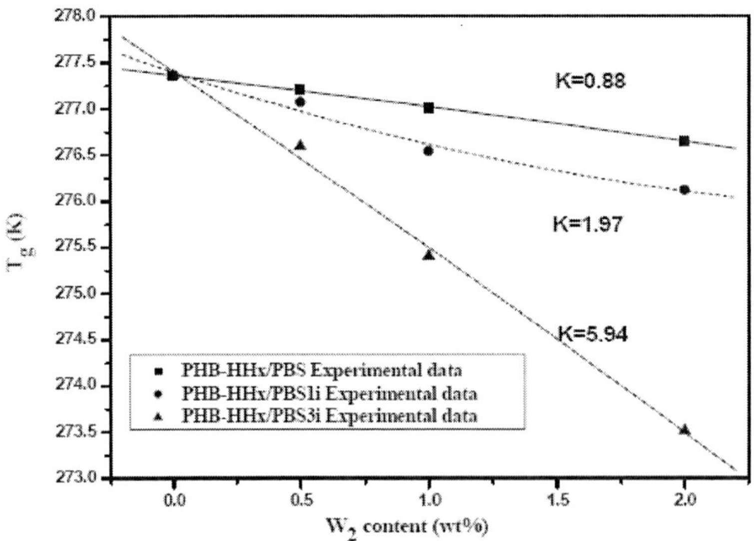

Figure 3-2. The Gordon-Taylor equation for the PHB-HHx/PBS and PHB-HHx/PBS ionomers.

3-3-2. Crystallization Rate

Figure 3-3 shows the DSC cooling curves for the PHB–HHx/PBS3i blends obtained at cooling rates of 5°C min^{-1}. The cold crystallization temperature T_{cc} of pure PHB–HHx and PBS3i were observed near 55.4 and 83. 8°C. Regardless of the ionomer content, the entire blend showed only one T_{cc} peak, pointing to the crystallization of a single phase. With the increase of the ionomer content, this T_{cc} value of blends gradually shifted to a lower temperature, indicating that the overall crystallization time increases with the addition of ionomer. In the previous paper, we investigated the effect of intermolecular interaction on the crystallization behavior of PHB-HHx/silica composites and concluded that their crystallization rates decreased because of the relatively strong intermolecular interaction between components [14]. On the same basis, it seems the presence of PBS ionomer apparently hinders the formation of PHB–HHx nuclei as a result of the relatively strong sodium metal ion–carbonyl interaction, thereby requiring more time to yield PHB–HHx crystals.

The cold crystallization temperatures of PHB–HHx/PBS, PHB–HHx/PBS1i, and PHB–HHx/PBS3i, determined at a cooling rate of 5°C min^{-1} from the melted state, are shown in Table 5-2. For a given blend composition, T_{cc} gradually shifted to a low temperature with an increase in the ionic group concentration. Compared to the T_{cc} of PHB–HHx/PBS (2.0 wt%)

and PHB–HHx/PBS3i (2.0 wt%), the cold crystallization temperature decreased by as much as 7.6°C. This result indicates that the crystallization rate decreases in the order PHB–HHx/PBS > PHB–HHx/PBS1i > PHB–HHx/PBS3i, which concurs with the order of the increasing strength for the sodium metal ion–carbonyl interaction as seen in FT-IR. It is apparent that as the ionic group concentration increases, the segmental mobility of the PHB–HHx molecules becomes more restricted due to the greater degree of sodium metal ion–carbonyl interaction, leading to a slower crystallization rate in the blends.

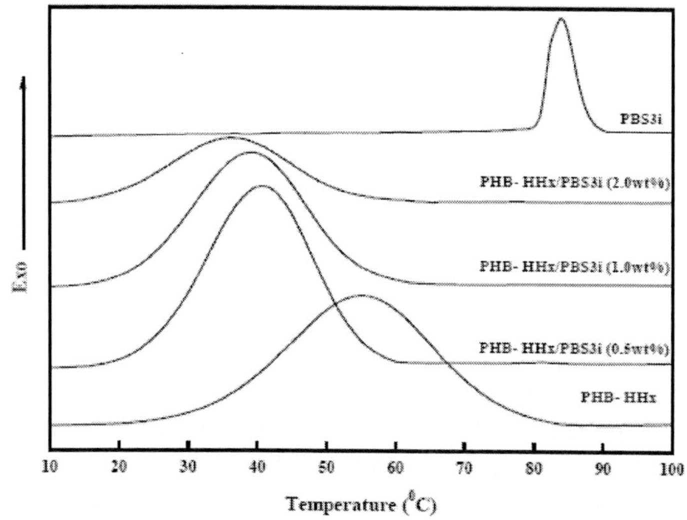

Figure 3-3. DSC cooling curves for the PHB-HHx/PBS3i blends.

Table 3-2. Cold crystallization temperature of PHB-HHx/PBS, PHB-HHx/PBS1i and PHB-HHx/PBS3i blends

Sample (°C) / wt%	PHB-HHx/PBS	PHB-HHx/PBS1i	PHB-HHx/PBS3i
0	55.4	55.4	55.4
0.5	46.1	40.6	40.4
1.0	45.1	39.9	39.2
2.0	43.7	37.8	36.1

The crystallization rate of PHB–HHx/PBS ionomer blends was systemically examined by isothermal crystallization analysis. Generally, the half time of crystallization $t_{1/2}$ was defined as the time required for half of the final crystallinity to develop. Table 3-3 provides $t_{1/2}$ for pure PHB–HHx, PHB–HHx/PBS, and the PHB–HHx/PBS ionomer blends at various crystallization temperatures. For a given crystallization temperature, $t_{1/2}$ showed a larger value with an increase in the ionic group concentration and ionomer content. This result lends strong support to the view that the greater sodium metal ion–carbonyl interaction makes the crystallization of PHB–HHx more difficult.

The crystallization kinetics of PHB–HHx/PBS ionomer blends was analyzed using the Avrami treatment [15, 16]. The Avrami exponent, n, is correlated with the nucleation mechanism and geometry of crystal growth. In turn, $\tau_{0.5}$ can be defined as the reciprocal of the crystallization half-time, commonly used to describe the overall crystallization rate. The values of the rate parameters obtained from the Avrami equation are given in Table 3-3.

Regardless of blend composition and isothermal crystallization temperature, the value of n is not changed in any significant way. This result indicates that the PBS ionomer has little effect on either the nucleation mechanism or the geometry of crystal growth of PHB–HHx. In contrast, for a given crystallization temperature, a progressive decrease in the $\tau_{0.5}$ value occurs not only with an increase in ionic concentration but also the ionomer content. The result confirms that the overall crystallization rate of PHB–HHx/PBS ionomer blend becomes slower in proportion to the strength of the sodium metal ion–carbonyl interaction.

3-3-3. DMA Analyses

Figure 3-4 (a) shows the temperature dependence of the storage modulus for PHB-HHx/PBS1i blend. Pure PHB-HHx has the highest modulus, which reflects its high crystallinity and stiffness. The storage modulus of blend was found to lie between that of pure PHB-HHx and PBS1i. The modulus value became lower with the increase of ionomer content, indicating that the crystallinity and stiffness of PHB-HHx decreases with the addition of ionomer. This phenomenon can be explained by the fact that PBS ionomer hinders the crystallization of PHB-HHx, which is consistent with the DSC data that indicated the crystallization rate decreased in proportion to the ionomer content.

Figure 3-4(b) displays the storage moduli of PHB-HHx/PBS and PHB-HHx/PBSi blends (1 wt%). At the same blend composition, an increase in the ionic group concentration caused a decrease in the storage modulus, confirming that the higher ionic group concentration leads to the dominant reduction of crystallinity in the PHB-HHx. By combining the FT-IR and the DSC results, it can be concluded that stronger sodium metal-carbonyl interaction within PHB-HHx/PBS3i makes the crystallization rate slower, consequently giving rise to the lower crystallinity and stiffness of PHB-HHx. We also analyzed the tan δ graphs for a series of PHB-HHx/PBSi blends, in which the maximum point of the tan δ peak is often associated with the glass transition temperature. All the blends show only a single peak between the peak positions of the pure components, and their values shift to lower temperatures as ionomer content increased. These results indicate that PHB-HHx forms a fully miscible blend over a range of PBS ionomer compositions (data not shown).

3-3-4. Equilibrium Melting Temperature and Interaction Parameter

Figure 3-5(a) and (b) show the DSC melting curves after the isothermal crystallization at 65°C for PHB-HHx/PBS1i and PHB–HHx/PBS3i. Pure PBS1i and PBS3i show the melting temperature at about 113.1°C. Independent of the ionomer content, none of the blends show the T_m peak of the added PBS ionomer, supporting PHB-HHx/PBSi blends are miscible. In addition, all the blend samples exhibited the double melting endotherms due to the melting and recrystallization process of pure PHB–HHx [17, 18]. On the other hand, as the ionomer content increases, melting points of both blends gradually shift to lower temperatures, implying that the PBS ionomer causes a depression in T_m. Taken into the morphological consideration, T_m depression may be attributed to the reducing crystal size or crystal defects induced in the crystalline lattice. Therefore, this result suggests that the inclusion of PBS ionomer causes less perfect PHB–HHx crystals, leading to the conclusion that an irregular molecular chain is present in PHB–HHx.

By using the DSC melting curves, the equilibrium melting temperature T^0_m, defined as the melting temperature of lamellar crystals with an infinite thickness, was calculated. In actual experiments, it is not possible to directly determine the T^0_m of polymer lamellar crystals since a chain-folded lamellar crystal usually grows with a finite thickness during the crystallization process. Therefore, to avoid any morphological contribution, T^0_m is usually extrapolated from the experimentally observed melting temperature. In the present case, the classical Hoffman–Weeks equation was used [19].

Table 3-3. $t_{1/2}$ (sec) and values of the rate parameters for pure PHB-HHx, PHB-HHx/PBS, PHB-HHx/PBS ionomer blends at various crystallization temperatures

Sample	T_c(°C)	$t_{0.5}$(sec)	n	$\tau^{0.5}$ (sec^{-1}x10^{-3})	Sample	T_c(°C)	$t_{0.5}$(sec)	n	$\tau^{0.5}$ (sec^{-1}x10^{-3})
	60	105.8	2.5	9.45		60	229.4	3.0	4.36
	65	130.5	2.7	7.66		65	270.6	3.0	3.70
PHB-HHx	70	142.8	2.7	7.00	PHB-HHx/ PBS3i (1.0wt%)	70	328.1	2.6	3.05
	75	198.4	2.7	5.04		75	375.1	2.7	2.67
	80	331.6	2.8	3.02		80	523.5	2.7	1.91
	60	120.5	2.6	8.30		60	144.9	2.6	6.90
	65	165.8	2.7	6.03		65	162.5	2.6	6.15
PHB-HHx/ PBS (0.5wt%)	70	191.7	2.6	5.22	PHB-HHx/ PBS (2.0wt%)	70	215.4	2.7	4.64
	75	236.9	2.6	4.22		75	244.9	2.6	4.08
	80	334.3	2.6	2.99		80	338.9	2.5	2.95
	60	172.8	2.6	5.79		60	242.6	2.9	4.12
	65	202.2	2.7	4.95		65	271.8	2.6	3.68
PHB-HHx /PBS1i (0.5wt%)	70	213.9	2.6	4.68	PHB-HHx/ PBS1i (2.0wt%)	70	313.2	2.8	3.19
	75	278.7	2.7	3.59		75	415.4	2.7	2.41
	80	359.9	2.7	2.78		80	532.1	2.5	1.88
	60	211.8	2.7	4.72		60	268.4	2.7	3.73
	65	247.1	2.8	4.05		65	297.8	2.6	3.36
PHB-HHx/ PBS3i (0.5wt%)	70	282.4	2.7	3.54	PHB-HHx/ PBS3i (2.0wt%)	70	386.8	2.5	2.59
	75	323.5	2.7	3.09		75	426.2	2.6	2.35
	80	464.7	2.7	2.15		80	623.0	2.8	1.61
	60	213.1	2.9	4.69					
	65	260.5	2.8	3.84					
PHB-HHx/ PBS1i (1.0wt%)	70	278.5	2.7	3.59					
	75	317.6	2.8	3.15					
	80	470.1	2.9	2.13					

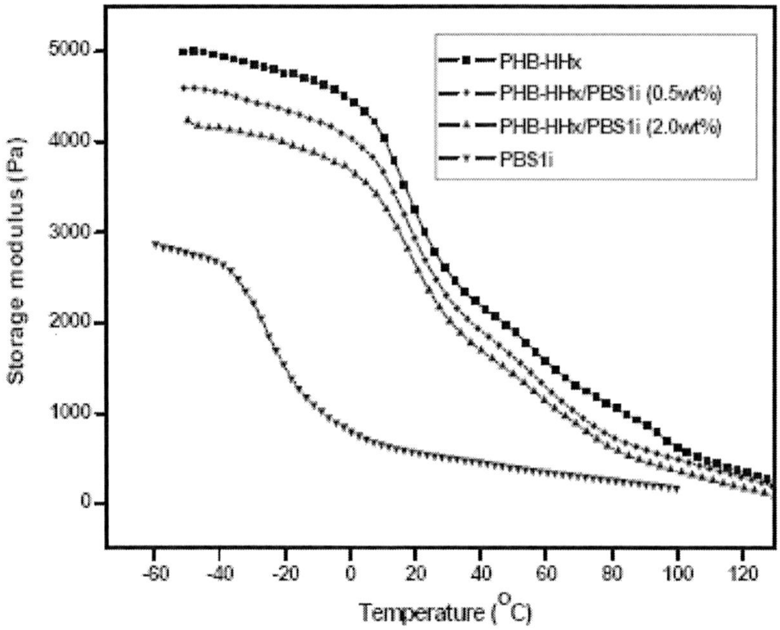

(a)

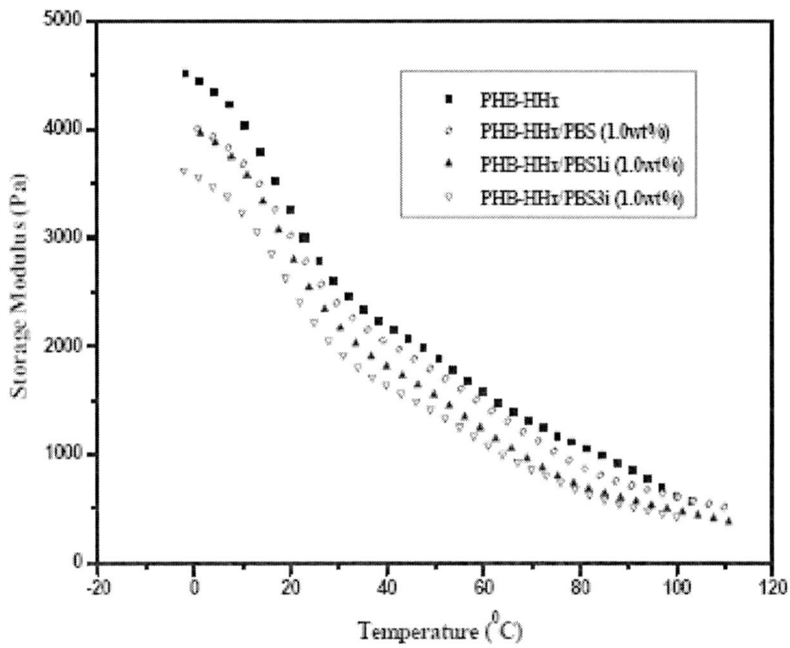

(b)

Figure 3-4. Temperature dependence of the storage modulus for the PHB–HHx/PBS ionomer blends; (a) PHB-HHx/PBS1i blends, (b) PHB-HHx/PBS and PHB-HHx/PBSi (1wt%).

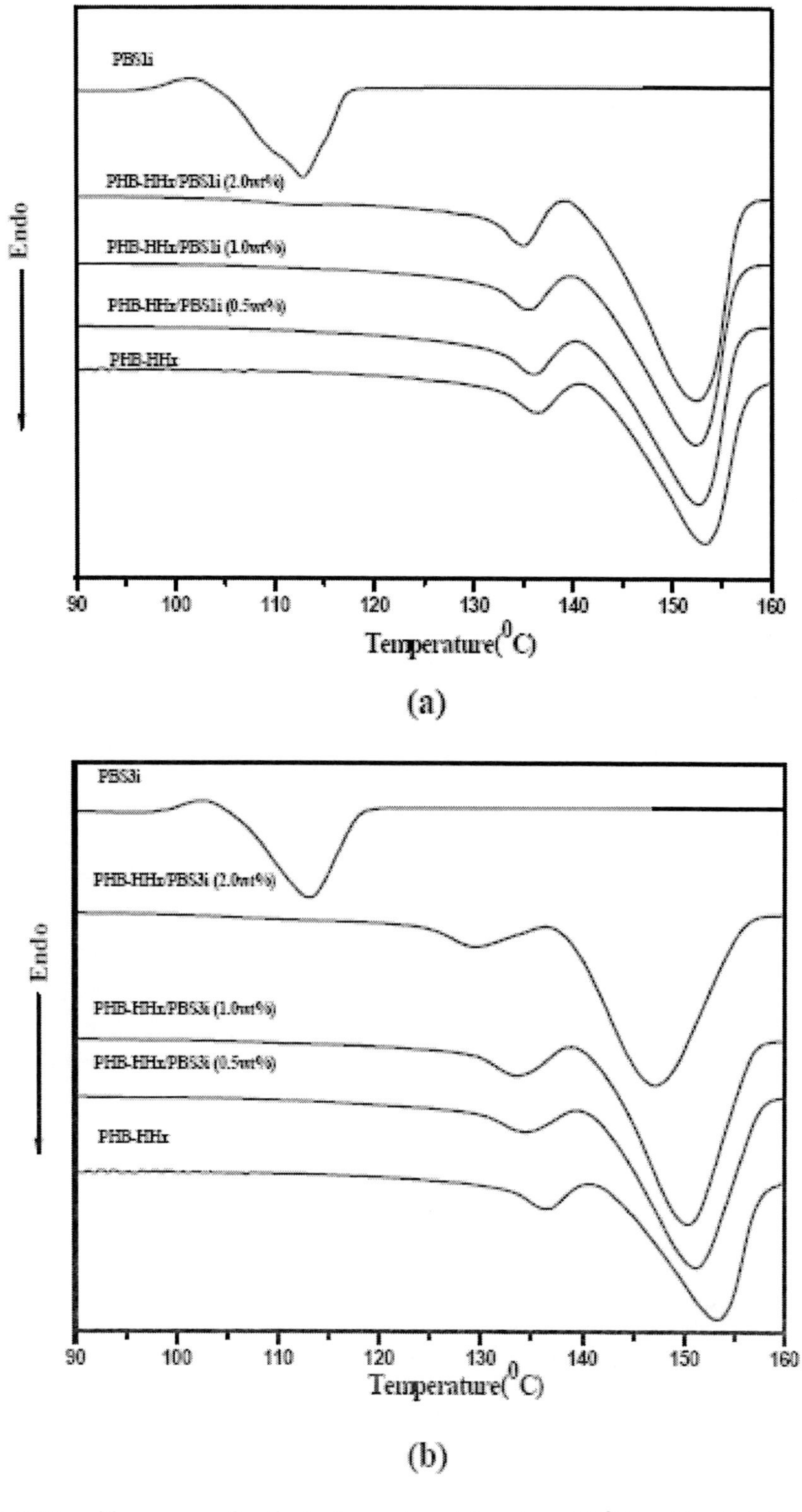

Figure 3-5. DSC melting curves after the isothermal crystallization at 65°C in PHB-HHx/PBS ionomer blends; (a) PHB-HHx/PBS1i, (b) PHB-HHx/PBS3i.

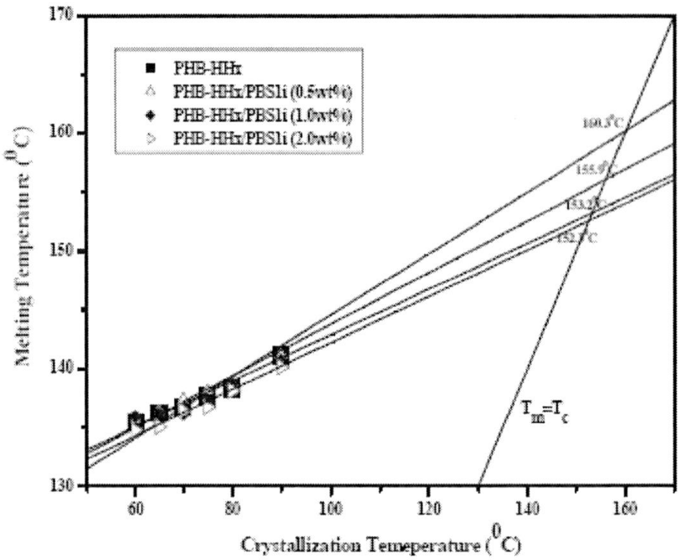

(a)

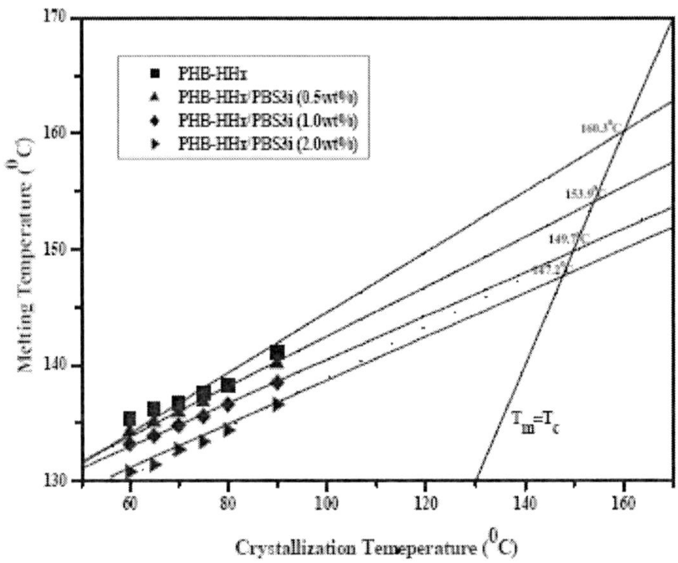

(b)

Figure 3-6. Plots T^0_m as a function of isothermal crystallization temperature of PHB-HHx/PBS ionomer blends; (a) PHB-HHx/PBS1i, (b) PHB-HHx/PBS3i.

Figure 3-6(a) and (b) plot T^0_m as a function of the isothermal crystallization temperature of PHB–HHx/PBS1i and PHB–HHx/PBS3i blends, where T^0_m of pure PHB–HHx occurs at about 160.3°C. Regardless of the ionic group concentration, T^0_m of both blends gradually decreased as the ionomer content increased. Compared to PHB–HHx/PBS1i, T^0_m of PHB–HHx/PBS3i decreased markedly. The maximum extent of this depression was about 13°C for the PHB–HHx/PBS3i (2wt%) blends, where T^0_m was approximately 147.2°C. This result indicates that a greater number of ionic groups within the PBS backbone lead to the large depression of T^0_m.

The miscibility of the PHB–HHx/PBS1i and PHB–HHx/PBS3i blends was systematically investigated by calculating their interaction parameter. In this study, the equilibrium melting temperature (T^0_m) depression was used to determine χ_{AB}. The data in Figure 3-7 were analyzed using the Nishi–Wang equation derived from Flory–Huggins theory [20]:

$$\frac{1}{T^0_m(blend} - \frac{1}{T^0_m(pure)} = -\frac{RV}{\Delta H_f^0 V_B}\left(\chi_{AB}\,\phi_B^2\right) \quad (5\text{-}2)$$

Using equation 3-(2), straight lines with negative slopes of -4.33×10^{-2} and -8.44×10^{-2}, equal to χ_{AB}, can be obtained, as shown in Fig. 5-7 by for both the PHB–HHx/PBS1i and PHB–HHx/PBS3i blends, respectively. This result provides a solid confirmation that PHB–HHx and the PBS ionomer are thermodynamically miscible. However, the χ_{AB} of the PHB–HHx/PBS3i shows a more negative value compared to that of PHB–HHx/PBS1i. Generally speaking, the presence of a relatively stronger intermolecular interaction between blend components can improve their miscibility, not only by creating a sufficient enthalpic interaction but also by making the chain configuration more ordered, thereby leading to a greater negative free energy of mixing [21]. Consequently, it is confirmed that the sodium metal-carbonyl interaction within the PHB–HHx/PBS3i blend, which is much stronger that that of PHB–HHx/PBS1i, results in improved miscibility.

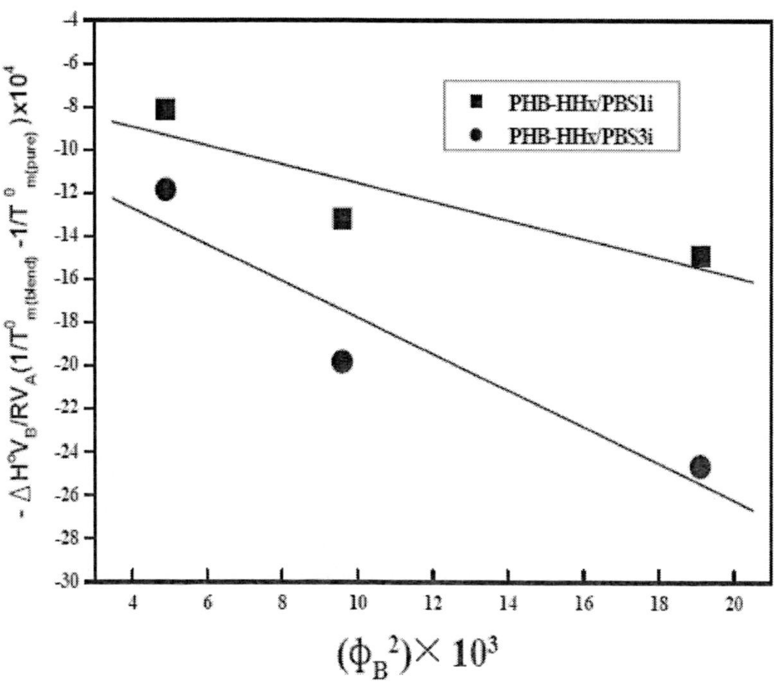

Figure 3-7. Application of Nishi - Wang equation for PHB-HHx/PBS ionomer blends.

3-3-5. Lauritzen–Hoffman secondary nucleation analysis

The Lauritzen–Hoffman equation was used to analyze the free energy of fold surface and the regime transition for the PHB–HHx/PBS ionomer blends [22]:

$$G = G_0\, exp\left(\frac{-U^*}{R\,(T_c - T_\infty)}\right) exp\left(\frac{-K_g}{f\,T_c(T^o{}_m - T_c)}\right) \tag{5-3}$$

the nucleation constant, K_g, can be expressed as [23]

$$K_g = \frac{n b_0 \sigma \sigma_e T^0{}_m}{\Delta H_f\, k_B} \tag{5-4}$$

where k_B is the Boltzmann's constant, ΔH_f is the heat of fusion per unit volume, b_0 is the thickness of molecular layer, σ is the lateral surface free energy, and σ_e is the fold surface free energy. The value of n may be 2 or 4, depending on the regime of crystallization and is determined by the relative rates of formation and spreading of new secondary nuclei at the growth front.

Prior to kinetic analysis, it was important to estimate the values of U^* and T_∞ in equation 3-(3). In many cases, these parameters in the literature have simply been assigned "universal" values of $U^* = 6300J\ mol^{-1}$ and $T_\infty = T_g - 30\ K$, which are appropriate in describing many polymers [22]. These values were adopted in our analysis. At the same time, the overall crystallization rate, $\tau_{0.5}$ was taken to represent the rate of crystallization, G, shown as [24]

$$G = \tau_{0.5} = (t_{0.5})^{-1}. \tag{3-5}$$

Combining equations (3) and (5) and rearranging

$$ln(t_{0.5})^{-1} + \frac{U^*}{R\,(T_c - T_\infty)} = lnG_0 - \frac{K_g}{f\,T_c(T^o{}_m - T_c)} \tag{5-6}$$

and by plotting the left-hand side of equation (6) versus $1/f\ T_c(T^0{}_m - T_c)$, the value of K_g is determined from the slope.

Figure 3-8(a) and (b) show the relationship of $ln(t_{0.5})^{-1} + U^*/[R(T_c - T_\infty)]$ versus $1/f\ T_c(T^0{}_m - T_c)^{-1}$ for both PHB–HHx/PBS ionomer blends. As a rule, with a sufficiently high undercooling, regime 3 may occur during the crystallization and n is then equal to 4 [23, 25]. This parameter was applied to the PHB–HHx/PBS ionomer. Regardless of the ionomer content and ionic group concentration, all the experimental data can be fitted by only one straight line without any point of inflection, thereby confirming that the presence of the PBS ionomer does not influence the regime of the PHB–HHx during crystallization process.

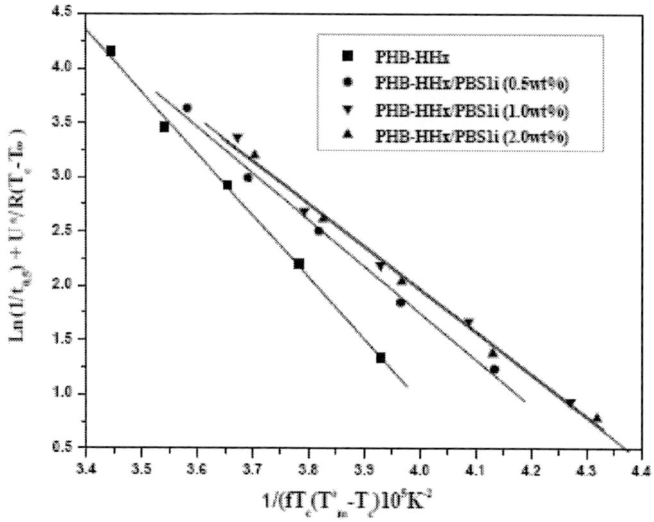

(a)

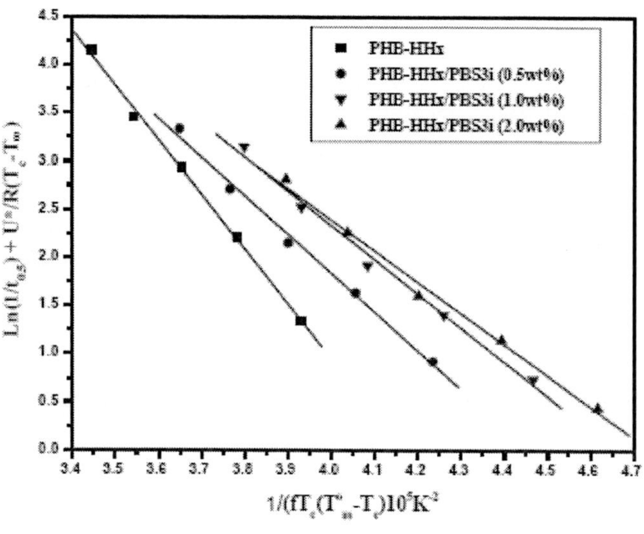

(b)

Figure 3-8. Relationship of $ln(t_{0.5})^{-1} + U^*/[R(T_c - T_\infty)]$ versus $f\,T_c(T^0{}_m - T_c)^{-1}$ for the PHB-HHx/PBS ionomer blends; (a) PHB-HHx/PBS1i, (b)PHB-HHx/PBS3i.

In other words, the value of K_g obtained from Figure 5-8 of 5.68×10^5 (K^2) is that of pure PHB–HHx. The K_gs so derived can then be used to calculate σ_e for the PHB–HHx/PBS ionomer blends using the relation [26]

$$\sigma = \alpha\, (\Delta H_f)\, (a_0 b_0)^{1/2} \qquad (5\text{-}7)$$

Taking $\alpha = 0.25$ for high-melting polyester and using the literature values of $a_0 = 0.66$ nm, $b_0 = 0.58$ nm, and $\Delta H_f = 1.85 \times 10^8$ J m^{-2} [27], the values of σ and σ_e for pure PHB–HHx are 28.6×10^{-3} J m^{-2} and 50.4×10^{-3} J m^{-2}, respectively.

The values of K_g and $\sigma\sigma_e$ for the PHB–HHx/PBS ionomer blends are given in Table 3-4. Both values decrease with an increase in ionomer content and are inversely proportional to the strength of sodium metal ion–carbonyl interaction. This phenomenon may occur because the sodium metal ion–carbonyl interaction strongly affects conformational properties between polymer chains segments. They become arranged within the crystal–liquid interphase, thereby leading to lower entropic change during the crystallization process [28, 29]. In other words, it can be inferred that the melt/crystal interfacial free energy of PHB–HHx decreases as the ionomer content increases.

For a given blend composition, a comparison of the magnitude of K_g and $\sigma\sigma_e$ for PHB–HHx and PHB–HHx/PBSi, showed that K_g and $\sigma\sigma_e$ values decrease with an increase in ionic group concentration. A drop in undercooling, resulting from melting point depression, is usually a major factor dictating a reduction in the crystallization rate of miscible polymer blends [30, 31]. In the previous section, the value of the interaction parameter for PHB–HHx/PBSi blends proved that the equilibrium melting point decreased with the increase in the metal ion–carbonyl interaction strength for a constant crystallization temperature. When two polymers are miscible, the chemical potential of the polymer blend is lowered by the addition of the second component, which leads to the depression of the equilibrium melting temperature. If the miscibility between blend components becomes much higher as a result of specific intermolecular interaction, the decrease in chemical potential may be prominent, leading to the large amount of T^0_m depression [32]. On this basis, it is concluded that a stronger sodium metal ion–carbonyl interaction leads to lower crystallization capability for the PHB–HHx/PBS ionomer blends. Furthermore, this result is consistent with the observations for other miscible blends, such as PHB/PVPh, PHB/PVAc, PHB/PMA, and PHBV/PECH-E, in which intermolecular interactions occur between blend components [25, 31, 33, 34].

From the calculated values of σ_e, the work of chain folding can be determined from

$$q = 2a_0 b_0 \sigma_e \tag{5-8}$$

where q is the work associated with chain folding, and provides a measure of chain stiffness[23, 26]. Smaller q values indicate more flexible polymer chains. The calculated q values are listed in Table 3-4, where the q of pure PHB–HHx is 3855J mol^{-1}. As the ionomer content increases, the q value gradually decreases. In addition, for the same blend composition, the value becomes smaller as the ionic concentration increases. This observation confirms that adding the PBS ionomer leads to a more flexible chain in PHB–HHx. Furthermore, this result agrees well with the storage modulus result from the previous DMA study.

Figure 3-9(a) and (b) show the relationship between the interaction parameter and fold surface free energy for PHB–HHx/PBS1i and PHB–HHx/PBS3i. As the χ_{AB} value becomes increasingly negative, the σ_e value exhibits a lower value in both blend systems. In addition, both variations display linear plots with confidence limits of 0.986 and 0.999, respectively:

It is clear that the greater the miscibility, the slower becomes the crystallization rate for the PHB–HHx/PBS ionomer, as a result of the varying strength of the sodium metal ion–carbonyl interactions.

Table 3-4. Values of K_g, $\sigma\sigma_e$, σ and q for PHB-HHx/PBS ionomer blends

Sample / Wt(%)	PHB-HHx/PBS1i				PHB-HHx/PBS3i			
	K_g (10^5K^2)	$\sigma\sigma_e$ ($10^{-6}J^2/m^4$)	σ_e ($10^3J/m^2$)	q (J/mol)	K_g (10^5K^2)	$\sigma\sigma_e$ ($10^{-6}J^2/m^4$)	σ_e ($10^{-3}J/m^2$)	q (J/mol)
0	5.68	1441	50.4	3855	-	-	-	-
0.5	4.29	1101	38.5	2946	4.02	1035	36.2	2771
1.0	3.92	1012	35.4	2708	3.55	924	32.3	2469
2.0	3.89	1003	35.1	2683	3.23	846	29.6	2264

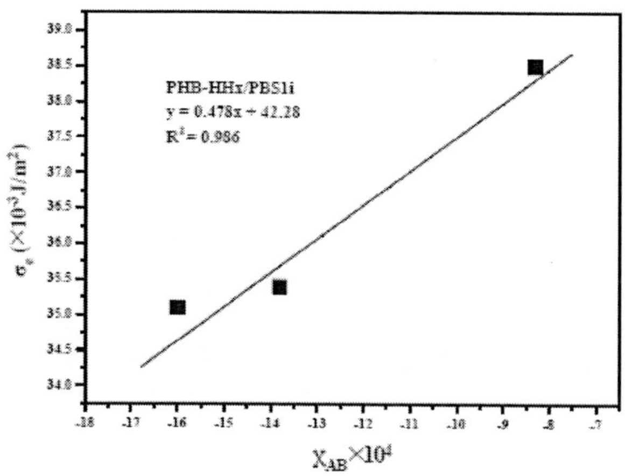

(a)

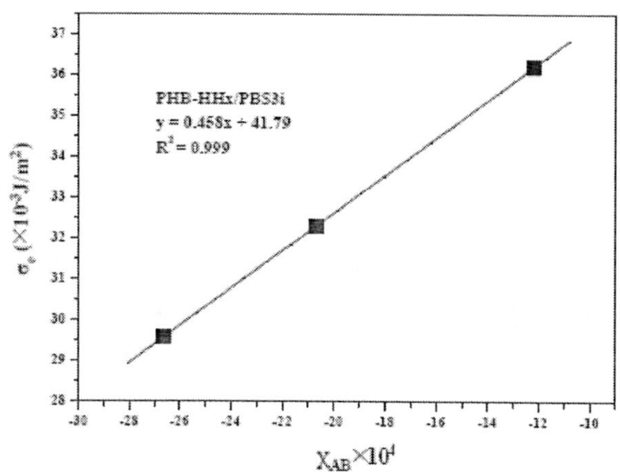

Figure 3-9. Relationship of the interaction parameter and fold surface free energy for the PHB–HHx/PBS ionomer blends; (a) PHB-HHx/PBS1i, (b) PHB-HHx/PBS3i.

3-4. Summary

1) FT-IR analysis demonstrated that sodium metal ion–carbonyl interactions occurred between the carbonyl group in PHB–HHx and the sodium cationic ion in the PBS ionomer; the extent of this interaction became stronger with an increase in the ionic group concentration. As a result, we concluded that a stronger sodium metal ion–carbonyl interaction facilitated miscibility in the PHB–HHx/PBS ionomer blends.
2) The crystallization rate of PHB–HHx decreased with an increase in the strength of the sodium metal ion–carbonyl interaction. Based on the Lauritzen–Hoffman secondary nucleation analysis, it can be demonstrated that the regime of the PHB–HHx/PBS ionomer was not altered during the crystallization process. Furthermore, the nucleation constant and surface free energy decreased with an increase in the ionomer content and ionic group concentration.
3) The greater the miscibility achieved, the slower became the crystallization rate of a PHB–HHx/PBS ionomer, as a result of the different extent of the sodium metal ion–carbonyl interaction.

References

[1] C. K. Chan and I. M. Chu, *Polymer*, 42, 6089 (2001).
[2] E. Kontou and M. Niaounakis, *Polymer*, 47, 1267 (2006).
[3] E. M. Woo and C. P. Chiang, *Polymer*, 45, 8415 (2004).
[4] H. Abe, Y. Doi, H. Aoki, and T. Akehata, *Macromolecules,* 31, 179 (1998).
[5] H. Sato, M. Nakamura, A. Padermshoke, H. Yamaguchi, H. Terauchi, S. Ekgasit, I. Noda, and Y. Ozaki, *Macromolecules*, 37, 3763 (2004).
[6] H. Sato, R. Murakami, A. Padermshoke, F. Hirose, K. Senda, I. Noda, and Y. Ozaki, *Macromolecules*, 37, 7203 (2004).
[7] J. D. Hoffman and J. J. Week, *J. Res. Natl. Bur. Stand.*, 66, 13 (1962).
[8] J. D. Hoffman and Regime-Iii, *Polymer*, 24, 3 (1983).
[9] J. D. Hoffman, G. T. Davis, and J. I. Lauritzen Jr., *In Treatise in Solid State Chemistry*; N. B. Hannay, Ed., Plenum Press: New York, (1976).
[10] J. F. Kadla and S. Kubo, *Macromolecules*, 36, 7803 (2003).
[11] J. Huang and H. Marand, *Macromolecules*, 30, 1069 (1997).
[12] J. I. Lauritzen Jr. and J. D. Hoffman, *J. Appl. Phys.*, 44, 4340 (1973).
[13] J. S. Lim, I. Noda, and S. S. Im, *Polymer*, 48, 2745 (2007).
[14] J. S. Lim, Y. I. Lee, and S. S. Im, *J. Polym. Sci. Part B: Polym. Phys.*, 46, 925, (2008).
[15] K. Nie, W. Pang, Y. Wang, F. Lu, and Q. Zhu, *Mater. Lett.*, 59, 1325 (2005).
[16] K. Vahik and J. P. Darrin, *Macromolecules*, 37, 6480 (2004).
[17] L. L. Zhang, S. H. Goh, S. Y. Lee and G. R. Hee, *Polymer*, 41, 1429 (2000).
[18] M. Avrami, *J. Chem. Phys.*, 7, 1193 (1939).
[19] M. Avrami, *J. Chem. Phys.*, 9, 177 (1941).
[20] M. Dosierem, *Crystallization of Polymer*; Kluwer Academic Publishers: Dordrecht, (1993).
[21] M. Gordon and J. S. Taylor, *J. Appl. Chem.*, 2, 493 (1952).

[22] N. Yoshie, Y. Okie, K. I. Kasuya, Y. Doi, and Y. Inoue, *Biomacromolecules,* 3, 1320 (2002).

[23] P. J. Barham, A. Keller, E. L. Otun and P. A. Holmes, *J. Mater. Sci.*, 19, 2781 (1984).

[24] P. X. Xing, L. A. Dong, and Z. Feng, *Macromolecules*, 31, 6898 (1998).

[25] P. Xing, L. Dong, Y. An, and Z. Feng, *Macromolecules*, 30, 2726 (1997).

[26] R. Xie and R. A. Weiss, *Polymer*, 39, 2851 (1998).

[27] S. W. Kuo and F. C. Chang, *Macromolecules*, 34, 5224 (2001).

[28] S. W. Kuo, S. C. Chan, and F. C. Chang, *Macromolecules*, 36, 6653 (2003).

[29] T. Nishi and T. T. Wang, *Macromolecules*, 8, 909 (1975).

[30] U. W. Gedde, *Polymer Physics*; Chapman & Hall, London, UK, Chapter 4

[31] X. Q. Liu, C. C. Li, G. H. Guan, X. P. Yuan, Y. N. Xiao, and D. Zhang, J. *Polym. Sci., Part B: Pol. Phys.,* 43, 2694 (2005).

[32] X. Y. Lu and R. A. Weiss, *Macromolecules*, 24, 4381 (1991).

[33] Y. An, L. Dong, G. Li, Z. Mo, and Z. Fng, *J. Polym. Sci., Part B: Pol. Phys.*, 38, 1860 (2000).

[34] Y. Feng, A. Schmidt, and R. A. Weiss, *Macromolecules*, 29, 3909 (1996).

In: Polymer Chains
Editors: Régine Penzkofer and Yasue Wu
ISBN: 978-1-62100-422-6

Chapter 4

Inorganic Polyphosphate: Biological Activities and Industrial Applications

Tatyana V. Kulakovskaya, Vladimir M. Vagabov and Igor S. Kulaev
Skryabin Institute of Biochemistry and Physiology of Microorganisms,
Russian Academy of Sciences, Pushchino, Moscow region, Russia

Abstract

High-polymeric inorganic polyphosphates are linear polymers containing a few to several hundred orthophosphate residues linked by energy-rich phosphoanhydride bonds. These inorganic compounds can be easily obtained by fusion of potassium or sodium phosphates at 700-800 °C for several hours followed by rapid cooling. These polymers are present in the cells of all living organisms from bacteria to mammals. Numerous functions of this compound in the cells, including phosphate and energy reservation, sequestration and storage of cations, formation of membrane channels, participation in phosphate transport, involvement in cell envelope formation and function, gene activity control, regulation of enzyme activities, stress response and stationary phase adaptation, are provided by the following properties: polyphosphate is an osmotically neutral form of phosphorus, the vital cell element necessary for the biosynthesis of nucleic acids and phospholipids, phosphorylation of proteins and sugars; the energy of phosphoanhydride bonds between phosphate residues is close to the energy of terminal phosphoester bond in ATP; these polyanions can form complexes both with inorganic cations and with positively charged organic compounds, modifying the properties of these compounds.

Polyphosphates are extensively used for sequestration of iron, manganese, calcium and magnesium in water treatment. Polyphosphates are also applied as fertilizers and food additives. Enhanced biological phosphorus removal (EBPR) based on polyphosphate accumulation by sludge bacteria is now an accepted and low-cost strategy for controlling eutrophication. Polyphosphates are promising as phosphodonors in enzymatic synthesis of biologically active compounds instead of ATP.

Great expectations are associated with their application in different fields of medicine. They display antiseptic, cytoprotective and antiviral activity. In view of the fact that polyphosphates are involved in the regulation of bone tissue metabolism, they are appropriate agents in treatment for bone diseases and in stomatology. Calcium polyphosphate bioceramics is being developed as a bone substitute. Polyphosphate plays a key role in both inflammation and formation of blood clots as an activator of Factor XII. Thus, these polymers are of great interest as a new type of treatment for inflammation, thrombosis, and other blood coagulation diseases.

INTRODUCTION

When Graham in 1833 [1] obtained a vitreous phosphate by fusion of NaH_2PO_4 at 700-800 °C for several hours followed by rapid cooling, nobody assumed that these simple inorganic polymers would be extensively used in various fields of industry and in agriculture. No data on the occurrence and functions of these polymers in living cells were available.

Since that time, the chemistry (specifically biochemistry and molecular biology) of these compounds has evolved, resulting in enhancement of practical application of these phosphorous compounds. The present review will consider the linear, unbranched polymers of orthophoshporic acid. These polymers have been applied in economy due to the low cost of production and a number of useful properties described below. In contrast to cyclic and branched polyphosphates, the linear polymers occur in living organisms from bacteria to humans. Biologically significant functions of these polymers revealed in the recent decade offer the new challenge of their applications in industry, medicine, and agriculture.

Inorganic polyphosphate, polyphosphate (PolyP) (formerly named as hexameta phosphate) has the general formula: $M_{(n+2)}P_nO_{(3n+1)}$. The anions are composed of chains, where each phosphorus atom is linked to its neighbors via two oxygen atoms, thus forming a linear unbranched structure (Figure 1). The degree of polymerization, n, may have values from 2 to 10^6.

The chemical properties and production of polyphosphates are well described in the literature on inorganic chemistry [2-5]. Hence, we will not dwell on these subjects in this review. The emphasis will be laid on application of inorganic polyphosphates in different fields of human economic activity, the new data on the functions of these polymers in living cells broadening the field of application of these polymers, and the prospects of using PolyP in industry, agriculture, environmental protection, and medicine.

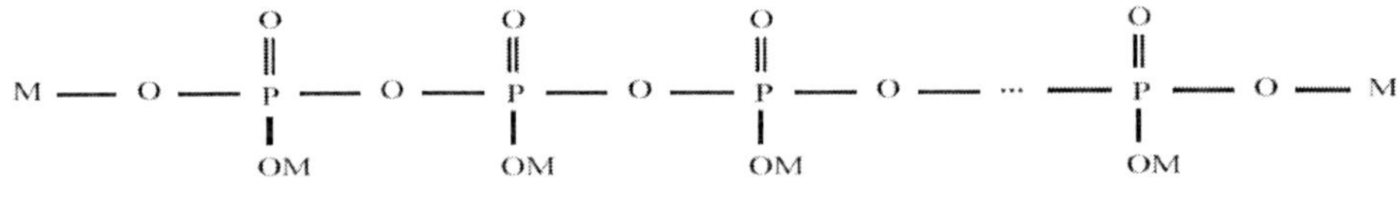

Figure 1. Structure of linear condensed phosphate (PolyP), where M is H^+ or monovalent metal cation.

1. INORGANIC POLYPHOSPHATES IN INDUSTRY AND AGRICULTURE

The basic properties determining the application of PolyP in different fields of economy are as follows:

- They are inorganic condensed phosphorous compounds;
- They are negatively charged polymers that can form complexes with different metal cations;

- They are able to bind and hold considerable amounts of water due to formation of a hydrogen bond between hydrogen atoms in water molecules and between oxygen atoms in polyphosphate molecules;
- They are nontoxic and biodegradable.

At present, PolyP are applied mainly as fertilizers, flame retardants, food additives, and water treatment agents.

1.1. Polyphosphates as a Fertilizer

Ammonium polyphosphate, an inorganic salt of polyphosphoric acid and ammonia, is one of the frequently used phosphoric fertilizers [4, 6]. Commercial ammonium polyphosphates are usually a mixture of ortho- and polyphosphate. The chain length of the polymeric compound is variable and can be greater than 1,000. The short chain compound ($n < 100$) is more sensitive to hydrolysis and less thermally stable than the longer chain one ($n >1000$).

Ammonium polyphosphate is a popular, effective fertilizer. It contains 34 % phosphate and 10 % nitrogen, but has higher water solubility and is more effective than Rock Phosphate and Superphosphates.

Plants absorb phosphorus mainly as orthophosphate. Polyphosphate fertilizers have to break down to the orthophosphate (P_i) form, and this conversion takes time. In contact with water, PolyP slowly hydrolyzes to P_i. Higher temperatures accelerate the hydrolysis. PolyP solutions were very stable when the pH was maintained close to neutral and the temperature was less than 25° [7]. The increasing temperature and decreasing pH have a deleterious effect on the stability of PolyP.

The conversion of PolyP to P_i by hydrolysis occurs in the soil. The time required for hydrolysis to occur varies with soil conditions. In some cases, 50% of the PolyP is hydrolized to P_i within two weeks [6] or longer under cool, dry conditions. PolyP hydrolysis in soil involves the enzymes of soil microorganisms and the enzymes secreted by plant roots, such as nonspecific phosphatases.

Application of PolyP, on the one hand, does not give such a quick response from plants under critical lack of phosphorus like P_i containing fertilizers but, on the other hand, provides for prolonged utilization. Application of some or other type of phosphorous fertilizers must conform to soil acidity, the content of metal cations in the soil, and soil moisture regime.

PolyP fertilizers are popular in agricultural industry due to easy application and yielding benefits in calcareous soils. PolyP is used for making the most types of liquid phosphate fertilizers. They are applied both in soil and in aquaculture for additional fertilizing of micro-alga, which are a feed for farmed fish [8].

1.2. Polyphosphate as a Flame Retardant

Ammonium polyphosphate is used as a non-halogen, high-efficient and nontoxic inorganic flame retardant. The principle of its action is as follows: under the influence of high temperatures, PolyP loses the bound water to produce vapor that swells up flameproof paint.

Non-inflammable PolyP is spread over the surfaces treated with such compositions. As a result, a thick foamy thermoisolation layer is formed, which isolates the material from the air and heating and prevents frame development. Its share may be up to 25% in the total composition of flameproof material. PolyP is used as a flame retardant for fiber materials (paper, wood, fireproof textile), all kinds of polymers, fire-refractory building board, coiled material, epoxy resin and unsaturated resin, cable and rubber, plastic material of electron device, etc. PolyP is a component of swelling up flameproof coating and is added as an antipyrene into flameproof paints, varnishes, impregnations, etc. PolyP is added into resins, plastics and provides low combustibility, inflammability, smoke-forming ability, toxicity of combustion products, and the low flame spread over the surface; it is also used in production of flameproof compounds based on rubbers, latexes, artificial leather, and lubricants.

1.3. Polyphosphates as a Food Additive

Phosphates are normally present in all living things, being an essential component of our diet. PolyP is easily degraded to orthophosphate by the alkaline phosphatase of the intestines and by other enzymes of human organism [9]; hence, they are not xenobiotics.

PolyP are legally permitted additives that are widely used to aid processing or to improve eating quality of many foods [10]. They are listed below.

E452 (Polyphosphates):

E452(i) Sodium polyphosphate (formerly E450c(i))
E452(ii) Potassium polyphosphate (formerly E450c(ii))
E452(iii) Sodium calcium polyphosphate
E452(iv) Calcium polyphosphate (also 544)
E452(v) Ammonium polyphosphate (formerly E450c(i))

All compounds are produced synthetically from the respective carbonates and phosphoric acid. In food industry, PolyP is used for different purposes, for instance, water holding in the product, emulsion stabilization, etc. The main value of PolyP lies in improving the retention of water. Its high buffering capacity, polyanion and sequestering properties, dry improvement, adhesion reduction ability and antibacterial effect are the technological properties, which make it an effective and multifunctional ingredient of foods such as ham, bacon, meat poultry, fish and shellfish. Besides, PolyP in food maintains the optimal pH level. It helps prolong the product lifetime, avoid undesirable change of product color and fat decomposition during storage.

Polyphosphate treatment of fish, meet and poultry before freezing reduces the amount of thaw drip, i.e., the liquid released when frozen fish is thawed. PolyP and pyrophosphate also help stabilize dispersions, emulsions and suspensions, favoring water binding capacity and gel formation. Many retail products such as fish fingers and portions are cut from the frozen blocks of bonefree flesh known as laminated blocks. The preparation of these blocks from boneless, skinless fillets is simplified by PolyP treatment. The PolyP swells some surface protein to give a lubricating layer, which enables the fillets to be packed easily into moulds, without leaving voids that could affect the shape and weight of products cut from the blocks. However, this additive is not an essential ingredient of laminated blocks; quantities of blocks have been manufactured without its addition. The first and universal effect of PolyP treatment is to increase the weight of fish and poultry by retaining water. The weight gain is of no technological benefit and clearly must not be the aim of PolyP treatment.

Most PolyP added to food are broken down to single phosphate units in the stomach after the food has been eaten; indeed, many are converted to single units in the food before it is eaten, for example in chill storage or during cooking. Nevertheless, since both phosphate and PolyP are important participants of metabolic processes, their consumption is to be controlled. Different Internet resources give the norm of daily phosphate consumption: 70 mg/kg of body weight per day.

The methods of PolyP detection in foods are rather labor-consuming, because foodstuff initially contains various phosphorous compounds; hence, different chromatographic and mass-spectrometric methods are used [11-13].

PolyP is nontoxic; nevertheless, it should be taken into account that the abundance of PolyP in food may have some unstudied effect on health. For example, if rats were fed on high phosphorus, especially polyP-rich diet, they got nephrocalcinosis [14].

The influence of PolyP in food on human health needs further investigation, especially in view of diseases associated with phosphate metabolism.

1.4. Polyphosphates in Water Treatment

PolyP is effective in sequestering iron, manganese, and alkaline earth metals such as calcium and magnesium. PolyP cleans or dissolves precipitated mineral scale that already exists in water distribution lines. It inhibits scale formation caused by calcium and magnesium through sequestration and crystal growth modification at a rate 20 times more effective than the pyrophosphates, but pyrophosphates can sequester iron and manganese at a rate 16 times more effective.

In order to sequester iron, manganese and calcium, it requires a 1:1 mole ratio with phosphate residue of PolyP. A dosage of only 2-4 mg/L is required to decrease the crystal growth of calcium carbonate. By modifying the crystalline structure, these compounds will not precipitate into scale and actually stay in solution through repelling and suspension. This crystal growth modification function prevents the formation of mineral scale within water distribution systems. PolyP will also dissolve already deposited mineral scale within the system, thereby increasing the carrying capacity of the water system. It does not remove iron from water but rather stabilizes and disperses it so that the water remains clear and does not produce iron stains.

As a cathodic inhibitor, PolyP prevents electrochemical corrosion and hence can be used in closed water systems, including the heating and cooling systems in power engineering [15]. They are more preferable than chromates as they are nontoxic. Their anticorrosion effect is considered to be based on absorption of calcium ions and formation of colloid particles, which then form a protective film on the cathode surface.

In Russia, because of the cold climate and necessity of heating, it has been proposed to use sodium polyphosphate for water softening in heating boilers and pipelines. The long-preserved chelation makes it possible to introduce them only once per the whole heating season. Additional advantages are their anticorrosion properties and the ability to reduce iron-oxide and copper scale formation on the internal heating surfaces of boilers. For prevention of pipe line corrosion, especially when using soft water, it is advisable to maintain the polyphosphate content at a level of 2-3 kg/m^3 of water or 0.2-0.3 % of flowing water weight and at pH 7-8. Decrease of Ca scale in pipelines with application of polyphosphates preserves the flow capacity of pipes and reliability of boilers [16].

PolyP-containing filters are sold for controlling scale formation in washers. When passing through these filters, water gradually dissolves them and is saturated with a certain amount of sodium polyphosphate that prevents scale formation as carbonate compounds, softens the water, and reduced detergent consumption.

PolyP, primarily sodium tripolyphosphate, is used in detergents for dishwashers and other cleaning products for households, hospitals and commercial or institutional laundries. PolyP getting into outflow and then into water bodies degrade to orthophosphate, among other things, due to microbial enzymes.

1.5. Enhanced Biological Phosphorus Removal

The presence of considerable amounts of phosphate in waste waters due to the runoff of fertilizers, industrial and household discharges is a great ecological problem, known as eutrophication. Under normal conditions, phosphate and nitrogen concentrations are a limiting factor for cyanobacteria growth [17]. The excess of these components typically promotes excessive growth of cyanobateria, which can accumulate a lot of PolyP and hence are resistant to phosphate excess in the medium. As they die and decompose, the high levels of organic matter deplete the water of available oxygen, causing the death of fish, shellfish, and plants. Eutrophication is a natural, slow-aging process for a water body, but human activity greatly speeds up the process.

For eutrophication control, an effective biotechnology has been developed as a tool for phosphorus removal from wastewater using polyphosphate-accumulating microorganisms. The currently available methods for removing P_i from waste water rely primarily on PolyP accumulation in sludge bacteria, which accumulate considerably more phosphate than is required for normal bacterial growth. This process is called enhanced biological phosphorus removal (EBPR) [18]. The cells of sludge bacteria take up the excess phosphate from the water phase. The P_i is stored in bacterial cells in the form of PolyP granules, and thus PolyP plays the key role in EBPR. Besides, some other extracellular microbial polymers associated with cell clusters bind and precipitate phosphate [19].

The process of EBPR is now an accepted and low-cost strategy for controlling eutrophication. Many detailed reviews summarize the data on technology, biochemistry, microbiology and molecular biology of this process [20-30]. EBPR involves the cycling of microbial biomass and influent wastewater through anaerobic and aerobic zones to achieve selection of microorganisms with the high capacity for PolyP accumulation in cells in the aerobic period. In the anaerobic zone of the treatment system, the cells are electron acceptor deficient and carbon rich. It has been proposed that PolyP is degraded to P_i, which is excreted from cells to increase the transmembrane proton gradient. Carbon is then taken up via a proton-symport pump and stored inside the bacterial cell as polyhydroxyalkanoate (PHA). In the subsequent aerobic zone of the treatment system, the environment is electron acceptor rich but carbon deficient. It has therefore been proposed that PHA is degraded and PolyP is synthesized from the ATP generated from PHA metabolism. Since more P_i is taken up during the aerobic phase than is secreted during the anaerobic phase, there is a net removal of P_i from waste water. A simplifying scheme of EBPR is represented in Figure 2.

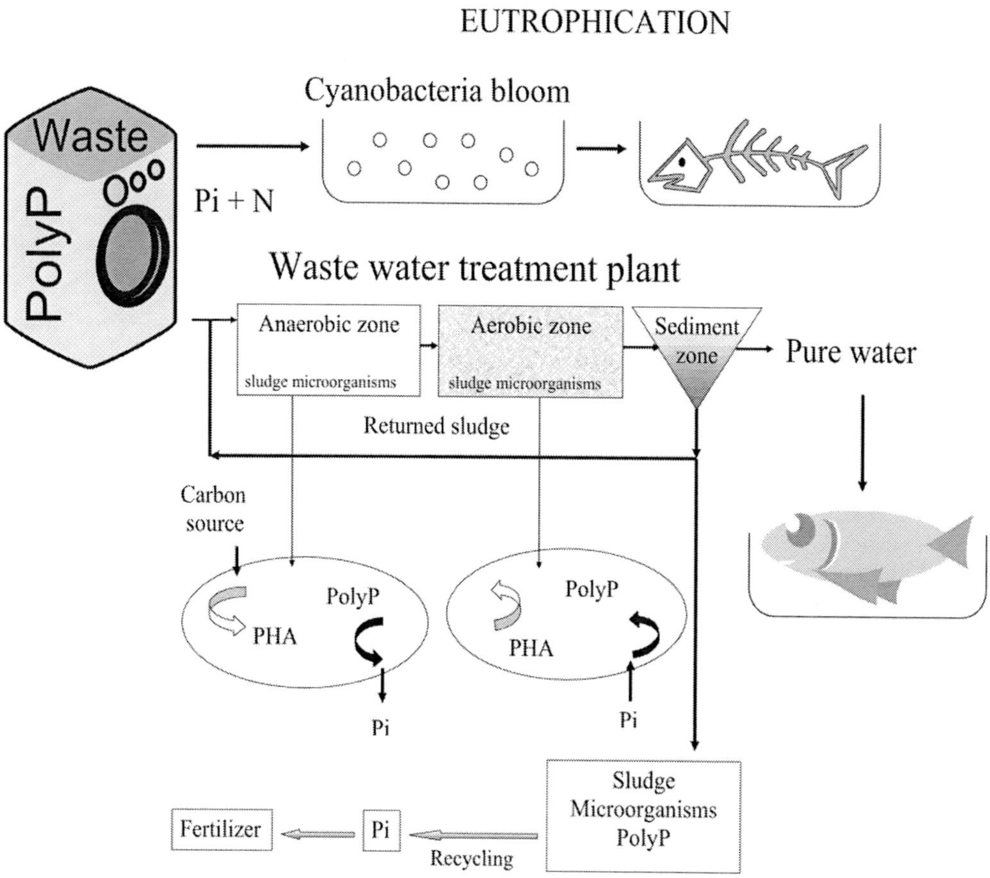

Figure 2. Eutrophication, EBPR and polyphosphate-accumulating microorganisms in activates sludge.

Although the anaerobic/aerobic process in EBPR is established from the engineering point of view, it has not been clearly defined in microbiological terms. According to the modern data, EBPR is implemented by complicate microbial communities [20-30]. The major part of bacteria of these communities is very difficult to isolate and cultivate in pure cultures. The microbial community structure depends on waste composition, organic substrates, and inorganic compounds. The EBPR sludge consists of several phylogenetical microbial groups, and their metabolic cooperation allows them to grow in special conditions such as waste water treatment plants. Microbial communities from technological plants are often analyzed by determination of polyphosphate kinase activity and by occurrence of the *ppk* genes encoding the enzyme [31].

In the EBPR process, one of the problems is utilization of accumulated phosphate of the sludge. It has been discovered that nearly all PolyP is released from activated sludge simply by heating it to 70°C for about 1 h [32]. The chain length of released PolyP ranged from 100 to 200 P_i residues. The addition of $CaCl_2$ precipitated approximately 75% of the total phosphorus without pH adjustment. Hence, in combination with EBPR, the present method has a potential for development of a simple process for recovering phosphorus in reusable form from wastewater [30, 32].

2. The Functions of Polyphosphates and Enzymes of Their Metabolism Offering New Pathways of Their Application

More than one hundred years ago L. Liberman found PolyP in the yeast [33]. It was difficult even to suggest that these polymers, having no ordered secondary and tertiary structure, were the most important regulatory molecules in living cells. Previously it was considered either as a "molecular fossil" and as a phosphorus and energy source providing the survival of microorganisms under extreme conditions. The studies of recent decades have shown that they are present in all living cells (from prokaryotic to human) and perform quite a number of most different functions [34-42].

The following set of physicochemical properties is responsible for the biological activity of these molecules:

- They are an osmotic neutral form of phosphorus, a vitally important element in the cell required for biosynthesis of nucleic acids and phospholipids, phosphorylation of proteins and sugars;
- Phosphate residues in PolyP molecules are linked with each other by phosphoanhydride bonds close by their energy to the energy of the terminal phosphoester bond in ATP molecule, due to which they participate in energy metabolism;
- They are polyanions and can form complexes both with inorganic cations and with positively charged organic compounds, modifying the properties of compounds with which they interact.

PolyPs emerged at the stage of prebiological evolution and now are present in all cell compartments: in membranes as a complex with poly-beta-hydroxybutyrate, in the nuclei, cytoplasm, vacuoles, mitochondria, and cell envelope. Their physicochemical properties, particularly the ability to form a complex with different cell components including proteins and nucleic acids, determine their involvement in various regulatory processes, from phosphorus reservation in microorganisms to the blood coagulation cascade in mammals.

The main functions of PolyP and enzymes for their metabolism in living cells may be as follows:

- They are a phosphorus reserve in microorganisms and participate in phosphorus homeostasis in cells and in the phosphorus cycle of the biosphere;
- PolyP are directly involved in the processes of energy reservation and consumption in cells;
- They participate in detoxification of heavy metals, including Cd2+ and Hg2+;

- They participate in the induction of rpoS, a RNA-polymerase subunit in bacteria, which is responsible for expression of a large group of genes involved in adjustments to the stationary growth phase and many stressful agents;
- They are important for cell motility, biofilm formation, and virulence;
- PolyP and exopolyphosphatases participate in regulation of the levels of the stringent response factor, guanosine 5'-diphosphate 3'-diphosphate (ppGpp), the second messenger in bacteria;
- The channels formed by PolyP and poly-β-hydroxybutyrate with Ca2+ are involved in the transport processes in organisms from different stages of evolution;
- PolyP participate in the regulation of activities of some important enzymes, including glucose transferase in yeasts, Lon-protease responsible for degradation of ribosome proteins under amino acid starvation in bacteria, and some protein kinases in animals;
- PolyP have water-retaining and antibacterial properties;
- PolyP participate in biomineralization processes, including the processes of bone tissue development;
- PolyP are involved in the blood coagulation cascade.

Let us consider in more detail the aspects of PolyP metabolism and function in living cells, which seem to be most important from the viewpoint of the closest future prospects of their practical application.

2.1. The Most Important Enzymes of Polyphosphate Metabolism and Their Functions

Polyphosphate kinase (PPK) (polyphosphate: ADP phosphotransferase, EC 2.7.4.1) catalyzes the reversible transfer of terminal phosphate from ATP to PolyP [37, 43].

$$PolyP_n + ATP \rightleftarrows PolyP_{n+1} + ADP$$

The most of bacterial species possess two PPK forms: PPK1 responsible for polyP synthesis and PPK2, an enzyme that favors the transfer of terminal phosphate residues from PolyP to GDP [42]. The PPK1 and PPK2 encoding genes are very conservative and have been identified in many species including major pathogens [42, 44]. The experiments with mutants in the PPK1 gene demonstrate that polyphosphate kinase and PolyP are essential for stationary phase responses and viability in *Escherichia coli* [45], *Vibrio cholerae* [46], *Pseudomonas aeruginosa* [47], *Helicobacter pylori* [48], Shigella and Salmonella [49]. These mutants demonstrate growth defects in the stationary phase, defective responses to stress and starvation, higher sensitivity to stress factors including heat, antibiotics, antiserum, UV-light and other effectors, impairment in motility, biofilm formation and virulence [42, 46, 47, 49]. Polyphosphate kinase is a colonization factor of *H. pylori*. The mutant with the enhanced level of this enzyme exhibited better capacity for colonizing mice, while the PPK-deficient mutants could not colonize the gastric mucosa of mice [50].

In mammals, no homologues of the *ppk* genes have been found. The importance of polyphosphate kinases in bacterial metabolism and virulence of pathogens is of interest for the development of new inhibitors that may be prospective antibacterial compounds. The importance of PPK1 for bacterial metabolism and virulence of pathogens makes it a target for the discovery of new antibiotics that inhibit this enzyme [42]. PPK2 is involved in alginate synthesis and biofilm formation, which are associated with virulence. Like PPK1, it may also be a target for design of novel antibiotics [42].

Exopolyphosphatase (polyphosphate phosphohydrolase EC 3.6.1.11) is one of the important enzymes involved in PolyP metabolism. The enzyme splits P_i from the end of the PolyP chain:

$$PolyP_n + H_20 \rightarrow PolyP_{n-1} + P_i$$

Two bacterial exopolyphosphatases [51, 52] and two yeast exopolyphosphatases [53-55] have been well studied. The latter are several times more active than and not homologous to the bacterial ones. One of the yeast exopolyphosphatases, PPX1, most effectively hydrolyzes short-chain polyphosphates and tripolyphosphate [53, 54]. The other one, PPN1, poorly hydrolyzes tripolyphosphate but is more active with long-chain polymers [55]. Such characteristic of substrate specificity is of interest for analyzing the content of polyphosphates of different length in the samples. It is important because recently it has been established that the regulatory effect of polyphosphates on some processes depends on chain length.

Due to obtaining of recombinant polyphosphate kinases and exopolyphosphatases it is possible to use these enzymes for analytical purposes such as polyphosphate detection in biological samples and as a component of ATP-regenerating systems.

2.2. PolyP-Poly-β-Hydroxybutyrate/Ca^{2+} Complexes

The discovery by R. Reusch [56-61] of PolyP-poly-β-hydroxybutyrate/Ca^{2+} complexes, which proved the involvement of PolyP in the formation of channels across cell membranes, broadened our view of the function of these compounds. The above channels formed by PolyP and poly-β-hydroxybutyrate with Ca^{2+} are involved in the transport processes in organisms from different stages of evolution. There are ion-selective, voltage-activated channels that facilitate ion selection, solvation, and transport. The complexes of two polymers, both isolated from bacterial plasma membranes and made of synthetic polymers, form voltage-dependent Ca^{2+}-selective channels in the planar lipid bilayers. These channels are selective for divalent over monovalent cations, permeant for Ca^{2+}, Sr^{2+}, and Ba^{2+}, and blocked by transition metal cations in a concentration-dependent manner. The *E. coli* PolyP/PHB complexes were investigated first of all [56]. PolyP/PHB complexes extracted from cell membranes into chloroform and premixed with the phospholipid solution form calcium-selective channels in artificial planar bilayers. The PHB/Ca^{2+}-PolyP complexes were reconstituted. PHB was recovered from *E. coli* and carefully purified, and Ca^{2+} ¯PolyP was prepared of commercial sodium PolyP and calcium chloride [56-61]. The channels formed in planar bilayers by synthetic complexes were similar to those formed by the PolyP/PHB complexes extracted from *E. coli*. The conductances of synthetic and E. coli channels were equivalent. The channels formed by the PolyP/PHB complexes, from *E. coli* or synthetic, show strong selectivity for divalent over monovalent cations. The complexes of PolyP and PHB were found in the membranes of the endoplasmic reticulum and mitochondria of animal cells [57] 1989), which suggests their participation in the processes of transmembrane transfer.

Patent No. 5650330, "Method of using polyhydroxybutyrate and polyphosphate membranes with channels" [62], describes lipid bilayers with incorporated poly-3-hydroxybutyrate/polyphosphate ion channels. The ion channels are preferably purified polyhydroxybutyrate polyphosphate complexes extracted from organisms or prepared of the individual constituents. The bilayers are useful for transporting of biological active molecules into living cells. The prospects of practical application of such complexes are associated with the possibility of injecting target compounds or ions into liposomes for delivering directly to the target cells. Besides, for the reason that the quantity of these complexes increases in the cells competent to transformation by genetic material, they have prospects for application in experimental practice for introduction of genetic material into cells.

2.3. PolyP in the Synthesis and Assay of Biologically Active Compounds

The cheapness of PolyP, which can be easily obtained by chemical synthesis from P_i, and the availability of recombinant high-purified and active polyphosphate kinase are the reason for using these components in assay and synthesis.

The methods for PolyP detection in different biological materials using polyphosphate kinase and polyphosphatases, especially the yeast exopolyphosphatase PPX1, have been developed and used in research practice.

PolyP and polyphosphate kinase can be used for creation of ATP-regenerating systems for research purposes and for the synthesis of biologically active compounds. PolyP and recombinant polyphosphate kinase of *E. coli* have been used for the synthesis of N-acetyllactosamine [63, 64].

Thermostable polyphosphate kinase of *Thermus thermophilus* has been cloned in *E. coli*. Thermal stability optimize ATP regeneration systems and other technologies for enzymatic synthesis of biologically active compounds [65].

Besides, various analytical methods with application of PolyP and PolyP-metabolizing enzymes are proposed. A method of glucose determination by immobilized polyphosphate glucokinase (EC 2.7.1.63, polyphosphate:glucose phosphotransferase) was elaborated [66]. A sensitive method was developed for detecting adenosine monophosphate (AMP) by using polyphosphate (polyP)–AMP phosphotransferase (PPT) and adenylate kinase (ADK) from *Acinetobacter johnsonii* in conjugation with firefly luciferase [67]. This method allowed us to detect AMP in the concentration range of 0.3–300 pmol per assay. Samples taken from various sources susceptible to bacterial contamination were subjected to AMP assay using the present method, because AMP is known as a reliable indicator in hygiene monitoring. The results showed that the present method was able to detect food residues with high sensitivity [67].

The PCR method with the polyphosphate kinase gene fragments is often used for analysis of sludge microbiota for its prospects in EBPR [68].

It may be expected that the potential of PolyP as a phosphodonor in enzymatic synthesis of biologically active compounds is not depleted. Besides, exopolyphosphatases with different substrate specificity may be of interest for analysis of the content of polyP with different chain length in blood and other biological samples and in foods.

2.4. PolyP as a Chelator of Heavy Metal Cations

The complexes of PolyP with common cations (Mg^{2+}, Ca^{2+} and K^{+}) are found in many organisms, but PolyP and the transport of metal-phosphate complexes could be involved in heavy metal tolerance in microorganisms. PolyP sequesters Ni^{2+} in *Staphylococcus aureus* [69]. The cells of cyanobacteria *Anacystis nidulans* with high intracellular PolyP levels show a greater tolerance to Cd^{2+} than those with small PolyP reserves [70]. In protozoa and some algae, cation sequestration is one of the functions of acidocalcisome [39]. It is an electron-dense acidic organelle, which contains pyrophosphate and PolyP bound with Ca^{2+} and other cations. Its membrane possesses a number of pumps and exchangers for the uptake and release of these elements.

The Cd^{2+} tolerance of *E. coli* depends on PolyP metabolism [71, 72]. The PolyP metabolic pathways in E. coli were genetically manipulated to test the effect of PolyP on the tolerance to cadmium. The mutant strain carrying multiple copies of exopolyphosphatase and polyphosphate kinase encoding genes showed the enhanced resistance to Cd^{2+} [71, 72]. The following mechanism of PolyP participation in detoxication of heavy metals has been proposed. PolyP sequesters heavy metals, on the one hand, and the entry of metal cations into cells stimulates exopolyphosphatase activity, which releases P_i from PolyP, on the other hand. The $MeHPO_4^-$ ions are then transported out of the cells [27].

Some data demonstrating the involvement of PolyP in copper resistance in the archae *Sulfolobus acidocaldarius, Sulfolobus metallicus*, and *Sulfolobus solfataricus* were obtained [73].

A plasmid was constructed for mercury bioaccumulation, using the fusion of the well-known mer-operon from Pseudomonas with the *ppk* gene from *Klebsiella aerogenes*. The *E. coli* strain with the above plasmid possessed more polyphosphates and accumulated more Hg^{2+} [74]. It was suggested that PolyP could play a direct role in mercury resistance, probably via chelate formation rather than precipitation of mercury by releasing P_i from PolyP. Based on the results obtained in the present study, the fusion plasmid may serve as a strategy for mercury remediation [74].

A strain of *Acinetobacter johnsonii*, accumulating many PolyP and found in activated sludge, removed La^{3+} from solution via precipitation of cell-bound $LaPO_4$ [75].

Experiments on expression of the genes responsible for the synthesis of PolyP, first of all bacterial polyphosphate kinase, in plant cells are of interest as well. The feasibility of transgenic tobacco engineered to express bacterial polyphosphate kinase for phytoremediation of mercury pollution was evaluated [76]. The transgenic tobacco produced a large amount of PolyP in leaves and showed a relatively high resistance to Hg^{2+} than the wild-type progenitors, probably via chelation formation with PolyP. The transgenic tobacco accumulated significantly more PolyP and mercury than the wild type and may be useful for phytoremediation of mercury pollution of soil [76].

As a whole, the PolyP-mediated accumulation from waste and environment could be a useful strategy for remediation of heavy metal containing pollutants.

2.5. Anticeptic and Antiviral Properties of PolyP

PolyP show antiseptic properties, particularly against pathogenic bacteria and the bacteria inhabiting foods. An example of such action is the mechanism leading to growth inhibition, morphological changes, and lysis of *Bacillus cereus* when challenged by a long-chain PolyP [77]. At a concentration of 0.1% or higher, PolyP had a bactericidal effect, inhibited spore germination and outgrowth, and a higher concentration (1.0%) was even sporocidal. The cells exposed to the action of polyP showed the impairment of septa formation during division. The bacteriostatic effect of PolyP on *Staphylococcus aureus* was observed as well [78].

The addition of Mg^{2+} and Ca^{2+} could almost completely block and reverse the antimicrobial activity of PolyP. Hence, it is assumed that the antibacterial effect of PolyP is associated with the fact that they form complexes with the Mg^{2+} and Ca^{2+} ions. As a result, these cations become inaccessible as co-factors or substrate components of the enzymes.

It was revealed that PolyP possessed antiviral activity by preventing the binding of a virus to a cell [79]. Sodium tetrapolyphosphate and the longer polymers, with average chain lengths of 15, 34, and 91 phosphate residues, significantly inhibited the infection of cells by the human immunodeficiency virus type 1 (HIV-1) *in vitro,* whereas $PolyP_3$ was ineffective. It is supposed that the anti-HIV effect of PolyP is caused by the binding of these compounds to both the host cell surface and the virus, thereby inhibiting attachment of the virus [79].

2.6. PolyP in Bone Tissues

The cells of mammals possess PolyP but in lower amounts than the cells of microorganisms. The role of PolyP in animal cells is still little studied, but some important data indicate their importance for development and regulatory processes. Thus, in spite of the fact that animal cells, probably with the exception of bone tissue cells, do not need PolyP as a phosphate reserve, the functions of these biopolymers are still quite significant. It is confirmed by the presence of these compounds nearly in all tissues and organs and practically in all subcellular fractions: nuclei, mitochondria, plasma membranes and microsomes [40-42]. One of the examples of probable regulatory function of PolyP in animals is the ability of this polymer to enhance the proliferation of human fibroblasts [80], to stimulated mammalian TOR, a kinase involved in the proliferation of mammary cancer cells [81]. PolyP plays an important role in the regulation of Ca^{2+} uptake and membrane transport in mitochondria [82].

All these fragmentary though interesting data demonstrate that PolyP may be very promising compounds for medical application.

The most impressive data are those obtained very recently on the role of PolyP in bone tissue and blood.

The vertebrate skeleton is composed of calcium phosphate mineral known as apatite. The skeleton is continually rebuilt, repaired and resorbed during the growth, normal remodeling, and rehabilitation after traumas and diseases.

PolyP was found in bone tissue nearly ten years ago [83, 84]. Significant amounts of PolyP and exopolyphosphatase activity were detected in human mandibule-derived osteoblast-like cells. The amounts of both soluble and insoluble long-chain PolyP in unstimulated osteoblast-like cells were higher than in human gingival cells, erythrocytes, peripheral blood mononuclear cells, and human blood plasma. The cellular content of PolyP in osteoblast-like cells significantly decreased after combined treatment of cells with the stimulators of osteoblast proliferation and differentation [83, 84]. The authors assumed that PolyP could be involved in the modulation of mineralization process in the bone tissue.

Later on, considerable advances were made in the study of PolyP contribution to bone tissue development. Relatively large amounts of PolyP have been found in normal osteoblasts [85]. The expression of both osteopontin and osteocalcin was induced by polyP (0.1 ~ 1 mM). The levels of alkaline phosphatase and polyphosphatase in these cells increased two and six-fold, respectively. It is postulated that the cells may utilize PolyP as a phosphate source for calcification [85]. Probably, PolyP have a potent role in differentiation of human mesenchymal stem cells into osteoblasts [86]. PolyP increases expression of the genes encoding osteocalcin, osterix, bone sialoprotein, and tissue non-specific alkaline phosphatase, resulting in low-chain-length PolyP, inorganic pyrophosphate, and inorganic phosphate production [87]. Differentiation of mesenchymal stem cells into mature osteoblasts under PolyP treatment was observed in cell cultures and in animal models. Poly(P) induces osteoblastic differentiation and bone mineralization and acts as a resource for mineralization [87]. Joint treatment of mesenchymal stem cells with polyphosphates and basic fibroblast growth factor is also efficient for differentiation [88].

Comparative experiments showed that PolyP- Ca^{2+} salt can efficiently replace β-glycerophosphate in the *in vitro* hydroxyapatite biomineralization [89]. In the presence of PolyP- Ca^{2+} salt, the cells extensively form hydroxyapatite, which remains intimately associated with or covered by the plasma membrane.

The presence of the so-called electron-dense granules in skeletal tissues can be detected under the treatment with organic solvents [41]. In dense granules, a considerable part of phosphate is PolyP, as is shown by specific staining with the DAPI dye. It has been demonstrated that formation of apatite is suppressed in the presence of PolyP in spite of the high local concentrations of calcium and phosphate. Enhanced quantity of PolyP has been found in calcifying cartilage where bone growth occurs [41, 90].

As a whole, modern concepts of the role of PolyP in bone tissue development come to the following [90]. The skeletal mineralization is associated with non-crystalline, calcium and phosphate containing electron-dense granules. These granules originate from mitochondria that have accumulated much calcium and lost their energy functions. The mitochondria take up P_i and condense them into PolyP and also sequester Ca^{2+}. The formed granules containing PolyP and Ca^{2+} may be transported out of the osteoclasts. Probably they can be transported into osteoblasts that build a new bone. The osteoblasts may embed the granules in the new unmineralized bone. Alkaline phosphatase releases P_i from PolyP, increasing local P_i and Ca^{2+} concentrations. This results in apatite mineral formation. In such a way, bone growth or posttraumatic repair and bone tissue renewal occur [90].

This mechanism is still clear only in general. So far, the enzyme responsible for PolyP synthesis in bone tissue cells has not been revealed and the signal molecules determining separate stages of this process are unknown. However, in spite of the above, the experiments with animals are already carried out and show that artificial introduction of PolyP into the bone growth or regeneration zone accelerates these processes. The *in vivo* experiments, when porous rods of calcium PolyP were implanted in the distal femur of rabbit, show that the rods can support bone ingrowth and give no adverse reaction [88]. PolyP/calcium hydroxyapatite composites enhances bone regeneration in animal models [91-93].

The finding that PolyP induces osteoblastic differentiation and bone mineralization and acts as a resource for mineralization is a basis for development of novel polyP-containing drugs for bone disease treatment. Besides, new polyP-containing materials for bone substitution and gels for dentistry and addition into toothpastes are developed. Calcium PolyP fiber has been synthesized [94], and new high-performance calcium polyphosphate bioceramics has been proposed as a bone substitute [95, 96]. Biodegradable PolyP matrix system was developed as a potential delivery vehicle for growth factors. PolyP was believed to be an effective carrier for morphogens, growth factors, or other classes of bioactive molecules [97]. Calcium PolyP fiber was used as a scaffold material for tendon tissue engineering *in vitro* [98]. Development of such materials is continued [99]. We do not issue the challenge of analyzing here the patent literature concerning new bone substituent materials; nevertheless, we will give reference to one of the patents as an example [100].

Based on the PolyP properties such as complexing ability, buffer capacity and antiseptic properties, new dentifrices with PolyP were elaborated. These dentifrices demonstrates anticaries potential and enhanced hard tissue safety [101, 102].

2.7. PolyP in Blood

PolyP was found in human blood more than ten years ago [84]. However, its role had remained unclear until it was shown that platelets contained PolyP-rich granules similar to acidocalcisomes [103]. PolyP of approximately 70-75 phosphate residues was found by ^{31}P NMR and by the urea-polyacrylamide gel electrophoresis of platelet extracts. Dense granules were isolated from platelets by means of metrizamide gradient centrifugation. Dense granules contained large amounts of calcium, potassium, V-ATPase, pyrophosphatase activities, and PolyP [103]. All of these characteristics of platelet dense granules, together with their known acidity and high density (both by weight and by electron microscopy), are similar to those of acidocalcisomes of bacteria and unicellular eukaryotes.

Platelets play a central role in thrombosis, hemostasis, and inflammation. These cells contain high levels of PolyP, so their functions and participation in many platelet functions have been extensively studied in recent years.

PolyP was shown to modulate blood coagulation and fibrinolysis [104]. It was also shown that PolyP is a potent hemostatic regulator accelerating blood clotting by activating the contact pathway and promoting the activation of factor V, which in turn results in abrogation of the function of the natural anticoagulant protein, tissue factor pathway inhibitor. PolyP was also found to delay clot lysis by enhancing a natural antifibrinolytic agent, thrombin-activable fibrinolysis inhibitor. It was supposed that PolyP released from platelets initially promotes clot formation and stability; subsequent degradation of PolyP by blood phosphatases fosters inhibition of clotting and activation of fibrinolysis during wound healing [105]. PolyP enhances fibrin clot structure in a calcium-dependent manner. Fibrin clots formed in the presence of PolyP had up to 3-fold higher turbidity, had higher mass-length ratios, and exhibited thicker fibers in scanning electron micrographs [105]. PolyP significantly shortened the clotting time in plasmas from patients with hemophilia A and B and that of normal plasmas containing a variety of anticoagulant drugs, including unfractionated heparin, enoxaparin (a low molecular weight heparin), argatroban (a direct thrombin inhibitor) and rivaroxaban (a direct FXa inhibitor), and antagonized the anticoagulant effect of these drugs via accelerating FV activation. PolyP also shortened the clotting times of plasmas from warfarin patients. These results suggest that PolyP may have utility in reversing anticoagulation and in treating bleeding episodes in patients with hemophilia [105-106].

PolyP modulates the blood coagulation cascade in 3 stages: it triggers the contact pathway, it accelerates factor V activation, and it enhances fibrin polymerization. PolyP exerts differential effects on blood clotting, depending on polymer length [107]. Very long polymers (≥ 500 phosphate residues) were required for optimal activation of the contact pathway, while shorter polymers (100 residues) were sufficient to accelerate factor V activation and abrogate the anticoagulant function of the tissue factor pathway inhibitor. Optimal enhancement of fibrin clot turbidity by polyphosphate required ≥ 250 residues. Pyrophosphate, which is also secreted by activated platelets, potently blocked polyphosphate-mediated enhancement of fibrin clot structure, suggesting that pyrophosphate is a novel regulator of fibrin function. In conclusion, polyphosphate of the size secreted by platelets is very efficient at accelerating blood clotting reactions but is less efficient at initiating them or at modulating clot structure [106].

PolyP interacts with thrombin's exosite II at a site that partially overlaps with, but is not identical to, the heparin-binding site. PolyP interactions with thrombin may be physiologically relevant [108]. In addition to its role in coagulation, polyP-induced Factor XII activation mediates the release of the inflammatory mediator bradykinin by activating the kallikrein-kinin system. This results in vascular leakage and edema formation in vivo. PolyP released from platelets has a physiological role in the thrombus-forming process, as well as in inflammatory reactions [108]. The interactions of PolyP with fibrin(ogen) and its effect on fibrin structure and fibrinolysis are under study as well [107]. Electrophoretic mobility and binding assays indicate that PolyP interacts with fibrinogen and soluble fibrin. Clots formed in the presence of PolyP exhibited the lower turbidity and permeability indicative of a tighter fibrin network, but these changes were not related to cross-linking or fibrinopeptide release. Microscopy showed a change in fibrin distribution in clots formed with PolyP, with formation of tight aggregates of fibrin fibers interspaced with large pores in contrast to homogenous fiber distribution in control clots. Lysis by tissue plasminogen activator (tPA) and plasminogen or plasmin was delayed in clots formed with PolyP and depended on both the activator and PolyP concentration. PolyP directly influences fibrin architecture and attenuates fibrinolysis through reduced binding of fibrinolytic proteins [108].

As a whole the role of platelet PolyP in the blood coagulation system is as follows [103-112]. Platelets release PolyPof 60-100 phosphate residues that directly bind to and activate factor XII. Platelet PolyP potently initiate fibrin formation via the factor XII-driven intrinsic pathway. PolyP are the endogenous factor XII activator in vivo linking platelet activation (primary hemostasis) and fibrin production (secondary hemostasis). The data identify PolyP as a new class of mediators having a key role in platelet-driven proinflammatory and procoagulant disorders. Targeting PolyP-mediated factor XII activation offers the exciting possibility for safe anticoagulation with minimal therapy-associated bleeding.

The role of PolyP in blood is not quite clear as yet. It is not improbable that these polymers per se are a factor that varies the activity of some enzymes due to their ion-exchange properties and a factor of nonspecific immunity due to antibacterial and antiviral properties. However, even now it may be supposed that these polymers, which can be easily obtained synthetically, may be a remedy in the therapy for some blood disorders. Particularly, it was shown that the addition of PolyP restored defective plasma clotting of Hermansky-Pudlak Syndrome patients, who lacked platelet PolyP [106, 112].

Conclusion

PolyP, due to their unique properties, particularly the low cost of synthesis, nontoxicity and biodegradability, are widely used in various fields of economy (Figure 3). At the same time, their extensive applications draw attention to the problem of environmental pollution with excessive phosphorus. In nature, phosphorous compounds are distributed unevenly. On the one hand, there are phosphorite and apatite fields and, on the other hand, soil often lacks phosphorus for high productivity of plants calling for the addition of fertilizers. The increasing polyphosphate pollution of waste results in the excess of phosphates in the environment, which leads to disturbance of ecological equilibrium.

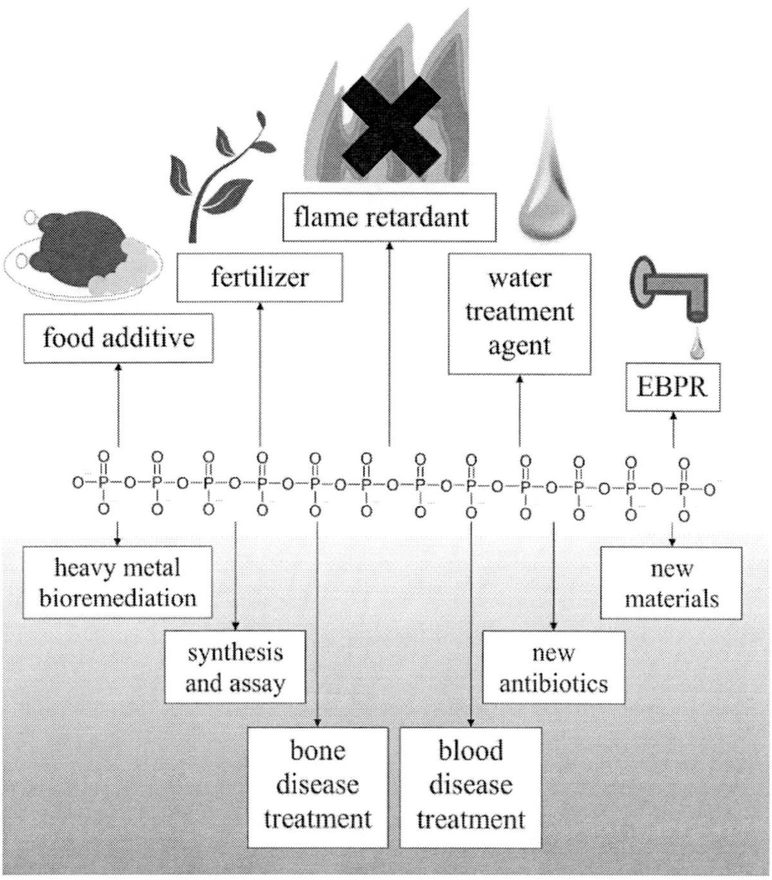

Figure 3. Modern and prospective applications of inorganic polyphosphates.

At the same time, on condition of construction of modern water treatment facilities where phosphate is taken up by sludge microorganisms and removed from effluents, PolyPs will not be hazardous for the environment.

Being a biodegradable material, they are less dangerous than other water softeners and anticorrosion agents.

With regard to foods, in view of the great role of PolyP in cell metabolism and in various physiological processes of human organism such as bone tissue development and blood clotting, it is necessary to control the quantity of these food additives and prevent exceeding the established normative values.

PolyP as regulators of different processes in living organisms are promising candidates in the therapy for bone tissue and blood diseases. These polymers are of interest as a basis for creation of novel bone substitute materials, as carriers for obtaining prolonged action drugs, and as a reagent for the synthesis of biologically active compounds. Taking into consideration the significance of polyphosphate kinase for pathogenicity, one might suggest the possibility of appearance of new antibiotics based on the inhibitors of this enzyme important for survival of pathogenic bacteria but absent in man. The studies in the field of PolyP biochemistry and cell biology have prospects for medicine, environmental protection, and agriculture.

Acknowledgments

The authors were supported by the Russian Foundation for Basic Research (grant 11-04-01009). We thank E. Makeeva for her help with preparation of the manuscript.

References

[1] Abramov, AY; Fraley, C; Diao, CT; Robert Winkfein, R; Colicos, MA; Duchen, MR; French, RJ; Pavlov, E. Targeted polyphosphatase expression alters mitochondrial metabolism and inhibits calcium-dependent cell death. *Proc. Natl. Acad. Sci. USA,* 2007, 13, 18091-18096.

[2] Akiyama, M; Crooke, E; Kornberg, A. An exopolyphosphatase of *Escherichia coli*. The enzyme and its *ppx* gene in a polyphosphate operon. *J. Biol. Chem.*, 1993, 268, 633-639.

[3] Andreeva, NA, Kulakovskaya, TV, Kulaev, IS. High molecular mass exopolyphosphatase from cytosol of the yeast *Saccharomyces cerevisiae* is encoded by the PPN1 gene. *Biochemistry (Moscow),* 2006, 71,975-977.

[4] Andreeva, NA; Okorokov, LA Purification and characterization of highly active and stable polyphosphatase from *Saccharomyces cerevisiae* cell envelope. *Yeast*, 1993, 9, 127- 139.

[5] Ayraud, S; Janvier, B ; Labigne, A ; Ecobichon, C ; Burucoa, C ; Fauchere, JL. Polyphosphate kinase: a new colonization factor of *Helicobacter pylori. FEMS Microbiol. Lett.*, 2005, 243, 45-50.

[6] Blackall, LL; Crocetti, GR; Saunders, AM, Bond, PL. A review and update of the microbiology of enhanced biological phosphorus removal in wastewater treatment plants. *Antonie Van Leeuwenhoek*, 2002, 81, 681-691.

[7] Boswell, CD; Dick, RE; Eccles, H; Macaskie, LE. Phosphate uptake and release by *Acinetobacter johnsonii* in continuous culture and coupling of phosphate release to heavy metal accumulation. *J. Ind. Microbial. Biotechnol.*, 2001, 26, 333-340.

[8] Busscher, HJ; White, DJ; van der Mei, HC; Baig, AA; Kozak KM. Hexametaphosphate effects on tooth surface conditioning film chemistry-in vitro and in vivo studies. *J. Clin. Dent.*, 2002, 13, 38-43.

[9] Caen, J; Wu, Q. Hageman factor, platelets and polyphosphates: early history and recent connection. *J .Thromb. Haemost.*, 2010, 8, 1670-1674.

[10] Cloete, TE; Oosthuizen, DJ. The role of extracellular exopolymers in the removal of phosphorus from activated sludge. *Water Res.*, 2001, 35, 3595-3598.

[11] Corbridge, DEC. Phosphorus. An Outline of its Chemistry, Biochemistry and Technology (Second edition) Elsevier Sci. Pub. Comp. Amsterdam-Oxford-New York, 1980.

[12] Cozzani, R; Barchi, D; Catalli, G; Cinquina, AL. Use of the microwave system for polyphosphate determination in meat products. *Industr. Aliment.,* 1996, 35, 515-516.

[13] Docampo, R; Moreno, SN. The acidocalcisome. *Mol. Biochem. Parasitol.*, 2001, 114, 151- 159.

[14] Fedenko, AN. Application of sodium polyphosphate in water treatment: peculiarities, application and economical effectiveness (In Russian). News of Water Heating (In Russian) , 2002, 11, 29-30. (www.ntsn.ru)

[15] Godd, GA; Bell, SG. Eutrophication in freshwaters. *J. Water. Pollut. Control. Fed.,* 1985, 84, 225-232.

[16] Gonzales, H; Jensen, TE. Nickel sequestering by polyphosphate bodies in *Stafilococcus aureus. Microbios,* 1998, 93, 179-185.

[17] Graham, T. Researches on the arsenates, phosphates and modifications of phosphoric acid. *Phil. Trans. R. Soc.*, 1833, 123, 253-356.

[18] Griffith, EJ. In search of safe mineral fiber. *Chemtech.*,1992, 22, 220-226.

[19] Griffith, EJ; Beeton, A; Spencer, JM; Mitchell DT. *Environmental Phosphorus Handbook.* John Wiley & Sons Ltd, 1973.

[20] Grynpas, MD; Pilliar, RM; Kandel, RA; Renlund, R; Filiaggi, M; Dumitriu, M. Porous calcium polyphosphate scaffolds for bone substitute applications in vivo studies. *Biomaterials*, 2002, 23, 2063-2070.

[21] Hartley, FR. *The chemistry of phosphorus compounds*. John Wiley & Sons Ltd, 1994.

[22] Hirota, R; Kuroda, A; Kato, J; Ohtake, H. Bacterial phosphate metabolism and its application to phosphorus recovery and industrial bioprocesses. *J. Biosci. Bioeng.*, 2010, 109, 423-432.

[23] Iwamoto, S; Motomura, K; Shinoda, Y; Urata, M; Kato, J; Takiguchi, N; Ohtake, H; Hirota, R; Kuroda, A.Use of an Escherichia coli recombinant producing thermostable polyphosphate kinase as an ATP regenerator to produce fructose 1,6-diphosphate, Appl. Environ. Microbiol., 2007, **73**, 5676–5678.

[24] Jen, CM; Shelef, LA. Factors affecting sensitivity of *Staphylococcus aureus* 196E to polyphosphates. *Appl. Environ. Microbiol.*, 1986, 52, 842-846.

[25] Kaufmann, M; Maden, K; Leisser, W; Matera, M; Gude, T. Analysis of polyphosphates in fish and strimps tissues by two different ino chromatography methods: implication of false- negative and –positive findings. *Food Addit. Contamin.*, 2005, 22, 1073-1082.

[26] Kawazoe, Y; Shiba, T; Nakamura, R; Mizuno, A; Tsutsumi, K; Uematsu, T; Yamaoka, M; Shindoh, M; Kohgo, T. Induction of Calcification in MC3T3-E1 Cells by Inorganic Polyphosphate. *J. Dent. Res.*, 2004, 83, 613-618.

[27] Keasling, JD., and Hupf, GA. Genetic manipulationof polyphosphate metabolism affects cadmium tolerance in *Escherihia coli. Appl. Enviroment. Microbiol.* 1996, 62, 743-746.

[28] Keasling, JD; Bertsh, L; Kornberg, A. Guanosine pentaphosphate phosphohydrolase of *Escherichia coli* is a long-chain exopolyphosphatase. *Proc. Natl. Acad. Sci. USA,* 1993, 90, 7029-7033.

[29] Keasling, JD; Van Dien, SJ; Pramanik, J. Engeneering polyphosphate metabolism in *Escherichia coli* - implication for bioremediation of inorganic contaminants. *Biotechnol. Bioeng.,* 1998, 58, 231-239.

[30] Keasling, JD; Van Dien, SJ; Trelstad, P; Renninger, N; McMahon, K. Application of polyphosphate metabolism to environmental and biotechnological problems. *Biochemistry (Moscow)*, 2000, 53, 324-333.

[31] Keyhani, S; Lopez, JL; Clark, DS; Keasling, JD. Intracellular polyphosphate content and cadmium tolerance in *Anacystis nidulans* R2. *Microbios,* 1996, 88, 105-114.

[32] Kim, KS; Rao, NN; Fraley, CD; Kornberg, A. Inorganic polyphosphate is essential for long-term survival and virulence factors in *Shigella* and *Salmonella* spp. *Proc. Natl. Acad. Sci. USA,* 2002, 99, 7675-7680.

[33] Kornberg, A; Rao, NN; Ault-Riché, D. Inorganic Polyphosphate: a molecule with many functions. *Ann. Rev. Biochem.*, 1999, 68, 89-125.

[34] Kornberg, SR. Adenosine triphosphate synthesis from polyphosphate by an enzyme from *Escherichia coli. Biochem. Biophys. Acta*, 1957, 26, 294-300.

[35] Kortstee, GJJ; Appeldoorn, KJ; Bonting, CFC; Van Niel, EWJ; Van Veen, HW. Biology of polyphosphate accumulating bacteria, involved in enchanced biological phosphorus removal. *FEMS Microbiol. Rev.* 1994, 15, 137-153.

[36] Kortstee, GJJ; Appeldoorn, KJ; Bonting, CFC; Van Niel, EWJ; Van Veen, HW. Recent developments in the biochemistry and ecology of enhanced biological phosphorus removal. *Biochemistry (Moscow)*, 2000, 65, 332-341.

[37] Kowalczyk, TH; Szymona, O. Glucose determination using immobilized polyphosphate glucokinase. *Anal. Biochem.*, 1991, 197, 326-332.

[38] Kulaev, IS. *The Biochemistry of Inorganic Polyphosphates.* John Wiley & Sons Ltd Wiley, 1979.

[39] Kulaev, IS; Kulakovskaya, TV. Polyphosphate and phosphate pump *Ann. Rev. Microbiol.*, 2000, 54, 709-735.

[40] Kulaev, IS; Vagabov, VM. Polyphosphate metabolism in microorganisms. *Adv. Microbiol. Physiol.,* 1983, 24, 83-171.

[41] Kulaev, IS; Vagabov, VM; Kulakovskaya, TV. *The Biochemistry of Inorganic Polyphosphates,* John Wiley & Sons Ltd 2004, Wiley.

[42] Kuroda, A; Takiguchi, N; Gotanda, T; Nomura, K; Kato, J; Ikeda, T; Ohtake, H. A simple method to release polyphosphate from activated sludge for phosphorus reuse and recycling. *Biotechnol. Bioeng.*, 2002, 78, 333-338.

[43] Leyhausen, G; Lorenz, B; Zhu, H; Geurtsen, W; Bohnensack, R; Müller, WEG; Schröder HC. Inorganic polyphosphate in human osteoblast-like cells. *J. Bone Miner. Res.,* 1998, 13, 803-812.

[44] Lieberman, L. Uber das Nuclein der Hefe und Kunstliche Darstellung eines Nucleus Eiweiss und Metaphosphatsaure. *Ber. Chem-Ges.* 1888, 21, 598-607.

[45] Lorenz, B; Leuck, J; Kohl, D; Müller, WEG; Schröeder, HC. Anti-HIV-1 activity of inorganic polyphosphates. *J. Acquir. Immun. Defic. Syndr. Hum. Retrovirol.,* 1997, 14, 110-118.

[46] Lorenz, B; Schröder, HC. Mammalian intestinal alkaline phosphatase acts as highly active exopolyphosphatase. *Biochim. Biophys. Acta,* 2001, 1547, 254-261.

[47] Mackman, N; Gruber, A. Platelet polyphosphate: an endogenous activator of coagulation factor XII. *J. Thromb. Haemost.*, 2010, 8, 865-867.

[48] Maier, SK; Scherer, S; Loessner, MJ. Long-chain polyphosphate causes cell lysis and inhibits *Bacillus cereus* septum formation, which is dependent on divalent cations. *Appl. Environ. Microbiol.*, 1999, 65, 3942-3949.

[49] Matsuzaki, H; Masuyama, R; Uehara, M; Nakamura, K; Suzuki, K. Greather effect of dietary potassium tripolyhosphate than potassium dihydrogenphosphate on the nephrocalcinosis and proximal tubular function in female rats from the intake of a high-phosphorus diet. *Biosci. Biotechnol. Biochem.* 2001, 65, 928-934. http://sperchemical.com/html /corrosion_inhibition.html

[50] McBeath, TM; Lombi, E; McLaughlin, MJ; Bünemann, EK. Polyphosphate- fertilizer solution stability with time, temperature, and pH. *J. Plant Nutr. Soil Sci.*, 2007, 170, 387-391.

[51] McGrath, JW; Quinn, JP. Microbial phosphate removal and polyphosphate production from wastewaters *Adv. Appl. Microbiol.*, 2003, 52, 75-100.

[52] McMahon, KD; Dojka, MA; Pace, NR; Jenkins, D; Keasling, JD. Polyphosphate kinase from activated sludge performing enhanced biological phosphorus removal. *Appl. Environ. Microbiol.*, 2002, 68, 4971-4978.

[53] McMAhon, KD; Yilmaz, S; He, S; Gall, DL; Jenkins, D; Keasling, JD. Polyphosphate kinase genes from full-scale activated sludge plants. *Appl. Microbiol. Biotechnol.*, 2007, 77,167-173.

[54] Mino, T. Microbial selection of polyphosphate accumulating bacteria in activated sludge wastewater treatment process for Enhanced Biological Phosphate Removal. *Biochemistry (Moscow)*, 2000, 65, 541-549.

[55] Mino, T; Kawakami, T; Matsuo, T. Location of phosphorus in activated sludge and function of intracellular polyphosphates in biological phosphorus removal process. *Water. Sci. Technol.*, 1985, 17, 93-106.

[56] Morimoto, D; Tomita, T; Kuroda, S; Higuchi, C; Rato, S; Shiba, T; Nakagami, H; Morishita, R; Yoshikawa, H. Inorganic polyphosphate differentiates human mesenchymal stem cells into osteoblastic cells. J. Bone Miner. Metab., 2010, 28, 418-423.

[57] Morita, K; Doi, K; Kubo, T; Takeshita, R; Kato, S; Shiba, T; Akagawa, Y. Enhanced initial bone regeneration with inorganic polyphosphate-adsorbed hydroxyapatite. *Acta Biomater.,* 2010, 6, 2808-2815.

[58] Müller, F; Mutch, NJ; Schenk ,WA; Smith, SA; Esterl, L; Spronk, HM; Schmidbauer, S; Gahl, WA; Morrissey, JH; Renné, T. Platelet polyphosphates are proinflammatory and procoagulant mediators in vivo. *Cell* 2009,139,1143-1156.

[59] Müller, F; Renné, T, Platelet polyphosphates: The nexus of primary and secondary hemostasis. *Scand. J. Clin. Lab. Invest.* 2011, 71, 82-86.

[60] Müller, WE; Wang, X; Diehl-Seifert, B; Kropf, K; Schloßmacher, U; Lieberwirth, I; Glasser, G; Wiens, M, Schröder, HC. Inorganic polymeric phosphate/polyphosphate as an inducer of alkaline phosphatase and a modulator of intracellular Ca(2+) level in osteoblasts (SaOS-2 cells) in vitro. *Acta Biomater*., 2011, Mar 30.

[61] Mutch, NJ; Myles, T; Leung, LL; Morrissey, JH. Polyphosphate binds with high affinity to exosite II of thrombin. *J. Thromb. Haemost*., 2010, 8, 548-555.

[62] Nagata, T; Ishikawa, C; Kivono, M; Pan-Hou, H. Accumulation of mercury in transgenic tobacco expressing bacterial polyphosphate. Biol. Pharm. Bull., 2006, 29, 2350-2353.

[63] Nelson, SR; Wolford, LM; Lagow, RJ; Capano, PJ; Davis, WL. Evaluation of a new high- performance calciumpolyphosphate bioceramics as a bone graft material. *J. Oral Maxilofacial Surgery*, 1993, 51, 1363-1371.

[64] Noguchi, T; Shiba, T. Use of *Escherichia coli* polyphosphate kinase for oligosaccharide synthesis. *Biosci. Biotechnol. Biochem.,* 1998, 62, 1594-1596.

[65] Ogawa, N; Tzeng, CM; Fraley, CD; Kornberg, A. Inorganic polyphosphate in V*ibrio cholerae*: genetic, biochemical, and physiologic features. *J. Bacteriol.* 2000, 182, 6687-6693.

[66] Ohtake, H; Kuroda, A. Molecular analysis of polyphosphate accumulation in Bacteria. *Biochemistry (Moscow)*, 2000, 65, 304-309.

[67] Omelon, S; Georgiou, J; Henneman, ZJ; Wise, LM; Sukhu, B; Hant, T; Wynnyckyj, S; Holmyard, D; Bielecki, R; Grynpas, MD. Control of Vertebrate Skeletal Mineralization by Polyphosphates. *PLoS ONE*, 2009, 4(5), e5634. doi:10.1371/journal.pone.0005634.

[68] Omelon; Grynpas, MD. Relationships between polyphosphate chemistry, biochemistry and apatite biomineralization. *Chem. Rev.,* 2008, 208, 4694-4715.

[69] Pan-Hou, H; Kiyono, M; Omura, H; Omura, T; Endo, G. Polyphosphate produced in recombinant *Escherichia coli* confers mercury resistance. *FEMS Microbiol. Lett.,* 2002, 10325, 159-164.

[70] Parker, R. *Aquaculture Science.* Cengage Learning, 2011.

[71] Patent 5650330 Method of using polyhydroxybutyrate and polyphosphate membranes with channels. http://www.patentstorm.us/patents/5650330.html

[72] Patent 6537589. Calcium phosphate artificial bone as osteoconductive and biodegradable bone substitute material.

[73] http://www.patentstorm.us/patents/6537589/claims.html

[74] Pfarrer, AM; White, DJ; Rapozo-Hilo, M; Featherstone, JD. Anticaries and hard tissue abrasion effects of a "dual-action" whitening, sodium hexametaphosphate tartar control dentifrice. *J. Clin. Dent.*, 2002, 13, 50-54.

[75] Pilliar, RM; Filiaggi, MJ; Wells, JD; Grynpas, MD; Kandel RA. Porous calcium polyphosphate scaffolds for bone substitute applications –in vitro characterization. Biomaterials, 2001, 22, 963-972.

[76] Pilliar, RM; Kandel, RA; Grynpas, MD; Zalzal, P; Hurtig, M. Osteochondral defect repair using a novel tissue engineering approach: sheep model study. *Technol. Health Care*, 2007, 15, 47-56.

[77] Rao, NN; Gómez-García, MR; Kornberg, A. Inorganic polyphosphate: essential for growth and survival. *Ann. Rev. Biochem*., 2009, 78, 605-647.

[78] Rao, NN; Kornberg, A. Inorganic polyphosphate support resistance and survival of stationary-phase *Escherichia coli. J. Bacteriol.*, 1996, 178, 1394-1400.

[79] Rashid, MH; Rumbaugh, K;Passador, L; Davies, DG; Hamood, N; Iglewski, BH; Kornberg, A. Polyphosphate kinase is essential for biofilm development, quorum sensing, and virulence of *Pseudomonas aeruginosa*. *Proc. Natl. Acad. Sci. USA*, 2000, 97, 9636-9641.

[80] Remonsellez, F; Orell, A; Jerez, CA. Copper tolerance of the thermoacidophilic archaeon *Sulfolobus metallicus*: possible role of polyphosphate metabolism. Microbiology, 2006, 152, 59-66.

[81] Renier, ML; Kohn, DH. Development and characterization of a biodegradable polyphosphate. *J. Biomed. Mater. Res.*, 1997, 34, 95-104.

[82] Reusch, RN. Biological complexes of poly-□-hydroxybutyrate. *FEMS Rev.*, 1992, 103, 119- 130.

[83] Reusch, RN. Biological complexes of poly-β-hydroxybutyrate. *FEMS Rev.*, 1992, 103, 119-130.

[84] Reusch, RN. Poly-beta-hydroxybutirate/calcium polyphosphate complexes in eukaryotic membranes. *Roc. Soc. Exper. Biol. Med.*, 1989, 191, 377-381.

[85] Reusch, RN. *Streptomyces lividans* potassium channel contains poly-(R)-3-hydroxybutyrate and inorganic polyphosphate. *Biochemistry,* 1999, 38, 15666-15672.

[86] Reusch, RN. Transmembrane ion transport by Polyphosphate/Poly-(R)-3-hydroxybutyrate complexes. *Biochemistry (Moscow)*, 2000, 65, 280-296.

[87] Reusch, RN; Huang, R; Bramble, LL. Poly-3-hydroxybutyrate/polyphosphate complexes from voltage-activated/Ca^{2+} channels in the plasma membranes of *Escherichia coli. Biophys. J.*, 1995, 69, 754-766.

[88] Reusch, RN; Sadoff, HL. Putative structure and functions of poly-beta-hydroxybutirate/calcium polyphosphate channel in bacterial plasma membranes. *Proc. Natl. Acad. Sci. USA,* 1988, 85, 4176-4180.

[89] Ruiz, FA; Lea, CR; Oldfield, E; Docampo, R. Human platelet dense granules contain polyphosphate and are similar to acidocalcisomes of bacteria and unicellular eukaryotes. *J. Biol. Chem.*, 2004, 279, 44250-44257.

[90] Schröder, HC; Kurz, L; Müller, WEG; Lorenz, B. Polyphosphate in bone *Biochemistry (Moscow)*, 2000, 65, 296-304.

[91] Sekiguchi, Y; Matsunaga, A; Yamamoto, A; Inoue, Y. Analysis of condensed phosphates in food products by ion chromatography with an on-line hydroxide eluent generator. *J. Chromatogr.*, 2000, 881, 639-644.

[92] Seviour, RJ; Blackall, LL. (Eds.) *The Microbiology of Activated Sludge*. Kluwer Academic Publishing, Boston, 1999.

[93] Shanjani, Y; De Croos, A; Pilliar, RM ; Kandel, RA.; Toyserkani, E. Solid Freeform Fabrication and Characterization of Porous Calcium Polyphosphate Structures for Tissue Engineering Purposes. *J. Biomed. Mater. Res. Part B - Application,* 2010, 93, 510-519.

[94] Shiba, T; Nishimura, D; Kawazoe, Y; Onodera, Y; Tsutsumi, T; Nakamura, R; Ohshiro, M. Modulation of mitogenic activity of fibroblast growth factors by inorganic polyphosphate. *J. Biol. Chem.,* 2003, 278, 26788-26792.

[95] Shiba, T; Tsutsumi, K; Ishige, K; Noguchi, T. Inorganic polyphosphate and polyphosphate kinase: their novel biological functions and applications. *Biochemistry(Moscow)*, 2000, 65, 315-324.

[96] Shirai, M; Kakada, J; Shibata, K; Morshed, MG; Matsushita, T; Nakazawa T. Accumulation of polyphosphate granules in *Helicobacter pylori* cells under anaerobic conditions. *J. Med. Microbiol.*, 2000, 49, 513-519.
[97] Smith, J; Hong-Shum, L. (Eds.) *Food additives data book*. Wiley-Blackwell, 2003.
[98] Smith, SA; Choi, SH; Davis-Harrison, R; Huyck, J; Boettcher, J; Reinstra, CM; Morrissey, JH. Polyphosphate exerts differential effects on blood clotting, depending on polymer size. *Blood*, 2010, 116, 4353-4359.
[99] Smith, SA; Morrissey, JH. Polyphosphate as a general procoagulant agent. *J. Thromb. Haemost.*,2008, 6, 1750-1756.
[100] Smith, SA; Mutch, NJ; Baskar, D; Rohloff, P; Docampo, R; Morrissey, JH. Polyphosphate modulates blood coagulation and fibrinolysis. *Proc. Natl. Acad. Sci. USA*, 2006, 103, 903-908.
[101] Sun, ZY; Zhao, L. Feasibility of calcium polyphosphate fiber as scaffold materials for tendon tissue engineering in vitro (in Chinese). *Zhongguo Xiu Fu Chong Jian Wai Ke Za Zhi,* 2002, 16, 426-428.
[102] Tanaka, S; Kuroda, A; Kato, J; Ikeda, T; Takiguchi, N; Ohtake, H.A sensitive method for detecting AMP by utilizing polyphosphate-dependent ATP regeneration and bioluminescence reactions, Biochem. Eng. J., 2001, 9, 193–197.
[103] Toerien, DF; Gerber, A; Lötter, LH; Cloete, TE. Enhanced phosphorus removal systems in activated sludge systems. *Adv. Microb. Ecol.,* 1009, 11, 173-230.
[104] Tzeng, CM; Kornberg, A. Polyphosphate kinase is highly conserved in many bacterial pathogenes. *Mol. Microbiol.*, 1998, 29, 381-382.
[105] Usui, Y; Uematsu, T; Uchihashi, T; Takahashi, M; Takahashi, M; Ishizuka, M; Doto, R; Tanaka, H; Komazaki, Y; Osawa, M; Yamada, K; Yamaoka, M; Furusawa, K. Inorganic polyphosphate induces osteoblastic differentiation. *J. Dent. Res.*, 2010, 89, 504-509.
[106] van der Meijden, PE; Heemskerk, JW. 2010). Polyphosphates: a link between platelet activation, intrinsic coagulation and inflammation? *Expert Rev. Hematol.,* 2010, 3. 269-272.
[107] Van Loosdrecht, MCM; Hooijmans, CM; Brdjanovic, D; Heijnen, JJ. Biological phosphate removal processes. *Appl. Microbiol. Biotechnol.* 1997, 48, 289-296.
[108] Van Wazer, JR. *Phosphorus and its Compounds.* John Wiley & Sons Ltd, 1958.
[109] Vasant Gowariker; Krishnamurthy, VN; Sudha Gowariker; Manik Dhanorkar; Kalyani Paranjape; Borlaug, N. *The Fertilizer Encyclopedia* John Wiley & Sons Ltd, 2009.
[110] Wang, L; Fraley, CD; Faridi, J; Kornberg, A; Roth RA. Inorganic polyphosphate stimulates mammalian TOR, a kinase involved in the proliferation of mammary cancer cells. *Proc. Natl. Acad. Sci. USA,* 2003,100, 11249-54.
[111] Wurst, H; Shiba, T; Kornberg, A. The gene for a major exopolyphosphatase of *Saccharomyces cerevisiae. J. Bacteriol.,* 1995, 177, 898-906.
[112] Yuan, Q; Kubo, T; Doi, K; Morita, K; Takeshita, R; Katoh, S; Shiba, T; Gong, P; Akagawa, Y. Effect of combined application of bFGF and inorganic polyphosphate on bioactivities of osteoblasts and initial bone regeneration. *Acta Biomater.*, 2009, 5, 1716-1724.

In: Polymer Chains
Editors: Régine Penzkofer and Yasue Wu

ISBN: 978-1-62100-422-6

Chapter 5

ALIPHATIC POLYESTER-BASED IONOMERS

Jung Seop Lim
Department of Fiber and Polymer Engineering, Hanyang University,
College of Engineering, Haengdang-dong, Seongdong-gu, Seoul, South Korea

1. OVERVIEW

An ionomer is a polymer that is partially ionized by the metallic ion. Research involving ionomers have received substantial attention in the past decades. As originally proposed by Eisenberg, the interaction between ionic groups leads to the formation of multiplets containing a small number of ion pairs, and ionic clusters, which constitute a second phase and contain many multiplets as well as portions of the hydrocarbon chains. These ionic structures have been shown to act as strong electrostatic cross-links and enhance the mechanical properties of the polymer. The improvement in mechanical properties of the ionomer has been reported to be related to the strength of the electrostatic cross-links, which can be varied by changing the size and type of the counterion present in the ionomer. Among several types of ionomers, poly(ethylene-*co*-methacrylic acid) (EMAA) is the most widely used base polymer for metal cation ionomers in industrial applications. Structural and mechanical properties of ethylene ionomers have been extensively investigated in both solid and melt states.

In ionomers, ionic constituents on the polymer backbone have a dramatic effect on the crystallization kinetics. Therefore, characterization of these effects holds considerable academic interest as well as practical importance. Since most industrial processing is performed under non-isothermal conditions, the crystallization behavior of ionomers under these circumstances is especially important and requires a detailed understanding.

In this chapter, we synthesited novel two types of aliphatic polyester based inomoer and investigated the effects of ionic substitution on physical properties, and crystallization rates of these ionomers. In addition, we also checked that these novel polyester-based ionomers have the infinitive potential as a superior biopolymer as well as industrial application.

2. Influence of Ionic Association on the Non-Isothermal Crystallization Kinetics of Sodium Sulfonate Poly(Butylene Succinate) (PBS) Ionomers

2-1. Introduction

In ionomers, ionic substituents on the polymer backbone have a dramatic effect on crystallization behavior [1, 2]. Therefore, characterization of these effects holds considerable academic interest as well as practical importance. Since most industrial processing is performed under non-isothermal conditions, the crystallization behavior of ionomers under these circumstances is especially important and requires a detailed understanding.

Orler *et al.* [3] studied the effect of alkali metal counterions on the isothermal crystallization kinetics of lightly sulfonated syndiotactic polystyrene ionomers. Their findings show that the crystallization rate was inversely proportional to the ionic radii of the counterions. Page and coworkers [4] investigated the influence of ionic aggregation on the isothermal crystallization rate in poly(butylene terephthalate) (PBT) ionomers and concluded that the crystallization rate of the ionomer decreased as the ionic monomer content increased.

Most studies to date have investigated the isothermal crystallization kinetics of ionomers. The relationship between the presence of ionic substituents and the non-isothermal crystallization kinetics, however, has received little attention despite its importance in industrial applications.

In this study, PBS-based ionomers (PBSis) containing a sodium sulfonate ionic group were synthesized by a two-step polymerization, and we investigated the effect of the ionic group on the non-isothermal crystallization behavior. The presence of ionic substituents in the PBS backbone markedly influences the non-isothermal crystallization kinetics, which has considerable impact on the development of process controls for new PBS ionomers and provides fundamental information on how ionic groups regulate non-isothermal crystallization behavior.

2-2. Experimental

2-2-1. PBS Ionomer Synthesis

Sulfonated dimethyl fumarate (SDMF) containing a sodium sulfonate ionic group was synthesized from dimethyl fumarate (DMF) and sodium hydrogen sulfite ($NaHSO_4$) [5]. PBS and PBS ionomers with 1.0, 3.0, and 5.0 mol% SDMF were prepared by a two-step polycondensation. First, calculated amounts of succinic acid, SDMF, and 1,4-butanediol were introduced into a three-neck flask equipped with a condenser, nitrogen inlet tube, and magnetic stirrer. The molar ratio of the diacid (including SDMF) to diol was 1:1.2. Titanium tetrabutoxide (0.3 wt% of total monomer) was then added to the reactor as an initiator.

The monomer mixture was melted and stirred at 190^0C for 2 h with nitrogen purging to evolve water condensate. The reaction mixture was then transferred to a glass reactor equipped with an overhead stirrer and temperature control device, and the reaction temperature was raised to 215^0C over a period of 30 min with gradual application of reduced pressure (<0.1 Torr) to remove low-molecular weight condensates. To obtain molecular weight comparable to that of the parent PBS, the reaction temperature was maintained for 4 to 7 h depending on the feed composition. The resulting PBS ionomers are denoted as follows: e.g., PBS1i represents PBS ionomer containing 1.0 mol% SDMF. The reaction route and structure of the PBS ionomer are shown in Figure 2-1.

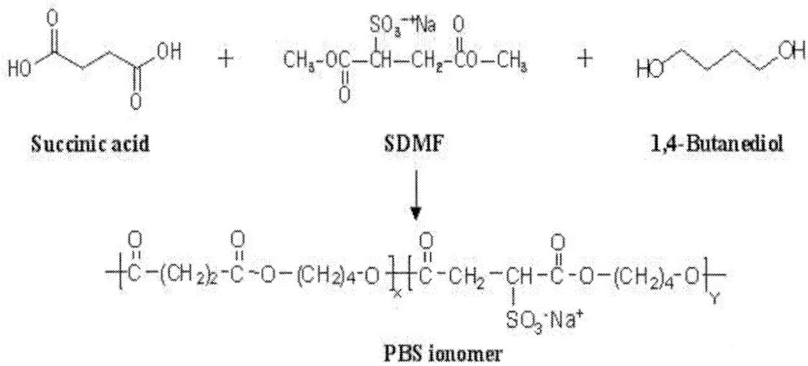

Figure 2-1. Synthetic scheme for PBS ionomers.

2-2-2. Measurement

The structure of the PBS ionomer was confirmed with ^{1}H-NMR spectra using a Fourier transform-NMR (300 MHz) spectrometer (Varian, Salt Lake City, UT, USA) with $CDCl_3$ containing tetramethylsilane (TMS) as an internal standard.

The molecular weight and molecular weight distribution of PBS and its ionomers were measured using a gel permeation chromatograph (GPC, RID-10A, Shimaz) equipped with an HPLC pump, refractive index detector using PS standards, and Styragel HR series (eluent: $CHCl_3$, 35^{0}C, flow rate: 0.3 ml/min). The molecular weights of pure PBS and PBS1i ionomer were 47,000 and 46,000 g/mol, respectively, while the molecular weights of the PBS3i and PBS5i could not be measured because of the adsorption of ionic groups on the column. Therefore, instead of using a GPC, we measured the inherent viscosity of the pure PBS and its ionomers using a Cannon-Ubbelohde microviscometer (Cannon Instrument Co., State College, PA, USA) at 30°C. The samples were dissolved in chloroform to give a concentration of 0.2 g dL^{-1} and their inherent viscosity, η_{Inh} was calculated from:

$$\eta_{rel} = \frac{t}{t_0} = \frac{\rho}{\rho_0} \qquad \eta_{inh} = (\ln \eta_{rel})/C$$

where,

t_0, ρ_0 : flow time and viscosity of the solvent

t, ρ : flow time and viscosity of the solution

C : concentration of the solution

The data are summarized in the Table 2-1. As shown in the Table 2-1, in the case of PBS ionomers, the value for IV was similar to that of PBS regardless of the added ionic group content, implying that the molecular weights of the PBS ionomers are comparable to that of pure PBS.

^{23}Na solid-state NMR spectra were obtained using a Super-conduction FT-NMR spectrometer (Bruker, Avance DSX-400) operating a 7-mm probe with the probe circuit tuned to 106 MHz for ^{23}Na. The frequency axis was set using an external NaCl reference. The speed of rotation for the analysis was $4000s^{-1}$. A pulse delay of 5 s was applied to obtain fully relaxed spectra. The analyses were carried out at 25°C to determine the extent of aggregation.

The rheological properties of PBS and its ionomers were measured using an advanced rheometric expansion system (ARES; Rheometric Scientific, Piscataway, NJ, USA) with a parallel plate geometry. The frequency ranged from 0.05 to 500 rad/s with a 12.5 mm diameter parallel plate with a 1.0-mm gap at 150°C.

To provide identical samples and to ensure consistent DMA measurements, the synthesized polymers were compression molded using a Carver Press at 150°C and 10,000 psi. The molded films were quenched in ice water and dried in a vacuum oven at 40°C for 48 h. The dynamic mechanical properties of PBS and its ionomers were determined using a dynamic mechanical analyzer (DMA 2980; TA Instruments, New Castle, DE, USA) set in tension mode at a frequency of 3 Hz under a nitrogen atmosphere. Samples (15 × 5 × 0.5 mm) were cooled to –50°C and then heated gradually to 100°C at a rate of 2°C min^{-1}.

The structure of PBS and its ionomers were determined using wide-angle x-ray diffraction (WAXD). Spectra were taken at room temperature at a scan speed of 10°/min over the range of 5° – 40° using an X-ray diffractometer equipped with nickelfiltered Cu Ka radiation (40 kV, 100 mA) (Rigaku Denki, Model DMAX 2000).

A PerkinElmer DSC7 (PerkinElmer, Wellesley, MA, USA) was used to analyze the thermal properties of 5mg samples of PBS and its ionomers sealed in aluminum sample pans. All measurements were performed in a high-purity nitrogen atmosphere to minimize degradation. For non-isothermal crystallization studies, the samples were heated at 10°C min^{-1} from 20 to 150°C, maintained at 150°C for 3 min, and cooled to 20°C at 5, 10, 20, 30, and 40°C min^{-1}.

Table 2-1. GPC and inherent viscosity of PBS and its ionomers

Sample	GPC results (g/mol)		[η](g/dL)
	M_n x 10-4	M_w/M_n	
PBS	4.7	1.7	0.99
PBS1i	4.6	1.8	0.97
PBS3i	-	-	0.94
PBS5i	-	-	0.93

The spherulitic morphology of PBS and ionomer samples was investigated with a Nikon polarized light microscope (Nikon, Tokyo, Japan). First, a Mettler-Toledo FP82HT hot stage (Mettler-Toledo, Greifensee, Switzerland) was equilibrated at the desired crystallization temperature before the sample was transferred to the hot stage. Then, each sample was melted on a hot stage at 200°C for 5 min and then quickly transferred to the second hot stage.

Crystallization occurred over 30 min and the spherulitic morphology of the samples was examined when crystallization was complete.

2-3. Results and Discussion

2-3-1. ^{1}H-NMR

Figure 2-2 represents the ^{1}H-NMR spectra of pure PBS and PBS ionomers, respectively. The major peaks of pure PBS can be assigned at 4.1, 2.6, 1.7 ppm and their integration ratio is 1: 1: 1, which is reflective of the complete synthesis of PBS. Relative to pure PBS, PBS1i exhibits the new peaks from 3.7 ppm to 3.65 ppm and the peak intensity grows gradually. These peaks may be attributed to the methine hydrogen in the sulfonate succinate unit of SDMF [5]. In addition, the integration ratio between the signal at these peaks and the resonance at 2.6ppm for the methylene hydrogens (-OCOCH$_2$CH$_2$COO-) was found to be proportional to the added SDMF content, confirming that the ionic group may be efficiently introduced to the PBS backbone.

2-3-2. Solid-State ^{23}Na NMR

Figure 2-3 shows the solid-state ^{23}Na NMR spectra of pure SDMF and the PBS ionomers. Before analyzing the results, a brief introduction to the chemical shift and the peak-broadening phenomena in ^{23}Na NMR follows. Since ^{23}Na is a quadrupole nucleus, a upfield chemical shift is usually observed when the quadrupolar coupling interaction between ^{23}Na nuclei becomes much stronger [6]. Conversely, the peak breadth can be strongly influenced by the degree of mobility for the ion group and the symmetry of its surrounding electric field: the lower both values, the broader the peak width. Therefore, ^{23}Na NMR peaks shift upfield and become broader as ionic interactions become stronger.

An intense peak can be detected in the ^{23}Na NMR spectra of SDMF at about –26 ppm due to the sodium sulfonate ion groups. In the case of PBS ionomers, as the ionic groups become denser, a narrow and sharp peak at –9 ppm gradually develops, which may be attributable to isolated sodium ions in lone sodium sulfonate groups. This assignment is consistent with those of O’Connell *et al.* [7], who examined sulfonated polystyrene ionomers using solid-state ^{23}Na NMR.

In addition, as for all ionomers, a broad peak near the isolated ion peak can be observed, which is attributable to aggregated sodium ions formed between numerous ionic groups [6, 7]. Interestingly, as the content of ionic substituents increased, the peak corresponding to the aggregated sodium ion was found to shift upfield (toward more negative ppm values). The aggregated ion peak exhibits a major upfield shift of 5 ppm from –18 to –23 ppm for PBS1i and PBS5i, respectively. This spectral behavior indicates that the quadrupolar coupling interaction between ^{23}Na nuclei increases in proportion to the ionic group content. In ionomers with a high ionic group content, this observation may be caused by stronger electrostatic interactions due to a greater number of ion pairs, thereby inducing the formation of ionic aggregates, *i.e.*, doublets, triplets, and multiplets. In addition, the aggregated ion peak gradually broadens with an increase in mole percentage of ion groups. This effect may also be due to stronger electrostatic interactions that restrict the environment of sodium ion resonance, by not only immobilizing the ion groups but also lessening the symmetry of the ion pair electric field. Therefore, as the percentage of ionic groups in the ionomer increases, the electrostatic cross-linking due to sodium ion interactions gradually becomes richer and leads to abundant multiplet structures.

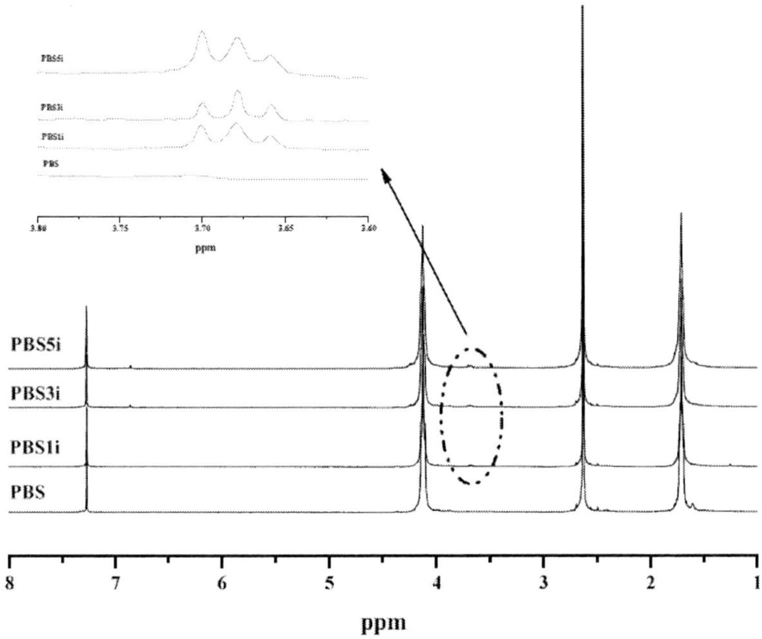

Figure 2-2. ^{1}H-NMR spectra of PBS and its ionomers.

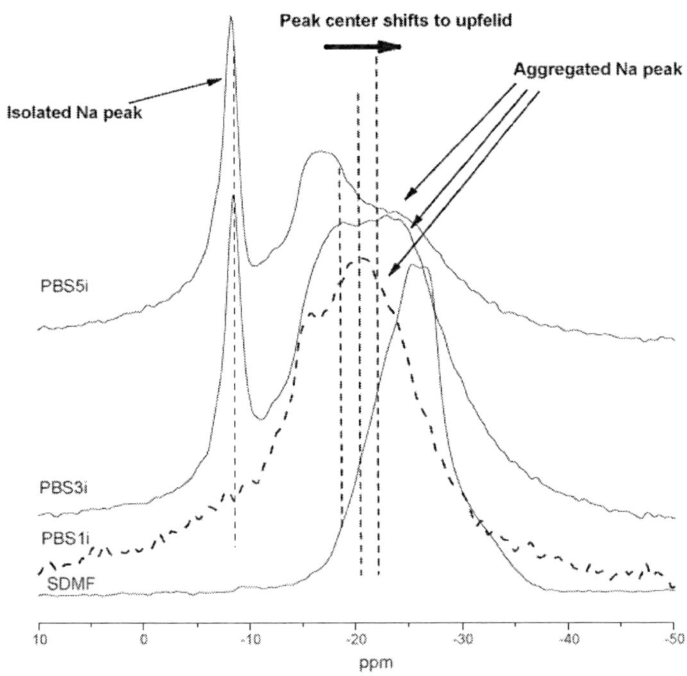

Figure 2-3. Solid-state ^{23}Na NMR spectra of pure SDMF and PBS ionomers.

2-3-3. Rheological Properties

Figure 2-4 shows the melt viscosity versus frequency of PBS and PBS ionomers measured at 150°C, at which pure PBS exhibits the typical shear thinning behavior at the high frequency region. In the low ω region, compared with pure PBS, PBS ionomers have higher η values and this value is gradually higher with the increase of the ionic group concentration. Generally, the high melt viscosity means that the polymer matrix may give more resistance to flow. If the there is a specific interaction or chemical bonding within polymer, this behavior is predominating [8]. Thus, higher η values of PBS ionomer can be clearly interpreted that with the increase of the ionic group content, the degree of cross-linkage between the ionic groups predominates as a result of abundant intermolecular ion-pair aggregation, classified as multiplets, within the PBS matrix. On the other hand, the value of melt viscosity decreases with increasing ω value in the PBS ionomer, while that of the parent PBS remained almost constant, indicating that the cross-linkage due to multilplets is broken down resulting from an increase in the shear force. The Cole–Cole plots of pure PBS and PBS ionomers are shown in Figure 2-5. Pure PBS has a linear curve with a slope of 1.57. The PBS ionomers yield similar curves, but their slopes become gradually lower as the ionic group concentration increases. The maximum slope reduction was observed in PBS5i, with a slope of 1.19. An isotropic and homogeneous polymer melt usually exhibits a slope of 2.0 on the Cole–Cole plot [9]. Therefore, this result demonstrates that sodium ion multiplets, created as a result of stronger ion-pair interactions, may give rise to the pronounced heterogeneity within the PBS chain, leading to lower slopes in the Cole–Cole plot.

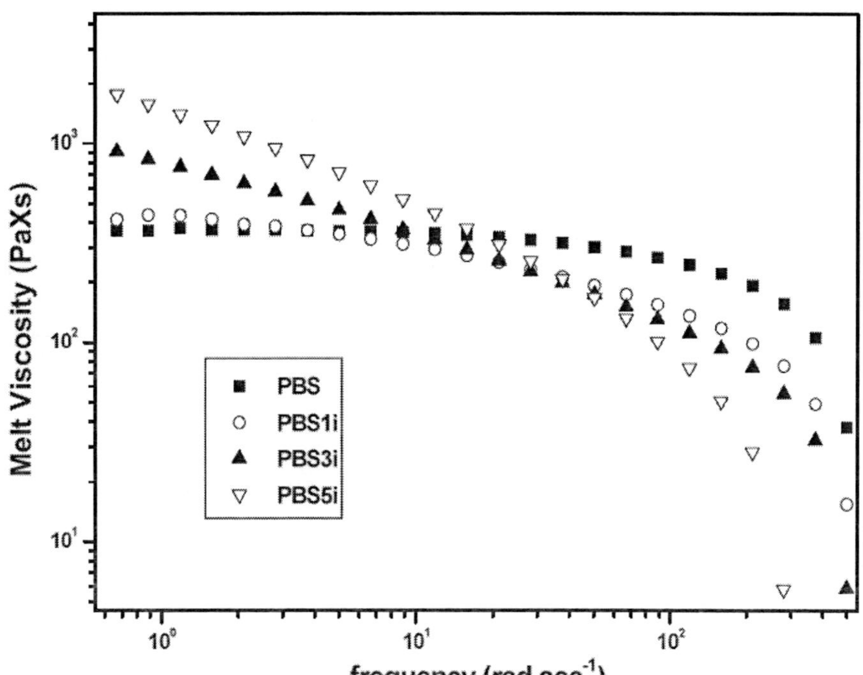

Figure 2-4. Melt viscosity versus frequency for PBS and the PBS ionomers measured at 150°C.

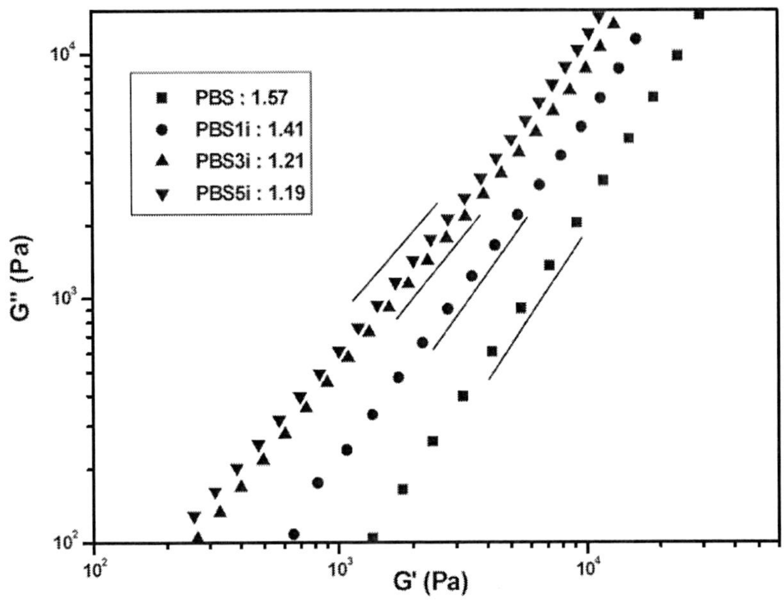

Figure 2-5. Cole–Cole plots for pure PBS and PBS ionomers.

2-3-4. Degree of Clustering

Figure 2-6 (a) plots the tan δ curves of pure PBS and its ionomers against temperature. Pure PBS shows a maximum at about -16.7°C relative to the glass transition temperature T_g. As summarized in the Table 2-2, the T_g of the PBS ionomers gradually shifts to higher temperatures with the increase in the ionic group concentration. The maximum shift of this T_g value is about 5.9°C for PBS5i, where T_g is approximately -10.8°C. This indicates that as the ionic group contnet increases, strong intermolecular ionic association effectively restricts the chain mobility in the PBS molecular backbone.

Interestingly, a vague shoulder peak, reflecting the T_g corresponding to the ion cluster region [10], is found near 53.9°C for the tan δ curve for PBS 1i. As the added ion group content increases, this peak progresses gradually and its position shifts to higher temperatures. Figure 2-6 (b) shows the relationship between the ionic group and the value of the cluster T_g, where the slopes of the regression line have a reliability of 0.9994.

Table 2-2. Glass transition temperatures for PBS and its ionomers

Sample	Matrix T_g (°C)	Cluster T_g (°C)
PBS	-16.7	-
PBS1i	-13.9	53.9
PBS3i	-12.5	55.4
PBS5i	-10.8	57.0

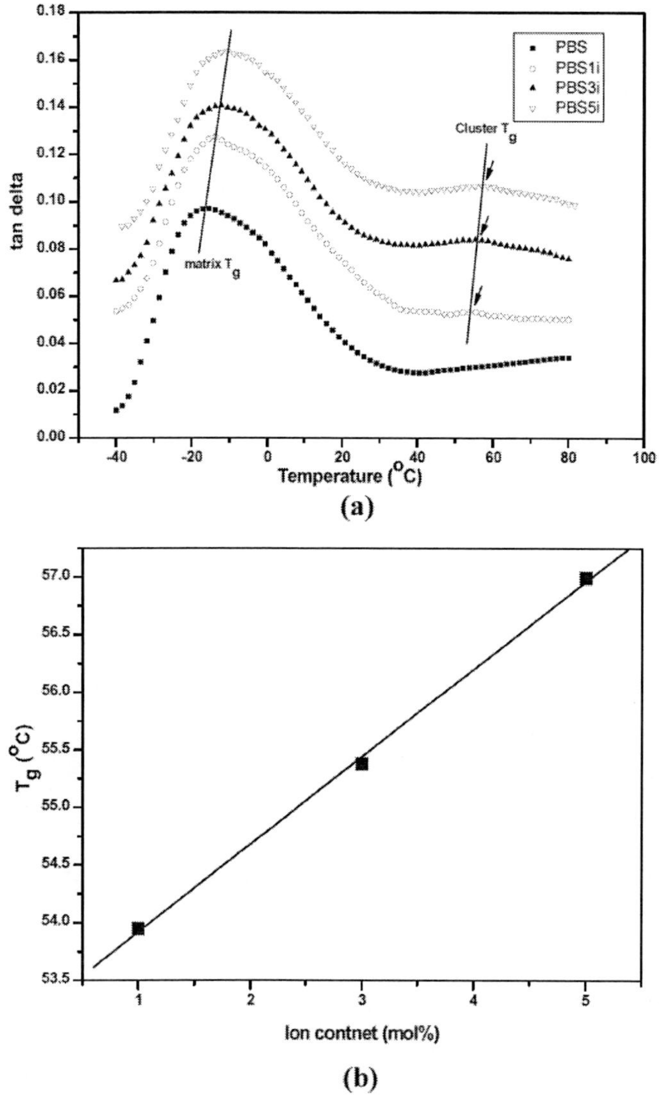

Figure 2-6. (a) tan δ curves for PBS and PBS ionomers, and (b) the relation of cluster T_g and ionic group concentration for PBS ionomers.

On the whole, a cluster can be defined as a contiguous region formed by interactions between sufficient numbers of multiplets and shows a second glass transition temperature. Kim et al. [11] investigated properties of poly(styrene-co-acrylate) and poly(vinyl cyc1ohexane-coacrylate) ionomers and reported that higher levels of clustering correspond to higher values of cluster T_g. Combined with the solid-state ^{23}Na NMR data, we conclude that the electrostatic ionic interaction increases in proportion to the degree of sodium sulfonate ionic group, which leads to the formation of a large number of multiplets and contributes favorably to the formation of clusters in a PBS molecular matrix.

Wide-angle x-ray diffraction patterns of PBS and its ionomers are shown in Figure 2-7. The pattern for pure PBS shows distinct peaks at 19.3, 21.7, 22.4, and 28.7°, which can be assigned to (020), (021), (110), and (111), respectively [12]. Regardless of the ionic group content, no new peaks occur and there are no significant changes in the d-spacing, reflecting that the PBS ionomers have the same monoclinic crystal structure as pure PBS. This result may be attributed that the ionic unit in the PBS ionomer is excluded from the crystal regions of the butylene succinate sequence [10].

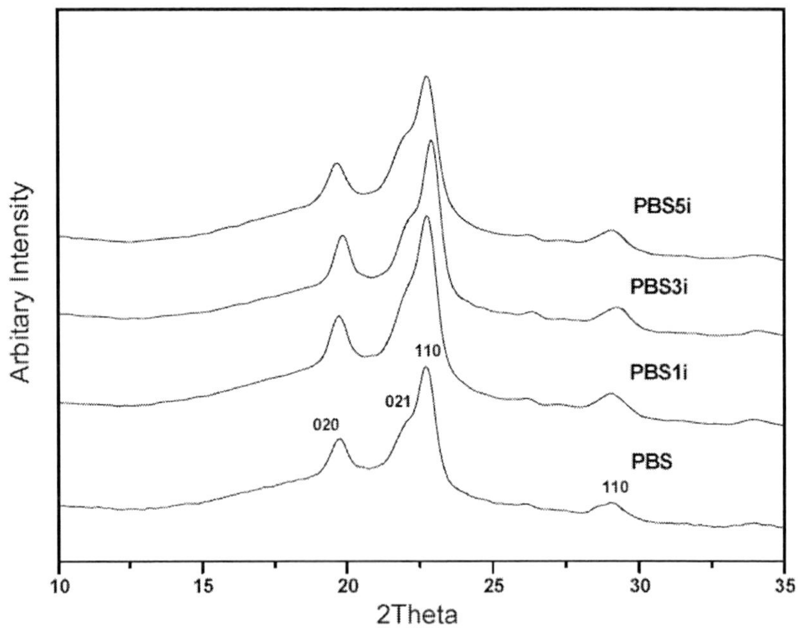

Figure 2-7. Wide-angle x-ray diffraction patterns for PBS and PBS ionomers.

2-3-5. Non-Isothermal Crystallization Analysis

Figures 2-8 shows the DSC cooling curves for pure PBS and its ionomers, respectively, obtained at various cooling rates. All samples show only one exothermic peak related to the crystallization of a single phase. In addition, as the cooling rate is increased, the maximum crystallization temperature decreases. Generally, when a specimen is cooled quickly, a greater degree of supercooling is required to initiate the crystallization, presumably as the motion of polymer molecule is unable to follow the cooling rate. At a given cooling rate, the crystallization peaks of pure PBS shift to the lowest temperature, suggesting that the presence of an ionic group in the PBS backbone may facilitate the crystallization process.

The relative crystallinity of polymers at various cooling rates as a function of temperature can be calculated as shown in Eq. (1):

$$X(T) = \frac{\int_{T_0}^{T} (dH_c/dT)dT}{\int_{T_0}^{T_\infty} (dH_c/dT)dT} \qquad (4\text{-}1)$$

where $X(T)$ is the degree of relative crystallinity as function of temperature, and T_0 and T_∞ are the initial and final crystallization temperatures, respectively. Generally, during the non-isothermal crystallization process, the crystallization temperature T can be transformed into the crystallization time t according to Eq. (2): where R is the cooling rate, and T is the temperature at time t. Figures 2-9 (a) and (b) show the relative crystallinity as a function of time for pure PBS and PBS3i, respectively. As the cooling rate increases, the sigmoidal curves shift to shorter times, indicating that the faster cooling rate is, the shorter the crystallization time requires. At a given cooling rate, the relative crystallinity curves of PBS3i shift toward shorter crystallization times than pure PBS, demonstrating that the non-isothermal crystallization rate of PBS3i is faster than that of pure PBS.

$$t = \frac{T_0 - T}{R} \tag{4-2}$$

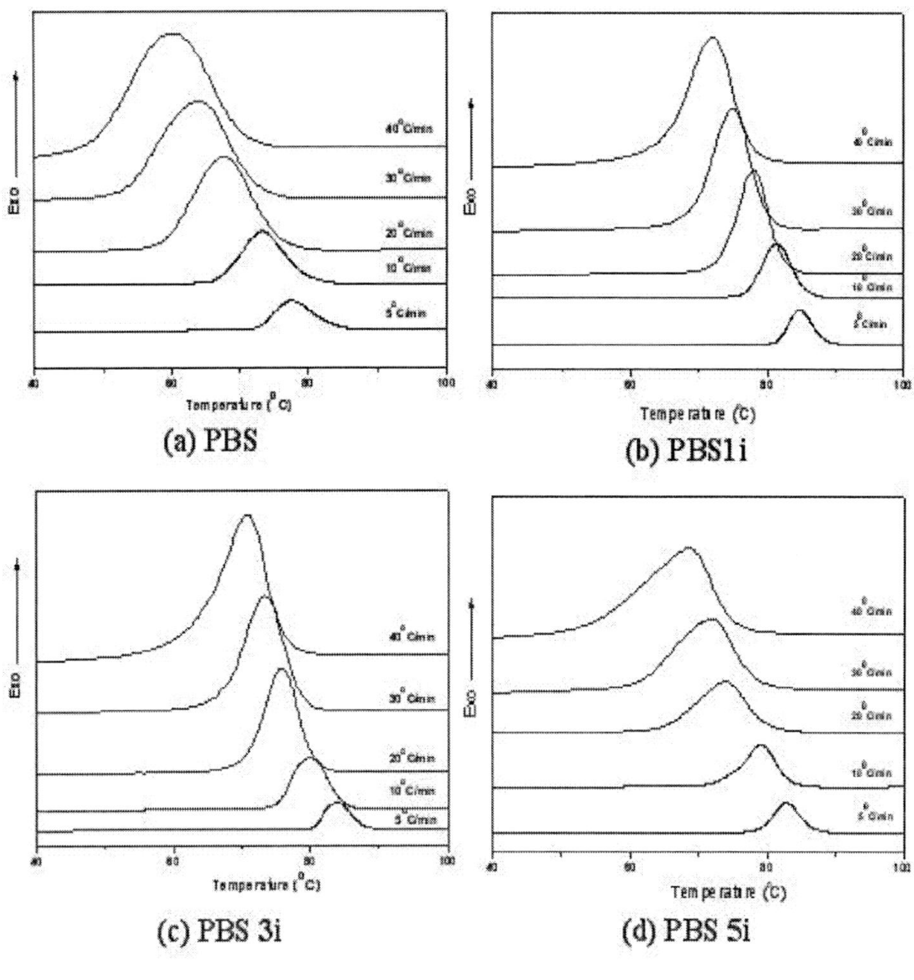

Figure 2-8. DSC cooling curves for pure PBS and its ionomers at various cooling rates; (a) PBS, (b) PBS1i, (c) PBS3i, (d) PBS5i.

We systematically investigated the non-isothermal crystallization rate for PBS and its ionomers by calculating the half time of crystallization, $t_{1/2}$, defined as the time required for half of the final crystallinity to develop. Figure 2-10 shows the variation in $t_{1/2}$ for PBS and its ionomers acquired at different cooling rates. Pure PBS has the largest value of $t_{1/2}$ at all cooling rates, indicating that pure PBS requires the greatest amount of time to produce crystalline phase when cooling from a molten state. This suggests that sodium sulfonate ionic groups can function as nucleation sites, promoting the crystallization of the PBS ionomer. However, $t_{1/2}$ decreases in the order PBS5i > PBS3i > PBS1i, implying that the enhancement of the crystallization rate is not directly proportional to the ionic group concentration.

The observed T_c values were used to determine the fastest crystallization time ($t_{\max}$) at various cooling rates. The results are summarized in Table 2-3. $t_{\max}$ can be expressed as a function of temperature (T) and the cooling rate (R) as [122]:

$$t_{\max} = \frac{T_o - T_c}{R} \quad (4\text{-}3)$$

where T_0 is the onset temperature of crystallization. For a given cooling rate, the lowest value of t_{max} was observed in samples of PBS1i. This strongly supports that during the non-isothermal crystallization process, the nucleation effect of the ionic group is most significant at the addition of 1 mol% ionic group. This result is further clarified below.

The cluster usually leads to the restriction of chain mobility in polymer molecules near itself as a result of the reduction in the number of chain configuration accessible to each chain, which may lead to a slower overall crystallization. The probability of a cluster being created in PBS1i is relatively weak since the ion pairs are evenly dispersed throughout the polymer matrix. As a result, the rapid crystallization rate of PBS1i can be attributed to dispersed ion pairs acting as effective nucleation sites and reducing the surface free energy during crystallization. A similar result was reported by Yu *et al.*, who examined the crystallization kinetics of polypropylene (PP) ionomers and showed that ionic groups along the PP chains significantly increased the crystallization rate [14]. As the ionic substituent content increases, the number of multiplets inevitably increases, eventually forming a cluster as seen in the DMA result above. Therefore, the relatively slower crystallization rate of PBS3i is likely due to a dominant cluster phase, which prevents the diffusion of ionomer chains during crystallization. This phenomenon is even more striking upon the incorporation of 5.0 mol% ionic substituents. A proposed mechanism for the ionic group concentration dependence of the crystallization rate is illustrated in Figure 2-11.

Figure 2-12 shows the spherulitic morphology of PBS and its ionomers obtained with a polarizing optical microscope at 80°C from the melt (magnification, 50×). In pure PBS, the spherulites exhibit a well-defined fibrilliar pattern with an apparent Maltese cross, which is true to many polymers [12, 15]. Upon the addition of 1 mol% ion content, the nuclei density increases and the spherulites become smaller. This result emphasizes the enhancement of nucleation by distributed ionic substituents.

In contrast, the spherulites of PBS5i display an irregular ring pattern without a distinct center. In our previous study, we confirmed that this irregular pattern was fundamentally attributable to the clusters [16]. Generally, banded/ringed spherulites are formed through lamellar twisting along the radial direction during crystal growth. Therefore, this result can be interpreted as the ionic group reducing the probability of forming a fold and reentering the crystal lamella, which consequently leads to formation of the ringed pattern observed through lamellar twisting. In summary, we can clearly conclude that the ionic cluster strongly influences the spherulite pattern.

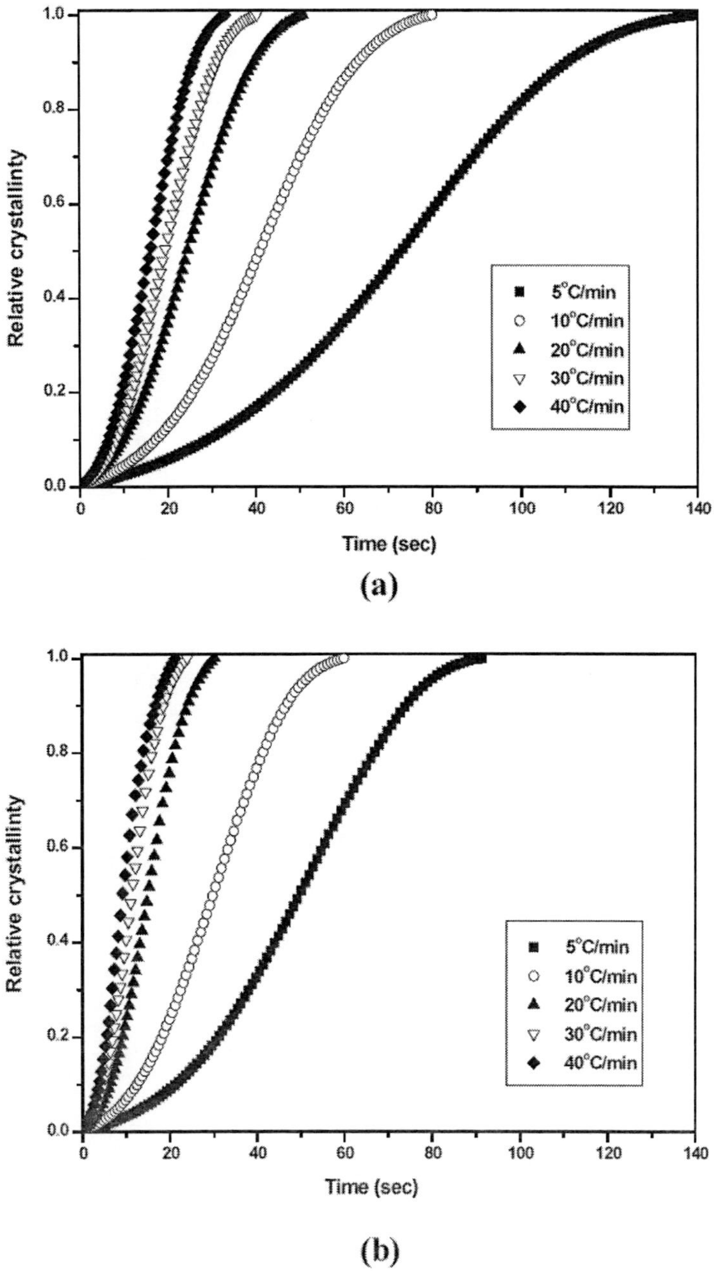

Figure 2-9. Relative crystallinity as function of time for pure PBS (a) and PBS3i (b).

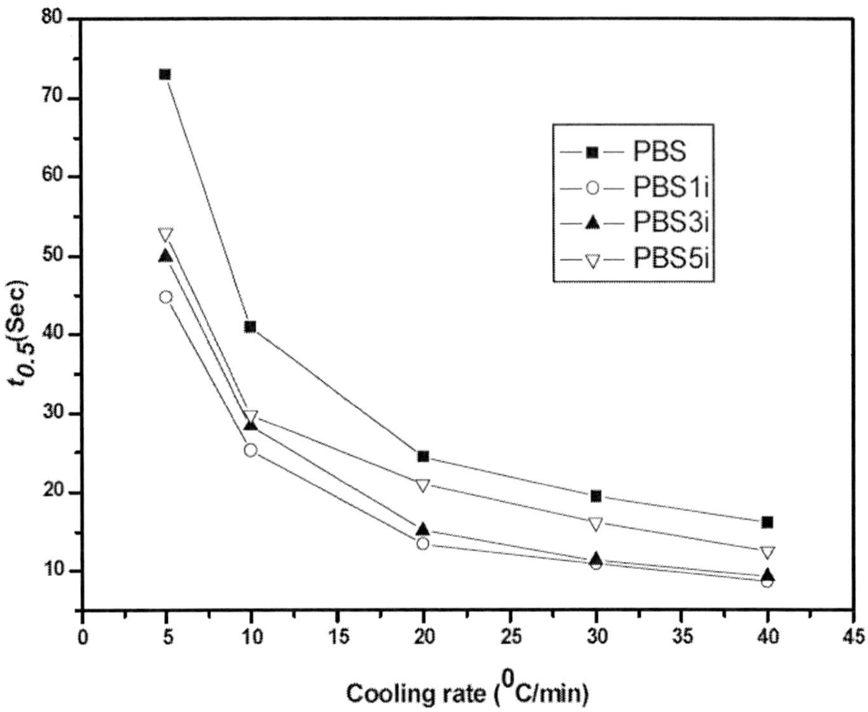

Figure 2-10. Variation in the crystallization half time, $t_{1/2,}$ for PBS and PBS ionomers at different cooling rates.

Table 2-3. Fastest crystallization time (t_{max}) for PBS and PBS ionomers at various cooling rates

Sample	Cooling rate (^oC/min)	t_{max}(sec)	Sample	Cooling rate (^oC/min)	t_{max}(sec)
PBS	5	77.8	PBS3i	5	46.4
	10	38.2		10	25.2
	20	23.2		20	13.7
	30	16.8		30	10.7
	40	14.9		40	8.0
PBS1i	5	41.0	PBS5i	5	46.9
	10	24.5		10	28.1
	20	12.2		20	18.0
	30	9.4		30	12.0
	40	7.7		40	9.2

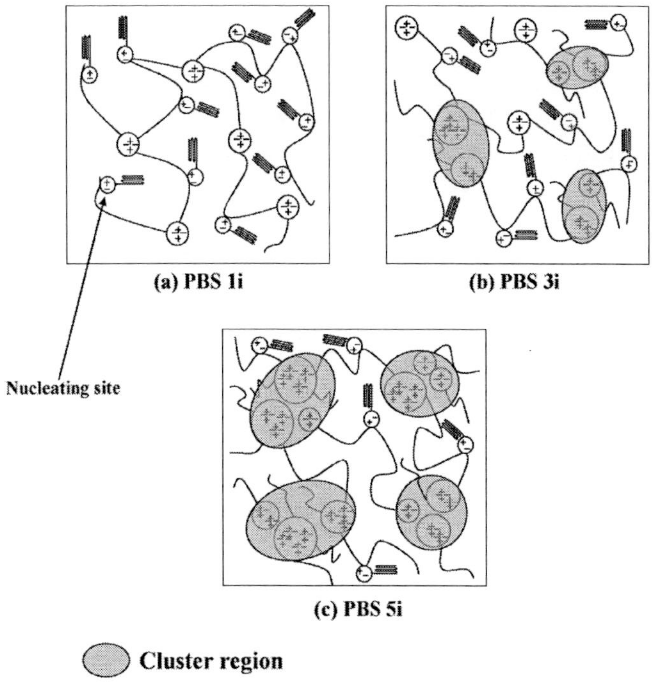

Figure 2-11. Proposed mechanism for the creation of clusters in PBS ionomers; (a) PBS1i, (b) PBS3i, (c) PBS51.

Figure 2-12. Spherulitic morphology of PBS and its ionomers at 80°C crystallized from the melt; (a) PBS, (b) PBS1i, (c) magnified PBS1i, (d) PBS5i.

2-3-6. Non-Isothermal Kinetics Equation

Ozawa [17] explained the effect of cooling rate on the non-isothermal crystallization of ionomers by modifying the Avrami equation, assuming that crystallization occurs at a constant cooling rate. According to Ozawa's theory, the relative degree of crystallization at temperature T, $X(T)$, can be calculated from the following equation:

$$1 - X(T) = exp[-P(T)/R^{m}] \quad (2\text{-}4)$$

where $P(T)$ is a function of the cooling rate, R is the cooling rate, and m is the Ozawa exponent, which is dependent on the dimension of crystal growth. Taking the double-logarithmic form, we obtain the following equation:

$$log\{-ln[1 - X(T)]\} = logP(T) - mlogR \quad (2\text{-}5)$$

A plot of $log\{-ln[1 - X(T)]\}$ versus $logR$ at a given temperature results in a straight line if the Ozawa method is applicable.

The results, based on the Ozawa method for PBS and its ionomers are displayed in Figure 2-13. An accurate analysis cannot be performed due to some curvature in the sample plots. This implies that the Ozawa exponent is not consistent with temperature during non-isothermal crystallization and for our PBS ionomers is therefore unable to estimate

$P(T)$ as it relates to the overall crystallization rate. The Ozawa theory seems not applicable because of the existence of a secondary crystallization process and/or the dependence of lamellar thickness on the crystallization temperature [18].

A method developed by Liu [19] was employed to describe the non-isothermal crystallization process using the physical variables of relative crystallinity [$X(T)$], cooling rate (R), and crystallization temperature (T). Both the Ozawa and Avrami equations give the following relationship:

$$logZ_t + nlogt = logP(T) - mlogR \quad (2\text{-}6)$$

In Eq. (6), the left term is a parameter of the Avrami equation. The coefficients n and m are the Avrami and Ozawa exponents, respectively, and depend on the nucleation mechanism and growth dimension. Z_t is a growth rate constant involving both nucleation and growth rate parameters. By rearranging Eq. (6) at a given crystallinity $X(T)$, the following equation may be obtained:

$$logR = logF(T) - alogt \quad (2\text{-}7)$$

where $a = n/m$, the ratio of the Avrami and Ozawa exponents. The expression $F(T) = [P(T)/Z_t]^{1/m}$ refers to the cooling rate, which may be chosen at unit relative crystallization time when the systems have obtained a certain degree of crystallinity. A high value of $F(T)$ indicates that a high cooling rate is required to reach $X(T)$ per unit time. $F(T)$ therefore provides a simple, practical measure of a polymer's ability to crystallize.

The plots of $logR$ versus $logt$ at a given degree of crystallinity are shown in Figure 2-14. Each plot shows a linear relationship, suggesting that this combined analysis may be more effective in describing the non-isothermal crystallization kinetics of PBS and its ionomers.

The values of *logF(T)*, determined from the *y*-intercept of the plots, is shown in Table 2-4. The value of *F*(*T*) increases with an increasing relative degree of crystallinity, indicating that at unit crystallization time, a higher degree of crystallinity can be obtained at a higher cooling rate. Interestingly, at the same relative crystallinity, the *F(T)* values of pure PBS and PBS1i exhibit the largest and lowest values, confirming the slowest and fastest crystallization rate, respectively. This result is consistent with the conclusions drawn from the non-isothermal crystallization analyses.

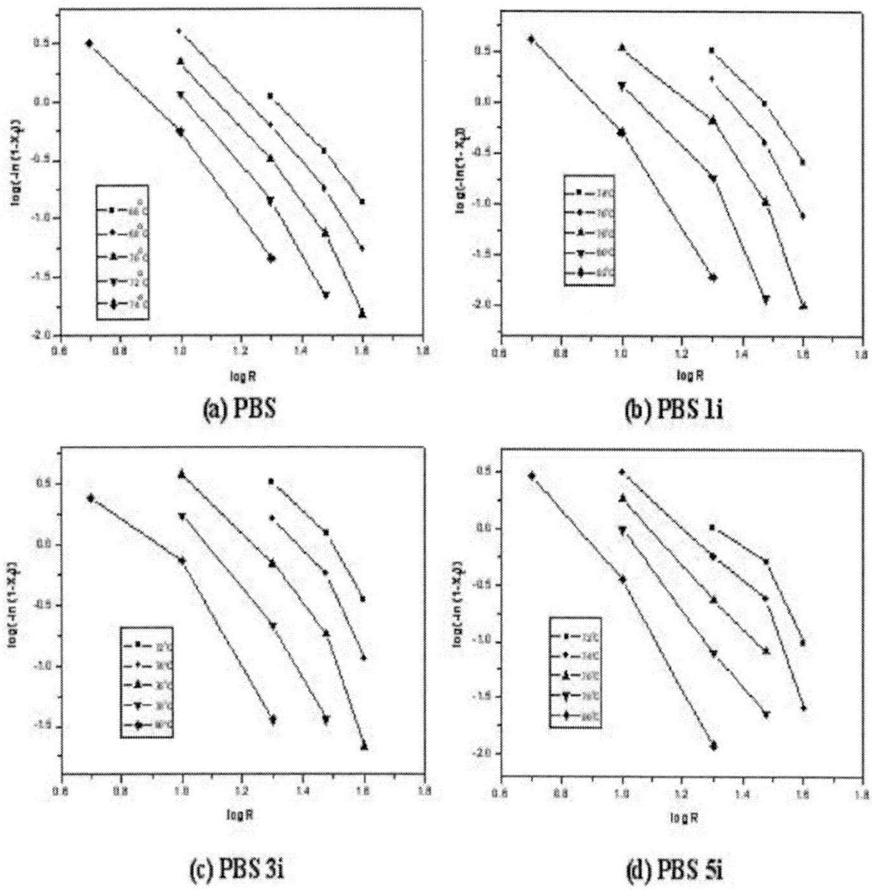

Figure 2-13. Ozawa plots of PBS and PBS ionomers during non-isothermal crystallization.

Table 2-4. Values of *F*(*T*) of PBS and PBS ionomers

Sample	Relative crystallinity (%)				
	10	30	50	70	90
PBS	1.94	2.25	2.35	2.44	2.51
PBS1i	1.79	2.07	2.17	2.22	2.31
PBS3i	1.84	2.07	2.18	2.26	2.33
PBS5i	1.92	2.17	2.29	2.35	2.40

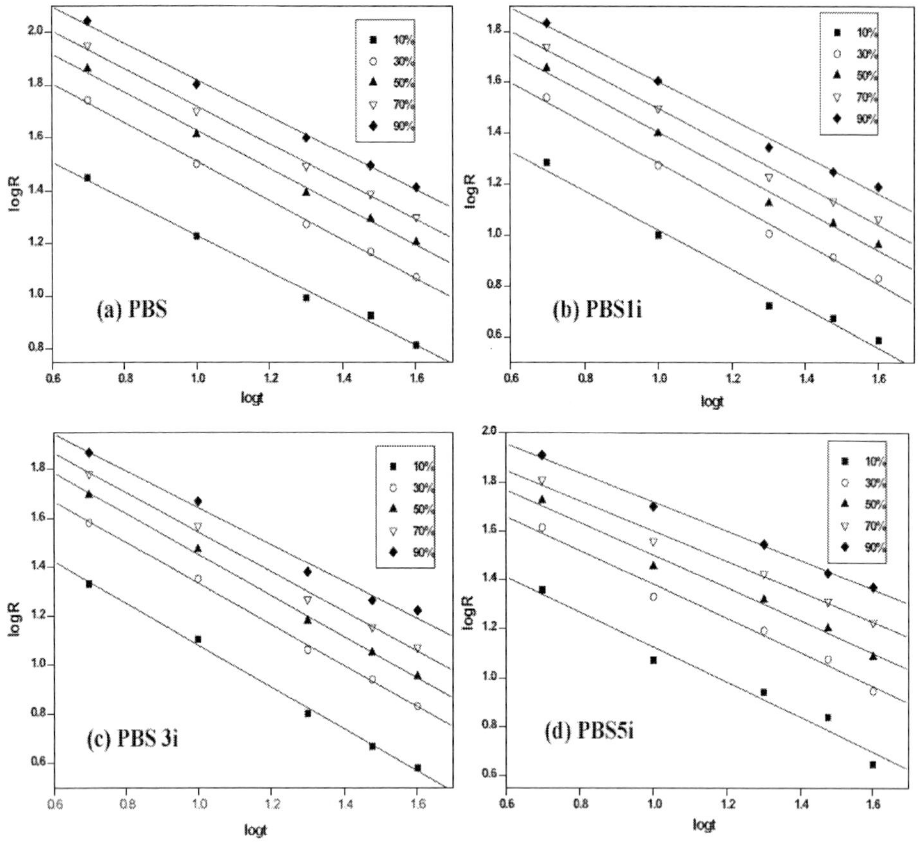

Figure 2-14. Plots of *logR* versus *logt* from the combined Avrami and Ozawa equation at different relative degrees of crystallinity for PBS and PBS ionomers.

2-3-7. Nucleation Activity

Dobreva *et al.* [20, 21] suggested a simple method for determining the nucleation activity of various additives during heterogeneous nucleation in a polymer melt. For homogeneous nucleation at temperatures near the melting temperature, the cooling rate is represented by Eq. (8):

$$logR = A - B/2.3\Delta T^2 \qquad (2\text{-}8)$$

In the case of heterogeneous nucleation, Eq. 2-(8) becomes

$$logR = A - B^*/2.3\Delta T \qquad (2\text{-}9)$$

where R is the cooling rate, A is a constant, ΔT is the degree of supercooling, and B is a parameter related to three-dimensional nucleation. The nucleating activity can then be calculated from the following relationship:

$$\phi = B^*/B \qquad (2\text{-}10)$$

The value of ϕ represents the amount by which the work of nucleation decreases with the addition of foreign nuclei. For example, if the foreign nucleus is extremely active, then ϕ will approach 0. The values of B and B^* can be obtained by using the slope of log$R$ versus $1/\Delta T^2$ [22], and ϕ can thereby be calculated according to Eq. 2-(10).

Figure 2-15 shows the *logR* versus $1/\Delta T^2$ plots for pure PBS and the PBS ionomers. Regressional analysis of all the slopes yields an R^2 correlation constant of 0.993. The three PBS ionomers display nucleation activities, ϕ, shown in order of increasing ion group concentration, of 0.79, 0.81, and 0.90. This observation shows unequivocally that the sodium sulfonate ion group assists the crystallization of PBS. However, this effect is not proportional to the concentration of ionic substituents. As the amount of ionic group increases, the ϕ value gradually increases; thereby supporting the idea that increased ionic group content can actually hinder PBS crystallization. The best nucleating efficiency was observed in PBS1i, which contained well-dispersed ion pairs, and also exhibited the fastest crystallization rate.

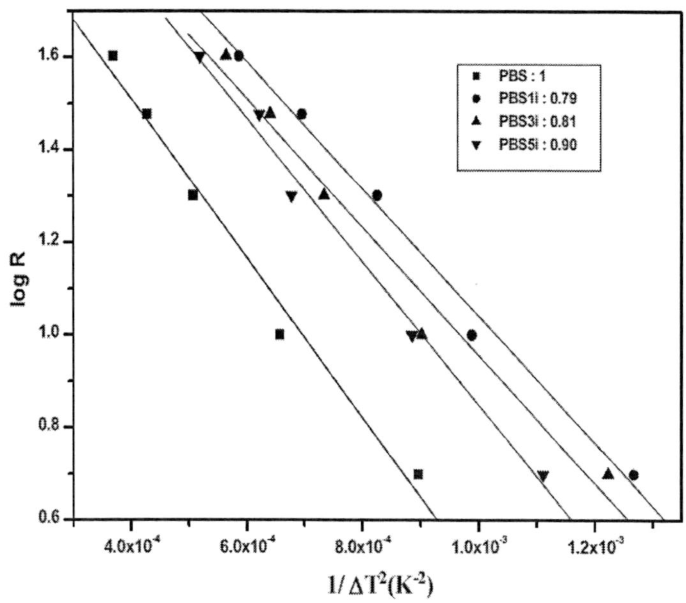

Figure 2-15. Plots of *logR* versus $1/\Delta T^2$ for pure PBS and the PBS ionomers.

2-4. Summary

1) Solid-state ^{23}Na NMR and DMA analyses confirmed that the formation of ion-pair multiplets increases with the concentration of ion groups, leading to a larger degree of clustering within the PBS matrix. Cole–Cole plot analysis confirmed that the multiplets induced a marked heterogeneity of PBS polymer chains.
2) Non-isothermal crystallization analysis confirmed that the crystallization rates of the PBS ionomers were greater than those of pure PBS, but the concentration of ionic substituents was not directly proportional to the increase in the crystallization rate.

3) A combined Ozawa and Avrami equation was more effective than the Ozawa equation in describing the non-isothermal crystallization kinetics of PBS and its ionomers. Moreover, an analysis of the nucleation activity showed that PBS1i, with the well-dispersed isolated ion pairs, showed the highest crystallization rate.

References

[1] A. Dobreva and I. J. Gutzow, *J. Non-Cryst. Solids*, 162, 1 (1993).
[2] A. Dobreva and I. J. Gutzow, *J. Non-Cryst. Solids*, 162, 13 (1993).
[3] A. K. Page, D. S. Greg, and R. B. Moor, *Polymer*, 45, 8425 (2004).
[4] B. K. Kim, C. H. Choi, *Polymer*, 37, 807 (1996).
[5] C. D. Han, J. Kim, and J. K. Kim, *Macromolecules*, 22, 383 (1989).
[6] E. B. Orler, B. H. Calhoun, and R. B. Moor, *Macromolecules*, 29, 5965 (1996).
[7] E. B. Orler, R. B. Moor, *Macromolecules*, 27, 4774 (1994).
[8] E. B. Orler, R. V. Gummaraju, B. H. Calhoun, and R. B. Moore, *Macromolecules*, 32, 1180 (1999).
[9] E. M. O'Connell, T. W. Root, and S. L. Cooper, *Macromolecules*, 27, 5803 (1994).
[10] J. S. Kim, G. Wu, and A. Eisenberg, *Macromolecules*, 27, 814 (1994).
[11] J. S. Lim, I. Noda, and S. S. Im, *Polymer*, 48, 2745 (2007).
[12] J. Yu and J. He, *Polymer*, 41, 891 (2000).
[13] K. J. Ihn, E. S. Yoo, and S. S. Im, *Macromolecules*, 28, 2460 (1995).
[14] M. Alonso, J. I. Velasco, and J. A. de Saja, *Eur. Polym. J.*, 33, 255 (1997).
[15] M. Eder and A. Wlochowicz, *Polymer*, 24, 1593 (1983).
[16] M. Takayanagi and T. Yamashita, *J. Polym. Sci.*, 22, 552 (1956).
[17] N. Lotti, V. L. Siracusa, L. Finelli, and A. Munari, *Polym. Eng. Sci.*, 42, 1590 (2002).
[18] S. I. Han, S. S. Im, and D. K. Kim, *Polymer*, 44, 7165 (2003).
[19] S. I. Han, W. D. Lee, D. K. Kim, and S. S. Im, *Macromol. Rapid. Commun.*, 25, 753 (2004).
[20] T. Liu, S. Mo, S. Wang, and H. Zhang, *Polym. Eng. Sci.*, 37, 568 (1997).
[21] T. Ozawa, *Polymer*, 12, 150 (1971).
[22] W. Zhu, L. Tian, T. Ren, X. Wang, X. Tang, and G. Li, *J. Appl. Polym. Sci.*, 84, 67 (2002).

3. Effects of Ionic Substitution on Physical Properties, and Crystallization Rates of, Poly(Oxyethylene-B-Butylene Adipate) Ionomers

3-1. Introduction

Ionomers, defined as ion-containing polymers that may bear their ionic groups either along the backbone chain or as pendant groups, have been useful in developing polymeric systems, because electrostatic cross-linking, arising from intermolecular ionic associations, can enhance the rheological and mechanical properties, relative to those of the neutral polymer [1–3]. However, the concentration of ionic substituents is not directly proportional to enhanced physical strength. In general, both the type and amount of ionic substituent, along with the nature of the polymer backbone, can affect the physical properties of an ionomer. The physical properties of semi-crystalline polymers, including ionomers, are dominated by their crystalline behavior. Not only does the choice of ionic substituent influence nucleation, growth rate, and characteristics of the individual folded chains, but it can also affect the extent of crystallinity and the crystalline morphology [4–7]. The crystallization rate must also be taken into account, especially when considering the implications for processing.

Recently, a series of biodegradable, aliphatic polyester-based poly(butylene succinate) ionomers (PBSi) was synthesized by the addition of sulfonated dimethyl fumarate (SDMF) to PBS. The nature of the ionic group was shown to influence the non-isothermal crystallization of the polymer [8]. A subsequent study described the synthesis of poly(oxyethylene-b-butylene adipate) ionomers (POBAis) by a two-step polycondensation and demonstrated the potential of this material in industrial applications [9]. However, the relationship between physical properties and crystallization rate needs to be carefully examined. Specifically, the effects of ionic substituents on the crystallization behavior and physical properties of polymer have received little attention. The current study demonstrates that the presence of ionic substituents in a novel poly(oxyethylene-b-butylene adipate) ionomer markedly affects the physical properties and crystallization rates of the polymer and would thus considerably impact the development of process controls.

3-2. Experimental

3-1-1. POBA and Its Ionomer Synthesis

Bis(poly(oxyethylene)) sulfonated dimethyl fumarate (SPEO), containing a sodium sulfonate ionic group, was synthesized from dimethyl fumarate (DMF), sodium hydrogen sulfite (NaHSO4), and poly(ethylene oxide; PEO, Mn = 400) according to a previously reported procedure [10]. Pure poly(oxyethylene-b-butylene adipate) (POBA) and POBA ionomer were also synthesized as described previously [8,9]. POBA ionomers are denoted as in the following example. POBA2.5i represents a POBA ionomer containing 2.5 mol% SPEO. The structures of SPEO and POBA ionomer are shown in Figure 3-1. ^{1}H NMR analyses (data not shown) showed that the ionic group was incorporated into the POBA backbone in proportion to the monomer feed concentration.

3-2-2. Measurements

2.2.1. ^{1}H NMR

The structure of the POBA ionomer was confirmed by interpretation of ^{1}H NMR spectra from a Fourier transform NMR (300 MHz) spectrometer (Varian, Salt Lake City, UT, USA). Samples were dissolved in CDCl3 containing tetramethylsilane (TMS) as an internal standard.

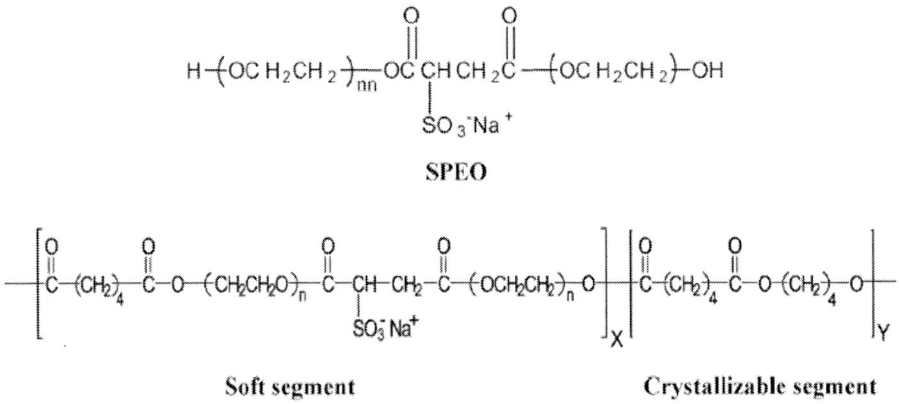

Figure 3-1. Chemical structures of SPEO and POBAi.

2.2.2. Size Exclusion Chromatography (SEC) and Inherent Viscosity

The molecular weight of POBA and its ionomers was measured with a SEC, equipped with an HPLC pump, a refractive index detector calibrated with PS standards, and a Styragel HR series column (eluent: CHCl3, 35◦C, flow rate: 0.3 mL/min). The molecular weight of pure POBA was about 30,000. The molecular weights of the ionomers could not be measured by SEC due to adsorption of the ionic groups to the column. Alternatively, as an indicator of molecular weight, the inherent viscosity of pure POBA and its ionomers was measured at 30 ◦C with a Cannon-Ubbelohde microviscometer (Cannon Instrument, State College, PA). The samples were dissolved in chloroform to give a concentration of 0.2 g dL−1. The data are summarized in Table 3-1. The inherent viscosity of the POBA ionomers was similar to that of POBA, regardless of the ionic group content. This indicated that the molecular weights of the POBA ionomers were comparable to those of pure POBA.

Table 3-1. Reaction time and Inherent Viscosity of POBA and its ionomers

Sample code	Molar ratio in feed (PBA/SPEO)	Reaction time (h)	η_{inh} (g dL^{-1})
POBAi-2.5	97.5/2.5	5.0+9.5	0.78
POBAi-5.0	95.0/5.0	5.0+11.5	0.76
POBA-2.5	97.5/2.5	5.0+7.0	0.75

2.2.3. Non-Isothermal Crystallization Analyses

A PerkinElmer DSC7 (PerkinElmer, Wellesley, MA) was used to analyze the thermal properties of 8-mg samples of POBA and POBAi sealed in aluminum sample pans. All measurements were performed in a high-purity nitrogen atmosphere to minimize degradation. For non-isothermal crystallization studies, samples were heated at 10 ◦C/min from −20 to 60 ◦C, maintained at 60 ◦C for 5min, and cooled to −20 ◦C at 2.5, 5, 10, and 20 ◦C/min.

2.2.4. Dynamic Mechanical Analyses (DMA)

To provide identical samples and to ensure consistent DMA measurements, the synthesized polymers were compression molded using a Carver Press at 60 ◦C and 69 MPa. The molded films were quenched in ice water and dried in a vacuum oven. The dynamic mechanical properties of POBA and its ionomers were determined using a dynamic mechanical analyzer (DMA 2980; TA Instruments, New Castle, DE) set in tension mode at a frequency of 3Hz under a nitrogen atmosphere. Samples (15mm×5mm×0.5mm) were cooled to −100 ◦C and then heated gradually to 30 ◦C at a rate of 2 ◦C/min. Mechanical tests were conducted with a universal tensile machine (Instron 4465, Instron Corp.) according to ASTM D882 at a crosshead speed of 20 mm/min. Each value from 20 successful measurements was averaged and expressed with standard deviation (S.D.)

3-3. Results and Discussion

3.1. Non-Isothermal Crystallization Analyses

Figure 3-2(a) and (b) shows the relative crystallinity as a function of time for pure POBA-2.5 and POBAi-2.5, respectively. All samples produced sigmoidal curves that shifted to shorter times with increasing cooling rates. These data indicate that faster cooling rates effectively shortened the time required for crystallization. At a given cooling rate, the relative crystallinity curves of POBAi- 2.5 shifted to longer crystallization times than those of pure POBA, demonstrating that the non-isothermal crystallization rate of POBAi-2.5 was slower than that of pure POBA-2.5. Non-isothermal crystallization rates for POBA and its ionomer were further characterized by their half-time of crystallization, $t_{1/2}$, defined as the time required for half of the final crystallinity to develop.

Table 3-2 shows the variation in $t_{1/2}$ for POBAand its ionomer at different cooling rates. Compared with POBAi-2.5, the $t_{1/2}$ values obtained with pure POBA were smaller at all cooling rates, indicating that pure POBA required less time to crystallize when cooling from the molten state. Thus, the presence of the sodium sulfonate ionic groups apparently prevents the diffusion of ionomer chains during crystallization, thereby slowing the crystallization process in POBA ionomers. In contrast, analyses of POBAi-2.5 and POBAi-5 (data not shown) revealed an increase in $t_{1/2}$ with increasing ionic group content. This result suggests that ionic substituents and clusters restrict chain mobility by limiting the number of accessible chain configurations, leading to slower overall crystallization. This result is consistent with a previous study [8].

Table 3-2. The crystallization parameters for the various cooling rates

Sample	Φ (°C/min)	$t_{1/2}$ (min)	n	Z_c
POBAi-2.5	2.5	1.62	3.30	0.461
	5	0.94	3.47	0.982
	10	0.46	3.55	1.250
	20	0.31	3.79	1.260
POBA-2.5	2.5	1.25	3.15	0.622
	5	0.61	3.23	1.177
	10	0.45	3.39	1.284
	20	0.30	3.47	1.296

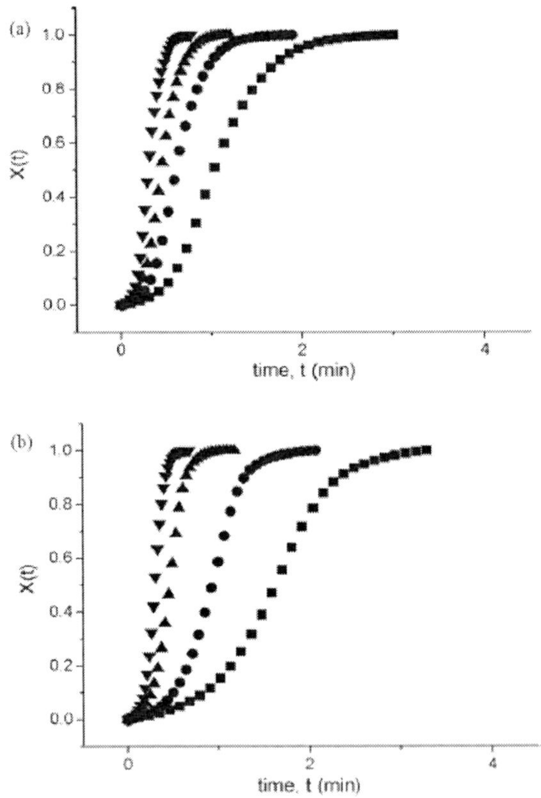

Figure 2. The development of relative crystallinity is shown as a function of time for(a) POBA-2.5 and (b) POBAi-2.5 at cooling rates of 2.5 ◦C/min (squares), 5 ◦C/min (circles), 10 ◦C/min (triangles), and 20 ◦C/min (inverted triangles).

3.2. Modified Avrami analyses

Generally, isothermal crystallization of polymers can be described by the Avrami equation [11,12], which assumes a relative degree of crystallinity proportional to the crystallization time, t. Taking the logarithm of both sides of the Avrami equation results in the following equation

$$\log[-\ln(1 - X(t))] = n \log t + \log Z \tag{1}$$

where n is the Avrami exponent, which is correlated with the nucleation mechanism and geometry of crystal growth, and Z is an overall kinetic rate constant that takes into account both nucleation and growth rate constants. Because the temperature is constantly changing in non-isothermal processes, the physical meanings of n and Z differ from those in isothermal crystallizations. Jeziorny [13] proposed that the kinetics of non-isothermal crystallization may be given as

$$\log Zc = \log Z \tag{2}$$

Eqs. (1) and (2) can be combined to yield a plot of log[−ln(1−X(t))] versus log t for both POBA and POBAi (Figure3-3). Note that the curve obtained for POBA contains a distinct nonlinear portion, which may be attributed to a secondary crystallization involving not only fibrillar growth between the primary lamellae of spherulite, but also the occurrence of spherulite impingements. The plot obtained for POBAi is much more linear, indicating a lesser degree of secondary crystallization processes. In semi-crystalline ionomers, the ionic aggregates have difficulty folding into crystalline lamellae during the crystallization process. The ionic group reduces the probability that a given chain will fold and reenter the same crystal lamella [8]. On this basis, it was concluded that the ionic groups in POBAi restrict the mobility of the polymer chain following primary crystallization. Consequently, secondary crystallization in POBAi is delayed relative to the same processes in POBA.

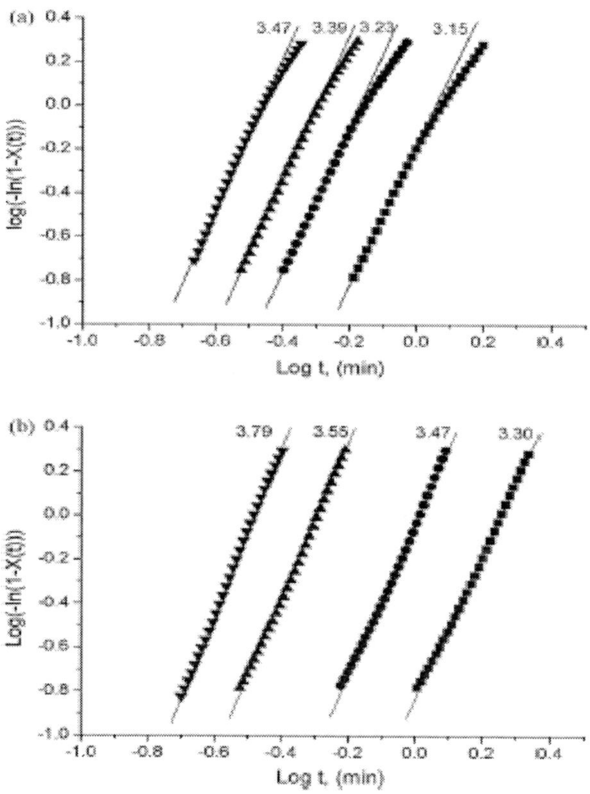

Figure 3. Avrami plots for the non-isothermal crystallization of (a) POBA-2.5 and (b)POBAi-2.5 are shown for cooling rates of 2.5 ◦C/min (squares), 5 ◦C/min (circles), 10 ◦C/min (triangles), and 20 ◦C/min (inverted triangles).

n and Zc for POBAi-2.5 and POBA-2.5 are given in Table 3-2. For two of the samples, n was approximately 3.4. This result indicates that crystallization in both POBA and POBAi occurred through a three-dimensional spherical growth mechanism with heterogeneous nucleation. Conversely, at the same cooling rate, the Zc of POBAi-2.5 was lower than that of POBA-2.5. This result demonstrates that crystallization of POBAi-2.5 was slower than that of POBA-2.5, providing further evidence that the ionic group on POBAi influenced crystallization kinetics.

3.3. The Combined Avrami and Ozawa Equation

A method developed by Liu et al. [14] was used to describe the non-isothermal crystallization process in terms of the relative crystallinity [X(T)], cooling rate (R), and crystallization temperature (T). Both the Ozawa and Avrami equations give the following relationship:

$$\log Zt + n \log t = \log P(T) - m \log R \tag{3}$$

where P(T) is a function of the cooling rate, R is the cooling rate, and m is the Ozawa exponent, which is dependent on the dimension of crystal growth. Rearranging Eq. (5) at a given crystallinity X(T) yields

$$\log R = \log F(T) - a \log t \tag{4}$$

where a = n/m, the ratio of the Avrami and Ozawa exponents. The expression F(T) = [P(T)/Zt]1/m refers to the cooling rate, which may be chosen at unit relative crystallization time when the systems have obtained a certain degree of crystallinity. A high value of F(T) indicates that a high cooling rate is required to reach X(T) per unit time. F(T) therefore provides a simple, practical measure of a polymer's ability to crystallize.

Plots of log R versus log t at a given degree of crystallinity are shown in Figure 3-4. Each plot shows a linear relationship, suggesting that this combined analysis may be effective in describing the nonisothermal crystallization kinetics of POBA and its ionomers. Values of log F(T), determined from the y-intercept of the plots in Figure 3-4, are shown in Table 3-3. The value of F(T) increases with an increasing relative degree of crystallinity, indicating that at unit crystallization time, a higher degree of crystallinity can be obtained at a higher cooling rate. At the same relative crystallinity, F(T) values of pure POBA were lower than those of POBAi. This result is consistent with a faster crystallization rate and corroborates the conclusions drawn from the non-isothermal crystallization analyses described above.

Table 3-3. Values of F(T) of POBA and POBA ionomers

Sample		X(T)			
		20%	40%	60%	80%
POBAi-2.5	F(T)	2.91	4.03	4.86	5.72
POBA-2.5	F(T)	1.25	2.03	2.93	4.37

3.4. DMA Analyses and Mechanical Properties

Figure 3-5 shows the storage moduli of POBA and POBAi as a function of temperature. Relative to POBA, POBAi-2.5 exhibited a higher storage modulus at all temperatures evaluated, indicating greater mechanical stiffness. This phenomenon can be explained by the presence of strong physical cross-links between polymer chains. Previous studies have verified that the presence of ionic groups in the ionomer may induce electrostatic cross-linking, because sodium ion interactions lead to abundant multiplet structures [8]. However, with increasing ionic group content, the trend reverses and the storage modulus begins to decrease. This may be attributed to a reverse effect of physical cross-linkage that results from preferential intramolecular ion pair associations [15,16].

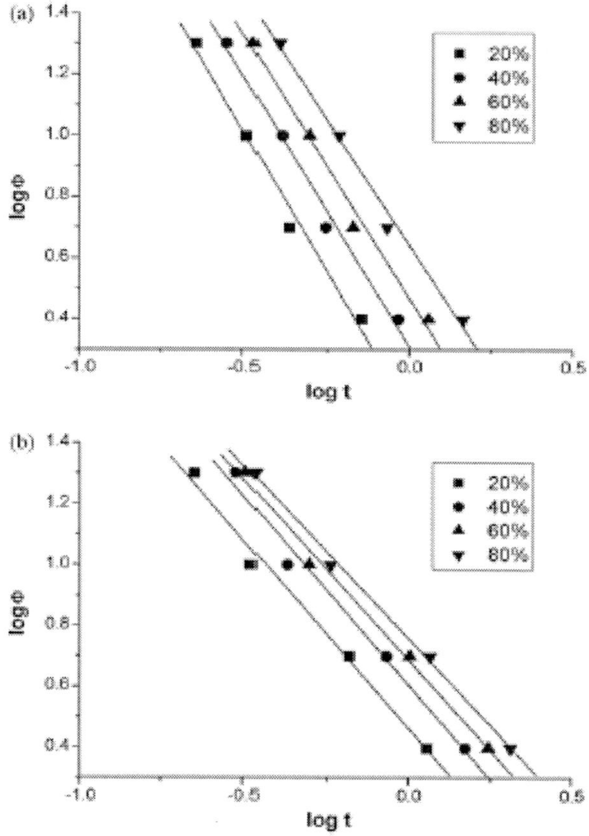

Figure 3-4. Plots from the combined Avrami and Ozawa equation to describe nonisothermal crystallization are shown for (a) POBA-2.5 and (b) POBAi-2.5.

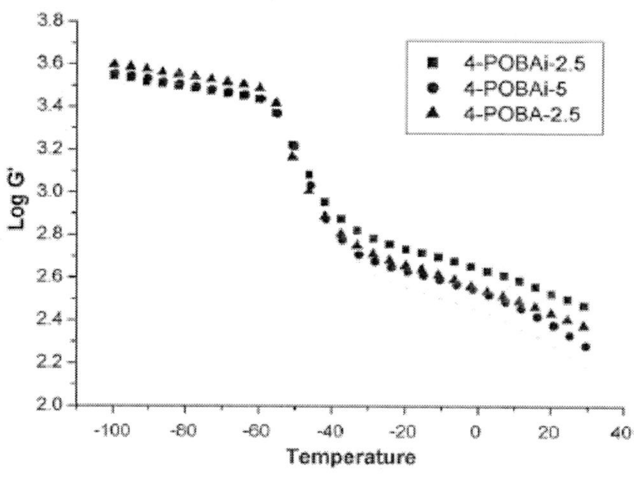

Figure 3-5. The storage modulus of each sample is given as a function of temperature.

The tensile properties of POBA and its ionomers are shown in Table 3-4. Compared with pure POBA, POBAi-2.5 exhibited greater mechanical strength. In general, tensile strength depends largely on the molecular weight of the polymer. However, in the current study, Inherent viscosity data confirmed that the molecular weight of POBAi was similar to that of pure POBA. Thus, physical cross- linking as a result of ionic groups may play a role in enhancing the mechanical strength of POBAi-2.5.

As the ionic group gradually enlarges, the ionomer becomes physically weaker. This suggests that there is a critical ionic group content at which the mechanical properties are optimal. Below this critical value, the ionomer chains are physically cross-linked to each other through intermolecular ion-pair associations. Above the critical ionic group content, an increasing degree of intramolecular ion-pair associations weakens the polymer. This result is consistent with the DMA analyses. In the current study, the optimal ionic group content in POBAi, as defined by enhancements in the physical strength of the ionomer, was at least 2.5 mol%.

Table 3-4. Tensile properties of POBA and its ionomers

Samples	Stress at break (MPa)	Elongation at break (%)	Modulus (MPa)
POBA-2.5	23.0 ± 0.35	400 ± 6.1	160 ± 2.4
POBAi-2.5	32.9 ± 0.46	599 ± 8.4	155 ± 2.2
POBAi-5.0	17.9 ± 0.32	209 ± 3.8	125 ± 2.3

3-4. Summary

Several poly(oxyethylene-b-butylene adipate) ionomers (POBAis) were synthesized and the role of the ionic group was evaluated in terms of physical properties and crystallization rate. Non-isothermal crystallization analyses confirmed that the crystallization rates of the POBA ionomers were slower than those of pure POBA. Based on a modified Avrami analysis, it was concluded that the ionic groups in POBAi restricted polymer chain mobility following primary crystallization, leading to a delay in the secondary crystallization processes relative to pure POBA. DMA analyses showed significant improvement in mechanical properties with an optimized ionic group content.

References

[1] A. Eisengerg, B. Hird, R.A. Moore, *Macromolecules.* 23 (1990) 4098.

[2] A. Eisengerg, K.S. Kim, *Introduction to ionomers,* Wiley, New York, 1998.

[3] A. Jeziorny, *Polymer.* 19 (1978) 1142.

[4] A. Molnar, A. Eisenberg, *Macromolecules.* 25 (1992) 5774.

[5] A.K. Page, D.S. Greg, R.B. Moor, *Polymer.* 45 (2004) 8425.

[6] E.B. Orler, B.H. Calhoun, R.B. Moor, *Macromolecules.* 29 (1996) 5965.

[7] E.B. Orler, R.B. Moor, *Macromolecules.* 27 (1994) 4774.

[8] J.S. Lim, Y.I. Lee, S.S. Im, *J. Polym. Sci. Part B: Polym. Phys.* 46 (2008) 925.
[9] M. Avrami, *J. Chem. Phys.* 7 (12) (1939) 1103.
[10] M. Avrami, *J. Chem. Phys.* 9 (2) (1941) 177.
[11] N. Lotti, V. Siracusa, L. Finelli, A. Munari, *Polym. Eng. Sci.* 42 (2002) 1590.
[12] R. Xie, R.A. Weiss, *Comput. Theor. Polym. Sci.* 7 (1997) 65.
[13] R.D. Lundberg, R.R. Phillips, *J. Polym. Sci., Polym. Phys. Ed.* 20 (1982) 1143.
[14] S.I. Han, S.S. Im, D.K. Kim, *Polymer.* 48 (2007) 1830.
[15] T. Liu, S. Mo, S.Wang, H. Zhang, *Polym. Eng. Sci.* 37 (1997) 568.
[16] X. Wang, L. Wang, H. Li, X. Tang, F. Chang, *J. Appl. Polym. Sci.* 77 (2000) 184.

4. New Application of Poly(Butylene Succinate) (PBS) Based Ionomer as Biopolymer: A Role of Ion Group for Hydroxyapatite (HAp) Crystal Formation

4-1. Introduction

A new type of thermoplastic resin, an ionomer, is a polymer that is partially ionized by the metallic ion [1]. As originally proposed by Eisenberg [2, 3], the interaction between ionic groups leads to the formation of multiplets containing a small number of ion pairs and ionic clusters, which constitute a second phase and contain many multiplets. These ionic structures have been shown to act as strong electrostatic crosslinks and enhance the mechanical properties of the polymer [2, 4].

Recently, we synthesized for the first time a series of poly(butylene succinate) (PBS), a typical biodegradable aliphatic polyester-based ionomer (PBSi), by the addition of sulfonated dimethyl fumarate (SDMF) with sodium sulfonate ionic group and extensively examined the relationship of the ionic group between nonisothermal crystallization behavior. The PBSi exhibited the enhanced nonisothermal crystallization rates than that of pure PBS, but the concentration of ionic substituents was not directly proportional to the increase in the crystallization rate [5]. Our subsequent study of the blend of it with poly(3-hydroxybutyrate-co-3- hydroxyhexanoate) (PHB-HHx), new PHA family, verified that PBSi induce to the superior miscibility with PHB-HHx as a result of sodium metal ion–carbonyl interaction between the ionic group and the carbonyl group [6]. As a result, we confirmed that the ionic group within the PBSi could help it apply as industrial fields.

Nowadays, investigation for hydroxyapatite (HAp), (Ca10(PO4)6(OH)2) has been greatly focused in biomaterial research area, since its chemical and crystallographic properties are similar to the inorganic component found in nature bone. In addition, this inorganic material has favorable advantages such as excellent biocompatibility, bioactivity, and osteoconductivity [7–9]. However, some drawbacks, like mechanical brittleness, low flexibility, and difficult process ability, are often encountered with its application as artificial bone materials. In order to improve the shortcoming of HAp and obtain intelligent biomaterials, fabrication of the composites of HAp with some organic polymers has been widely examined via biomimetic methods [10–14]. In these HAp/polymer composites, the key point in brilliant composites fabrication is how the functional group can be efficiently introduced into the polymer matrix, since this functional group can operate as a nucleus to the formation of hydroxyapatite, consequently not only leading to the increase the interfacial adhesion but also restricting the phase separation between the polymer matrix and HAp. Xiao et al. [15, 16] radiated with UV light to introduce –OOH groups onto the surface of the PLA and fabricated HAp/PLA composites by this modified PLA. Wang et al. [17, 18] have examined the silanation of HAp and acrylic acid grafting on polyethylene for improving bonding between HAp and polyethylene in HAp/HDPE composites.

However, theses studies have traditionally focused mainly monotonous alteration on the polymer surface. No systematic research has been published concerning the regulation of functional group, especially the ionic group, directly within the molecular chain. Based on this consideration, for the purpose of putting PBSi into biopolymer practice, we report the fabrication of novel PBS ionomer/hydroxyapatite composites. The ionic groups in the PBSi molecules can act as effective functional group of HAp crystal formation, generating that not only enhance the interfacial adhesion between HAp and PBSi but also control its formation and growth. Furthermore, it is anticipated that this novel polyester-based ionomer has the infinitive potential as a superior biopolymer as well as industrial application.

4-2. Materials and Methods

4-2-1. PBS Ionomer Synthesis

Pure PBS and PBS ionomers with 1.0, 3.0, and 5.0 mol% SDMF were synthesized in our previous study [5, 6]. The resulting PBS ionomers are denoted as follows: e.g., PBS1i represents PBS ionomer containing 1.0 mol% SDMF. PBS ionomer/hydroxyapatite composites fabrication PBSi/HAp composites were fabricated using a biomimetic methodology that involved immersing the polymers in simulated body fluid (SBF). The SBF solutions were prepared depending on references [19, 20]; NaCl, KCl, $MgCl_2$, $NaHCO_3$, K_2HPO_4, Na_2SO_4, and $CaCl_2$ were poured, in this order, into a 1000-mL volumetric flask that contained the nominated amount of 0.4 mol Tris [$(CH_2OH)_3CNH_2$] solution, 90% of the corresponding tabulated amount of 0.36 M HCl solution, and 800 mL of distilled water. The remainder of the HCl solution and 200 mL of distilled water were added dropwise to obtain a pH of 7.4 while stirring gently. The final SBF product was a transparent solution. Films of PBS and its ionomers measuring 3 cm 93 cm 9 0.25 mm were immersed in the SBF solution and maintained at body temperature 36.5 ^{o}C in an incubator for immersion times of 10, 20, 40, 80, 160, or 320 h. Then, the saturated films were removed and washed with distilled water. Each sample was dried and kept at 40 ^{o}C in an oven while awaiting further analyses.

4-2-2. Measurement

The structure of PBS and its ionomers were determined using wide-angle X-ray diffraction (WAXD) with a Rigaku Denki D-Max2000 X-ray diffractometer operated at 40 kV, 100 mA. The X-ray source consisted of an 18-kW rotating anode X-ray generator equipped with a rotating anode Cu target. The X-ray diffraction patterns of the powdered samples were recorded at room temperature over the range 2h = 5–40° at a scan speed of 5°/min. WAXD was also used to examine the HAp crystals grown on both PBS and its ionomers, but using film samples and a scan range of 2h = 10–60°. Thermogravimetric analysis (TGA) was used to determine on the dried samples to evaluate HAp content. The measurement was performed between 20 and 1000 ^{o}C at heating rate of 20 ^{o}C min^{-1} using a

PYRIS TGA (Perkin– Elmer, Norwalk, CT, USA). Fourier transform infrared (FT-IR) spectra were collected at room temperature on an IFS 88-IR spectrometer (Bruker AXS GmbH, Karlsruhe, Germany). The scanned wavenumber range was 4000–500 cm^{-1}. All spectra were recorded at a resolution of 4 cm^{-1} and 16 scans were co-added for each sample. Scanning electron microscopy (SEM) of the PBS ionomer/ HAp composite films was undertaken after drying the samples at room temperature under nitrogen. The samples were then Au–Pd coated by ion sputtering and examined by using field emission (FE) SEM (S-3000N, Hitachi).

4-3. Results and Discussion

Figure 4-2a shows the magnified WAXD diffraction of PBS5i films immersed in SBF solutions for various times. In addition, the WAXD pattern of HAp, which was synthesized in our laboratory, is displayed on the right, where several unique peaks, reflective of HAp crystal, can be arranged. Especially, the most dominant HAp peak appears at 32_, which corresponds to the (211) plane of the hexagonal HAp crystal [22]. At initial SBF immersion time, no new peak is observed for the PBS5i. After immersing in SBF for 80 h, the new peak at 32°, the remarkable peak of HAp, can be detected. Besides, this peak gradually intensifies as the immersion time of the PBSi in SBF solution becomes longer. This result indicates that more HAp is formatted on the surface of PBS ionomer films in proportion to the immersion time.

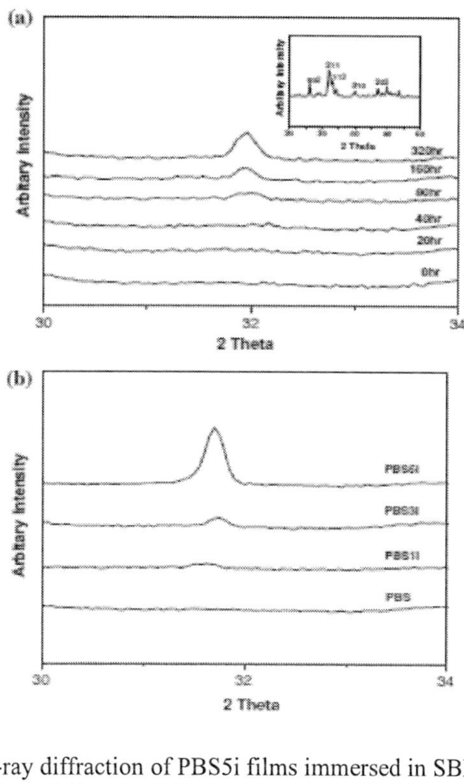

Figure 4-2. a Wide-angle X-ray diffraction of PBS5i films immersed in SBF solutions for different intervals (above right: the X-ray diffraction of hydroxyapatite). b Wide-angle X-ray diffraction images of PBS and PBS ionomers immersed in SBF solutions for 320 h.

This result is consistent with the data of SEM (as shown in Figure 4-3); at immersion time of 80 h, the formation of inorganic HAp particles was observed on the surface of PBSi films. As the immersion time increases, not only HAp particles grew with the larger size with sizes but also these particles were regularly dispersed. This result can be interpreted subsequently.

When the PBS ionomer was immersed in the SBF, it may release the positive Na^+ ions of $SO^{3-}Na^+$ ionic group into the surrounding fluid via exchange with H_3O^+ ions in the SBF fluid. As a result of this ion exchange, the PBS ionomer created $SO^{3-}H^+$ groups on its surface. In fact, as the immersion time increases, PBS ionomer liberated a large amount of Na^+ ion into the SBF, consequently generating a PH increase in the SBF fluid. These $SO^{3-}H^+$ groups immediately interacted with the calcium ions in the SBF fluid to form a calcium sulfonate. The calcium sulfonate thereafter incorporated the phosphate ions in the fluid to deposit the HAp on the surface of PBS ionomer. Here, the appearance of the calcium phosphate may be attributed to an electrostatic interaction between the positively charged calcium sulfonate and the negatively charged phosphate ions in the SBF. As the immersion time increases, HAp was shown to grow toward the outside of the PBS ionomer surface by continuously absorbing the calcium and phosphate ions in the SBF. As a result, it can be clearly summarized that the sodium sulfonate ionic group can play a main role in the formation of HAp crystalline on the surface of PBS ionomer. Furthermore, this observation is consistent with the work of Hiroaki et al., who investigated mechanism of biomineralization of HAp on a sodium silicate glass [23]. This mechanism is illustrated in Figure 4-4.

Figure 4-2b displays the magnified WAXD images of PBS and its ionomers immersed in SBF solution for 320 h. As the ionic group concentration increases, the intensity of the characteristic HAp peak at about 32° also increases, proving that HAp crystal formation and growth is proportional to the ionic group concentration. This observation demonstrates that the added sodium sulfonate ionic groups operate as the functional groups with negative charge densities which can bind relatively plentiful Ca^{2+} ions in SBF solution, generating that not only contributes to enhance the interface bonding strength between HAp and PBS but also serve as active sites allowing HAp crystals to grow on the surfaces of the polymer matrix.

We also checked TGA curves of pure PBS and its ionomer immersed in SBF solution for 320 h, where pure PBS exhibits about 0.3% residual weight. With the increase of added ionic group mol%, the residual weight of PBS ionomer shows the higher value. The maximum extent may be observed for the PBS5i, where its value is about 2.5%. This result confirms that a large amount of inorganic HAp may remain with the increase of the sodium sulfonate ion groups, since there is little possibility that organic polymer molecules can be survived at very higher temperature (data not shown). Based on this finding, we can conclude that the sodium sulfonate ionic groups may take active part in the production of HAp crystalline, leading to the more deposition of HAp on the surface of PBS ionomer.

Figure 4-5 shows the FT-IR spectra of PBS and its ionomers following immersion in SBF solutions for 320 h. In PBS5i, the characteristic band of PO_4^{3-}, corresponding asymmetrical stretching [24, 25], can be apparently detected at 1038 cm^{-1} in comparison with pure PBS. This characteristic band as well as TGA curves may be powerful evidence that the sodium sulfonate ionic groups promote the creation and growth of HAp crystals and then much amount of HAp is formed on the surface of the PBS ionomers.

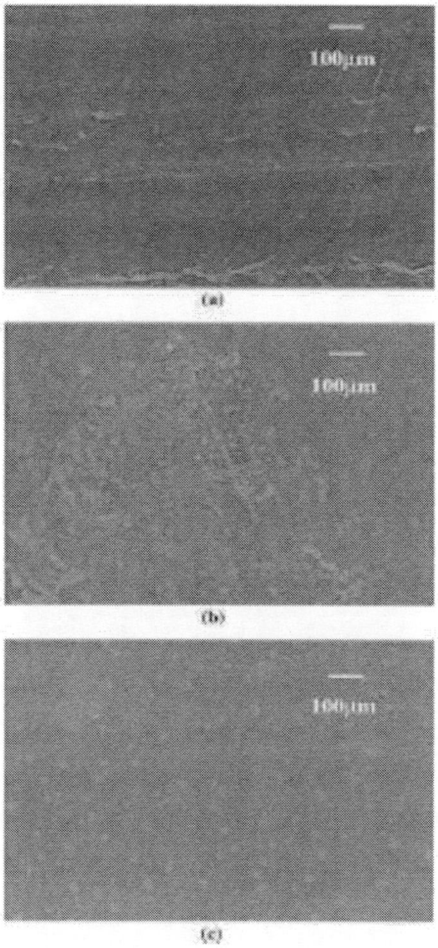

Figure 4-3. SEM micrographs of PBS5i: a before SBF immersion, b after SBF immersion (80 h), and c after SBF immersion (320 h).

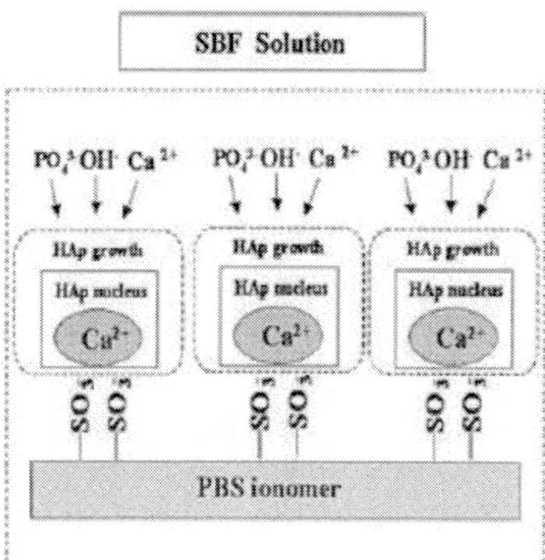

Figure 4-4. Schematic illustration of the formation and growth of a Hap crystal on a PBS ionomer.

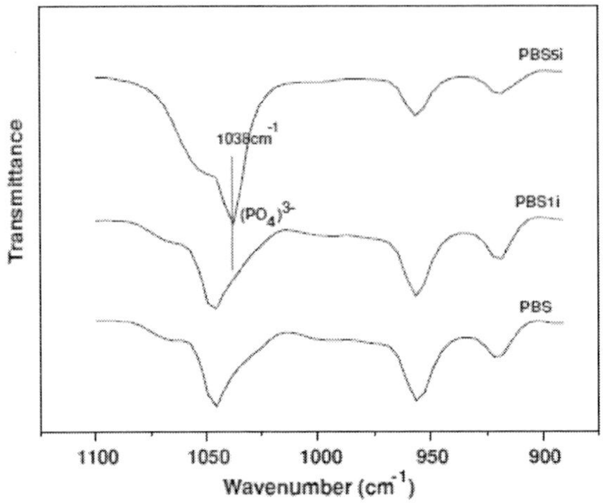

Figure 4-5. FT-IR spectra of PBS and its ionomers after immersion in SBF solutions for 320 h.

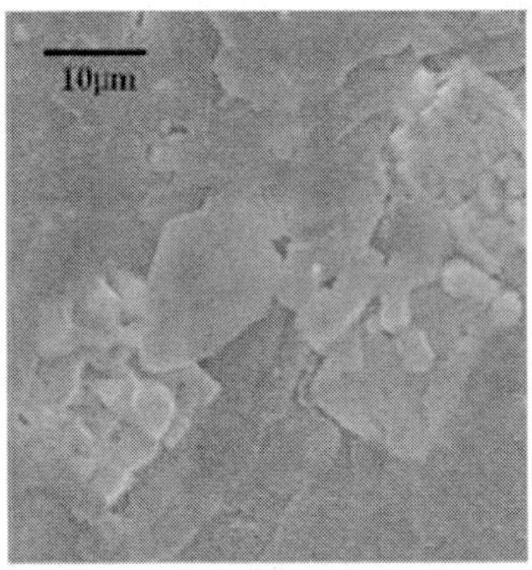

(a)

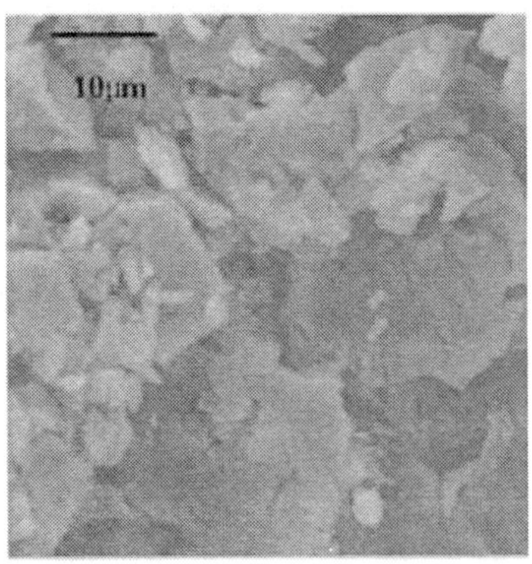

(b)

Figure 4-6. SEM images of PBS ionomers after SBF immersion for 320 h: a PBS1i and b PBS5i.

Figure 4-6 displays the SEM micrographs of PBSi films after SBF immersion (320 h). As for the PBS1i, the Hap crystal was successfully integrated into the matrix with no visible agglomerate formation at low particle amount. As the ionic group content grows richer, more and more Hap crystal deposits on the surface of PBSi. The reason for this phenomenon may be explained that large amount of sodium sulfonate ionic group interact favorably with the calcium ion and phosphate ion in the SBF solution, producing that the formation of HAp crystal may become easier. Additionally, in the SEM image of PBS5i, there are not only bulks of crystal deposits but also holes in the bulk of deposition. These holes are maybe very suitable for the ingrowths of the surrounding tissue and assist the bone formation. In summary, it can be clearly concluded that the ionic group in the PBSi may be decisive factors in growing HAp, and it is anticipated that this novel material can contribute to excellent biopolymer.

4-4. Summary

PBS-based ionomers (PBSi), incorporating the solfonated dimethyl fumarate (SDMF) with the sodium sulfonate ionic group, and hydoxyapatite (HAp) composites were fabricated by soaking in the SBF solution and we investigated the influence of ionic groups on the HAp crystal formation and growth. By WAXD, TGA and FT-IR analyses, relative to pure PBS, the biomimetic growth of HAp on the surface of PBS ionomer was remarkably increased and this ratio was predominated in proportion to the ionic group mol%. This demonstrated that the sulfonate groups, with the negative charge densities, interact favorably with the calcium ions and phosphate ions in the SBF solution, generating that the formation of HAp crystal may become more straightforward. The bulks of crystal deposits and holes, which were observed at PBS ionomer, may be anticipated to be very satisfactory for the ingrowths of the surrounding tissue and the bone formation.

References

[1] Beppu MM, Torres MA, Aimoli CG, Coulart GAS, Santana CC (2005) *J Mater Res* 20:3303

[2] Cho SB, Nakanishi K, Kokubo T, Soga N, Ohtsuki C, Nakamura T, Kitsugi T, Yamamuro T (1995) *J Am Ceram Soc* 78:1769

[3] Eisengerg A, Hird B, Moore RA (1990) *Macromolecules* 23:4098

[4] Eisengerg A, Kim KS (1998) *Introduction to ionomers.* Wiley, New York

[5] Guild FJ, Bonfield W (1993) *Biomaterials* 14:985

[6] Han SI, Kim DK, Im SS (2003) *Polymer* 44:7165

[7] Hao J, Liu Y, Zhou S, Li Z, Deng X (2003) *Biomaterials* 24:1531

[8] Huang M, Feng JQ (2003) *J Mater Sci Mater Med* 14:655

[9] Ihn KJ, Yoo ES, Im SS (1995) *Macromolecules* 28:2460

[10] JCPDS card no. 73-293 (2000) ICDD, PCPDFWIN v.2.1 JCPDS—International Centre for Diffraction Data

[11] Lim JS, Isao N, Im SS (2008) *Eur Polym J* 44:1428

[12] Lim JS, Lee YI, Im SS (2008) *J Polym Sci B Polym Phys* 46:925
[13] Liou SC, Chen SY, Liu DM (2003) *Biomaterials* 24:3981
[14] Molnar A, Eisenberg A (1992) *Macromolecules* 25:5774
[15] Murugan R, Ramakrishna S (2005) *Cryst Growth Des* 5:111
[16] Murugan R, Ramakrishna S (2006) *Acta Biomater* 2:201
[17] Petricca SE, Marra KG, Kutma PN (2006) *Acta Biomater* 2:277
[18] Shikinami Y, Okuno M (1999) *Biomaterials* 20:859
[19] Shikinami Y, Okuno M (2001) *Biomaterials* 22:3197
[20] Takadama H, Kim HM, Kokubo T, Nakamura T (2001) *Chem Mater* 13:1108
[21] Wang M, Bonfeld W (2001) *Biomaterials* 22:1311
[22] Wang M, Bonfiel W (2001) *Biomaterials* 22:1311
[23] Wang M, Deb S, Bonfelid W (2000) *Mater Lett* 44:119
[24] Xiao Y, Li D, Fan H, Li X, Gu Z, Zhang X (2007) *Mater Lett* 61:59
[25] Xiao Y, Xu Y, Lu J, Zhu X, Fan H, Zhang X (2007) *Mater Lett* 61:2601

In: Polymer Chains
Editors: Régine Penzkofer and Yasue Wu

ISBN: 978-1-62100-422-6

Chapter 6

INTERACTIONS OF SMALL PERMEANTS IN POLYMERIC MATRICES: A LOW-TEMPERATURE DIFFERENTIAL SCANNING CALORIMETRY STUDY

Shingjiang Jessie Lue*

Department of Chemical and Materials Engineering,
Chang Gung University, Taiwan

ABSTRACT

This chapter discusses the interaction between small permeant states in polymer-based membranes. The permeants, in the vapor and liquid phases outside the membranes, are sorbed into different membrane systems, which include hydrophilic, hydrophobic and organic/inorganic composite films. The permeants and membranes are characterized using low temperature differential scanning calorimetry (DSC). The permeant state can be classified into freezable, intermediate, or non-freezable bound states. We use three systems to represent these various permeant-polymer interactions. In the weak interaction system of water and organic solvents in poly(dimethyl siloxane) (PDMS), the sorbed permeant exists in both freezable and non-freezable states. The fusion temperature and enthalpy of the freezable permeant resembles those of the bulk, pure solvent. At low sorption levels, the permeant molecules are present in the non-freezable bound state and the swollen PDMS film behaves like a pseudo-homogeneous system. Beyond a critical sorption level, all of the excessive permeant molecules in the PDMS are in the freezable state and the films behave like a phase-separated solution. In medium interaction systems, such as water in polyvinyl alcohol (PVA), the water distributes into freezable (with freezing at 0°C), intermediate (with sub-zero freezing temperatures), and non-freezable bound water. However, the incorporation of fumed silica (FS) nano-particles into the PVA results in a complex matrix in which FS generates dilution, shielding and suppression of hydrogen bond formation, as well as a shift in the freezable and non-freezable water equilibrium. In the strong interaction system of water in perflurosulfonic acid (Nafion®), no free water is detected. Only non-freezable bound water and intermediate water with subzero fusion temperatures are found. Mathematical models are

* Tel: +886-3-2118800, fax: +886-3-2118700, email: jessie@mail.cgu.edu.tw (Prof. S.J. Lue).

proposed to express the permeant state distribution in these three systems. The interpretation of the corresponding plots of the solvent content of each state versus total solvent uptake is elucidated. These various micro-interactions may have different impacts on the transport properties of polymer-based membranes, hydrogels, and solid electrolytes used in separation, gas barriers, controlled drug release, fuel cells, and other applications.

NOMENCLATURE

ΔH_1	Polymer crystal melting enthalpy per gram of wet sample (J/g)
ΔH_2	Permeant fusion enthalpy per gram of wet sample (J/g)
ΔH_m	Polymer crystal melting enthalpy per gram of dry film (J/g)
ΔH_f	Permeant fusion enthalpy per gram of dry film (J/g)
ΔH_f^*	Permeant fusion enthalpy per gram of sorbed permeant (J/g)
ΔH_s^o	Pure solvent fusion enthalpy (J/g)
K	Equilibrium constant between bound and free (including freezable) water associated with filler
T_f	Permeant fusion temperature (°C)
T_m	PDMS crystal melting temperature (°C)
w	Permeant mass uptake per gram of dry membrane (g/g)
W	Water content (g/g membrane)
W'	Water content in composite when FS acts as a diluter (g/g composite)
W″	Water content in composite when FS acts as both a diluter and hydrogen bonding suppresser (g/g composite)

Greek letters

ϕ	Filler mass fraction
γ	Number of water molecules per SiO_2 molecule (mol/mol)
Λ	Number of free or non-freezable water molecules per hydrophilic functional group
ξ	Hydrogen-bonding hindrance coefficient due to filler

Subscript

f	in afreezable state
b	in a bound, non-freezable state
c	critical threshold

1. INTRODUCTION

Water and organic solvents (penetrants) in polymer networks exhibit specific interactions with the polymer chains. Water, for example, has been shown to be present in free and/or bound states, as well as in an intermediate state [1]. The free water molecules are considered to be far away from the polymer chains and surrounded by other water molecules. Therefore, free water has the same properties as the pure bulk water: it freezes at 0°C and evaporates at 100°C. On the other hand, bound water is situated immediately adjacent to the polymer chains and has strong interactions with the polymer matrix. The strong interactions between the bound water and the polymer suppress many water characteristics. The bound water does not freeze below 0°C, nor does it evaporate above 100°C [1]. The bound water is also called non-freezable, non-freezing, or unfrozen water. Intermediate water has moderate interactions with the polymer chains and can freeze at sub-zero temperatures.

The amounts of free and bound water are important characters in many applications. For example, these water states play different roles in conductivity improvement of solid electrolytes for fuel cells [2,3,4,5,6,7]. The electrolyte membranes with high freezable water contents exhibit significantly high ionic conductivity [6,7] while bound water helps maintaining conductivity at sub-zero temperatures and high temperatures above 80°C [8,9]. The water states in cellulose acetate membranes are related to the permeability in a reverse osmosis process, as shown by Taniguchi and Horigome [10]. In addition, water state distribution in hydrogels is associated with swelling kinetics [11], glass-transition temperatures [12], and degradability [13] of the polymers. Yamane et al. [14] found that enzyme (lipase) activity increased remarkably at optimal free water content by adding hydroscopic sugars in a water-immiscible organic solvent.

In addition to water, other small penetrants in polymers play a critical role on permeant transport properties and polymer chain packing and orientation characteristics. Membrane separation, food packaging, controlled drug release, barrier films, and conductive membranes in fuel cells are some examples utilizing these properties. The amount of penetrants sorbed in polymers depends on the penetrant activity, temperature of the surroundings, polymer characteristics, and the presence of other penetrants. The mechanical and physical properties of the polymer materials may be altered after sorption of these penetrants, including transport properties, optical, thermal, and electrical properties of the polymer films. Furthermore, the absorbed permeants in polymer systems can affect the chain flexibility and mobility, which in turn impacts the penetrant diffusivity. It is therefore necessary to understand the states of permeants in the polymer systems in order to elucidate the mass transfer mechanism and the material properties.

Many instruments have been used to measure the permeant states in polymer systems [15,16,17,18,19]. Among these several techniques, differential scanning calorimetry (DSC) has been used extensively to study the different states of penetrants in a polymer network. Not only can the solvent freezing temperature and the fusion enthalpy in the polymer systems be measured, but the polymer characteristics upon solvation can also be determined. In this study, we will use many systems to illustrate various types of permeant-polymer interactions and the consequences of penetrate states and polymer modifications.

2. Small Permeants in a Poly(Dimethylsiloxane) Membrane

2.1. Water States in a Poly(Dimethylsiloxane) Membrane

Poly(dimethylsiloxane) (PDMS) is a hydrophobic polymer with a number of exceptional physical properties that make it desirable for a range of industrial applications. It is an amorphous rubbery polymer and is used as a first illustration of a simple system. To test for distribution of permeant states, PDMS films were immersed in water for various lengths of time to achieve different levels of water uptake. The wet films were then sealed in a DSC pan and quenched in liquid nitrogen. The pan containing the frozen PDMS sample was heated in a DSC and a thermogram was recorded [1]. Typical DSC thermograms of PDMS with different amounts of sorbed water are shown in Fig. 1. The dry PDMS membrane exhibits only one endothermic peak in the low temperature region (approximately –47°C), which corresponds to the melting point of PDMS crystalline [1,20,21]. In the moistened PDMS films, an additional endothermic peak at 0°C is observed, which is due to the fusion of water molecules present inside the PDMS membrane (Fig. 1). As more water is sorbed into the PDMS, the endothermic peak at 0°C becomes larger.

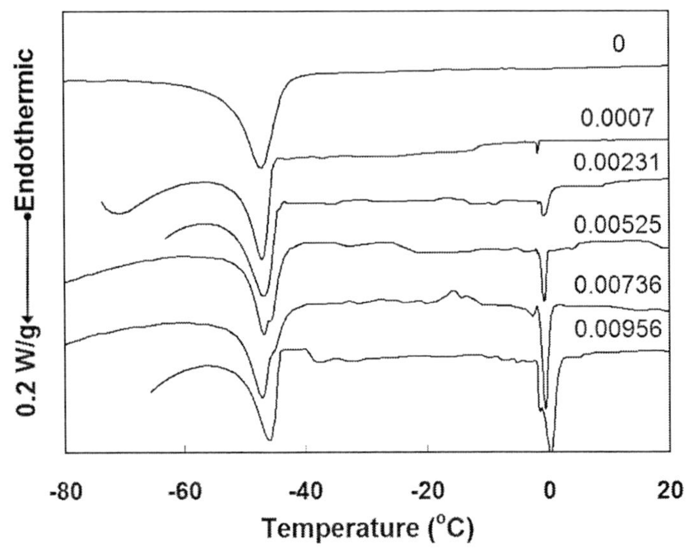

Figure 1. DSC thermograms of dry and moistened PDMS membranes with various water contents (in unit of g/g dry PDMS film) shown next to the curves (data from [1]).

When the PDMS film was suspended above water and allowed to take up water vapor, the sorbed PDMS films also showed similar thermograms for various vapor uptakes. There were two endothermic peaks: one for PDMS crystalline melting (–47°C) and one for water solvent fusion (at 0°C). Similarly, the higher vapor uptake sample exhibited a larger water fusion peak [1].

The DSC results, including the phase transition temperature and the associated enthalpy value, are useful indicators to determine water and PDMS characteristics. The polymer crystalline melting enthalpy from the DSC output (ΔH_1, in J/g wet sample) is converted to the dry weight base using the following equation:

$$\Delta H_m = \Delta H_1(1+w) \quad (1)$$

where ΔH_m is the polymer crystal melting enthalpy (in J/g dry PDMS) and w is the water uptake (in grams of solvent per gram of dry PDMS film). Similarly, water fusion enthalpy from the DSC output (ΔH_2, in J/g wet sample) was converted to the dry weight base:

$$\Delta H_f = \Delta H_2(1+w) \quad (2)$$

where ΔH_f is solvent fusion enthalpy, in J/g dry PDMS. We also converted this solvent fusion enthalpy based on the solvent weight:

$$\Delta H_f^* = \Delta H_2(1+w)/w \quad (3)$$

where ΔH_f^* is in J/g solvent.

The values of ΔH_m and ΔH_f can be plotted as a function of w, as shown in Fig. 2. This figure combines the data from the PDMS sorption measurements from water liquid and vapor phases. The data points for ΔH_m in Fig. 2 do not deviate significantly in the range of 11.1 to 11.5 J/g dry PDMS. This implies that the polymer crystallinity does not change significantly upon cooling after water sorption. The data points for ΔH_f in Fig. 2, however, seem to lie in a straight line. The slope of this straight line represents the enthalpy increase associated with one gram of sorbed water. Specifically, the slope was 333.8±9.0 J/g for the PDMS sorbed from both the liquid and vapor phases. This value is not significantly different from the pure water fusion enthalpy reported in the literature (333.5–333.6 J/g [22,23]).

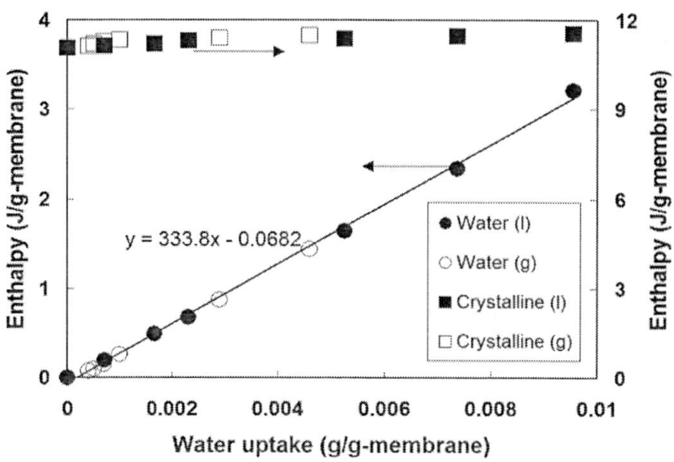

Figure 2. Water fusion enthalpy, ΔH_f (left axis), and PDMS crystalline melting enthalpy, ΔH_m (right axis), of a PDMS membrane containing penetrants sorbed from liquid (filled symbols) and vapor (open symbols) phases at a sorption temperature of 25°C (data from [1]).

The PDMS melting temperature (T_m) of the water-sorbed PDMS was in the range of –45.8 to –47.5 °C, (Fig. 1, [20]) implying that the sorbed water barely affects the polymer. Note, however, that the maximum water uptake was less than 0.01 g/g PDMS. At these low sorption levels, both water and PDMS do not demonstrate significant modification from their pristine states and almost all of the water molecules in the PDMS are in the freezable state. These sorbed water molecules thus behave like pure water, and the water present in the PDMS membrane do not alter the crystalline melting temperature or the fusion enthalpy.

2.2. Organic Solvent States in PDMS Membrane

The low-temperature DSC thermograms for PDMS sorbed with organic solvents were also determined. The tested solvents included cyclohexane (Cx), benzene (Bz), toluene (Tol), and m-xylene (Xyl), as they are known to have high solubilities in the PDMS film. The DSC thermograms of PDMS films with different amounts of Bz are shown in Fig. 3. When low levels of Bz were sorbed in the PDMS, only one endothermic peak was observed between –47°C and –49°C, which corresponds to the PDMS crystalline melting temperature. As the Bz content of the PDMS increases beyond a certain value (0.17 g/g dry PDMS in this case), an additional endothermic peak appears at 4°C, representing the fusion temperature of Bz [21,22,23]. In addition, the plot of ΔH_f versus w reveals a segment line of Y=0 at low sorption levels (less than 0.17 g/g dry PDMS) and a straight line with a positive slope at higher Bz sorption levels (as shown in the square symbols in Fig. 4).

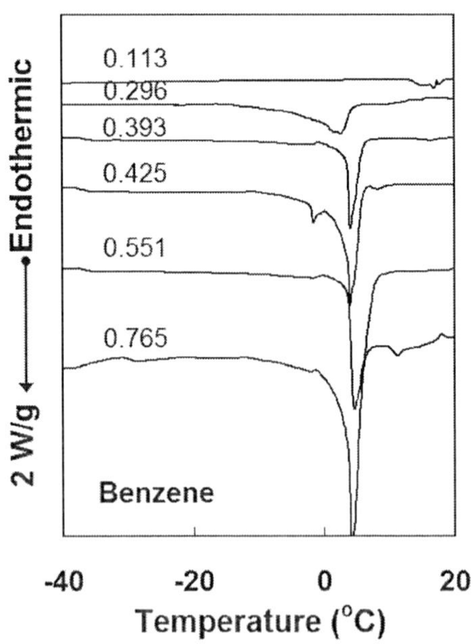

Figure 3. DSC thermograms of benzene sorbed PDMS membranes with various solvent uptakes (unit in g/g dry membrane) (data from [1]).

The DSC thermograms of Tol, Xyl, and Cx sorbed in PDMS membranes exhibited similar trends to those seen in Fig. 3. Below certain threshold values, there were no solvent fusion peaks present in the thermograms. Beyond these thresholds, the solvent fusion endothermic peaks were visible at temperatures corresponding to the permeant fusion, as follows: −95°C for Tol [22,23], −49°C for Xyl [22,23], and 5°C for Cx [22,23,24]. Although these threshold values vary for different permeants (Fig. 5), the solvent fusion enthalpies increased linearly with increasing amount of solvent uptakes in excess of the thresholds. The PDMS sorbed with the vapor phase of solvents followed similar patterns to those with liquid penetrants (as shown in the open symbols in Figs. 4–5).

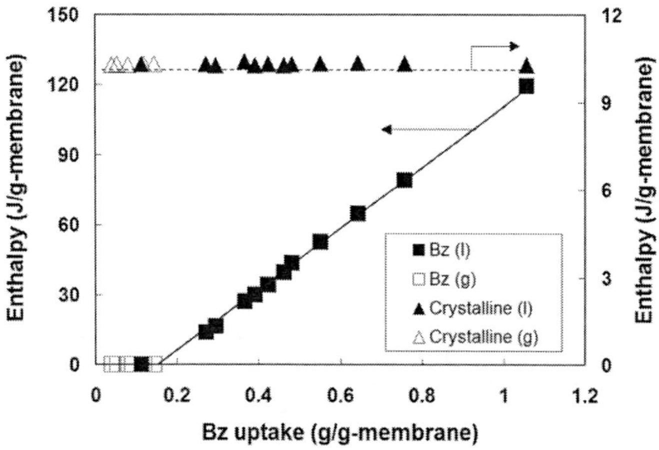

Figure 4. Solvent fusion enthalpy (left axis) and PDMS crystalline melting enthalpy (right axis) for PDMS containing benzene sorbed from the liquid (filled symbols) and vapor (open symbols) phases at a sorption temperature of 25°C (data from [1]).

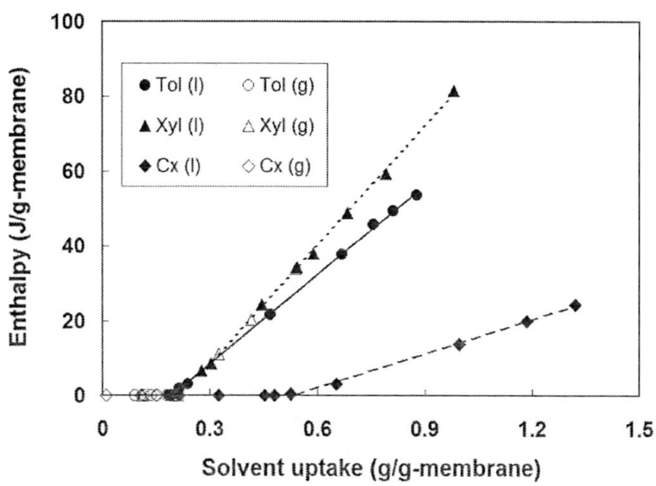

Figure 5. Solvent fusion enthalpy for PDMS membranes containing penetrants sorbed from the liquid (filled symbols) and vapor (open symbols) phases at a sorption temperature of 25°C (data from [1]).

The slopes of the lines in Figs. 4–5 represent the fusion enthalpies per gram of the solvents sorbed in the PDMS in excess of the threshold uptakes. The values were estimated to be 134.8, 80.0, 106.1, and 30.6 J/g for Bz, Tol, Xyl, and Cx in the PDMS membrane, respectively. These data and their 95% confidence intervals confirm that the sorbed solvent fusion enthalpy is not significantly different from that of pure bulk solvent (Table 1). From this analysis, we suggest that sorbed solvent in excess of the critical threshold displays similar characteristics to the pure solvent.

Table 1. Fusion temperature (T_f) and fusion enthalpy (ΔH_f) of freezable solvent in PDMS membrane [1]

Solvent	Water	Benzene	Toluene	Xylene	Cyclohexane
T_f, experimental [a] (°C)	−0.02±1.27	3.50±0.99	−95.55±0.36	−48.91±0.70	4.67±1.40
T_f, literature data (°C) [b]	0.00	5.53	−94.99	−47.87	6.72
ΔH_f of freezable solvent in PDMS, experimental [a] (J/g)	333.8±9.0	134.8±2.1	80.0±2.6	106.1±3.7	30.6±2.4
ΔH_f of pure solvent, experimental (J/g)	333.1	132.0	79.9	109.0	32.0
ΔH_f, literature data[b] (J/g)	333.5–333.6	125.9–126.3	71.8–72.0	109.0	31.8

[a] Mean ± 95% confidence interval.
[b] Obtained from [21,22,23,24].

Table 2. Maximum non-freezable contents, solubility parameters, and solubility estimates of various solvents and PDMS [1]

Component	PDMS	Water	Benzene	Toluene	Xylene	Cyclohexane
Maximum non-freezable content [a] (g/g membrane)	— [d]	0.00021 ±0.00011	0.17±0.01	0.20±0.01	0.22±0.01	0.54±0.04
Solubility parameter [b] ($J^{1/2}/cm^{3/2}$)	14.9	47.9–48.1	18.5–18.8	18.2–18.3	18.0	16.7
Solubility estimate [c] (g/g)	— [d]	0.0076	0.64	0.68	0.76	1.0

[a] Mean ± 95% confidence interval.
[b] Cited from reference [25].
[c] Estimated using the UNIquac group-Functional Activity Coefficient (UNIFAC) model [26].
[d] Not applicable.

A certain sorption level, however, has to be reached for a given solvent to respond like the pure solvent. This threshold can be estimated from the X-intercept of the rising line in Figs. 4–5. We called these thresholds “maximum non-freezable contents” and the values of 2.1×10^{-4}, 0.17, 0.20, 0.22, 0.54 g/g dry PDMS were obtained for water, Bz, Tol, Xyl, and Cx, respectively, in the PDMS membrane (Table 2). With any sorption level below these thresholds, the permeant fusion endothermic peaks were undetectable and only the PDMS crystal fusion peak was observed. The interpretation of this lack of permeant solvent fusion phenomena will be discussed in detail in the following sections.

2.3. Polymer Modifications

The PDMS crystal melting temperature and fusion enthalpy from the DSC thermograms also revealed useful information on polymer modification. At low temperatures, PDMS undergoes several transitions, such as chain-segment motion (glass transition) and ordering (crystallization), etc. [27]. In addition, the polymer molecular weight, preparative method [28,29], and cooling rate prior to DSC analysis [28,30,31,32,33] are known to influence the onset and presence of these transitions and the associated enthalpies [34,35,36,37]. The glass transition temperature (T_g) of the PDMS membrane is –110°C [20] and the crystalline melting temperature of PDMS (T_m) was about –46°C (Fig. 1) [38,39,40].

The heat of fusion enthalpy (ΔH_m) of the tested PDMS crystalline region was almost 11.1 J/g (Fig. 2), which is in the lower range of the literature data [34,35,37]. The ΔH_m for completely crystallized PDMS is reported as 36.7 J/g [30,31], but higher molecular weights [30], higher polydispersity, and/or a higher degree of crosslinking in PDMS lower crystallinity, resulting in a lower ΔH_m [30,37].

2.3.1. Polymer Crystallinity in Water and Solvent Sorbed PDMS Film

The DSC thermograms of PDMS crystal melting with different water uptakes are shown in Fig. 6. Additionally,Table 3 shows the PDMS crystal melting temperature (T_m) and ΔH_m for the moistened films at different water uptakes. The T_m and ΔH_m values of both the liquid- and vapor-sorbed membranes showed similar results (Table 3) and their values were close to that of the dry PDMS membrane. Although, the ΔH_m values of water-sorbed samples were slightly higher (up to 3.5%) than that of the dry PDMS, the differences were within the experimental error.

The DSC thermograms of the Cx pure solvent and Cx-sorbed PDMS samples are shown in Fig. 7. Upon inspection of the figure, the pure Cx solvent shows two endothermic peaks at –87°C and 5°C that correspond to the solid-solid transition temperature and the fusion temperature of Cx, respectively [20,24]. The broad peaks at –50 ~ –64°C are due to modified PDMS crystalline melting. The transition temperature of this peak varies with the Cx uptake but poses a fairly constant fusion enthalpy value (10.2–10.7 J/g dry PDMS) when compared with an experimental error of 0.70 J/g. Therefore, it was hypothesized that this peak may have originated from the PDMS membrane transition in the presence of Cx [24]. Moreover, Fig. 8 shows the crystalline melting temperature changes in the PDMS films with different Cx solvent uptakes. As can be seen, the crystalline melting temperature of the PDMS membrane decreased with increasing Cx solvent uptake. Extrapolating the curve to a zero sorption level yielded a transition temperature corresponding to that of pure PDMS, which was close to the experimental value (–46°C) of the PDMS crystalline melting temperature. This finding implies that the peak at –50 ~ –64°C originated from the modified PDMS crystal region.

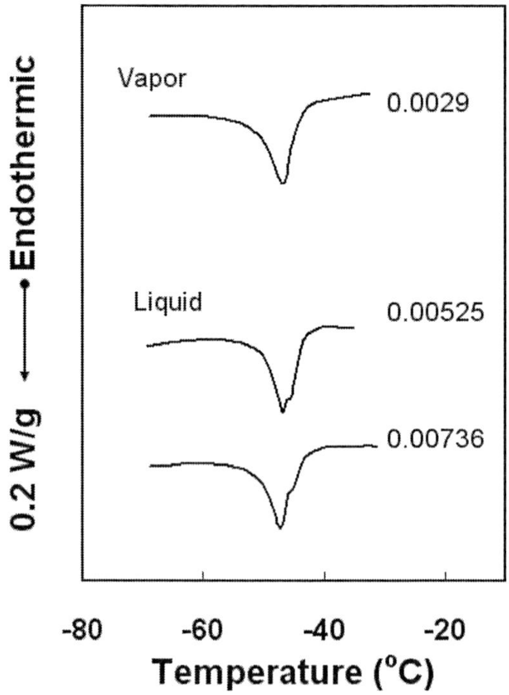

Figure 6. Typical thermograms of moistened PDMS films with water vapor and liquid uptakes [20].

Table 3. PDMS melting temperature and the melting enthalpy of dry and moistened PDMS films [20]

Water uptake (g/g dry membrane)	Liquid penetrant		Vapor uptake (g/g dry membrane)	Vapor penetrant	
	T_m (°C)	ΔH_m (J/g) [a]		T_m (°C)	ΔH_m (J/g) [a]
Dry PDMS	–46.0	11.1	Dry PDMS	–46.0	11.1
0.0007	–46.4	11.1	0.0004	–47.6	11.1
0.0017	–46.6	11.2	0.0005	–46.6	11.2
0.0023	–46.6	11.3	0.0007	–47.1	11.3
0.0053	–6.7	11.4	0.0010	–46.5	11.3
0.0074	–47.2	11.5	0.0029	–46.6	11.4
0.0096	–45.8	11.5	0.0046	–47.5	11.5

[a] Based on 1 g of dry membrane.

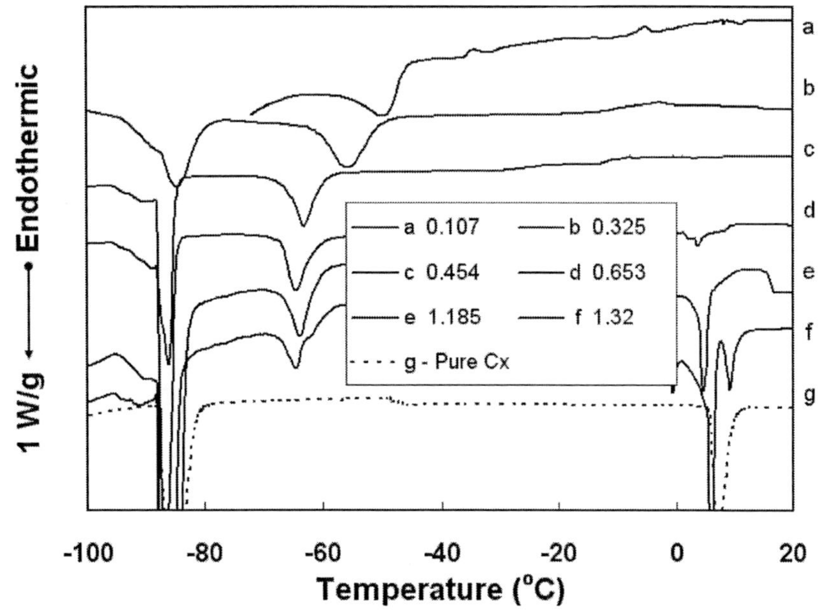

Figure 7. Typical thermograms of Cx-PDMS films at sorption temperature of 25°C. Data are normalized for dry PDMS weight; solvent content in g/g dry PDMS are shown in the legend [20].

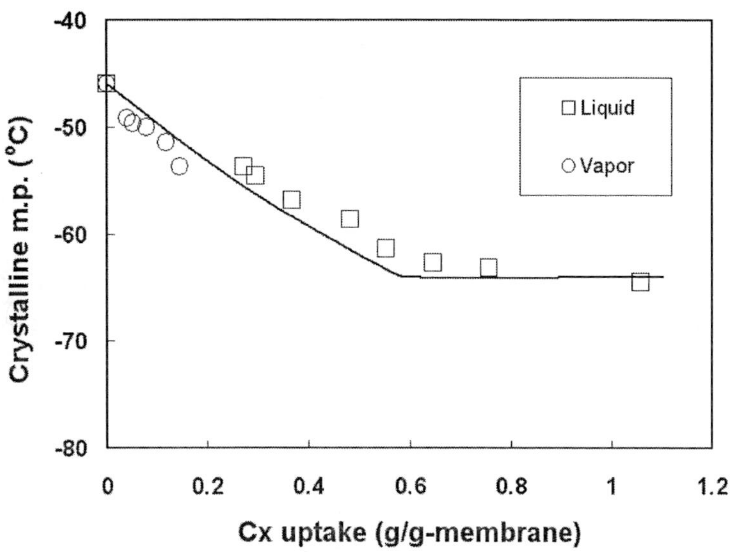

Figure 8. Crystalline melting temperature of PDMS samples with sorption from Cx vapor and liquid phases [20].

Similar trends for the changes in T_m and ΔH_m changes with respect to solvent uptake were also observed in the Tol-sorbed PDMS films. Figs.9 (a) and (b) show the thermograms between –100 and –40°C for the PDMS sorbed from liquid and vapor phases. The endothermic peak observed at –96°C originated from the Tol solvent and the other broad peak (–46 to –68°C) was due to the polymer crystallization transition. The ΔH_m of the Tol-sorbed

PDMS films was 10.5 J/g for Tol uptakes of 0.09–0.0875 J/g, regardless of the Tol phase penetrating into the PDMS [20]. Furthermore, the T_m of these samples significantly decreased from –46 to –68°C with increasing the Tol uptake. This decreasing trend for the toluene sorbed PDMS crystalline melting temperature was similar for both liquid and vapor permeants (Fig. 10).

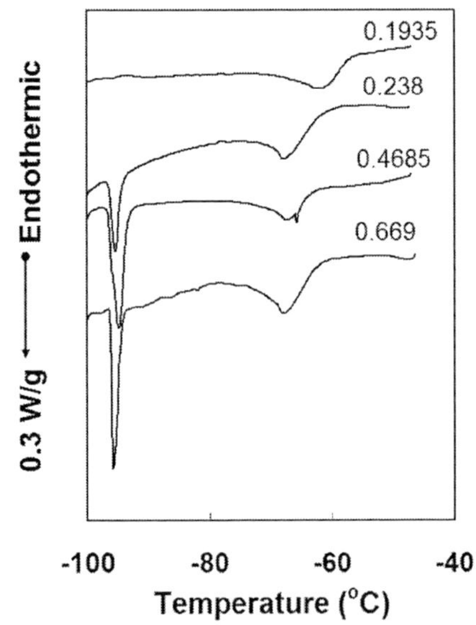

Figure 9. Typical thermograms of Tol-sorbed PDMS films (data normalized for dry PDMS weight; sorption contents shown next to the curves) (data from [20]).

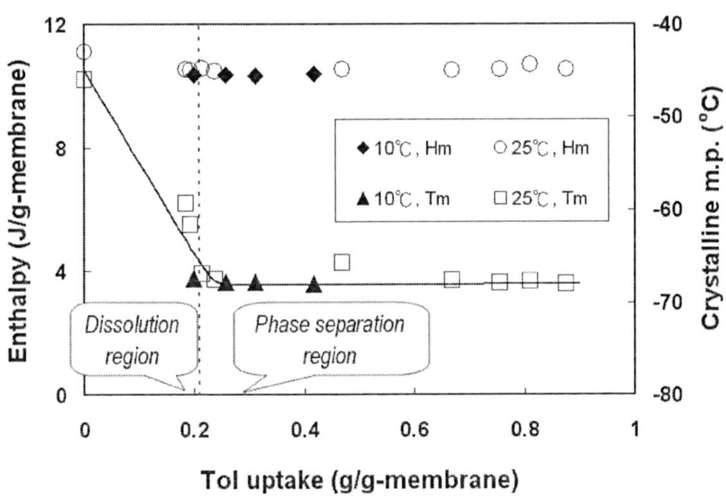

Figure 10. Crystalline melting enthalpy (◆,○ left axis) and crystalline melting temperature (▲,□ right axis) of Tol-sorbed PDMS samples at sorption temperatures of 10°C (filled symbols) and 25°C (open symbols) [20].

The polymer crystal modification of the Tol-sorbed PDMS was also carried out at 10°C of the sorption temperature (Fig. 10) for comparision with the findings at 25°C shown in the above sections. After sorption at either temperature, the Tol-sorbed PDMS film was cooled to "freeze" the structure present at the moment of sorption using liquid nitrogen. The resulting structure of the Tol-sorbed PDMS film remained unchanged until the solvent or polymer crystal was melted during the DSC heating cycle. Therefore, the melting point and the enthalpy reflect the physical properties of the solvent or polymer crystalline associated with the respective sorption temperature. Moreover, the T_m and ΔH_m depression for a comparable Tol content was similar at these same sorption temperatures (Fig. 10).

2.3.2. Phase Separation in a PDMS-Permeant System

PDMS is a rubbery polymer and the polymer segments exist in an amorphous phase. During the sorption of solvents at either 10 or 25°C, the penetrants are dispersed uniformly within the polymer matrix at low levels of sorption, as in a pseudo-homogeneous solution. When the samples are quenched, the PDMS chain mobility decreases and the polymer segments begin to form crystalline (Fig. 11 (a)). During the DSC heating cycle, the polymer crystalline is melted which is the only phase transition observed. The melting temperature is depressed because of the presence of the permeant, which is analogous to a melting point suppression of a true solution due to the presence of a soluble solute (left region in Fig. 10). The permeant does not show fusion character and is in a non-freezable state under this circumstance.

When the sorbed permeant uptake exceeds the solubility of the polymer (which depends on the degree of the chemical compatibility between permeant and polymer), the permeant then leads to cluster formation (Fig. 11 (b)), which results in phase separation from the polymer-rich domain. The permeant content in the polymer portion does not increase further and the T_m of the polymer crystal remains constant (right region of Fig. 10). The new permeant-rich phase demonstrates its own characteristics, such as solvent fusion above the critical threshold value (Figs. 4–5). The penetrant at this stage is in a freezable state. We therefore partition the solvent into these two states (non-freezable and freezable) at various uptake levels for these organic penetrants.

These critical thresholds are related to the chemical compatibility between the PDMS and the penetrant. A higher threshold is ascribed to higher penetrant solubility in the polymer region. Therefore, a penetrant with a solubility (Table 2) close to that of PDMS indicates that it is highly compatible with the polymer and results in a higher critical threshold. We will use the maximum non-freezable content to represent this threshold.

Table 2 summarizes the maximum non-freezable content, solubility, and estimated solubility using the UNIquac group-Functional Activity Coefficient (UNIFAC) model for each solvent [26]. The non-freezable content was the highest for Cx and lowest for Bz, following the order Cx > Xyl > Tol > Bz > water. The reverse order was observed for the difference in solubility between the solvent and PDMS. Cx had a solubility parameter (16.7 $J^{1/2}/cm^{3/2}$) closest to that of PDMS (14.9 $J^{1/2}/cm^{3/2}$), implying the highest degree of chemical similarity and mutual compatibility among the tested solvents. Hence, the greatest interaction was between the Cx solvent and the PDMS, as well as the highest amount of non-freezable solvent (Table 4). On the contrary, water has a solubility parameter of 48 $J^{1/2}/cm^{3/2}$, the least compatible with PDMS. The maximum non-freezable content determined was as low as

2.1×10^{-4} g/g, as determined from the *X*-intercept by linear regression of the enthalpy data in Fig. 2.

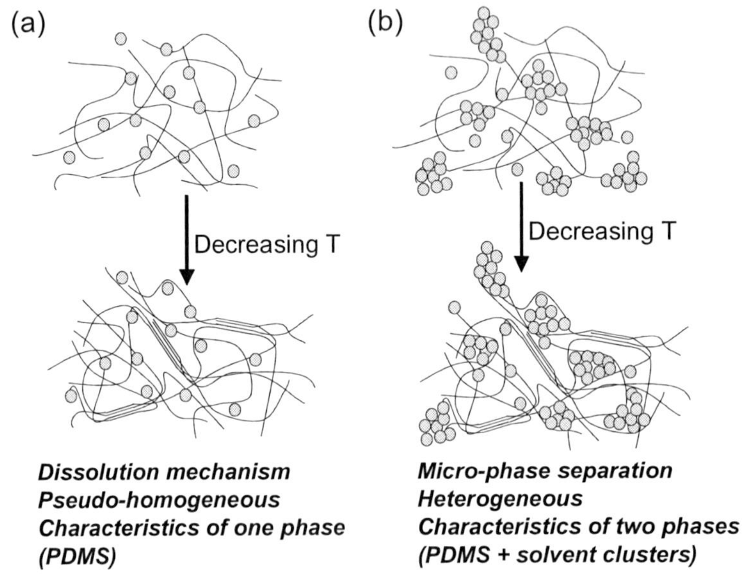

Figure 11. Illustration of polymer-penetrant interaction of (a) low sorption and (b) high sorption levels upon cooling [20].

2.4. Model Distribution of Permeant States in PDMS

The solvent enthalpy data can be used to calculate the distribution of the freezable solvent content (in g solvent/g dry film).

$$w_f = \Delta H_2(1+w)/\Delta H_s^o \tag{4}$$

where w_f is freezable solvent content and ΔH_s^o is the corresponding pure solvent fusion enthalpy. The sorbed solvent that does not show this fusion capability is considered to be "in the non-freezable state." The amount of solvent in the non-freezable state is calculated by subtracting the solvent in the freezable state from the total solvent uptake:

$$w_b = w - w_f \tag{5}$$

where w_b is non-freezable bound solvent content.

The sorbed solvent classified as freezable and non-freezable are shown in Fig. 12. The figure clearly illustrates that in the initial sorption stage, all solvents are in a non-freezable state. This holds true until the sorption level reaches the maximum non-freezable content. Beyond this threshold, the non-freezable solvent content remains constant and the freezable solvent content is detected and increases with solvent uptake. We can then write a set of mathematical equations to describe the freezable and non-freezable contents shown in Fig. 12:

$$w_b = w \quad for\ w \leq w_c \tag{6}$$

$$w_f = 0 \quad for\ w \leq w_c \tag{7}$$

For low sorption level.

$$w_b = w_c \quad for\ w \geq w_c \tag{8}$$

$$w_f = w - w_c \quad for\ w \geq w_c \tag{9}$$

Equations (6) and (7) are used for low sorption levels and Equations (8) and (9) for high (above threshold) sorption levels. The symbol w_c is the critical value of the maximum non-freezable solvent content in the above equations.

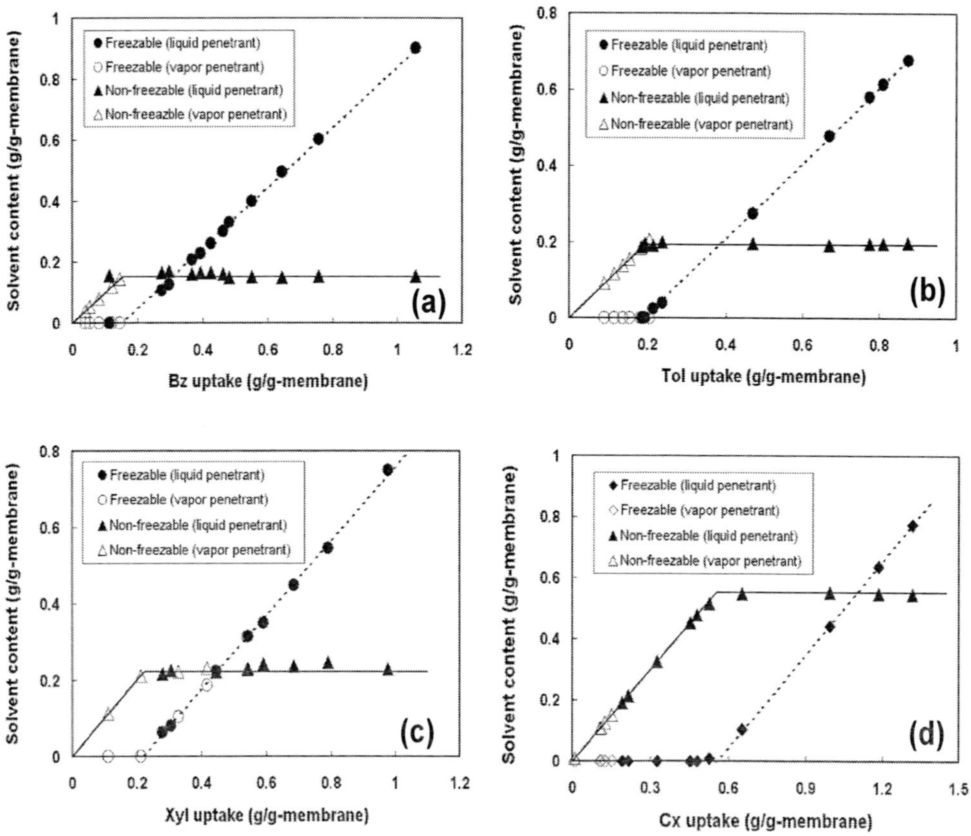

Figure 12. Relative quantities of freezable and non-freezable states of (a) Bz, (b) Tol, (c) Xyl, and (d) Cx in PDMS films (sorption temperature: 25°C) [1].

More specifically, in the plot of w_f and w_b versus w shown in Fig. 13, the slope of the non-freezable line is unity (Eq. (6)) in the initial stage and the freezable content is zero. This non-freezable line extends to the maximum non-freezable content, and then remains constant above the critical value. At this critical maximum non-freezable solvent content, freezable solvent content starts to increase with a slope equal to unity with respect to solvent content (Eq. (9)). This simple illustration is shown in Fig. 13 and can help identify the solvent state distribution in polymer systems.

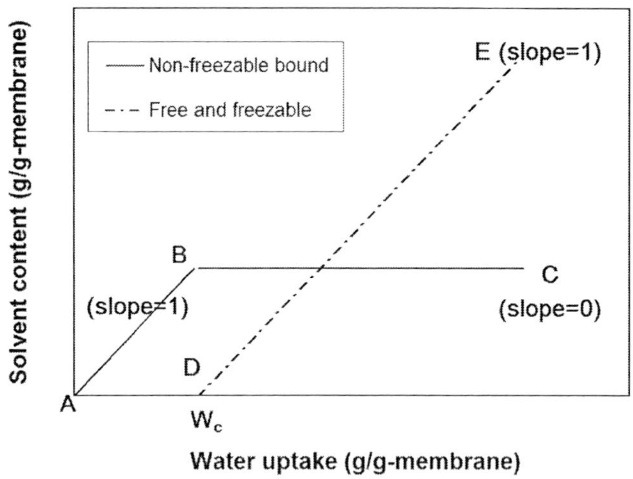

Figure 13. General permeant partition to free/freezable and non-freezable bound states in swollen PDMS film.

One may observe that the maximum non-freezable solvent content is a constant in the PDMS system. For example, Danjo *et al.* [41] studied the water states in polyvinyl pyrrolidone and reported that the maximum non-freezable content is approximately 0.7 g/g, independent of water content. We propose that the van der Waals interaction between the permeant and the PDMS helps the permeant occupy the first layer of free volume in the amorphous polymer segments. Increasing the solvent uptake further results in more non-freezable molecules distributing within the amorphous matrix, analogous to a relatively homogeneous phase (i.e., with only one endothermic peak present in the DSC thermograms) until the critical threshold w_c. The microstructure and morphology of the polymer matrix is not significantly affected upon solvation and the vacancy sites available for accommodating permeant molecules are solely related to the chemical compatibility, as mentioned before.

3. Interaction between Water and Polyvinyl Alcohol-Fumed Silica Nano-Composites

3.1. Water States in a Polyvinyl Alcohol Membrane

The hydrophilic polyvinyl alcohol (PVA) and its fumed silica (FS) nono-composites are illustrations of medium interaction systems. Fig. 14 (a) shows the DSC thermograms of moistened PVA films with different water contents. The PVA crystalline melting peak was absent in the scanned temperature range as it would occur only when the temperature reached 227°C [42]. No water fusion peak was observed in the scanned temperature range for sorption levels below 0.7 g/g. However, in the intermediate sorption levels (between 0.7 and 1.5 g/g membrane), two water fusion peaks were discovered at zero and subzero temperatures. The endothermic peak at ~0°C originates from ice crystal melting upon heating and it is attributed to free water [10]. The other broad endothermic peak appears at a subzero temperature range (−12°C) and is ascribed to intermediate freezable water. This indicates the presence of disordered, incomplete ice crystals [43] or liquid-like water confined in dynamic pores [44]. Such intermediate freezable water is also known as "freezable bound water" or "intermediate water" [45]. The water content in the PVA thus decreases the melting point depression of the freezable bound water. At higher water content, the melting peak of the freezable water shifts towards free water and these two melting peaks are merged above 1.5 g/g water content (Fig. 14 (a)).

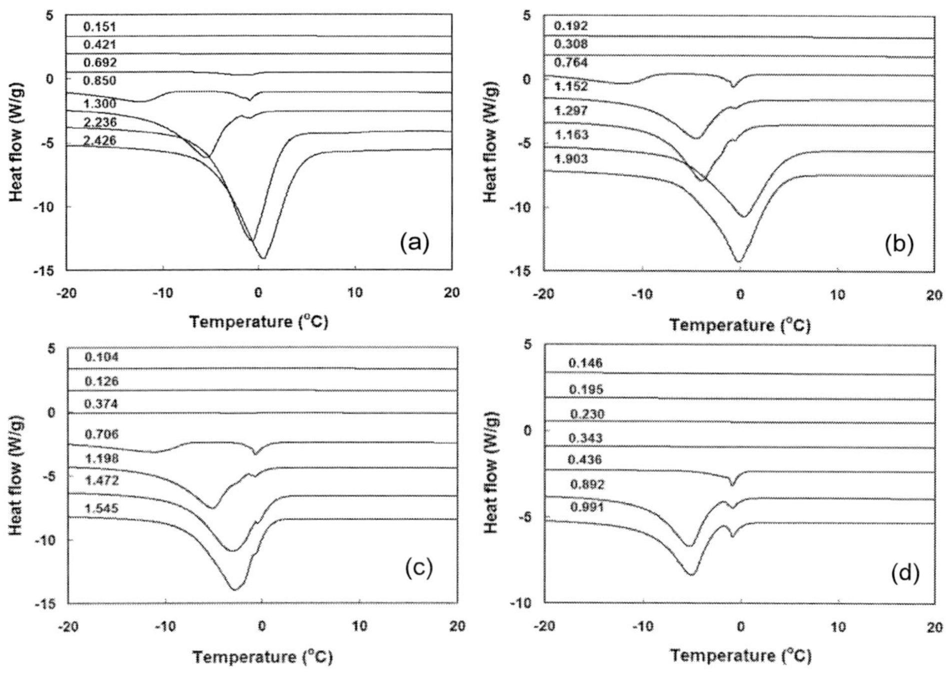

Figure 14. DSC thermograms of moistened (a) PVA membranes, (b) PVA-FS (10%) composite, (c) PVA-FS (20%) composite, and (d) PVA-FS (30%) composite at various water uptakes (in g/g-membrane) (data from [46]).

We sum up the fusion enthalpies of the free and intermediate water and refer to it as freezable in this section. Fig. 15 (a) shows that the sorbed water in PVA has partitioned into bound non-freezable (W_b) and freezable (W_f) water contents using Eqs. (4) and (5). When water uptake is less than the sorption capacity of the PVA (0.675 g/g-membrane), all sorbed water is in non-freezable bound state. Above the critical value, the excess water exists only in a free or freezable bound state, with no change in the amount of non-freezable water content. Statistical analysis indicated that in this region, non-freezable bound water content was not related to water uptake and the slope (right portion of the solid line in Fig. 13) was not significantly different from zero (with a p-value greater than 0.52 using statistical analysis).

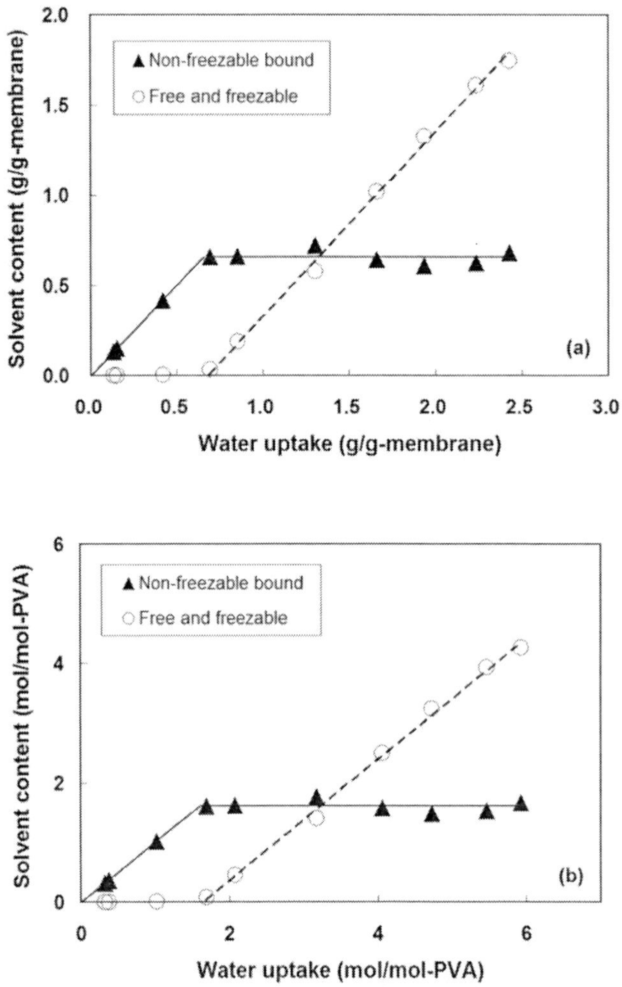

Figure 15. Distribution of water states into non-freezable bound and freezable states in PVA using mass unit (a) and molar unit (b) [46].

The following equations are used to convert the freezable and non-freezable water contents to molar units i.e., the number of free and non-freezable water molecules per PVA repeating unit, λ:

$$\lambda = W_{t,PVA} \cdot 44/18 \tag{10}$$

where $W_{t,PVA}$ is the water content (g/g-PVA) in pure PVA film and 18 and 44 are the molecular weights of water and the PVA repeating unit, respectively. Similarly, the number of non-freezable bound water molecules sorbed in one PVA repeating unit (λ_b) is calculated as:

$$\lambda_b = W_b \cdot 44/18 \tag{11}$$

The number of free and freezable bound water molecules sorbed in one mole of PVA repeating units (λ_f) is then determined by subtraction:

$$\lambda_f = \lambda - \lambda_b \tag{12}$$

The distribution between λ_b and λ_f as a function of λ is expressed in Fig.15 (b). The critical λ value (λ_c) corresponding to the point at which free and freezable water start to appear was extrapolated to be 1.65. The PVA used in this study was 99% hydralysed, meaning that 99 hydroxyl groups were hydrolyzed for every 100 PVA repeating units. If non-freezable bound water originates from the interaction between water molecules and polymer chains, including hydrogen bonding between water molecules and hydroxyl groups in the PVA polymer chains [47], then the critical value means that each PVA hydroxyl group can form 1.67 hydrogen bonds. This number is in line with the findings of Zhang *et al.* [45] and with those of Puffr and Sebenda [47]. Zhang *et al.* [45] reported 1.5 water molecules bound to each PVA unit. Puffr and Sebenda [47] found three water molecules (one tightly and two loosely) bound to every two amide groups in Nylon 6. However, other researchers have arrived at lower values of 0.5–0.66 [43,44,48]. The maximum number of bound water molecules that can be accommodated by each PVA unit therefore appears to depend on many factors, including polymer crystallinity, molecular weight, polymer free volume, and the chemical environment of the polymer matrix.

A mathematical expression for the lines in Fig. 15 (b) can be written as the following:

$$\begin{aligned} \lambda_b &= \lambda; \lambda_f = 0 \quad for\ \lambda < \lambda_c \\ \lambda_b &= \lambda_c; \lambda_f = \lambda - \lambda_c \quad for\ \lambda \geq \lambda_c \end{aligned} \tag{13}$$

where subscripts *b*, *c*, and *f* stand for non-freezing bound state, critical value, and freezable states respectively.

3.2. Water States in Polyvinyl Alcohol-Fumed Silica Membranes

The DSC thermograms of water sorbed PVA-FS composites are shown in Fig. 14 (b)–(d), while water contents of free/freezable and non-freezable bound states in the PVA-FS composites are shown in Figs. 16 (a)–(c). As can be seen, the PVA-FS composites absorbed water as a non-freezable bound state below a critical value. This critical value depends on the FS loading in the polymer. For PVA-FS composites, the critical water content was 0.565, 0.459 and 0.352 g/g, for 10%, 20% and 30% FS loading, respectively. These values are significantly lower than that of PVA (0.675 g/g). The lower binding capacity is assumed to be due to the dilution effect and suppressed hydrogen binding ability from the FS particles. A detailed discussion on this critical value is given in the following sections.

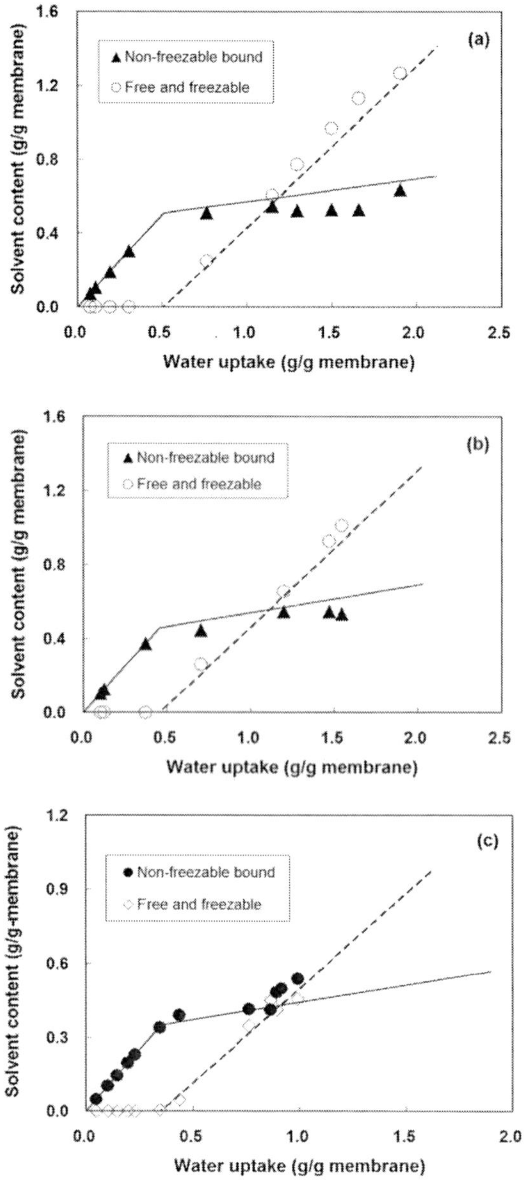

Figure 16. Distribution of water states into non-freezable bound and free/freezable states for (a) PVA-FS (10%), (b) PVA-FS (20%), and (c) PVA-FS (30%) composites. Solid and dashed lines represent fitted results using Eq. (25) [46].

Above the corresponding critical value, free and freezable water appears. However, non-freezable bound water content shows a different trend. Unlike the constant non-freezable bound water value for the PVA system, non-freezable bound water content in PVA-FS is not constant beyond the critical water content: it increases with increasing water uptake, especially for the PVA/30% FS composite. Furthermore, the maximum water uptake for the PVA membrane was approximately 2.426 g/g-membrane, and the maximum water uptake decreased to 1.904, 1.545, and 0.991 g/g-membrane for PVA-FS composites containing 10%, 20%, and 30% FS, respectively.

3.2.1. Inert Filler Dilution Effect on W_b

The dilution effect on overall water sorption for PVA-FS composites is proposed when FS is considered to be an inert filler. Therefore, for a water content of W' in one gram of composite membrane containing ϕ grams of FS, the number of sorbed water molecules is λ (1 - ϕ), where λ is the number of water molecules per hydroxyl group in PVA. Non-freezable bound and free/freezable water contents (W_b' and W_f', respectively) in this system can then be expressed as follows:

$$\begin{aligned} W_b' &= 18\lambda(1-\phi)/44; W_f' = 0 \quad \text{for } \lambda < \lambda_c \\ W_b' &= 18\lambda_c(1-\phi)/44; W_f' = W' - 18\lambda_c(1-\phi)/44 \quad \text{for } \lambda \geq \lambda_c \end{aligned} \tag{14}$$

where ϕ is the FS mass fraction of FS and W' the mass of water molecules per gram of composite when FS acts as a diluter. Equation (14) clearly shows the linear reduction of non-freezable bound water content (in water mass per gram of composite) as FS is introduced.

3.2.2. Suppressed Hydrogen-Bonding Ability of PVA and Reduced W_b

In addition to the dilution effect of FS, FS may act as a physical cross-linker and introduce steric hindrance. This can restrict polymeric chains from forming hydrogen bonds between PVA and water, as well as other inter-molecular interactions. Because the nano-particles are tightly embedded in the PVA polymer matrix [49], their shielding effect is then proportional to the weight ratio of filler to polymer, which equals $\phi/(1-\phi)$. Assuming the hindrance coefficient (ξ) is applied to suppress interaction and hydrogen-bonding ability of the polymer, one can obtain Equation (15) for the maximum non-freezing bound water content:

$$\begin{aligned} W_b'' &= 18\lambda(1-\phi)(1-\xi\phi/(1-\phi))/44; W_f'' = 0 \quad \text{for } \lambda < \lambda_c \\ W_b'' &= 18\lambda_c(1-\phi)(1-\xi\phi/(1-\phi))/44; W_f'' = W'' - 18\lambda_c(1-\phi)(1-\xi\phi/(1-\phi))/44 \quad \text{for } \lambda \geq \lambda_c \end{aligned} \tag{15}$$

where W'' is the water uptake (water mass in one gram of composite) when FS acts as both a diluter and hydrogen-bonding suppresser. Equation (15) indicates a further reduction in the critical non-freezable bound water content.

3.2.3. Equilibrium of Bound Water to Free Water in the Presence of FS

Equations (14)–(15) describe constant non-freezing bound water contents above critical values. The gradual increase in non-freezing bound water content with respect to water uptake, as shown in Fig. 16, must be taken into consideration. We therefore propose that FS could also adsorb water present in both free and bound states. In addition, we assume that an equilibrium is established between these two states:

$$\text{water}_f\text{-FS} \rightleftharpoons \text{water}_b\text{-FS}$$

$$K = [water_b - FS]/[water_f - FS] \tag{16}$$

where subscripts b and f represent bound and free/freezable water states, respectively, K is a partition constant between the bound and free/freezable water states and can be fitted experimentally, and water-FS represents the interaction between water and FS. The parameter γ is the number of water molecules associated with each FS (SiO_2) unit. Thus one can obtain:

$$K = \gamma_b / \gamma_f = W_{b,FS} / W_{f,FS} \tag{17}$$

$$\text{or } W_{b,FS} = KW_{f,FS} \tag{18}$$

where subscripts b, FS and f, FS are the bound and free water associated with the FS, respectively.

The free/freezable water content with respective to the polymer is the same amount available to FS. In every gram of composite, there are ϕ grams of FS and $(1 - \phi)$ grams of PVA. In addition, each mole of constituent has γ_f and λ_f moles of free water. Therefore:

$$\lambda_f (1-\phi)/44 = \gamma_f \phi / 60 \tag{19}$$

where 44 and 60 are the molecular weights of the PVA repeating unit and SiO_2, respectively.

Substituting Equation (19) into Equation (12) yields:

$$\gamma_b = K\lambda_f (1-\phi) 60 / (44\phi) \tag{20}$$

This equation implies that there is no free water associated with PVA ($\lambda_f = 0$) for low sorption levels with all water molecules present in a non-freezable state in the PVA-FS composites. Meanwhile, no bound water is associated with FS ($\gamma_b = 0$) and there is no free water from FS ($\gamma_f = 0$). Under these circumstances, FS does not adsorb any water because the sorbed water molecules are hydrogen bound to the PVA chains. Above the critical sorption level, the excessive water is in a free/freezable state. This water can be adsorbed by FS and a portion of it becomes water bound to FS. The remaining water is in a free state, which is neither bound to FS nor to PVA.

3.2.4. Overall Water Sorption

One can obtain the overall bound water content at low sorption levels (below critical λ_c) from the previous paragraphs:

$$\begin{aligned} W_b &= 18\lambda(1-\phi)(1-\xi\phi/(1-\phi))/44 \\ W_f &= 0 \\ W_t &= W_b + W_f \quad \textit{for } \lambda < \lambda_c \end{aligned} \tag{21}$$

At high sorption levels ($\lambda \geq \lambda_c$), the overall bound water content and free/freezable water content can be expressed as:

$$W_b = W_b'' + W_{b,FS}$$
$$W_f'' = W_{f,FS} \tag{22}$$

where W is water uptake in the mass unit per g of composite when FS acts as both a diluter and hydrogen bonding suppresser, and when the free and bound water reaches an equilibrium on the FS surface. In addition, a mass balance over water states applies, apart from Equations (18) and (22):

$$W_{b,FS} = KW_{f,FS} = KW_f'' = K(W_t - W_b'' - W_{b,FS}) \tag{23}$$

Rearranging Equation (23) and combining it with Equation (15) yields the following equation:

$$W_{b,FS} = K/(1+K)\cdot(W_t - W_b'') = K/(1+K)[W_t - 18\lambda_c(1-\phi)(1-\xi\phi/(1-\phi))/44] \tag{24}$$

Combining Equations (15), (21), and (24), the overall bound water and free/freezable water contents at high sorption levels then becomes:

$$W_b = 18\lambda_c(1-\phi)(1-\xi\phi/(1-\phi))/44 + K/(1+K)\cdot[W_t - 18\lambda_c(1-\phi)(1-\xi\phi/(1-\phi))/44]$$
$$W_f = 1/(1+K)\cdot[W_t - 18\lambda_c(1-\phi)(1-\xi\phi/(1-\phi))/44] \quad \text{for } \lambda \geq \lambda_c \tag{25}$$

3.2.5. Model Adequacy and Implications

The data in Fig. 16 were used to fit the unknown parameters ξ and K, as shown in Equation (25). The ξ value was found to be 0.6, whereas the K value varied with the FS content. These K values were 0.08, 0.13, and 0.42 for PVA-FS (10%), PVA-FS (20%), and PVA-FS (30%), respectively [46]. The predicted W_b and W_f values using Eq. (25) with these parameters were plotted as solid and dashed lines, respectively, in Fig. 16. The predicted results were in excellent agreement with the experimental data, confirming the adequacy of the proposed model.

A general plot employing Equation (25) is illustrated in Fig. 17. First, the threshold at which the free/freezable water appears (which also represents the critical amount of non-freezable bound water) is dependent on the critical value for the polymer to form a bound state (λ_c), the filler mass fraction (ϕ), and the hindrance coefficient (ξ). In addition, this threshold is proportional to λ_c. At a predetermined λ_c, an increase in either diluting filler content (ϕ) or hindrance coefficient (ξ) results in a reduced threshold, as shown in the shorter line AB′ in Fig. 17 (a). Under some circumstances, a negative hindrance coefficient (ξ) is observed for salts, urea, and thiourea additives. The composites containing these fillers can enhance the maximum non-freezable water content by bonding more water molecules than the pristine polymer films [45] due to their strong hydration (hygroscopic) effect [45]. Thus, the line AB is extended to line AB″ in Fig. 17 (a).

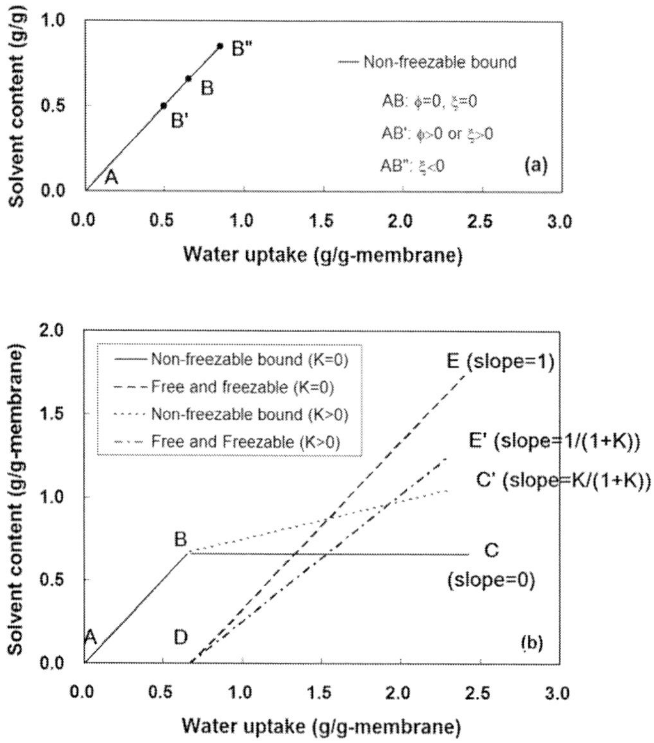

Figure 17. Illustration of the effects of (a) FS filler content (ϕ) and hydrogen-bonding hindrance coefficient (ξ) on maximum bound state, and (b) equilibrium constant (K) between bound water and free water associated with FS on water states in PVA-FS composites. The values of ϕ and ξ influenced the maximum bound state (point B, B', or B"), while the K value changed the slopes of BC and DE to BC' and DE' for bound water and freezable water, respectively[46].

Secondly, the slope may change for the bound and free/freezable water contents with respect to total water content. At low sorption levels ($\lambda < \lambda_c$), all water in the membrane exists in a bound state. The relationship $Y = X$ holds true (i.e., with a slope of unity) in the W_b versus W_t plot, as shown in line AB in Fig. 17 (b). At high sorption levels ($\lambda \geq \lambda_c$), free/freezable water starts to appear and increase in rate depends on K. If K equals zero, as shown in the case of the PVA membrane (Fig. 15 (a)), all sorbed water in excess of λ_c is in a free/freezable state, and a slope of unity is obtained (line DE) in the W_f versus W_t plot and the X-intercept corresponds to the maximum bound water content $[18\lambda_c(1-\phi)(1-\xi\phi/(1-\phi)/44)]$ from Eq. (21). The bound water remains constant with water uptake (i.e., with a slope of zero) in this high sorption range, as illustrated in line BC (Fig. 17 (b)).

For a positive K value, as in the PVA-FS composites, the relationship between W_b and W_t is similar to the PVA system at low sorption levels ($\lambda < \lambda_c$), but with a lower λ_c value. When water content is above the critical bound water content, the W_b versus W_t plot displays an increase in the slope equal to $K/(1+K)$ as shown in the dotted line BC' in Fig. 17 (b). In addition, the slope of W_f versus W_t decreases to $1/(1+K)$, as shown in the dash-dotted line DE' (Fig. 17 (b)). Therefore, Fig. 17 (b) offers a fast and simple analysis tool to determine the effect of high water-binding additives on the water states sorbed into composite membranes.

Table 5. Critical water content, water fusion enthalpy (ΔH_f) based on one additional gram of sorbed water above the critical content, partition coefficient (K) between bound and free water, and number of water states for work presented here and in the literature [46]

Polymer matrix	Critical water (g/g) [a]	ΔH_f (J/g) [a]	K [b]	# water states	Source
PVA [c]	0.675±0.066	339.9±20.5	≈ 0	3	Lue *et al.* [46]
PVA-FS [d] (10%)	0.565±0.117	308.8±37.3	0.08	3	Lue *et al.* [46]
PVA-FS (20%)	0.459±0.174	297.3±59.9	0.13	3	Lue *et al.* [46]
PVA-FS (30%)	0.352±0.085	235.1±51.9	0.42	3	Lue *et al.* [46]
PVA	0.387	– [e]	0.055	3	Zhang *et al.* [45]
PVA in aqueous salt solutions	0.385–0.590	– [e]	0.0074–0.160	3	Zhang *et al.* [45]
PDMS [f]	0.000208± 0.000105	333.8±8.9	≈ 0	2	Lue and Yang [1]
Polyhydroxystyren e	0.083	330	≈ 0	3	Nakamura *et al.* [50]
Nafion	0.0785± 0.0128	189.7±24.0	0.755	2	Lue and Shieh [52]
PVA hydrogel	0.26–0.40	234–327	– [e]	3	Nagura *et al.* [43]
Cellulose acetate	0.12–0.13	– [e]	Varied from 0 to 0.081 as water content increased from 0.1 to 0.58.	4	Taniguchi and Horigome [10]
CS [g]	0.401	304.3	0.094	3	Guan *et al.* [11]
CS/polyether IPN	0.201	272.8	0.221	3	Guan *et al.* [11]
CS/gelatin	0.361	306.0	0.088	3	Guan *et al.* [11]
CS/pectin	0.436	201.4	0.654	3	Guan *et al.* [11]
CS	0.244	289.1	0.055	3	Qu *et al.* [51]
CS/LA [h] (1:2–2:1)	0.601–0.546	144.4–178.0	1.112–0.713	3	Qu *et al.* [51]
CS/GA [i]/LA (1:1:1)	0.424	192.3	0.586	3	Qu *et al.* [51]
CS/GA (1:2–2:1)	0.333–0.377	201.0–231.7	0.517–0.316	3	Qu *et al.* [51]

[a] Mean ± 95% confidence interval.
[b] Calculated from the data reported in the corresponding reference.
[c] Polyvinyl alcohol.
[d] Fumed silica.
[e] Not available.
[f] Polydimethyl siloxane.
[g] Crosslinked chitosan.
[h] Lactic acid.
[i] Glycolic acid.

The model presented in Fig. 17 is not only useful in polymer-filler composite matrices, but can also be extended to other homogeneous, heterogeneous or micro-phase separated polymer systems. For example, organic solvents in poly(dimethyl siloxane) [1] and water in polyhydroxystyrene and polyvinylpyrrolidone [41,50] show constant maximum non-freezable solvent contents. Their partitioning into freezable and non-freezable states can thus be demonstrated as in the BC curve shown in Fig. 17 (b). On the contrary, water in cellulose acetate [10], chitosan (CS)-based hydrogels [11,51], perfluorosulfonic acid membranes [52], and hydrofluorocarbon polymer electrolytes [8] exhibits a similar trend to that illustrated in the BC' curve in Fig. 17 (b). For comparision, the critical non-freezable bound water content and partition coefficient (*K*) results from the literature are summarized in Table 5. These characteristic parameters confirmed adequacy of the model and can be used to demonstrate the distribution of water into freezable and bound states in various polymer matrices.

4. Interaction between Water and a Perfluorosulfonic Acid Membrane (Nafion)

4.1. Water States in Nafion

Fig. 18 shows the DSC thermograms of wet Nafion membranes with various water vapor uptakes. Water fusion peaks are not observed for the low water sorption Nafion membrane (e.g. below 0.080 g/g dry membrane), as shown in Fig. 18 (a). At high sorption levels (greater than 0.088 g/g dry membrane), the sorbed water melts at a temperature between –17 and –9 °C, as shown in Figs. 18 (a) and (b). Only freezable intermediate water [52] is present in the Nafion membrane; there is no free water in the Nafion membrane, judging by the absence of a fusion peak at 0°C. Fig. 19 shows the fusion temperature of sorbed water in the Nafion membranes from water vapor, liquid water and methanol aqueous solutions. The fusion temperature increased with increasing water uptake in the membranes approaching a higher value of –5°C at higher water contents above 0.241 g/g. Similarly, only intermediate freezable and non-freezable bound water states are present in the Nafion film. The absence of free water indicates the strong interaction between the water molecules and the sulfonate groups in the Nafion film.

Fig. 20 shows the relationship between water fusion enthalpy and water uptake into the Nafion membranes from liquid water and vapor phases. Up to a certain water uptake the solvent enthalpies are not visible. Beyond this value, the solvent fusion enthalpy increases linearly with increasing water uptake. Furthermore, the freezable water content starts to emerge above the critical value, which is obtained from the *X*-intercept of this straight line at ΔH=0. The critical value is estimated to be 0.0785 ±0.0128 (mean ± 95% confidence interval) g/g membrane. The slope (189.7 ± 24.0 J/g) represents the enthalpy change per unit mass of the solvent [52].

The slopes of the water fusion enthalpy data per unit of water content in polyvinyl alcohol (PVA), poly(dimethylsiloxane) (PDMS), and Nafion are shown in Fig. 21 for comparison. The critical values were 0.675 g/g and 0.000208 g/g for PVA and PDMS, respectively. The slopes beyond the respective critical water contents were estimated to be 344.97 ± 40.75 J/g in PVA and 333.3 ± 17.0 J/g in PDMS (Fig. 4 (b)) [1,46]. These numbers do not deviate from the water fusion enthalpy value [1,48,50] and confirm that the water content in the PVA and PDMS above the critical level behaves like pure bulk water (free water). Every gram of additional water sorbed in the PVA and PDMS is consider to be in the freezable state, and the increase in water fusion enthalpy per gram of sorbed water matches the pure water fusion enthalpy.

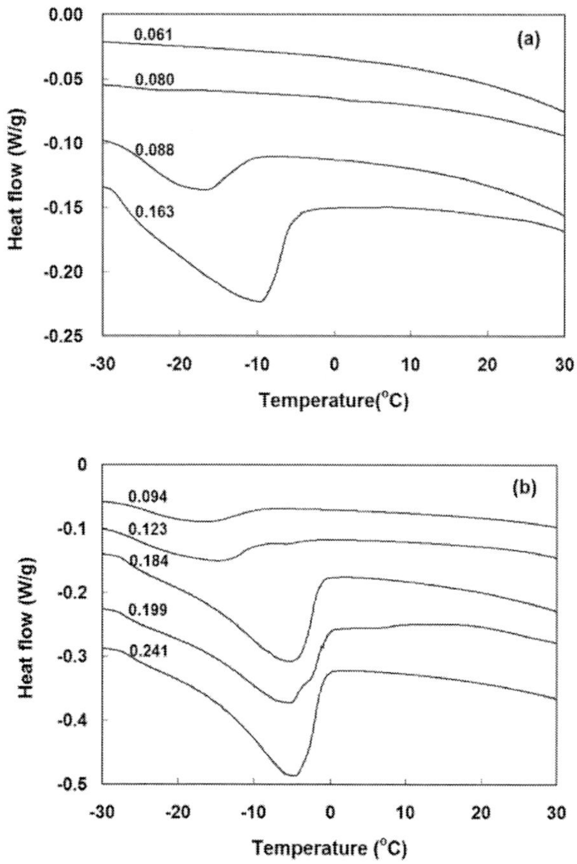

Figure 18. (a) DSC thermograms of Nafion membranes containing various (a) water vapor contents and (b) liquid water uptakes [52].

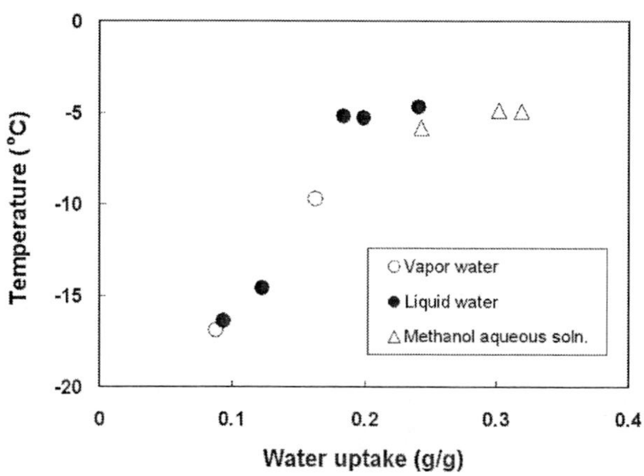

Figure 19. Fusion temperature of water in Nafion membranes as a function of water content from vapor, liquid, or methanol aqueous solutions [52].

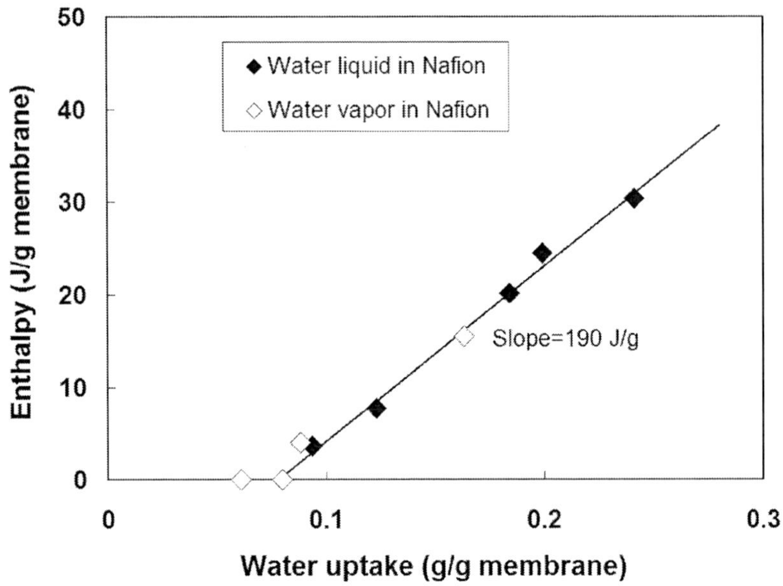

Figure 20. Water fusion enthalpy of water in wet Nafion containing water sorbed from the liquid (filled symbols) and vapor (open symbols) phases [52].

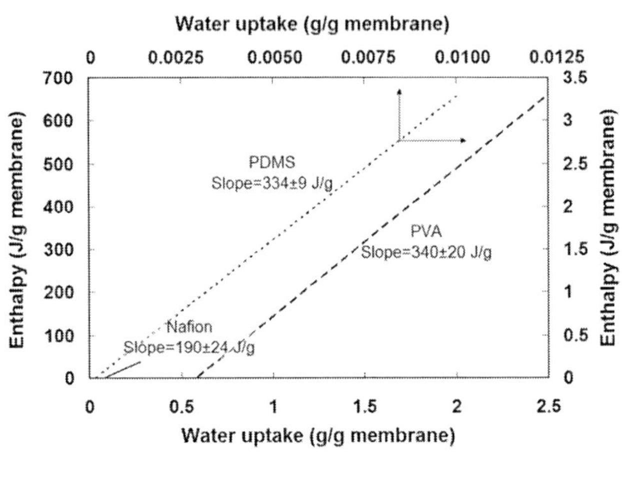

Figure 21. Water fusion enthalpy dependance on water uptakes in Nafion, PVA, and PDMS membranes (sorption temperature: 25°C). The fitted line for PDMS data has different *X* (upper) and *Y* (right) axes [52].

The Nafion samples, however, exhibited much lower changes in enthalpy per unit mass of water uptake (189.7 J/g) than pure water. This enthalpy change is also significantly lower than the water fusion enthalpies in PVA and PDMS polymers (Fig. 21) due to the strong interaction of water in the Nafion membrane. This phenomenon suggests that all the water molecules sorbed above the critical value (0.0785 g water/g Nafion) in the Nafion membrane do not completely behave like free water. Therefore, only a portion of the sorbed water is in the freezable state and the remaining part is in a non-freezable state. Thus, with every additional gram of water uptake in the Nafion, only a portion is freezable and the resulting enthalpy increase is smaller than the fusion enthalpy of pure water. The implication of this phenomenon will be detailed in the next two paragraphs.

This finding (lower enthalpy per g of sorbed water) is not an isolated event. For example, Qu *et al.* [51] reported different slopes in ΔH vs. water content in physically crosslinked chitosan hydrogels obtained by grafting D,L-lactic acid (LA) and/or glycolic acid (GA). The slope of the graphed data corresponded to 144.4–289.1 J/g, significantly lower than that of the pure water fusion enthalpy (333–334 J/g [50,52]). Guan *et al.* [11] also reported that the slopes ranged from 48.14 to 73.14 cal/g (equivalent to 201.2–305.7 J/g) for crosslinked chitosan and chitosan hydrogels crosslinked with interpenetrating polyether, gelatin, and ionic pectin complexes.

Eq. (2) can be used to calculate the amount of non-freezable water content by subtracting the freezable state amount from the total solvent uptake. The amount of freezable and non-freezable water states in the Nafion membrane with various amount of water uptake (from the liquid and vapor phases) are shown in Figs. 22 (a) and (b). During the initial sorption, the water liquid or vapor sorbed in the Nafion was in a non-freezable state until a characteristic threshold value (0.0785 g/g, determined from the intercept of the regression line in Fig. 20). Beyond this threshold, the non-freezable water content increased with water content, but with a gentle slope (Fig. 21).

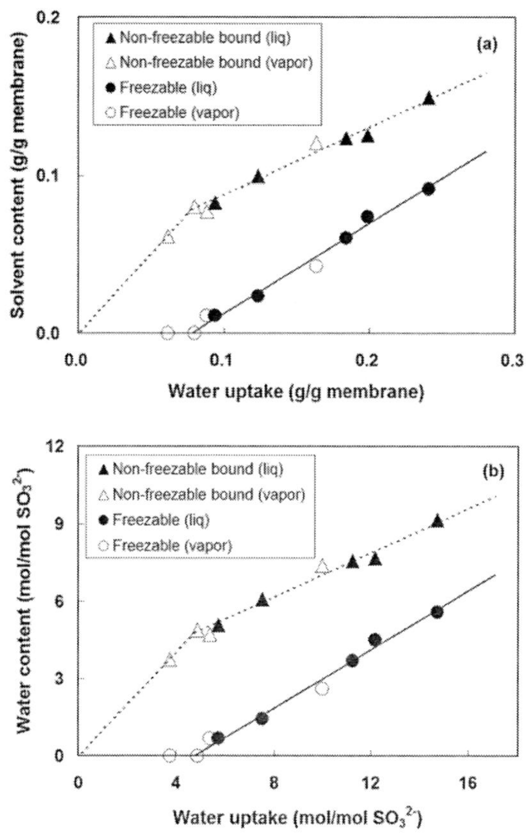

Figure 22. Freezable and non-freezable states of water in Nafion films as a function of water uptake in (a) mass units and (b) number of water molecules per sulfonate group [52].

In the Nafion membrane, the distribution pattern of the freezable and non-freezable water content is different from that in PVA [46] and polyvinylpyrrolidone (PVP) [41] films. In PVA and PVP membranes, there is no change in the non-freezable solvent content beyond the critical value and the freezable water content increases linearly with solvent uptake. The saturation of hydrogen bonding ability between the PVA network and sorbed water molecules are responsible for maintaining the amount of non-freezable water and the saturation capacity is governed by the chemical structures of the water permeant and the PVA matrix. Below the critical region, the polymer matrix acts as a pseudo-homogeneous system with the low level of sorbed water evenly dispersed in the PVA matrix (polymer-rich phase). The excess penetrants formed multiple layers (i.e., aggregates, a solvent-rich phase) after occupying all vacant sites surrounding the polymer chains, creating a second pseudo-phase. This phase acted like a bulk state and the fusion property could then be detected by DSC [1,46].

Nafion has a rather complicated structure [53]. In the Nafion membrane, the water clusters and channels can be easily created by sulfonate groups and associated water molecules [54,55]. However, the relative humidity and amount of water uptake may change these cluster sizes [54]. The clusters can then coalesce or evolve to form a lamellar structure [56] when the Nafion membrane uptakes more water molecules. The extent of water molecule interaction with sulfonate groups then dynamically changes according to the cluster or lamella size. The isolated ionic clusters are dispersed in a dry or limited swollen fluorocarbon matrix with a diameter of ~1.5 nm [57]. Few water molecules are able to enter into the clusters because of the steric volume of the sulfonate groups in the Nafion membrane [53]. This stage is similar to the non-freezable state below the critical water content (0.0785 g/g) in this study. As more water is absorbed to expand the cluster [54] or enlarge the lamella space, more water molecules can be accommodated and more hydrogen bonded water is formed. These aggregated water clusters are then viewed as being in the freezable water state. Meanwhile, more non-freezable bound water has the ability to interact with sulfonate groups since the steric restriction no longer plays a limiting factor due to the enlarged hydrophilic clusters. Under these circumstances, additional sorbed water molecules may be present in these two states at different equilibria. Therefore, the ability of the sulfonate group to induce interactions with water molecules is affected by changes in morphology and microstructure of the Nafion membrane, resulting in an increase in the non-freezable bound water content with increasing total water uptake.

The trend in water state distribution in Fig. 22 is also analogous to that of the chitosan hydrogels reported by Guan et al [11]. They showed that in the initial stage all sorbed water molecules were in a bound state. The free water only starts to emerge above a certain critical water content (ranging from 0.15 to 0.4 g/g dry membrane) and increases linearly with water content. At the same time the non-freezable bound water content increases gradually, but at a slower rate with respect to the water uptake. It is noted that the liquid and vapor penetrants followed a similar sorption pattern and partitioned into freezable and non-freezable states, regardless of the original permeant phase outside the Nafion film during the sorption process (Fig. 22). This trend is consistent with the finding for organic vapors in PDMS [1], as shown in Figs. 4–5.

4.2. Model for Water States in Nafion

Now we apply the mathematical model described in the previous section for PVA-FS composites to the wet Nafion system. It is assumed that an equilibrium exists between non-freezable bound water and freezable water states inconjunction with the sulfonate groups in the Nafion. Below a critical water content, all water is in the non-freezable state. Equation (26) describes this situation.

$$\begin{aligned} W_b &= W_t \\ W_f &= 0 \quad \text{for } W_t < W_c \end{aligned} \tag{26}$$

where subscripts *b*, *c*, *t*, and *f* stand for bound (non-freezable) state, critical value, total, and free/freezable states, respectively.

Above the critical value, the additional sorbed water molecules partition into bound and freezable states. Water molecules in these two states are in equilibrium with each other, as shown in the following equations [46]:

$$\begin{aligned} &water_f - Nafion \longleftrightarrow water_b - Nafion \\ &K = [water_b - Nafion]/[water_f - Nafion] \end{aligned} \tag{27}$$

$$\begin{aligned} W_b &= W_c + \frac{K}{1+K}(W_t - W_c) \\ W_f &= \frac{1}{1+K}(W_t - W_c) \quad \text{for } W_t \geq W_c \end{aligned} \tag{28}$$

The general illustration in Fig. 17 (b) can be used to fit the data in Fig. 22. At low sorption levels ($W_t < W_c$) all water in the membrane exists in the bound state. At high sorption levels ($W_t \geq W_c$), however, free/freezable water starts to appear and the increase rate depends on the K value. In the Nafion system, the K value is obviously positive. The W_b versus W_t plot exhibits an increase with respect to water content with a slope of $K/(1+K)$ as indicated in the line BC' in Fig. 17 (b). This slope deviates from a horizontal line (with a K value of zero). In addition, the slope of W_f versus W_t plot decreases to $1/(1+K)$ as shown in the line DE' in Fig. 17 (b). As K increases, however, this slope moves farther away from unity in the sorption regime above the critical threshold. From the slopes in Fig. 22 (a), we can obtain the K value of 0.755 for this water-Nafion system.

Fig. 22 (a) can be converted into the number of water molecules per sulfonate group using Eq. (29):

$$\lambda = W \cdot 1100/18 \tag{29}$$

where λ is the number of water molecules sorbed on one Nafion sulfonate unit, W is the water content (g/g) in the Nafion film, and 18 and 1100 are the water molecular weight and Nafion equivalent weight, respectively.

The molar ratios of the non-freezable and freezable water states in the Nafion system are shown in Fig. 22 (b). The critical value at which the freezable water begins to appear corresponds to 4.797 water molecules per sulfonate functional group, with little dependence on whether a liquid or vapor penetrant was present. As the number of water molecules per sulfonate group increased above this critical value, both the non-freezable and freezable water components increased. In addition, the equilibrium water uptake (i.e., water solubility) at 25°C was about 14.7 water molecules per sulfonate group. At this equilibrium water uptake, 9.1 water molecules per sulfonate group were in the non-freezable state and 5.6 water molecules per sulfonate group were in the freezable state. Therefore, about 38% of the sorbed water molecules were in the freezable state.

Similar results are found in the literature. For instance, the number of non-freezable bound water molecules per sulfonate group in Nafion 117 were reported as 9 and 10 by Siu *et al.* [8,58] and Walsby *et al.* [59], respectively. Furthermore, Gupta *et al.* [60] reported 8.5–8.9 bound water molecules per sulfonate group in grafted FEP-g-polystyrenesulfonic acid membranes. Moreover, Saito *et al.*[61] found that 20.8 water molecules per sulfate group were sorbed in H-type Nafion, in which 8 water molecules per sulfate group were in the freezable state. The ratio of freezable water to total water content was 0.39 here. Finally, Chen and Kuo [62] reported that freezable water accounted for 40.3% of sorbed water in wet Nafion film.

The interpretation of the partition coefficient (K) can be explained in another way. With a K value of 0.755, one additional mole of water sorbed in Nafion above the critical salvation value is composed of 0.57 mol in the freezable state and 0.43 mol in the non-freezable state. The ratio of non-freezable water molecules to the freezable water molecules is thus equal to 0.43/0.57, or 0.755. Additionally, the K value is less than unity, indicating that more freezable water is associated with the sulfonate group than the non-freezable water as the water uptake increases. Therefore, the slope in the freezable water content (solid line in Fig. 22 (a)) is greater than that for the non-freezable water content (right segment of the dotted line in Fig. 22 (a)). The K value, in conjunction with the critical water content (W_c), can then serve as a quantitative indicator for to characterizing water states in ionomers.

Conclusion

In this chapter, low temperature DSC thermograms are used to discover the interactions between permeating molecules and a polymer matrix. The presence and numerical values of the fusion temperature and the associated enthalpy of each constituent are indicators of the number of phases present and the chemical compatibility between these constituents. From the permeant fusion enthalpy data, one can then extract quantitatively the individual contents of free/freezable and bound non-freezable solvent states. The distribution into these states is a useful characteristics in many applications, including membrane separation, hydrogels for controlled drug release, and solid electrolytes for fuel cells.

ACKNOWLEDGMENTS

The author would like to express the gratitude for the technical assistance from Mr. Song-Jiang Shieh and Mr. Shi Wei Yang. Financial support from Chang Gung University (BMRP 326) is also greatly appreciated.

REFERENCES

[1] Anantaraman, A.V., Gardner, C.L. Studies on ion-exchange membranes. Part 1. Effect of humidity on the conductivity of Nafion. *J. Electroanal. Chem.* 1996*, 414, 115-120.*

[2] Andrianov, K.A., Slonimskii, G.L., Zhdanov, A.A., Levin, V.Y., Godovskii, Y.K., Moskalenko, V.A. Some Physical Properties of Polyorganosiloxanes. I. Linear Polyorganosiloxanes. *J. Polym. Sci. A1, Polym. Chem.* 1972, *10, 1–22.*

[3] Aranguren, M.I. Crystallization of polydimethylsiloxane: Effect of silica filler and curing. *Polyme*r 1998, *39, 4897–4903.*

[4] Bordeaux, D., Cohen-Addad, J.P. Silica-filled polydimethylsiloxane: Crystallization process of the adsorbed layer. *Polymer.* 1990, *31, 743–748.*

[5] Chen, W.F., Kuo, P.L. Covalently cross-linked perfluorosulfonated membranes with polysiloxane framework. *Macromolecules.* 2007, 40, *1987–1994.*

[6] Clarson, S.J., Dodgson, K., Semlyen, J.A. Stydies of cyclic and linear poly(dimethylsiloxane): 19. Glass transition tempera`tures and crystallization behaviour. *Polymer.* 1985, *26, 930–934.*

[7] Clarson, S.J., Mark, J.E., Dodgson, K. Effect of network chain length distribution and in situ precipitated silica on the low-temperature behaviour of poly(dimethylsiloxane) model networks. *Polym. Commun.* 1988, *29, 208–212.*

[8] Danjo, K., Nishio, F.. Zhow, B.D., Otsuka, A. Interactions between Water and Pharmaceutical Polymers Determined by Water-Vapor Sorption Measurement and Differential Scanning Calorimetry. *Chem. Pharm. Bull.* 1995, *43, 1958–1960.*

[9] Feio, G., Cohen-Addad, J.P. NMR Approach to the kinetics of polymer crystallization. II. Polydimethylsiloxane solutions. *J. Polym. Sci. B, Polym. Phys*. 1989, *27, 1–24.*

[10] Felder, R.M., Rousseau, R.W. (1986). Elementary principles of chemical processes, 2nd Ed.; Wiley: New York.

[11] Fushimi, H.; Ando, I.; Iijima, T. States of water in cationically charged poly(vinyl alcohol) membranes. *Polymer* 1991, *32, 241–248.*

[12] Gavach, C., Pamboutzoglou, G., Nedyalkov, M., Pourcelly, G. AC impedance investigation of the kinetics of ion transport in Nafion® perfluorosulfonic membranes. *J. Membr. Sci.* 1989*, 45, 37-53.*

[13] Gebel, G. Structural evolution of water swollen perfluorosulfonated ionomers from dry membrane to solution. *Polymer.* 2000, 41, *5829–5838.*

[14] Gebel, G., Moore, R.B. Small-angle scattering study of short pendant chain perfuorosulfonated lonomer membranes. *Macromolecules. 2000*, 33, 4850–4855.

[15] Gierke, T.D., Munn, G.E., Wilson, F.C. The morphology in Nafion perfluorinated membrane products, as determined by wide- and small-angle X-ray studies. *J. Polym. Sci., Polym. Phys. Ed.* 1981, *19, 1687–1704.*

[16] Gref, R., Nguyen, Q.T., Rault, J., Neel, J. Etats des solvants absorbes dans des films semi-cristallins en alcool polyvinylique—I. Gonflement dans l'eau Europ. *Polym. J.* 1992, 28, 1007–1014.

[17] Guan, Y.L., L. Shao, L., Yao, K.D. A study on correlation between water state and swelling kinetics of chitosan-based hydrogels, *J. Appl. Polym.Sci. 1996, 61, 2325-2335.*

[18] Gupta, B., Büchi, F.N., Staub, M., Grman, D., Scherer, G.G. Cation exchange membranes by pre-irradiation grafting of styrene into FEP films. II. Properties of copolymer membranes. *J. Polym. Sci. Part A: Polym. Chem.* 1996, *34, 1873–1880.*

[19] Hashimoto, Y., Sakamoto, N., Iijima, H. Influence of the water and cluster structure on the PFSA membrane conductivity. *Kobunshi Ronbunshu.* 2006, *63*, *166–173.*

[20] Helmer, J.D., Polmanteer, K.E. Supercooling of polydimethylsiloxane. *J. Appl. Polym. Sci.* 1969, *13, 2113–2118.*

[21] Kawahara, M., Rikukawa, M., Sanui, K. Relationship between absorbed water and proton conductivity in sulfopropylated poly(benzimidazole). *Polym. Adv. Technol.* 2000, *11, 544-547.*

[22] Ke, B. Differential thermal analysis of high polymers. VII. Some low-temperature transitions, *J. Polym. Sci. B, Polym. Lett.* 1963, *1, 167–170.*

[23] Kim, D.S.; Bum Park, H.O., Jang, J.Y., Lee, Y.M. Synthesis of sulfonated poly(imidoaryl ether sulfone) membranes for polymer electrolyte membrane fuel cells. *J. Polym. Sci. Part A: Polym. Chem.* 2005, *43, 5620-5631.*

[24] Kim, S.J., Lee, C.K., Kim, S.I. Characterization of the water state of hyaluronic acid and poly(vinyl alcohol) interpenetrating polymer networks, *J. Appl. Polym. Sci.* 2004, *92, 1467-1472.*

[25] Kitamura, K., Sakaguchi, Y., Takase, S., Nakao, J. Synthesis and characterization of ion-conducting polybenzoxazoles as thermostable proton exchange membrane for fuel cells. *Kobunshi Ronbunshu.* 2006, *63, 174-181.*

[26] Lebedev, B.V., Mukhina, N.N., Kulagina, T.G. Thermodynamics of polydimethylsiloxane in the range 0-350K. *Polym. Sci. USSR* 1978, *20, 1458–1466.*

[27] Lee, C.L., Johannson, O.K., Flaningam, O.L., Hahn, P. Calorimetric studies on the phase transitions of crystalline polysiloxanes. Part I. Polydimethylsiloxane. *Polym. Prepr.* 1969, *10, 1311–1318.*

[28] Li, J.; Yu, H. Synthesis and characterization of sulfonated poly(benzoxazole ether ketone)s by direct copolymerization as novel polymers for proton-exchange membranes. *J. Polym. Sci., Polym. Chem.* 2007, *45, 2273–2286.*

[29] Li, W., Xue, F., Cheng, R. States of water in partially swollen poly(vinyl alcohol) hydrogels, *Polymer.* 2005, *46, 12026–12031.*

[30] Litt, M. H. Reevaluation of Nafion morphology. *Polym. Prepr.* 1997, 38, 80–81.

[31] Liu, W.G.; Yao, K.D. What causes the unfrozen water in polymers: Hydrogen bonds between water and polymer chains? *Polymer.* 2001, *42*, *3943–3947.*

[32] Lue S. J, Lee D.T., Chen J.Y., Hu C. C., Jean Y. C., Lai J. Y. Diffusivity enhancement of water vapor in poly(vinyl alcohol)–fumed silica nano-composite membranes: Correlation with polymer crystallinity and free-volume properties. *J. Membr. Sci.* 2008, *325, 831–839.*

[33] Lue S.J., Hsieh SJ. Water states in perfluorosulfonic acid membranes using differential scanning calorimetry, *J. Macromol. Sci. Phys.* 2009, *48, 114–127.*

[34] Lue, S.J., Hsieh, S.J. Modeling water states in polyvinyl alcohol-fumed silica composites, *Polymer* 2009, *50, 654–661*
[35] Lue, S.J., Yang, S.W. Sorption of organic solvents in poly(dimethyl siloxane) membrane. I. Permeant states quantification using differential scanning calorimetry. *J. Macromol. Sci. Phys*. 2005, *44, 711–725.*
[36] Lue, S.J., Yang, S.W., Shin, C.Y. Sorption of organic solvents in poly(dimethyl siloxane) membrane. II. Effect of sorption on the polymer crystalline upon cooling. *J. Macromol. Sci. Phys*. 2006, *45, 417–430.*
[37] Lue,S.J., Chen, J.Y., Yang,J.M. Crystallinity and stability of poly(vinyl alcohol)-fumed silica mixed matrix membranes. *J. Macromol. Sci. Phys*. 2008, *47, 39–51.*
[38] Mauritz, K.A., Moore, R.B. State of understanding of Nafion. *Chem. Rev*. 2004, *104, 4535–4586.*
[39] Nagura, M., Saitoh, H., Gotoh, Y., Ohkoshi, Y.. States of water in poly(vinyl alcohol)/poly(sodium L-glutamate) blend hydrogels, *Polymer* 1996, *37, 5649–52.*
[40] Nagura, M., Takagi, N., Katakami, H., Gotoh, Y., Ohkoshi, Y., Koyano, T., States of water in poly(vinyl alcohol) hydrogels. *Polym. Gels Network.* 1997, *5*, *455–68*
[41] Nakamura, K., Hatakeyama, T., Hatakeyama, H. Relationship between hydrogen bonding abd bound water in polyhydroxystyrene derivatives. *Polymer.* 1983, *24, 871–876.*
[42] Puffr, R., Sebenda, J. On the structure and properties of polyamides. XXVII. The mechanism of water sorption in polyamides, *J. Polym. Sci., Part C: Polym. Symp*. 1997, *16, 79-93.*
[43] Qu, X., Wirsén, A., Albertsson, A.C. Novel pH-sensitive chitosan hydrogels: Swelling behavior and states of water. *Polymer.* 2000, *41, 4589–4598*
[44] Reid, R.C., Prausnitz, J.M., Poling, B.E. The Properties of gases and liquids. McGraw-Hill: Singapore 1988.
[45] Riddick, J.A., Bunger, W.B., Sakano, T.K. Organic solvents, 4th Ed.; Wiley: New York, 1986.
[46] Robeson, L.M., Hwu, H.H., McGrath, J.E. Upper bound relationship for proton exchange membranes: Empirical relationship and relevance of phase separated blends. *J. Membr. Sci.* 2007, *302, 70-77.*
[47] Saito, M., Hayamizu, K., Okada, T. Temperature dependence of ion and water transport in perfluorinated ionomer membranes for fuel cells. *J. Phys. Chem. B* 2005, *109, 3112–3119.*
[48] Siu, A., Pivovar, B., Horsfall, J., Lovell, K.V., Holdcroft, S. Dependance of methanol permeability on the nature of water and the morphology of graft copolymer proton exchange membranes. *J. Polym. Sci. Part B: Polym. Phys*. 2006, *44, 2240–2252.*
[49] Siu, A., Schmeisser, J., Holdcroft, S. Effect of water on the low temperature conductivity of polymer electrolytes. *J. Phys. Chem. B* 2006, *110, 6072-6080.*
[50] Taniguchi, Y., Horigome, S. The states of water in cellulose acetate membranes. *J. Appl. Polym. Sci.* 1975, *19, 2743-2748.*
[51] Urayama, K., Yokoyama, K., Kohjiya, S. Low-temperature behavior of deswollen poly(dimethylsiloxane) networks. *Polymer.* 2000, *41, 3273–3278.*
[52] Van Krevelen, D.W., Hoftyzer, P.J. Properties of polymers-their estimation and correlation with chemical structure, 2nd Ed.; Elsevier Science: Amsterdam, 1976

[53] Walsby, N., Hietala, S., Maunu, S.L., Sundholm, F., Kallio, T., Sundholm, G. Water in different poly(styrene sulfonic acid)-grafted fluoropolymers. *J. Appl. Polym. Sci.* 2002, *86, 33–42.*

[54] Wan, L.S.; Huang, X.J.; Xu, Z.K. Diffusion and structure of water in polymers containing N-vinyl-2-pyrrolidone. *J. Phys. Chem. B* 2007, *111, 922–928.*

[55] Wang, B., Krause, S. Properties of dimethylsiloxane microphases in phase-separated dimethylsiloxane block copolymers. *Macromolecules.* 1987, *20, 2201–2208.*

[56] Xiao, C., Zhou, G. Synthesis and properties of degradable poly(vinyl alcohol) hydrogel, *Polym. Degrad. Stab.* 2003, *81, 297-301.*

[57] Yagfarov, M.S., Ionkin, V.S. Thermographic study of the low-temperature behaviour of polydimethylsiloxanes. *Polym. Sci. USSR.* 1968, *10, 1867–1873.*

[58] Yamane, T., Ichiryu, T., Nagata, M., Ueno, A., Shimizu, S. Intramolecular esterification by lipase powder in microaqueous benzene: Factors affecting activity of pure enzyme. *Biotechnol. Bioeng.* 1990, *36, 1063-1069.*

[59] Yang, H., Nguyen, Q.T., Ding, Y., Long, Y., Ping, Z. Investigation of poly(dimethylsiloxane) (PDMS)-solvent interaction by DSC. *J. Membr. Sci.* 2000, *164, 37–43.*

[60] Yim, A., St. Pierre, L.E. The effect of interfacial energy on heterogeneous nucleation in the crystallization of polydimethylsiloxane. *J. Polym. Sci. B, Polym. Lett.* 1970, 8, *241–245*, and references therein.

[61] Yoshikawa, M., Cooney, D., Matsuura, T. Anomalous physico-chemical behaviour of cyclohexane in a poly(dimethylsiloxane) membrane. *Polym. Commun.* 1991, *32, 555–557.*

[62] Zhang, W.Z., Satoh, M., Komiyama, J. A differential scanning calorimetry study of the states of water in swollen poly(vinyl alcohol) membranes containing nonvolatile additives. *J. Membr. Sci.* 1989, *42, 303–314.*

In: Polymer Chains
Editors: Régine Penzkofer and Yasue Wu
ISBN: 978-1-62100-422-6

Chapter 7

ROLE OF POLYMER CHAIN ENDS IN PLASMA SURFACE MODIFICATION

Norihiro Inagaki*
Laboratory of Polymer Chemistry, Shizuoka University; Hamamatsu, Japan

ABSTRACT

Plasma treatment is a useful technique for modification of polymeric surface properties. In the plasma treatment process, two different reactions, implantation of functional groups (chemical modification) and etching of polymer surface (physical modification) occur on polymer surfaces. These reactions do not occur separately but occur simultaneously during the plasma treatment process. How to control the etching reactions on polymer surfaces is an important factor to accomplish successful surface modification by plasma.

Plasma modification reactions are restricted at a thin topmost layer of a few nanometers on polymer surfaces, and never occur at deep layers (a few micrometers or more) from the surfaces. Therefore, polymer segments located at a topmost layer of polymer surfaces are subjected to be modified by plasma. What segments of polymer chains are located at the polymer surfaces?

Polymer chain ends do not distribute homogeneously in polymer bulks, but tend to aggregate at surface layers of the polymers. Therefore, surface functionalities such as adhesion with Cu metal are mainly due to polymer chain ends aggregated at the polymer surfaces.

From above concepts, plasma surface modification is discussed on following items:

1) Effects of polymer chain composition on plasma modification: implantation and degradation reactions
2) Polymer chain end groups at polymer surfaces and glass transition temperature at polymer surface
3) Role of polymer chain end groups in adhesion with copper metal

---

* Norihiro Inagaki: CBL06881@nifty.com.

1. INTRODUCTION

Plasma treatment is a conventional technique for modification of polymeric surface properties. Plasma which contains chemically-active species such as radicals, ions, electrons, etc. makes strong interactions with polymer surfaces, when polymer surfaces are exposed to plasma. As a result, chemical and physical modifications occur simultaneously on the polymer surfaces [1]. For example, polyolefin film surfaces are modified from hydrophobic to hydrophilic by oxygen plasma (chemical modification), and also the surfaces are roughed up to adhere strongly with cupper metal by argon plasma (physical modification). The chemical modification is due to implantation reactions of hydrophilic or hydrophobic groups on the polymer surfaces, and the physical modification is due to degradation reactions of polymer chains on polymer surfaces. The implantation and degradation reactions do not occur separately but occur simultaneously in the plasma modification process. We can never avoid the degradation reactions as long as plasma is used for the modification. Degradation reactions on polymer surfaces are triggered plasma bombardment. Plasma bombards polymer surfaces to make C-C or C-O bond scission in polymer chains. As a result, carbon or oxygen radicals will be formed at the end of the polymer chains, and then, degradation reactions will start from these radicals. This degradation process is a firstly-important issue to understand surface modification by plasma.

The plasma modification process is restricted at a topmost layer of the surface of polymeric materials, and properties of bulk polymers are never changed by the plasma treatment. Therefore, only polymer chains which locate near the surface of polymer materials take part in the plasma modification, and polymer chains locating at deep layers from the surface never contribute to the modification. Surface properties such as wettability and water-repellency are modified dramatically, but bulk properties such as tensile strength and elongation are never modified by the plasma modification. This heterogeneity of the area to be modified is an important issue to understand effects by plasma modification.

Surface of polymeric materials is an extended layer of bulk polymer and contacts with air. Is the surface is the same in physical and chemical properties as the bulk polymer? Although the surface is a part of the bulk polymer, the surface is remarkably different in physical and chemical properties from the bulk polymer. Polymer chain ends tend to aggregate at the surface rather in the bulk of polymers. As a result, chain ends tend to concentrate at the surface rather than in the bulk, and glass transition temperature at the surface is lower by a few tens degrees than that for the bulk polymer. Therefore, plasma modification process is due to modification of chain end groups aggregated at the surface. From such viewpoints, effects of plasma modification are closely related to the structure of polymer chains at surfaces.

In this chapter, from above aspects, plasma surface modification is discussed on following items:

1) Effects of polymer chain composition on plasma modification: implantation and degradation reactions
2) Polymer chain end groups at polymer surfaces and glass transition temperature at polymer surface
3) Role of polymer chain end groups in adhesion with copper metal

2. Effects of Polymer Chaim Composition on Plasma Modification: Implantation and Degradation Reactions

Three polyester films, poly(lactic acid) film, PLA;poly(ethylene terephthalate) film, PET; and poly(oxybenzoate-co-oxynaphthoate) film, PBN; were used for plasma surface modification experiments. PLA is composed of aliphatic carbon chains and ester groups, PBN consists of aromatic carbon chains and ester groups, and PET is composed of a mixture of aliphatic and aromatic carbon chains and ester groups (see Figure 1).

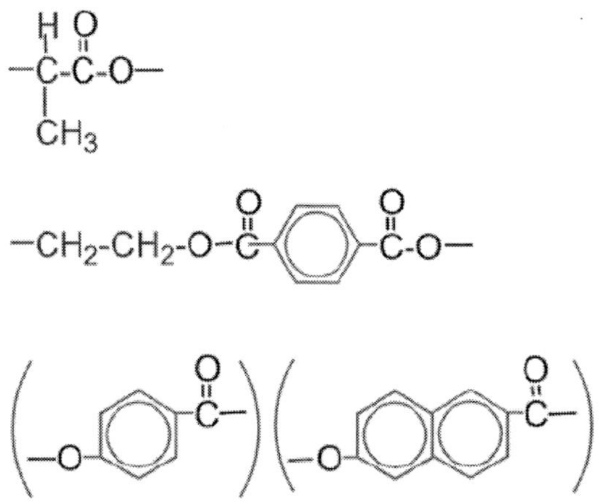

Figure 1. Chemical composition of PLA, PET, and PBN.

A special plasma reactor was used for argon and oxygen plasma treatments of the three polyester films (Figure 2) [2]. The plasma reactor consisted of a cylindrical Pyrex glass tube (45 mm diameter, 1000 mm long) with a gas inlet for the injection of argon or oxygen gas and a pair of electrodes for the energy input of rf power (13.56 MHz frequency). The polyester films were positioned between the electrodes, and exposed to the argon and oxygen plasmas. The argon and oxygen plasmas were operated at rf powers of 25 to 100 W at 13.56 MHz frequency at a system pressure of 13.3 Pa for given times (5 - 180 s). After finishing the plasma exposure, the films were rinsed with ethanol or acetone to remove deposited degradation products from the film surfaces, and then dried in vacuum. Weight changes for the polyester films before and after the plasma exposure were evaluated as a function of the plasma exposure. From a linear relationship between the weight loss and plasma exposure time, we calculated the weight loss rate for PLA, PET, and PBN by the argon and oxygen plasmas. The weight loss rate is summarized in Table 1 as functions of a kind of the plasmas and the rf power. To compensate large difference in the repeating units of the three polyesters, 78, 192, and 290 g/mol for PLA, PET, and PBN, respectively; the weight loss rate in Table 1 is shown in a dimension of mol/cm^2-s. We believe that the weight loss rate is an index of susceptibility to plasma, and that the rate was in the order of PLA > Pet > PBN. PLA showed the highest rate for weight loss process of the three films, even when either of argon or oxygen plasmas was used at rf powers of 25 - 100 W. Its weight low rate was higher by about 15 times than PBN. From this comparison, we conclude that PLA is easy to be subjected to degradation reactions, and that PBN is difficult to be subject to degradation reactions. This conclusion means that the plasma susceptibility is related to the chemical composition of the polymer chains.

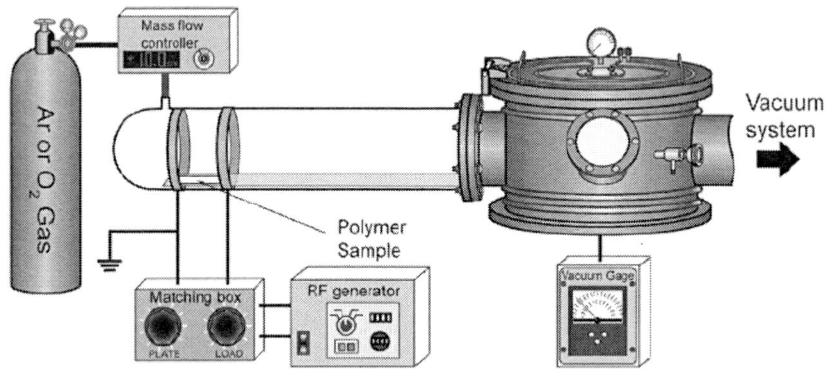

Figure 2. Schematic illustration of plasma reactor used for surface modification.

Table 1. Weight loss rate of PLA, PET, and PBN by plasma

Plasma treatment		Weight loss rate (10^{-9} mol/cm^2-s)		
Plasma	RF power (W)	PLA	PET	PBN
Ar plasma	25	2.78	0.443	0.080
	50	3.94	0.677	0.242
	100	4.17	0.641	0.280
O_2 plasma	25	5.22	1.50	0.768
	50	16.4	2.75	1.52
	100	30.9	3.49	2.17

Contact angle of water on the plasma-treated PLA, PET, and PBN surfaces was measured by the sessile drop method [3-5]. PLA, PET, and PBN surfaces were exposed to argon or oxygen plasma at rf powers of 25 to 100 W for 10 to 180 s, and then the surfaces were rinsed with ethanol or acetone to remove degradation products formed on the surfaces. Afterwards, the water contact angle was measured as functions of the plasma exposure time and rf power. Figures 3 and 4 show typical results of water contact angle on the plasma-treated PLA, PET, and PBN surfaces as functions of the argon and oxygen plasma exposure time and rf power. PET and PBN, as shown in Figure 3 and 4, showed large decreases in contact angle by the argon or oxygen plasma treatment. When exposed to argon plasma at 75 W, as shown in Figure 3, the contact angle for PBN surface decreased from 80 degrees (untreated) to 58 degrees (treated for 10 s). Afterwards, the contact angle decreased slightly with increasing the treatment time, and reached a constant of 54 degrees at a treatment time of 120 s. On the other hand, PLA showed no decrease even when treated with argon and oxygen plasma for 120 s. The contact angle for the PLA, PET, and PBN surfaces treated with argon and oxygen plasmas for 120 s are summarized in Table 2 as functions of the plasma and rf power.

Table 2. Contact angle of water on PLA, PET, and PBM surfaces treated with Ar and O_2 plasmas for 120 s

Plasma treatment conditions			Contact angle of water (degree)		
Plasma	RF power (W)	Treatment time (s)	PLA	PET	PBN
None	-	-	74	78	80
Ar plasma	25	120	74	32	64
	75	120	74	28	54
	100	120	74	24	36
O_2 plasma	25	120	74	26	56
	75	120	74	16	52
	100	120	74	16	44

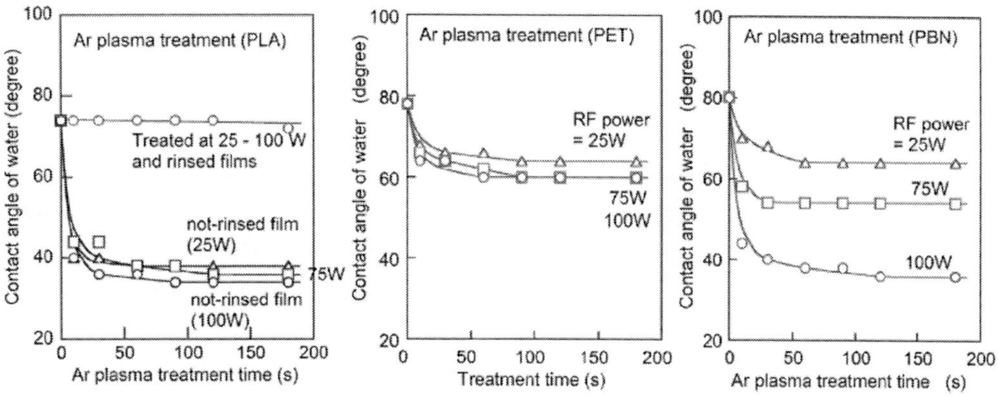

Figure 3. Contact angle of water on Ar plasma-treated PLA, PET, and PBN as functions of the Ar plasma treatment time and rf power.

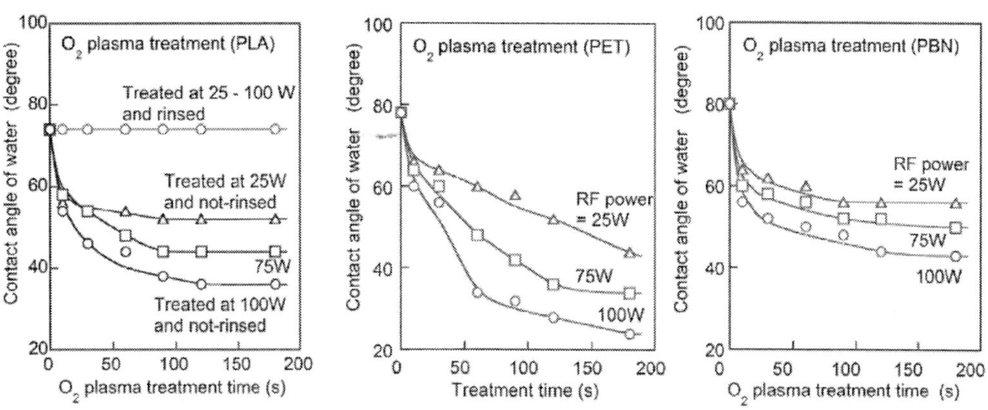

Figure 4. Contact angle of water on O_2 plasma-treated PLA, PET, and PBN as functions of the O_2 plasma treatment time and rf power.

PET and PBN showed large decreases in contact angle from 78 to 16 – 32 degrees and from 80 to 34 – 64 degrees, respectively. These decreases in contact angle indicate that some hydrophilic groups were introduced on their surfaces by the argon and oxygen plasmas. On the other hand, PLA showed no decrease in contact angle even when exposed to argon and oxygen plasma at 100 W for 120 s. This comparison shows that PET and PBN surfaces were effectively modified into hydrophilic by either of argon- and oxygen-plasmas. However, PLA surface was never modified even by argon- and oxygen-plasmas. Why was PLA never modified by the plasma exposure, although PLA was easy of weight-loss reactions? Specimens used for the contact angle measurement were ones rinsed with alcohol to remove degradation products formed on the PLA surfaces after finishing the plasma treatment. What contact angle do the specimens show? The contact angle for the plasma-treated PLA without the alcohol rinsing, as shown in Figures 3 and 4, showed large decreases by the argon and oxygen plasma treatments. For example, the contact angle decreased from 74 degrees to 44 and 58 degrees by the Ar and O_2 plasma treatments at 75 W for 10 s, respectively. The contact angle for Ar and O_2 plasma-treated PLA surfaces for 120s was 36 and 44 degrees, respectively. These results show that PLA surfaces were modified by Ar and O_2 plasmas into hydrophilic, but at the same time, degradation reactions occurred on the surfaces. Degradation products with hydrophilic groups were rinsed off completely from the surface, and as a result, not-modified PLA surfaces appeared on the surface. This points out an important concept for surface modification by plasma. Modification reactions, which are initiated by plasma, proceed always with degradation reactions together. A schema of the concept is shown in Figure 5. How to minimize degradation reactions is a key factor for effective surface modification.

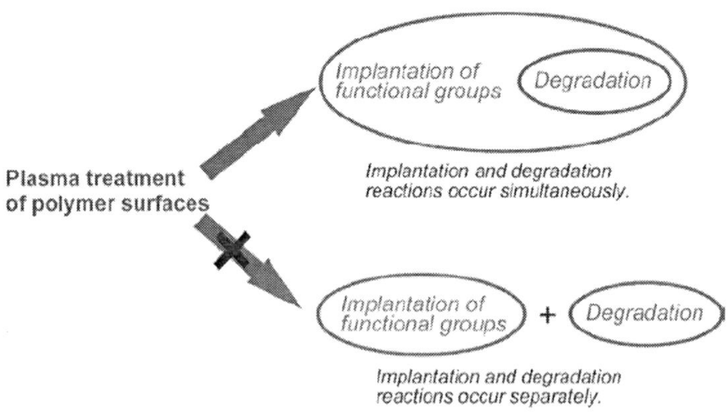

Figure 5. A concept of plasma treatment of polymer surfaces.

From the weight loss rate and contact angle measurements, we observed that PLA was distinguished from PET and PBN with respect to hydrophilic modification by argon and oxygen plasmas. Even when PLA was exposed to argon and oxygen plasma, hydrophilic surface were not obtained. On the other hand, PET and PBN surfaces were effectively modified into hydrophilic. We assumed that this difference in hydrophilic modification may be due to degradation reactions. To substantiate the assumption, the PLA, PET, and PBN surfaces which were treated with the Ar and O_2 plasmas and then rinsed with alcohol or acetone were analyzed with XPS. Figure 6 shows typical C1s spectra for these surfaces. There was less difference in the C1s spectra between the Ar plasma-treated PLA and untreated PLA. The C1s spectra were decomposed into four components: Comp. 1 at 285.0 eV due to CH_3, CH_2, and CH components; Comp. 2 at 286.6 – 287.0 eV due to C-O component; Comp. 3 at 288.9 -289.1 eV due to C=O component; and Comp. 4 at 291.4 -291.8 eV due to $\pi-\pi^*$ component [6]. Their relative concentration of Comp. 1, 2, and 3 was 41, 30, and 29 % for the Ar plasma-treated PLA, respectively. The concentration for the untreated PLA was 36, 31, and 33 %, respectively. There was less change in Comps. 1, 2, and 3 between the Ar plasma-treated and untreated PLA. On the other hand, the Ar plasma-treated PBN showed increase in Comp. 2 and 3, and decrease in Comp. 1, compared with the untreated PBN: The Comp. 2 and 3 increased from 13 to 19 % and from 12 to 14 %, respectively. The Comp. 1 decreased from 75 to 69 %. Conclusively, this composition shows that there was less change in chemical composition on the PLA surfaces by the Ar plasma treatment. Changeless modification may be due to chemistry of repeating unit between PLA and PBN. PLA is composed of aliphatic carbon chains and ester groups, and contains no aromatic group in the polymer chains. PBN contain aromatic groups in the polymer chains. Now we discuss why PLA surface could not be modified into hydrophilic surface but PBN surface could be modified into hydrophilic surface. It is well known that when polyesters are exposed to high-energy radiation such as gamma and ionizing rays, C-O bonds in the ester groups are broken to form radicals at the ends of polymer chains [7]. Degradation reactions are initiated from these radicals, and degradation products with low-molecular weight are formed on the surface. A similar bond scission will occur in exposing PLA and PBN surfaces to Ar plasma, and radicals will be formed at the end of the polymer chains. Typical degradation reactions are shown in Figure 7 for PLA and in Figure 8 for PBN. First, C-O bond scission for PLA occurs to form two radicals, A1 and A2, at polymer chain ends (Figure 7). Subsequently, A1 and A2 radicals will transform into A3 and A4 radicals, respectively, with elimination of small molecules such as CO_2 and CH_2=CH_2. Furthermore, A3 and A4 radicals will transform into A5 and A2 radicals, respectively. A5 radical will transform into A3 radical to eliminate CO.

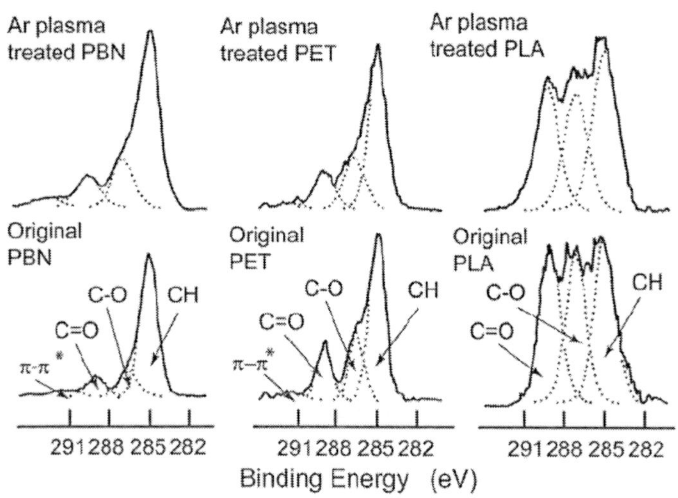

Figure 6. C1s spectra for PLA, PET, and PBN surfaces treated with Ar plasma.

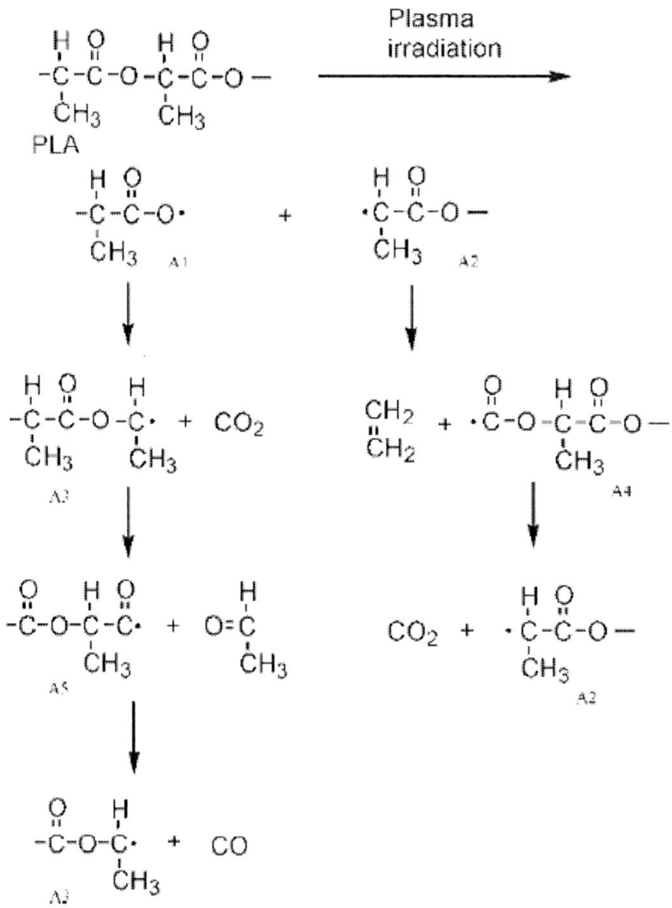

Figure 7. Possible degradation reactions for PLA.

Therefore, once the C-O bond scission for PLA occurs, two chain reactions of A1-A3-A5-A1, and A2-A4-A2 start from the polymer chain ends. These chain reactions continue as long as the radicals exist at the polymer chain ends. On the other hand, PBN also undergo scission of ester groups to form two radicals, B1 and B2 (Figure 8). Each of these radicals will subsequently convert into B3 radicals with elimination of small molecules such as CO_2. Some of B2 and B3 radicals will remain still on the PBN surface because of the stabilization of these radicals due to delocalization in aromatic rings. B2 and B3 radicals remaining on the surface will be oxidized to form oxygen functional groups, when exposed to air after finishing the Ar plasma treatment. As a result, PBN surface is effectively modified into hydrophilic. Therefore, aromatic rings in polymer chains of PBN contribute greatly to hydrophilic modification.

Consequently, hydrophilic modification by plasma is closely related to radicals formed at polymer chain ends. Instable radicals formed at aliphatic carbon atoms lead to unceasing degradation process. Stable radicals formed at aromatic carbon atoms contribute effectively to hydrophilic modification.

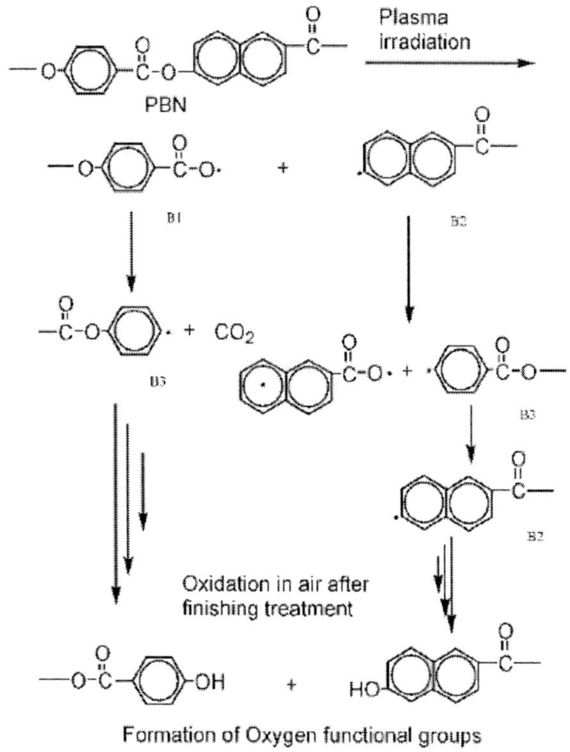

Figure 8. Possible degradation reactions for PBN.

3. Polymer Chain End Groups at Polymer Surfaces and Glass Transition Temperature at Polymer Surface

Polymer surface is a face contacting with the air, and is the extension of polymer bulk in topography. What segments of polymer chains occupy at the surface? Are segments occupied at the surface middle segments of polymer chains or end groups of polymer chains? Kajiyama, Takahara, and Tanaka investigated the distribution of polymer chain end groups in thin films using dynamic secondary ion mass spectroscopy, SIMS [8]. They used a special polystyrene (Mn=15,000; Mw/Mn=1.09) as a specimen for SIMS measurement. The special polystyrene was composed of deuterated styrene ($-CD_2-CD(Ph)-$) unit as a main chain and un-deuterated C_4H_9- and $(Ph)_2CH$- groups as polymer chain ends (Figure 9). Thin films were casted from a chloroform solution of the deuterated polystyrene by a spin-coating technique. Figure 10 shows a typical SIMS spectrum, where $^1H^+$, $^{12}C^+$, and $^2D^+$ atom intensities are plotted as a function of the depth from the thin film surface. The ^1H atom intensity means the concentration of polymer chain ends, and the $^2D^+$ atom intensity means the concentration of polymer main chains. The ^1H atom intensity, as shown in Figure 10, did not distribute homogeneously in the direction from the film surface to the inside of the film, but concentrated near the film surface. At more than 10 nm deep from the film surface, the ^1H atoms distributed homogeneously. On the other hand, the $^2D^+$ atoms distributed heterogeneously. The $^2D^+$ atom intensity was low at the film surface, and high at more than

10 nm deep from the film surface. From this SIMS spectrum, it is reasonable conclusion that the polymer surface was rich in polymer chain ends rather than polymer main chains.

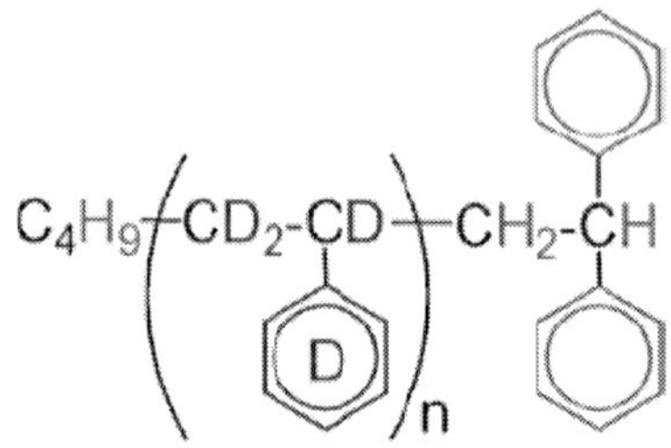

Figure 9. Poly(deuterated styrene) used for SIMS analysis.

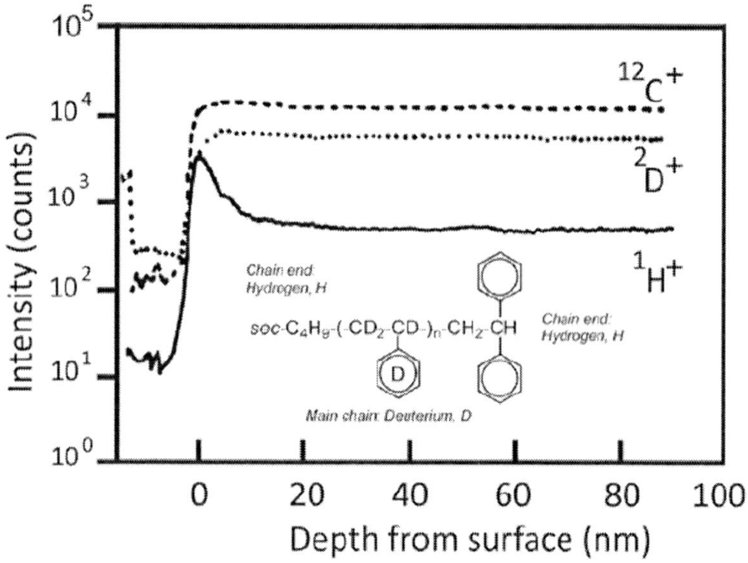

Figure 10. SIMS spectrum of poly(deuterated styrene).

Although the surface is a part of bulk polymers, the surface is remarkably different in polymer chain distribution from the polymer bulk. Polymer chain ends tend to aggregate near the surface rather than in the polymer bulk. How does such heterogeneous distribution of polymer chain ends influence on physical and chemical properties? Figure 11 is a duplicate of glass transition temperature (T_g) for surface and bulk of polystyrene films as a function of the molecular weight, which has been investigated by Kajiyama, Tanaka and Kawaguchi [9]. T_g for the film surface, as shown in Figure 11, is lower by a few tens degrees than that for the film bulk. Especially, lowering of T_g was remarked for polystyrene with low-molecular weight. This may be due to large free volume of polymer chain end groups aggregated at the film surface layer. Such aggregation of chain end groups near surface has been observed also by means of other techniques such as neutron reflectivity, angular XPS spectroscopy, and Monte Carlo simulation [10-16].

How do other polymer chain ends such as C(O)OH and NH_2 groups influence on T_g? Figure 12 compares T_g for surface and bulk of polystyrene derivatives having C(O)OH groups (PS-C(O)OH) and NH_2 groups (PS-NH_2) at polymer chain ends as well as normal polystyrene (PS-H) [17, 18]. As shown in Figure 12, PS-C(O)OH and PS-NH_2 also showed lower T_g at their surface layers than in their bulks. This indicates that C(O)OH and NH_2 chain end groups also aggregated near the surfaces. A magnitude of the difference was in the order of PS-H > PS-NH_2 > PS-C(O)OH. This order may be related closely to hydrophilicity of the chain end groups. Hydrophilicity of C(O)OH residue is the highest of the three residues, C(O)OH, NH_2, and H residues. Hydrophilicity of NH_2 residues is higher than that of H residues. We assume that highly-hydrophilic C(O)OH residues did not like contacting to hydrophobic air, and escaped from surfaces into deep layers. As a result, difference in T_g at the surface and in the bulk for PS-C(O)OH was smaller than that for PS-H.

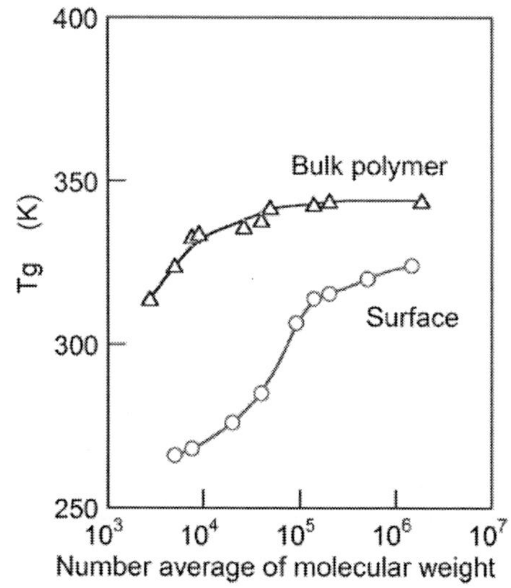

Figure 11. Glass transition temperature (T_g) for surface and bulk of polystyrene as a function of molecular weight.

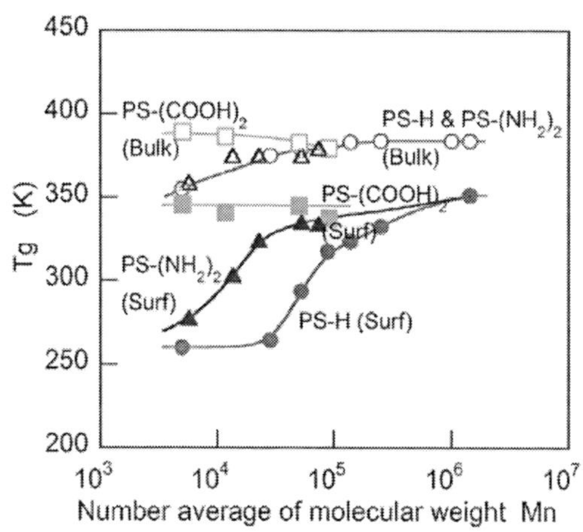

Figure 12. Glass transition temperature (Tg) for surface and bulk of polystyrene derivatives as functions of functional chain end groups and molecular weight.

Consequently, chemistry of polymer surface is not same as that of bulk, but polymer chain end groups aggregate predominately on the surface. From this viewpoint, it is an interesting issue that how polymer-chain end groups contribute to plasma modification. Contribution of the polymer-chain end groups on plasma modification will be discussed in a next chapter.

4. Role of Polymer Chain End Groups in Adhesion with Copper Metal

Aromatic polyester is often called one of the third-generation engineering plastics because of its excellent electrical and thermal properties as well as mechanical properties [19]. Vecstar OC film which is a trade name of Kuray Company, Ltd., Japan is an aromatic polyester made from acetoxy benzoic acid and acetoxy naphthoic acid (Figure 13), and possesses excellent electrical properties as well as mechanical and thermal properties (Table 3). The dielectric constant and dissipation factor for Vecstar OC film is comparable to those for polyimide film (Kapton H), and the water absorption for Vecstar OC film (0.04%) is much lower than that for Kapton H film (2.9%). From a viewpoint of electrical properties, Vecstar OC film is expected as a new material in application for microelectronics. Flexible printed circuit board (FPCB) is an example of promising applications of Vecstar OC films. An FPCB is a composite of polymer film and copper metal, which is deposited directly on the surface of the polymer film by conventional techniques such as electroless plating, vacuum deposition, sputtering, ion plating, etc. How strongly the deposited copper metal adheres to Vecstar OC film surface is an important item in the preparation of FPCB.

Table 3. Essential properties of Vecstar FA and OC and Kapton H

Properties	Vecstar FA	Vecstar OC	Kapton H
Tensile strength (MPa)	333/304*	196/196*	340
Tensile modulus (GPa)	70./6.7*	3.9/3.8*	3.4
Elongation at fracture (%)	16/14*	30/28*	80
Melting point (°C)	280	325	-
Glass transition temperature (°C)	205	285	-
Thermal expansion coefficient (ppm)	-20~0	-10~10	27
Water absorption (%)	0.04	0.04	2.9
Resistivity (Ωcm)	$7x10^{-15}$	$7.7x10^{-15}$	$1x10^{-15}$
Dielectric constant at 1 MHz	3.0	3.0	3.1
Dissipation factor at 1 MHz	0.022	0.022	0.089

* Machine direction/transverse direction.

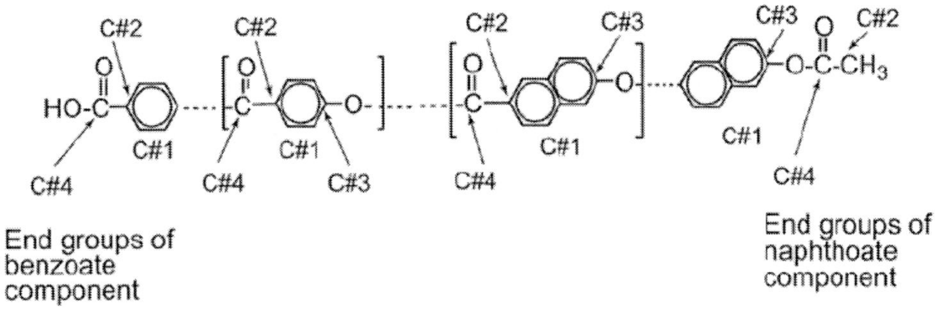

Figure 13. Chemical composition of repeating unit of Vecstar.

There are two types of Vecstar films, Vecstar OC and FA. Vecstar OC and FA are composed of chemically same repeating unit, which are copolymerized with acetoxy benzoic acid and acetoxy naphthoic acid. Vecstar FA is just fabricated from molten Vecstar by a blow molding technique, and Vecstar OC is subjected to be annealed at about 230°C after the film fabrication. In the film fabrication process, polymer chains near film surfaces are oriented in the direction of the film fabrication, but the polymer chains in inner layers of the films are distributed randomly. As a result, heterogeneous structure (skin-core structure) is formed in the film [19, 20]. In order to eliminate the skin-core structure, fabricated films are subjected to be annealed at about 230°C. During the annealing treatment, condensation reactions between carboxylic acid and acetoxy groups occurs. As a result, the concentration of polymer chain ends, acetoxy (CH_3-C(O)-O) and carboxylic acid (HO-C(O)-) groups decreases (Figure 14).

Aliphatic carbons of these polymer chain ends can be distinguished in XPS (C1s) spectra from aromatic carbons which are composed of polymer main chains. Aliphatic carbons appeared at 285.0 eV, and aromatic carbons in benzoate and naphthoate appeared at 284.7 eV [6]. C1s spectra for Vecstar FA and OC at a take-off angle of 20 degrees were decomposed into five components: Comp. 1 at 284.7 eV, Comp. 2 at 285.6 eV, Comp. 3 at 286.2 eV, Comp. 4 at 288.7 eV, and Comp. 5 at 291.0 eV, which were assigned to $\underline{C}$H in aromatic groups, $\underline{C}$-C(O) groups, $\underline{C}$-O groups, $\underline{C}$(O)-O groups, and π–π* shake-up satellite, respectively [6]. These spectra are not presented here for the sake of contraction. From the relative concentration of these C1s components, a concentration of the polymer chain end groups for Vecstar FA and OC could be estimated to be 47 % and 40 % of the total of carbon atoms, respectively.

Vecstar OC and FA were modified by Ar, O_2, N_2, and NH_3 plasmas, and on the plasma-modified Vecstar surfaces, copper metal (30 μm thick) was deposited by a combination of vacuum deposition and electroplating processes. The peel strength for the copper metal/Vecstar film systems was evaluated (Figure 15). There were large differences in adhesion between Vecstar OC/Cu and Vecstar FA/Cu composites. For example, the N_2 plasma-modified Vecstar OC/Cu composite showed peel strength of 0.70 N/5mm; on the other hand, the N_2 plasma-modified Vecstar FA/Cu composite showed peel strength of 1.3 N/5mm, which was 1.9 times larger than that for the N_2 plasma-modified Vecstar OC/Cu composite. Similar difference also was observed in the modification by Ar, O_2, and NH_3 plasmas. This indicates that the plasma modification is not always effective in improving the adhesion. Vecstar FA was effectively modified, but Vecstar OC film was not effectively improved. Why was Vecstar OC not modified effectively by plasma?

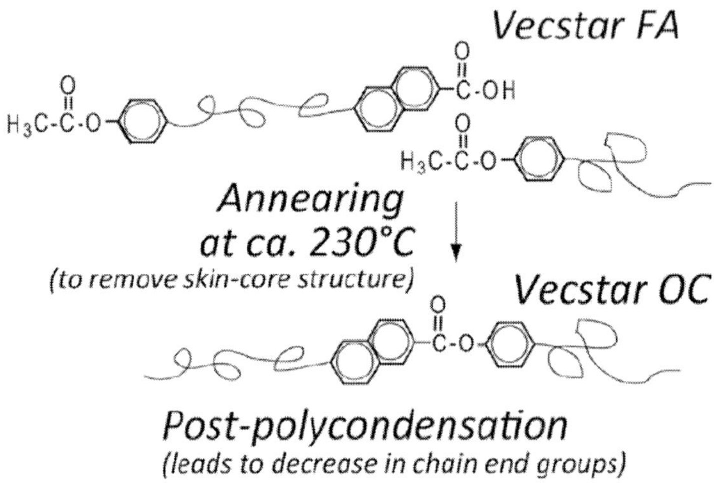

Figure 14. Annealing process of Vecstar film.

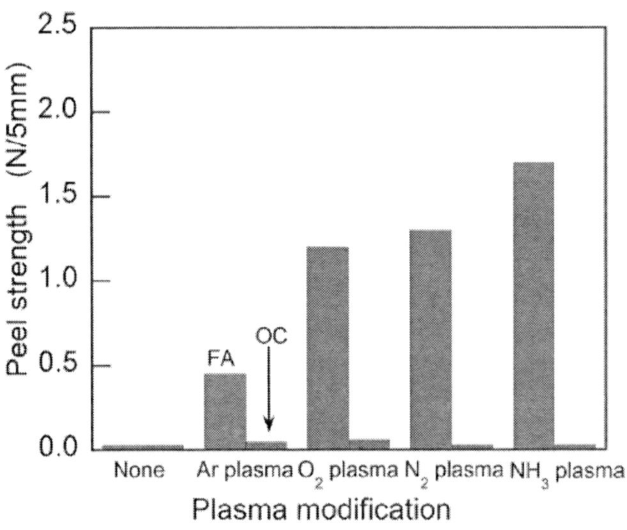

Figure 15. Peel strength for Vecstar/Cu composites.

Table 4 compares water contact angle between the Vecstar OC and FA surfaces which were plasma-modified at an rf power of 100W for 60s. Not only Vecstar FA film surfaces but also Vecstar FA surfaces showed some decrease in contact angle by the plasma modification. The atomic composition of the plasma-modified Vecstar OC and FA (Table 4), which were analyzed by XPS, also showed similar increase in O/C atomic ratio: The O/C atomic ratio increased from 0.18 to 0.35, and that for Vecstar FA increased from 0.19 to 0.33. Furthermore, both N_2 and NH_3 plasma modifications led to the formation of nitrogen functional groups at the Vecstar OC and FA surfaces. From these contact angle date and atomic composition results, it is reasonable to conclude that both Vecstar OC and FA surfaces were successfully modified by Ar, O_2, N_2, and NH_3 plasmas, and oxygen and nitrogen functional groups were formed on their surfaces by the Ar and O_2 plasma and the N_2 and NH_3 plasma treatments, respectively.

Table 4. Water contact angle and atomic composition of plasma-modified Vecstar films

Plasma	Contact angle (degree)*1		O/C atomic ratio*2		N/C atomic ratio*2	
	Vecstar FA	Vecstar OC	Vecstar FA	Vecstar OC	Vecstar FA	Vecstar OC
Ar plasma	41	36	0.23	0.32	0	0
O2 plasma	50	23	0.33	0.33	0	0
N2 plasma	63	30	0.25	0.35	0.07	0.09
NH3 plasma	73	66	0.22	-	0.13	-
None	85	84	0.19	0.18	0	

*[1]: modified at a plasma discharge power of 100 W for 60s.
*[2]: estimated from XPS at a take-off angle of 45 degrees.

It is well known that two mechanisms, chemical bond effect and mechanical interlocking effect, operate in adhesion between a metal layer and a polymer surface [21]. The chemical bond effect is the formation of Metal-O-C or Metal-N bonds at the interface between metal layer and polymer surface, and contributes to the adhesion. The formation of Metal-O-C and Metal-N bonds has been observed by XPS in the case of oxygen-containing groups such as carbonyl, carboxyl, and ether groups; and in the case of nitrogen-containing groups such as amide, cyano, and imidazole groups [21]. On the other hand, the mechanical interlocking effect is due to mechanical locking at the interface between the metal layer and polymer surface.

The plasma-modified Vecstar surfaces were scanned with a scanning probe microscope, SPM, to evaluate changes in surface topography. Figure 16 shows typical SPM pictures of the plasma-modified Vecstar OC. The specimens used for SPM analyses were the same as those used for XPS and water contact angle measurements. Topography of the plasma-modified Vecstar OC surfaces, as shown in Figure 16, was very different from that of the pristine Vecstar OC surface. These changes in topography were evaluated in terms of surface roughness (R_a), which was an arithmetic average of the profile height deviations. R_a for the plasma-modified Vecstar OC and FA surfaces are summarized in Table 5. This table indicates that 1) The pristine Vecstar OC is different in surface roughness from the pristine Vecstar FA. The surface roughness, R_a, for the pristine Vecstar OC is smaller by 6.4 times than that for the pristine Vecstar FA. 2) The plasma surface modification, independently of the kind of plasma used for the modification, caused to roughen Vecstar surfaces. For example, the Ra roughness for Vecstar OC surfaces increased from 2.31 nm (the pristine Vecstar OC film) to 5.39 - 10.0 nm. Ra for Vecstar FA surface also increased from 14.8 nm (the pristine film surface) to 11.8 - 23.4 nm. 3) A kind of plasma used for the surface modification influenced strongly the surface roughness of the plasma-modified Vecstar surfaces. The NH_3 plasma surface modification showed the highest Ra value (10.0 and 23.4 nm for Vecstar OC and FA , respectively), and the Ar plasma showed the lowest Ra value.

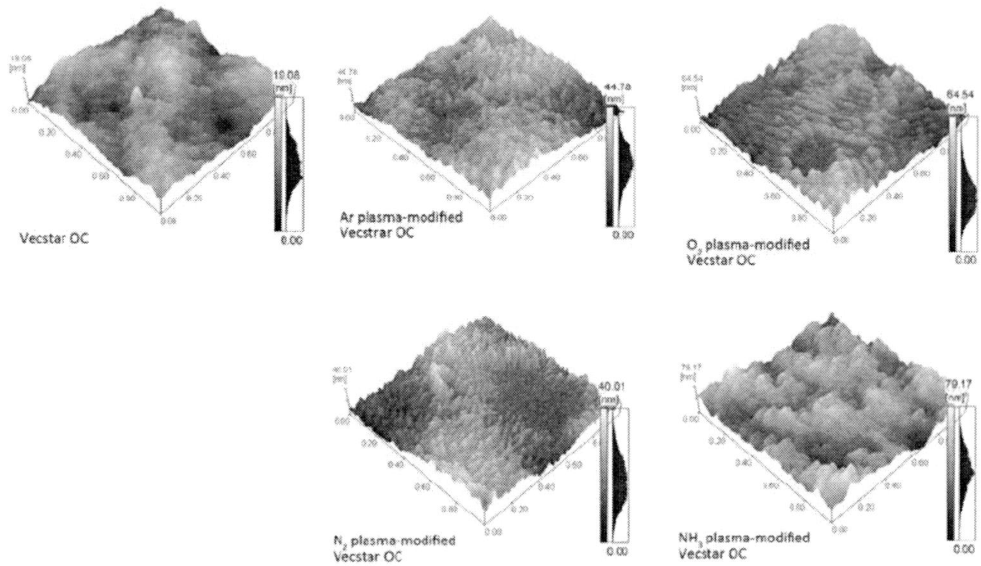

Figure 16. SPM pictures of the plasma-modified Vecstar OC surfaces.

Table 5. Surface roughness (Ra) of plasma-modified Vecstar surfaces

	Surface roughness (R_a) (nm)	
Plasma*	Vecstar FA	Vecstar OC
Ar plasma	11.8	5.41
O_2 plasma	16.4	7.70
N_2 plasma	19.0	5.39
NH_3 plasma	23.4	10.0
None	14.8	2.31

* Operated at a plasma discharge power of 100W for 60 s.

Figure 17 shows peel strength for copper metal/plasma-modified Vecstar OC and FA composites as a function of the surface roughness (R_a) of the plasma-modified Vecstar surfaces. The peel strength was poor in linear relationship with surface roughness, and the correlation coefficient for the linear relationship was 0.77. Such low correlation coefficient indicates that although the surface roughness could contribute to adhesion between copper metal and plasma-modified Vecstar, the surface roughness may be not the primary factor for the adhesion. We will discuss in the next section what factor operates effectively at the interface between the copper metal layer and plasma-modified Vecstar surface.

2. Effects of Polymer Chaim Composition on Plasma Modification: Implantation and Degradation Reactions

Three polyester films, poly(lactic acid) film, PLA;poly(ethylene terephthalate) film, PET; and poly(oxybenzoate-co-oxynaphthoate) film, PBN; were used for plasma surface modification experiments. PLA is composed of aliphatic carbon chains and ester groups, PBN consists of aromatic carbon chains and ester groups, and PET is composed of a mixture of aliphatic and aromatic carbon chains and ester groups (see Figure 1).

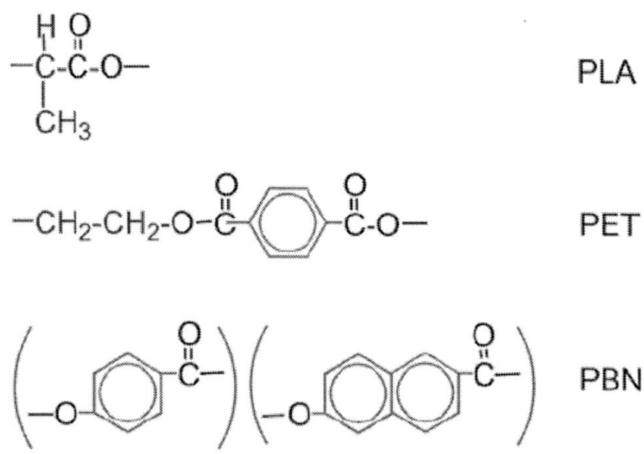

Figure 1. Chemical composition of PLA, PET, and PBN.

A special plasma reactor was used for argon and oxygen plasma treatments of the three polyester films (Figure 2) [2]. The plasma reactor consisted of a cylindrical Pyrex glass tube (45 mm diameter, 1000 mm long) with a gas inlet for the injection of argon or oxygen gas and a pair of electrodes for the energy input of rf power (13.56 MHz frequency). The polyester films were positioned between the electrodes, and exposed to the argon and oxygen plasmas. The argon and oxygen plasmas were operated at rf powers of 25 to 100 W at 13.56 MHz frequency at a system pressure of 13.3 Pa for given times (5 - 180 s). After finishing the plasma exposure, the films were rinsed with ethanol or acetone to remove deposited degradation products from the film surfaces, and then dried in vacuum. Weight changes for the polyester films before and after the plasma exposure were evaluated as a function of the plasma exposure. From a linear relationship between the weight loss and plasma exposure time, we calculated the weight loss rate for PLA, PET, and PBN by the argon and oxygen plasmas. The weight loss rate is summarized in Table 1 as functions of a kind of the plasmas and the rf power. To compensate large difference in the repeating units of the three polyesters, 78, 192, and 290 g/mol for PLA, PET, and PBN, respectively; the weight loss rate in Table 1 is shown in a dimension of mol/cm^2-s. We believe that the weight loss rate is an index of susceptibility to plasma, and that the rate was in the order of PLA > Pet > PBN. PLA showed the highest rate for weight loss process of the three films, even when either of argon or oxygen plasmas was used at rf powers of 25 - 100 W. Its weight low rate was higher by about 15 times than PBN. From this comparison, we conclude that PLA is easy to be subjected to degradation reactions, and that PBN is difficult to be subject to degradation reactions. This conclusion means that the plasma susceptibility is related to the chemical composition of the polymer chains.

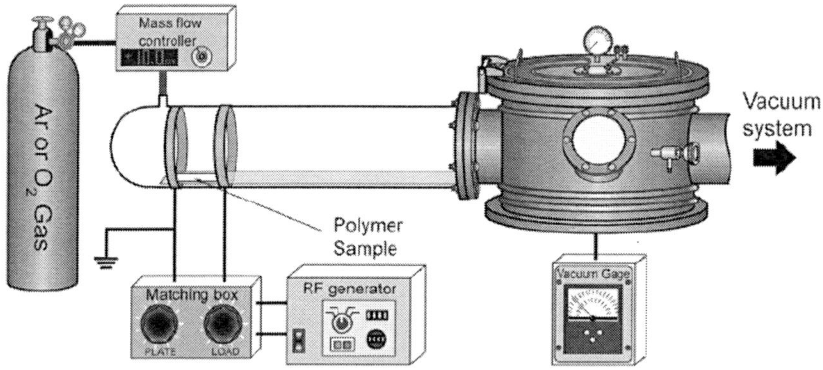

Figure 2. Schematic illustration of plasma reactor used for surface modification.

Table 1. Weight loss rate of PLA, PET, and PBN by plasma

Plasma treatment		Weight loss rate (10^{-9} mol/cm^2-s)		
Plasma	RF power (W)	PLA	PET	PBN
Ar plasma	25	2.78	0.443	0.080
	50	3.94	0.677	0.242
	100	4.17	0.641	0.280
O_2 plasma	25	5.22	1.50	0.768
	50	16.4	2.75	1.52
	100	30.9	3.49	2.17

Contact angle of water on the plasma-treated PLA, PET, and PBN surfaces was measured by the sessile drop method [3-5]. PLA, PET, and PBN surfaces were exposed to argon or oxygen plasma at rf powers of 25 to 100 W for 10 to 180 s, and then the surfaces were rinsed with ethanol or acetone to remove degradation products formed on the surfaces. Afterwards, the water contact angle was measured as functions of the plasma exposure time and rf power. Figures 3 and 4 show typical results of water contact angle on the plasma-treated PLA, PET, and PBN surfaces as functions of the argon and oxygen plasma exposure time and rf power. PET and PBN, as shown in Figure 3 and 4, showed large decreases in contact angle by the argon or oxygen plasma treatment. When exposed to argon plasma at 75 W, as shown in Figure 3, the contact angle for PBN surface decreased from 80 degrees (untreated) to 58 degrees (treated for 10 s). Afterwards, the contact angle decreased slightly with increasing the treatment time, and reached a constant of 54 degrees at a treatment time of 120 s. On the other hand, PLA showed no decrease even when treated with argon and oxygen plasma for 120 s. The contact angle for the PLA, PET, and PBN surfaces treated with argon and oxygen plasmas for 120 s are summarized in Table 2 as functions of the plasma and rf power.

As shown in Figure 15, there was large difference in peel strength between the plasma-modified Vecstar OC/Cu composites and plasma-modified FA/Cu composites. The plasma-modified Vecstar OC was poor in adhesion (peel strength of 0.03 N/5 mm) with copper metal. On the other hand, the plasma-modified Vecstar FA showed good adhesion (peel strength of 1.7 N/5 mm) with copper metal. What factor does difference in adhesion come from? There is no difference in chemistry between Vecstar OC and FA: poly (oxybenzoate-co-oxynaphthoate). Vecstar FA is just fabricated from molten Vecstar resin by a blow molding technique. On the other hand, after fabrication into a form of film, Vecstar OC is subjected to be annealed at about 230°C after the film fabrication. In the annealing process, the skin-core structure is destroyed. At the same time, condensation reactions between polymer chain end groups, acetoxy (CH_3-C(O)-O) and carboxylic acid (HO-C(O)-) groups, occur. As a result, Vecstar OC has lower in the concentration of polymer chain ends than Vecstar FA. Is there a possibility that polymer chain end groups may be closely related to adhesion in the copper metal/plasma-modified Vecstar OC composite?

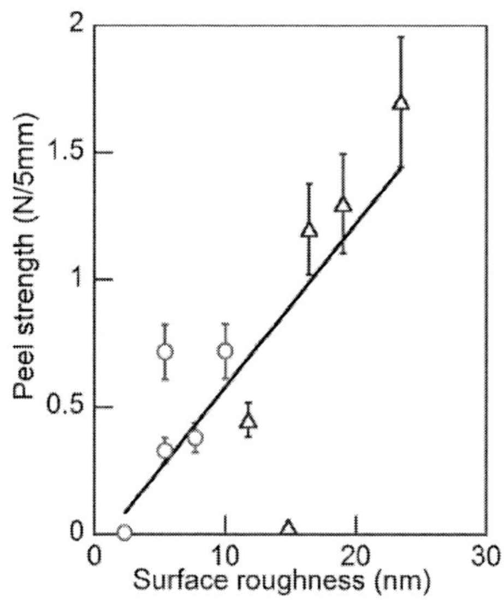

Figure 17. Peel strength for the plasma-modified Vecstar FA/Cu composite (Δ) and plasma-modified Vecstar OC/Cu composite (○) as a function of surface roughness.

Vecstar OC surface was irradiated by electron beam radiation at nitrogen atmosphere at room temperature or hydrolyzed in supercritical carbon dioxide containing methanol (2.5 mol%) at 25 MPa at 200°C for 30 min, in order to make polymer chain ends on the surface. Afterwards the modified Vecstar OC was treated by O_2 plasma and then, metallized with copper metal by a combination of vacuum deposition and electroplating. Figure 18 shows effects of the pretreatments by electron beam radiation or hydrolysis in super critical carbon dioxide solution on the adhesion to copper metal layer. The pretreatment by the electron beam irradiation led to great improvement of the adhesion for the Vecstar OC/Cu composites. The peel strength for the plasma-modified Vecstar OC/Cu composites increased from 0.03 to 1.9 N/5mm, which was corresponding to the adhesion for the plasma-modified Vecstar FA/Cu composite. Furthermore, a failure mode changed from the interface between the copper metal and Vecstar OC (interfacial failure) to failure within the Vecstar OC (cohesive failure). From these experimental results, we believe some contribution of polymer chain end groups to adhesion with Cu metal.

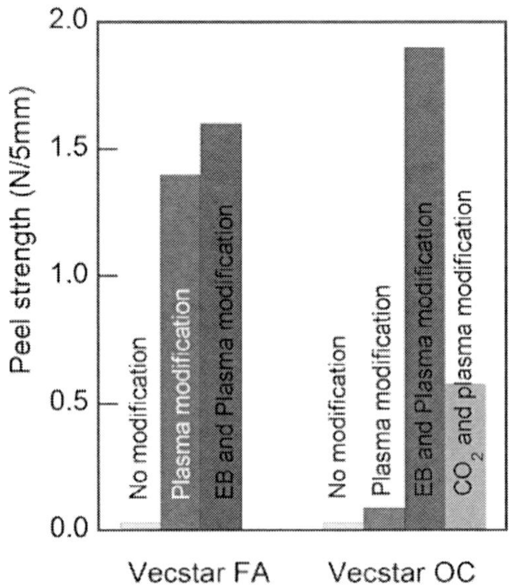

Figure 18. Adhesion improvements by a combination of electron beam irradiation and plasma modification or by a combination of hydrolysis in supper critical CO_2 solution and plasma modification.

In order to move oxygen- and nitrogen-containing functional groups formed by the plasma irradiation toward the topmost layer of the films, the plasma-modified Vecstar OC was heated at 200 – 300°C for 60 min, and then, copper layer was deposited on the Vecstar OC surface. The heating process of the plasma-modified Vecstar OC was done in two different environments: hydrophilic and hydrophobic conditions. In the hydrophilic condition, the plasma-modified Vecstar OC surface was contacted with glass plates during the heat treatment at 200 – 300°C for 60 min (heat treatment on glass plates). In the hydrophobic condition, the plasma-modified Vecstar OC surface was contacted to nitrogen atmosphere (heat treatment in N_2). After finishing these heat treatments, the Vecstar OC surface was metallized by a combination of vacuum deposition and electroplating of copper metal. Figure 19 shows the peel strength for copper metal/plasma-modified and heated Vecstar composite as a function of the heating temperature. The Ar and O_2 plasma-modified Vecstar OC films without the heat treatments, as shown in Figure 15, showed poor adhesion with copper metal, and their peel strength with copper metal was 0.3 and 0.4 N/5mm, respectively. The heat treatment on glass plates contributed significantly to the adhesion. The peel strength, as shown in Figure 19, increased with increasing heating temperature, and showed a maximum of 0.7 – 0.8 N/5mm at about 250 – 275°C. The heat treatment on glass plates was effective at 250 – 275°C, which was close to the glass transition temperature (285°C) of Vecstar OC. N_2 and NH_3 plasma-modified Vecstar OC surfaces also showed remarked improvement from 0.7 to 1.2 N/5mm (Figure 20). On the other hand as shown in Figure 20, the heat treatment in nitrogen atmosphere did not entirely contribute to the adhesion improvement. Therefore, we can conclude that the heat treatment alone does not contribute to the adhesion improvement, but a combination of heat treatment and hydrophilic atmosphere play an important role in the adhesion improvement. Furthermore, we believe that the adhesion improvement by the heating treatment may be due to some changes at the topmost layer of the plasma-modified Vecstar OC.

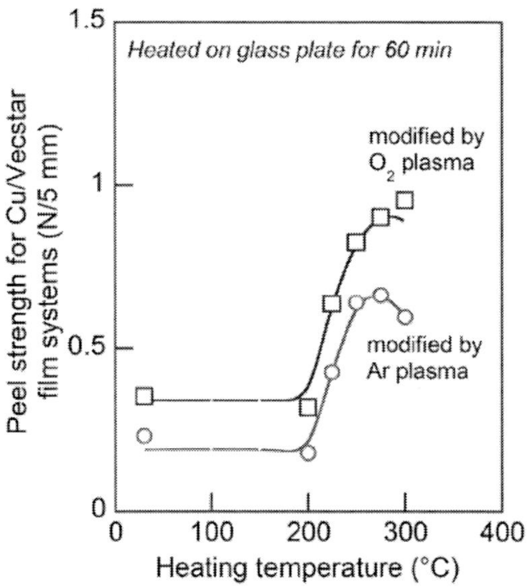

Figure 19. Peel strength for the Cu/plasma-modified and heat-treated Vecstar OC composites as a function of the heat temperature.

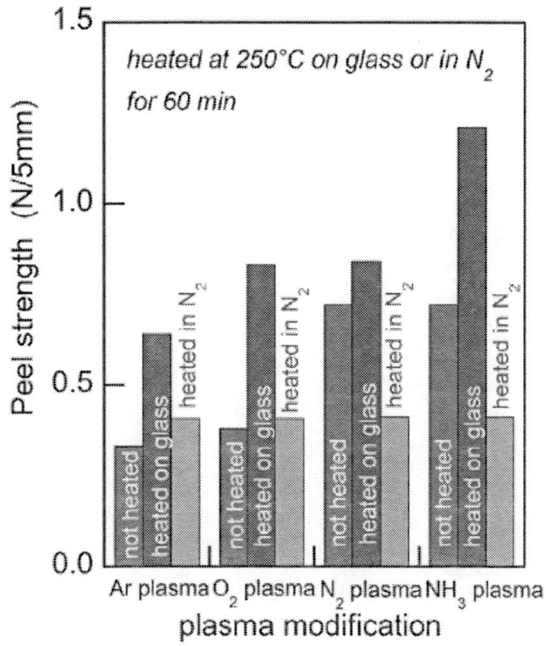

Figure 20. Effects of heat treatment of the plasma-modified Vecstar films on adhesion with copper metal.

In order to realize what polar components appeared at the topmost layer of the plasma-modified Vecstar OC by the heat treatment, XPS spectra were measured at a take-off angle of 20 degrees, which corresponds to an analysis depth of 3 nm. Figures 21 and 22 show the C1s spectra for the Ar plasma-modified and heat-treated, and N_2 plasma-modified and heat-treated Vecstar surfaces, respectively. These C1s spectra were decomposed into six components: C#1 at 284.70 eV (aromatic C), C#2 at 285.36 eV (aliphatic C), C#3 at 286.36 eV (C-O and C-N groups), C#4 at 287.94 eV (C=O groups), C#5 at 288.96 eV (C(O)-O groups), and C#6 at 290.7 – 291.91 eV (π–π* shake-up satellite) [6, 22]. There was no large difference in the C1s spectra between no-heat-treated and heat-treated surfaces. On the other hand, the O1s spectra as shown in Figures 23 and 24 were decomposed into three components, O#1 at 531.98-532.10 eV (O=C groups), O#2 at 523.38-532.76 eV (O=C(O) groups), and O#3 at 533.50-533.78 eV (O-C groups) [21, 24]. There was large difference in the relative concentration of the three O#1, O#2 and O#3 components between the plasma-modified Vecstar OC before and after the heat treatment. This difference changes indicate that some of the O1s components moved from the inner layer of the plasma-modified Vecstar OC to the topmost layer during the heat treatment process and/or another O1s component escaped from the topmost layer to the inner layer. In order to inspect such movement and escape of the O1s components by the heat treatment, peak intensity ratios of the O1s components (O#1, O#2, and O#3 components) at take-off angles of 20 and 45 degrees were measured and the peak intensity ratios were determined (Figure 25). When the Ar plasma-modified and N_2 plasma-modified Vecstar OC were heat-treated at 250°C on glass plates, the ratio of O=C peak intensities was more than one, however, the ratios of O=C-O- and O-C peak intensities were less than one. On the other hand, the ratios of all O=C, O=C-O-, and C-O peak intensities were less than one, when the Ar plasma-modified and N_2 plasma-modified Vecstar OC were heat-treated at 250°C in N_2 atmosphere. This comparison indicates some important aspects regarding the dynamics of plasma-modified polymer surfaces.

1) Some of oxygen functional groups formed on Vecstar OC by plasma modification moved toward the topmost layer of the plasma-modified films and other components escaped from the topmost layer to the inner layer of the plasma-modified films, when the plasma-modified surfaces were heat-treated at 250°C.
2) Such movement of oxygen functional groups was closely related to what substrate contacts to Vecstar OC film surfaces. A hydrophilic substrate, glass plate, accelerated the migration of O=C functional groups to the topmost layer of the films, however, O=C(O) and O-C groups migrated from the topmost layer to the inner layer. A hydrophobic substrate, N_2 atmosphere, accelerated the escape of O=C, O=C(O) and O-C functional groups from the topmost layer to the inner layer.

Such movement of O=C groups by the heat treatment on glass plate may be related to improvement of adhesion with copper metal. All plasma-modified Vecstar OC, as shown in Figure 25, showed adhesion improvement by the heat treatment, and the failure mode changed from copper metal/Vecstar interface to within the Vecstar film. Therefore, we believe that the O=C groups migrating to the topmost layer of the plasma-modified Vecstar OC contribute to the adhesion with copper metal [23, 24].

In conclusion, the adhesion between Vecstar OC and copper metal can be improved by a combination of plasma modification and heat treatment. This adhesion improvement is due to interaction between copper metal and O=C groups formed by the plasma modification. How to aggregate the O=C groups to the interface of the copper metal/Vecstar OC is a key factor for good adhesion. The heat treatment of the plasma-modified Vecstar OC on glass plates is effective in the aggregation at the top layer.

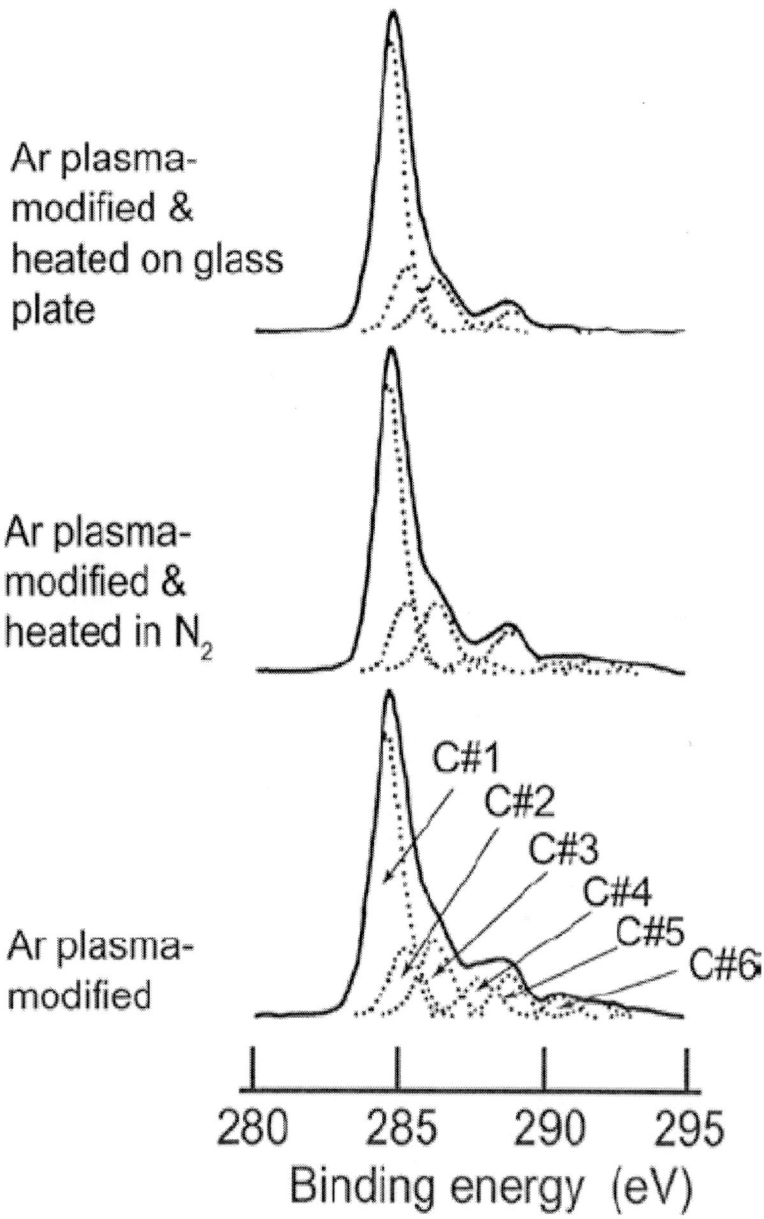

Figure 21. XPS (C1s) spectra at a take-off angle of 20 degrees for the Vecstar OC films modified by Ar plasma and heated on a glass plate and in N_2 atmosphere.

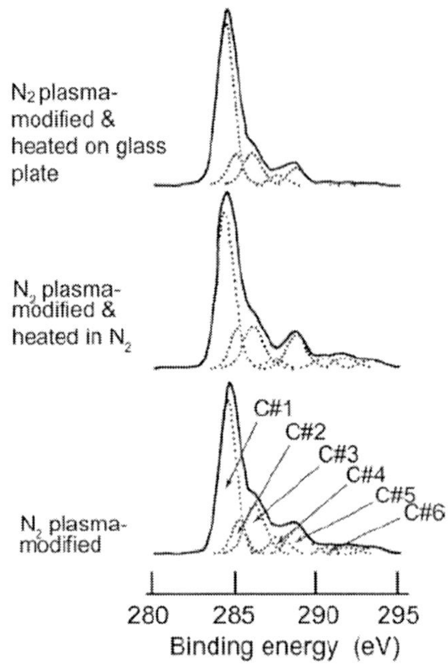

Figure 22. XPS (C1s) spectra at a take-off angle of 20 degrees for the Vecstar OC films modified by N_2 plasma and heated on a glass plate and in N_2 atmosphere.

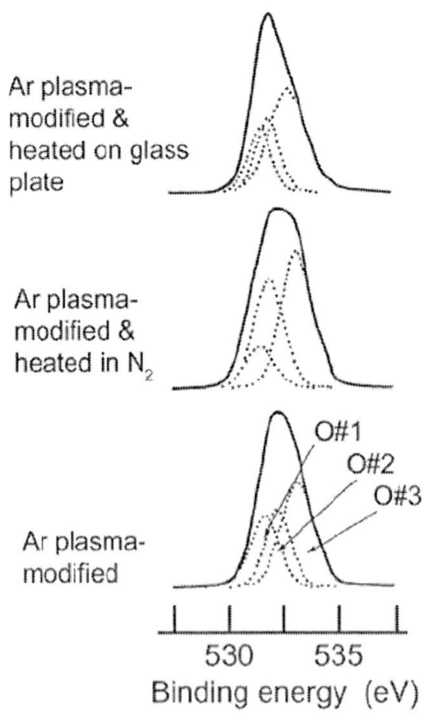

Figure 23. XPS (O1s) spectra at a take-off angle of 20 degrees for the Vecstar OC films modified by Ar plasma and heated on a glass plate and in N_2 atmosphere.

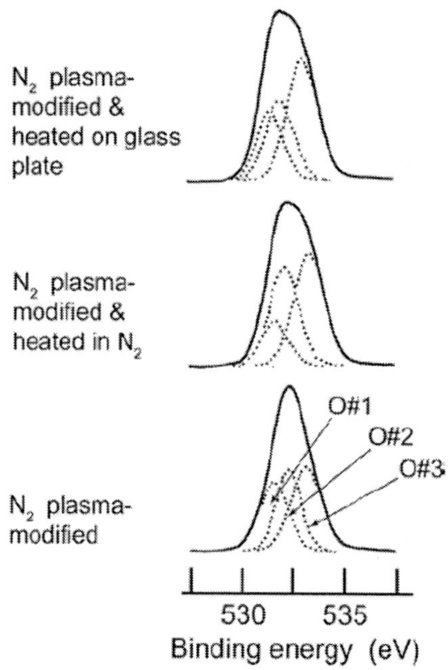

Figure 24. XPS (O1s) spectra at a take-off angle of 20 degrees for Vecstar OC films modified by N_2 plasma and heated on a glass plate and in N_2 atmosphere.

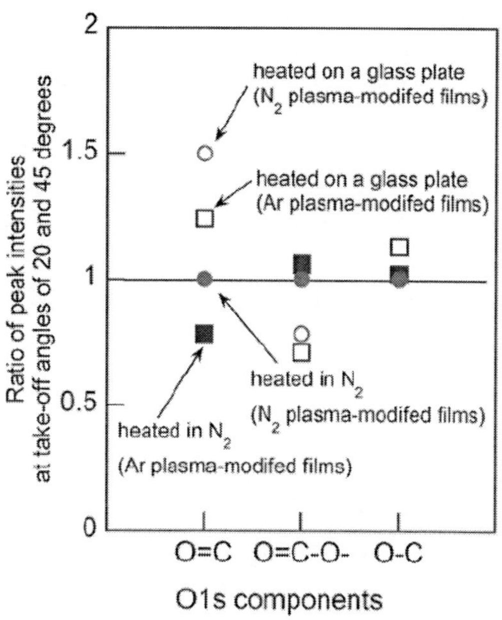

Figure 25. Ratio of peak intensities at take-off angles of 20 and 45 degrees for different oxygen functional groups: N_2 plasma-treated and heated on a glass plate (○), N_2 plasma-treated and heated in N_2 atmosphere (●), Ar plasma-treated and heated on a glass plate (□), and Ar plasma-treated and heated in N_2 atmosphere (■).

Conclusion

Plasma treatment is a useful technique for modification of polymeric surface properties. In the plasma treatment process, two different reactions, implantation of functional groups (chemical modification) and etching of polymer surface (physical modification) occur on polymer surfaces. These reactions do not occur separately but occur simultaneously in the plasma treatment process. Three polyester films composed of different repeating units, PLA, PET, PBN, were used as specimens for discussion of implantation and etching reactions by plasma. PLA is composed of aliphatic carbon chains and ester groups, PBN consists of aromatic carbon chains and ester groups, and PET is composed of a mixture of aliphatic and aromatic carbon chains and ester groups. Hydrophilic modification by plasma was closely related to radicals formed at polymer chain ends. Instable radicals formed at aliphatic carbon atoms led to unceasing degradation process. Stable radicals formed at aromatic carbon atoms contributed effectively to hydrophilic modification. Therefore, PBN were effectively modified into hydrophilic surface, however, PLA were never modified. Chemical composition of polymer chains is an important factor for effective plasma modification.

Plasma modification reactions are restricted at a thin topmost layer of a few nanometers from polymer surfaces, and never occur at deep layers (a few micrometers or more) from the surfaces. Therefore, polymer segments located at a topmost layer of polymer surfaces are subjected to be modified by plasma. What segments of polymer chains are located at the polymer surfaces? SIMS analyses for deuterated polystyrene showed that polymer chain ends do not distribute homogeneously in polymer bulks, but tend to aggregate at surface layers of the polymers. Therefore, surface functionalities such as adhesion with Cu metal are mainly due to polymer chain ends aggregated at the polymer surfaces.

From above concepts, plasma surface modification was discussed on following items:

1) Effects of polymer chain composition on plasma modification: implantation and degradation reactions
2) Polymer chain end groups at polymer surfaces and glass transition temperature at polymer surface
3) Role of polymer chain end groups in adhesion with copper metal

References

[1] A. J. Wagner, G, M. Walde, and D. H. Fairbrothe, Reactivity of Vapor-Deposited Metal Atoms with Nitrogen-Containing Polymers and Organic Surfaces Studied by in situ XPS, 2003, *Applied Surface Science,* 219, 317-328.

[2] A. R. Shultz, Chapter IX Intermolecular Reactions, in E. M. Fettes, Editor, *Chemical Reaction of Polymers,* New York, Wiley, 1964, pp. 723-810.

[3] C. E. Sroog, Polyimide, in J. C. Salamore, Editor; *Polymeric Materials Encyclopedia,* Boca Raton, FL; CRC, 1996, pp. 6253-6255.

[4] C. Perruchot, M. L. Abel, J. F. Watts, C. Lowe, J. T. Maxted, and R. G. White, High-Resolution XPS Study of Crosslinking and Segregation Phenomena in

Hexamethoxymethyl Melamine-Polyeste Resins, 2002, Surface Interface Analysis, 34, 570-574.

[5] C. Y. Liu and T. –K. Chung, The Diffusion of Polymer Chains with a Surface-Active End Groups, 2001, *Macromolecular Rapid Communications,* 22, 1278-1289.

[6] F. Xie, H. F. Zhang, F. K. Lee, B. Du, O. K. C. Tsui, Y. Yokoe, K. Tanaka, A. Takahara, and T. Kajiyama, Effect of Low Surface Energy Chain Ends on the Glass Transition Temperature of Polymer Thin Films, 2002, *Macromolecules,* 35, 1491-1492.

[7] G. Beamson and D. Briggs, Hight Resolution XPS of Organic Polymers, New York, Wiley, 1992.

[8] H. Chen, L. Ki, A. G. Merzlikine, Y. –T. Hisia, and M. S. Jhon, Surface Energy and Adhesion of Perfluoro-Polyether Nanofilms on Carbon Overcoat: The End Group and Backbone Chain, 2006, *Journal of Applied Physics,* 99, 08N103/1-08N101/3.

[9] J. Bachman, A. Ellies, and K. H. Hartgo, Development and Application of a New Sessile Drop Contact Angle Method to Assess Soil Water Repellency, 2000, Journal of Hydrology, 231/232, 55-75.

[10] J. D. Malcolm and H. M. Payneter, Simultaneous Determination of Contact Angle and Interfacial Tension from Seeile Drop Method, 1981, *Journal of Colloid and Interface Science,* 82, 269-275.

[11] J. Klos and T. Pakula, Computer Simulation of Chains End-grafted onto Spherical Surface, Effect of Matrix Polymer, 2004, *Macromolecules,* 37, 814-815.

[12] K. Tanaka, A. Taura, S. –R. Ge, A. Takahara, and T. Kajiyama, Molecular Weight Dependence of Surface Dynamic Viscoelastic Properties for the Monodisperse Polystyrene Film, 1996, Macromolecules, 29, 3040-3042.

[13] L. Li, K. –M. Ng, C. –M. Chan, J. –Y. Feng, and X. –M. Zeng, Surface Studies of the Rearrangement of End Groups of a Polymer, 2000, *Macromolecules,* 33, 5588-5592.

[14] L. M. Lander, L. M. Siewierski, W. J. Brittain, and E. A. Vogler, 1993, A Systematic Comparison of Contact Angle Methods, 1993, *Langmuir,* 9, 2237-2239.

[15] M. Strobel, C. S. Lyons, and K. L. Mittal, Plasma Surface Modification of Polymers: Relevance to Adhesion, Utrecht, VSP, 1994.

[16] N. Inagaki, K. Narushima, S. K. Lim. Y. W. Park, and Y. Ikeda, Surface Modification of Ethylene-co-Tetrafluroroethylene Films by Remote Plasma, 2002, *Journal of Applied Polymer Science,* 40, 2871-2882.

[17] N. Inagaki, Plasma Surface Modification and Plasma Polymerization, Lancaster, PA; Technomic Pub., 1996.

[18] N. Satomi, K. Tanaka, A. Takahara, and T. Kajiyama, Surface Molecular Motion of Monodisperse –Diamino-Terminated and –Dicarboxyl-Terminated Polystyrene, 2001, *Macromolecules,* 34, 8761-8767.

[19] P. Bébin and R. E. Prud'homme, Comparative XPS Study of Copper, Nickel, Aluminum Coating on Polymer Surfaces, 2003, *Chemistry Materials,* 15, 965-973.

[20] S. T. Chow, The Glass Transition Nanoscale Polymeric Films, 2002, *Journal of Physics: Condensed Materials,* 14, L333-L339.

[21] T. Inoue and S. Okita, Liquid Crystalline Polymers (Polyesters), in J. C. Salamore, Editor; *Polymeric Materials Encyclopedia,* Boca Raton, FL; CRC, 1996, pp. 3706-3711.

[22] T. Kajiyama, D. Kawaguchi, and K. Tanaka, Surface Mobility in Monodisperse Polystyrene Films, 2003, *Macromolecular Symposia,* 201, 95-102.

[23] T. Kajiyama, K. Tanaka, and A. Takahara, Surface Thermal Molecular Motion of Chain End-Modified Polystyrene, 2003, *Macromolecular Symposia,* 192, 265-270.
[24] X. Jiang, C. Z. Yang. K. Tanaka, A. Takahara, and T. Kajiyama, Effect of Chain End Group on Surface Glass Transition Temperature of Thin Polymer Film, 2001, *Physics Letters,* A281, 363-367.

In: Polymer Chains
Editors: Régine Penzkofer and Yasue Wu
ISBN: 978-1-62100-422-6

Chapter 8

Composition Effects on Thermodynamic and Conformational Properties of Ethylene/Propylene Copolymers

Inmaculada Suárez and Baudilio Coto*
Department of Chemical and Energy Technology,
ESCET, Universidad Rey Juan Carlos, Madrid, Spain

Abstract

Copolymerization reactions are widely used for the modification of many important end use properties of common polymers. In these copolymers, the composition and the chemical heterogeneity appear as main factors to understand polymer behaviour, and have to be determined in addition to the common characterization properties for polymer systems as molecular weight, polydispersity index, melting temperature, crystallinity, etc.

In this chapter, a set of ethylene/propylene copolymers covering the whole composition range was analyzed in order to fully describe the strong composition effects on thermodynamic and conformation properties in solution. Ethylene/propylene copolymers were synthesised by using the heterogeneous catalytic system MAO/rac-$Me_2Si[2Me\text{-}Ind]_2ZrCl_2$ and the polymerization process was carried out with controlled monomers feed that lead to copolymers covering a wide composition range. Details on that synthesis process leading to controlled compositions are also included.

Introduction

Plastics are a global success story as the industry has grown continuously for over 50 years. Production increased from 1.5 million tonnes in 1950 to 230 million tonnes in 2009. This growth is around 9% a year on average. This phenomenal growth of polymers is due to their versatility with respect to structure and properties and hence their applications. In

* To whom correspondence should be addressed: Phone: 34 91 4887089 ; Fax: 34 91 4887068 ; E-mail: baudilio.coto@urjc.es.

performance characteristics and applications prospects, they offer novelty and versatility, which are not found in other categories of materials like ceramics and metals [1].

There are around 20 distinct groups of plastics, each with numerous grades available to help deliver specific properties for each different application. There are five high-volume plastic families: polyethylene (including low density, PE-LD, linear low density, PE-LLD, and high density, PE-HD), polypropylene (PP), polyvinyl chloride (PVC), polystyrene (solid PS and expandable PS) and polyethylene terephthalate (PET). Together, that big five families account for around 75 % of all European plastics demand. The most used resin types are polyolefins (PE-LD, PE-HD, PE-LLD and PP) which account for around 50% of all plastics demand [2].

Polypropylene (PP) is a semi crystalline thermoplastic polymer widely used (8.6 Mt in 2009 for Europe) [2]. Polypropylene (PP) is widely used in packaging, textile, and automobile industries because of its good processability and properties. Nevertheless, its applications are limited because PP toughness is low, especially at room and low temperatures [3]. This is the reason why thermoplastic polymer production with elastomeric properties has been the focus of scientists and manufacturers in the last decade [4]. When a comonomer as ethylene is introduced in the polypropylene backbone, the crystallinity is reduced, as well as the melting and glass transition temperature, whereas the impact resistance and the transparency are increased. Thus, the copolymerization of propylene with small amounts of ethylene allows controlling the crystallization process [5] and therefore, the mechanical properties of the final copolymer. In the last years, ethylene/propylene (E/P) copolymers and terpolymers with 1-hexene or 1-butene have become elastomers with important commercial applications. A random monomer distribution along the copolymer chains together with high molecular weight and narrow molecular weight distribution (MWD) are the targeted structural characteristic in order to improve the elastomeric properties of the polymer and its potential applications [6].

The elastomeric properties of the copolymer are closely dependent on comonomer composition and considerable efforts were done to elucidate and control this aspect during copolymerization. Several influencing factors have been reported that allow control over the composition of the copolymer: 1) operating conditions [7], 2) chemical composition and physical state of the catalyst [8, 9, 10], 3) physical state of the copolymer being formed, and 4) structure of the monomers. Some of these are closely interdependent [11].

Final copolymer composition and chemical heterogeneity at various composition and configuration levels [12] are usually analyzed by ^{13}C-NMR, which is considered the best tool to carry out that study. Nevertheless, the properly application of this technique requires high spectral acquisition time to carry out quantitative intensity measurements and careful setup of experimental parameters.

In addition, the big influence of composition in some polymer properties are the key to study them by analyzing the samples through different techniques from two main different points of view: thermodynamic properties and conformation properties in solution.

Between the most common and used techniques to determine thermodynamic properties for polymer systems, it is of interest the use of Differential Scanning Calorimetry (DSC) to measure transition temperatures and crystallinity percentage, and the Temperature Rising Elution Fractionation (TREF) and Crystallization Analysis Fractionation (CRYSTAF) to qualitative estimate the Chemical Composition Distribution (CCD) which is also related to crystallinity.

The study of conformation of the macromolecular chains in solution used to be carried out by Gel Permeation Chromatography (GPC) with Multi-Angle Light Scattering Detector (MALS) which provides quantitative information on molecular weight distribution and radius of gyration of the chains. The addition of an Infrared detector to a GPC (GPC-IR) allows measuring the comonomer distribution along the molecular weight.

Ethylene/propylene copolymers have been previously widely studied by different authors [13, 14, 15] and by using most of the cited techniques above. However the covered composition ranges were limited to a narrow region with only small content of comonomer and thus close to both limiting homopolymers. In order to show composition effects in the whole composition range, a main topic is the synthesis of copolymers with a widely variable and controlled composition [16].

First part of this chapter will be then focussed from a general point of view in the procedures to obtain copolymers as well as some practical details about the synthesis for the copolymer set whose properties against composition are the topic of this chapter. Next sections will be related with the discussion of the characterization properties for such a set of copolymer mainly in terms of composition.

SYNTHESIS

The polymerization reactions can be carried out in several ways:

- bulk polymerization
- solution polymerization
- gas-phase polymerization
- slurry polymerization
- suspension polymerization
- emulsion polymerization

Slurry polymerization is often used in the manufacture of polyolefins. Initially, the reaction system involves the catalyst dispersed (or dissolved as in the case of soluble metallocene catalysts) in a continuous medium, which may be a solvent with solved monomer or the pure monomer. As the final polymer is insoluble in that continuous medium, it precipitates on the catalyst forming a slurry [17].

In spite of the processes, the efficient ways of producing the polyolefins are based on the catalysts too. The discovery of the catalyst based on titanium tetrachloride and diethylaluminum chloride as cocatalyst was made by Karl Ziegler [18]. The conventional Ziegler-Natta catalysts used in industrial production of polyolefins are heterogeneous systems, referring not only to the insolubility of the catalyst in the polymerizing medium, but also to its multisite nature. The results are catalysts which are difficult to study and understand that produce complex polymer mixtures with widely varying molecular weights and microstructures [19].

Metallocene/Methylaluminoxane (MAO) catalyst systems for olefin polymerization were discovered about 25 years ago in Hamburg laboratories [20, 21, 22]. MAO has three roles in metallocene polymerization. First, it acts as an alkylating agent for the halogenated metallocene precursors. Second, it behaves as a Lewis acid, abstracting an alkyl anion from the metallocene molecule and generating both a cationic active species and a non-coordinating alkyl-MAO anion. Lastly, MAO also acts as a scavenger, reacting with nucleophilic impurities that could otherwise deactivate the active species [23].

In comparison with Ziegler-Natta, metallocene catalysts are soluble in hydrocarbons, show only one type of active site and their chemical structure can be easily changed. These properties allow one to predict accurately the properties of the resulting polyolefin by knowing the structure of the catalyst used during their manufacture and to control the resulting molecular weight and distribution, comonomer content and tacticity by careful selection of the appropriate reactor conditions [24].

It is of great importance [25] to immobilize the homogeneous metallocene catalysts on inorganic carriers [26] for applying them to gas or slurry phase [27] polymerization processes. Supported catalysts can avoid reactor fouling, allow the reduction of the amount of cocatalyst during the reaction [28], and control the morphology of the produced polymer more easily [29]. The anchorage of the catalytic system can be carried out by several well-known techniques [19, 30]. The synthetic methods of supported metallocene catalysts can be classified as follows [19]: 1) immobilization of cocatalysts on a carrier, 2) immobilization of metallocene complex on a carrier and 3) simultaneous immobilization of metallocene complex and cocatalyst on a carrier. Another method of impregnation developed is the incipient wetness. This method provides an alternative route for depositing a dissolved substance onto a solid support. The principle of this method is to add only enough of the solution to fill the pores in the support [31].

Nevertheless, the immobilization is often accompanied by significant decreases in catalyst activity, sometimes even an order of magnitude respect to homogeneous polymerization [32]. Supports like SiO_2, Al_2O_3, $MgCl_2$, etc, can be used, all of them characterized by high activity and good control of the morphology of the final product [29].

Metallocene supported catalysts have allowed the preparation of copolymers which became a new generation of elastomers [33] with controlled structure and properties [34]. The polymers obtained show a random incorporation of monomer in the polymer backbone, narrow molecular weight distribution [35], especially when they are compared to conventional Ziegler-Natta catalysts [36].

An important issue in the metallocene catalyzed reactions for copolymer synthesis is the rate of monomer incorporation. For instance, ethylene incorporation in the polymer chain is about 100 times faster than that of propylene [37] in slurry polymerization because the solubility of both monomers is different and dependent of the solvent used, the pressure and temperature during the reaction [38]. Therefore if the ratio between the monomers in the feeding is kept constant in batch or semi-batch reactors, the final composition and the properties of the copolymer will vary during the reaction and the composition of the polymer could be quite different from that expected according to the feed. The final properties and the performance of such copolymer are determined not only by the monomer content but very often by their microstructure. The insertion of each monomer depends on the catalyst and the conditions of the reaction (pressure and temperature), which define the kinetics of the process. Due to the determinant effect of the monomer composition in the polymer properties, in the synthesis method the ethylene/propylene ratio in the gas phase must be kept constant during the polymerization process.

The selection of the best catalyst to a certain copolymerization process requires testing at different scales: laboratory, pilot plant and industrial scale. The set-up and conditions used in each scale could be very different and they must be considered carefully in order to compare results between them. The polymerization at the laboratory is usually carried out in batch or semi-batch reactors while at industrial scale reactors operate in continuous mode. Usually, these reactors show low monomer conversion per pass due to short residence time at high monomer flow.

The economy of the process leads to the recycle of the monomers that have not been consumed during the reaction, for which a separation process is needed. This procedure is not easily adapted at the laboratory scale. Besides, the rate of both monomers incorporation to the copolymer chain could differ in orders of magnitude, changing the gas phase composition of the slurry with the time of reaction, obtaining copolymers with variable composition depending on the kinetics of monomer incorporation. Several alternatives have been studied in order to get results in semi-batch reactors at laboratory scale easily extrapolated to continuous processes. Thus, copolymerization reactions with set-up involving continuous feeding and discharge of the monomers at very high space velocities of the circulating gases and low conversions have been described [39], as well as the operation with short polymerization experiments [40].

An alternative procedure can be carried out in order to run polymerization in semi-batch laboratory reactors by continuously modifying the feed to keep constant the gas phase composition. The control system includes an online Micro-Gas Chromatograph (Micro-GC), which analyzes both monomers in the gas phase, modifying the propylene flow to keep the monomers ratio constant in the gas phase during the reaction. Vapour liquid equilibrium conditions are considered to estimate the liquid composition.

This last procedure is used in order to obtain the set of copolymers covering the whole composition range. During the synthesis procedure the average gas phase composition, i.e. the gas ethylene/propylene (C_2/C_3) ratio, is controlled within the range 0.010 to over 100. Obtained copolymers are listed in Table 1, named as PP and PE for homopolymers and EP + number for copolymers, where the number represents the copolymer composition as will be explained later. Involved C_2/C_3 ratio is also listed in Table 1.

The first point of the synthesis method was the impregnation of metallocene catalyst (rac-dimethyl-silylbis (2-methylindenyl) zirconium dichloride) on silica modified with methylaluminoxane, MAO. All the air-sensitive compounds have to be handled in a dry nitrogen atmosphere inside a glove box and by using Schlenk techniques.

The conditions of the support pre-treatment, the determination of the hydroxyl content and the physical properties of the silica in these conditions have been previously reported [41]. The immobilization method is carried out in two steps: 1) reaction between silica material and MAO using the incipient wetness procedure; 2) anchorage of the metallocene on silica/MAO sample previously synthesized.

Once the catalytic system is prepared is carried out the synthesis of the range of copolymers. Ethylene and propylene monomers must be deoxygenated and dried before to be introduced in the polymerization reactor, because the traces of water can reduce or avoid the catalytic activity.

In these kinds of reactions, where the catalytic system is supported, the co-catalyst used is triisobuthylaluminium (TIBA) also added into the reactor as scavenger.

Table 1. Average gas phase composition of polypropylene, polyethylene and ethylene/propylene copolymers synthesized using Micro-GC for controlling the feed

Sample	C_2/C_3 (mol/mol) gas phase composition
PP	
EP 03	0.010
EP 06	0.013
EP 13	0.029
EP20	0.126
EP 30	0.390
EP 59	0.970
EP 81	5.046
EP 87	6.740
EP 94	19.000
EP 99	107.690
PE	

Copolymerization reactions are carried out in a stirred glass semi-batch reactor, where the pressure and the temperature are controlled for all the runs. The solvent (n-heptane) is saturated with a ethylene/propylene gas mixture previously to begin the copolymerization reaction.

During the reaction, monomers are continuously fed with an ethylene/propylene ratio varied during the experiments to keep constant the gas phase composition. To control the gas phase composition, real-time monitoring of the gas phase is performed by means of a Micro-GC. Ethylene and propylene signals are analyzed separately and the ratio is calculated, achieving a C_2/C_3 analysis within 60 seconds. The C_2/C_3 mass ratio is fitted to the set point value by modifying the propylene mass flow valve. The reactions are carried out during 30 minutes in order to avoid an influence of monomer gradients on catalyst activity and copolymer composition [40].

As an example, Figure 1 shows the results obtained during the synthesis process for copolymer EP 30. Initially, the reactor solvent is saturated with the same C_2/C_3 ratio (0.4); when reaction begins in the first stage (2 minutes) the consumptions of the monomers are higher and the effect of the faster insertion of one of the monomers by the activity centre of the metallocene catalyst can be seen due to the C_2/C_3 ratio increases.

If the control system is used the propylene feed is modified in order to keep constant the ratio between the monomers in the gas phase during the reaction time. Thus, the composition in the gas phase, and therefore in the liquid phase, is kept around certain value during the whole reaction time to ensure the final control on the copolymer composition and its properties next to the set point (S.P.).

Following this procedure it is possible to synthesize copolymers in whole composition range which allows establishing the correspondence between this property and all the others, which is the main topic of this chapter.

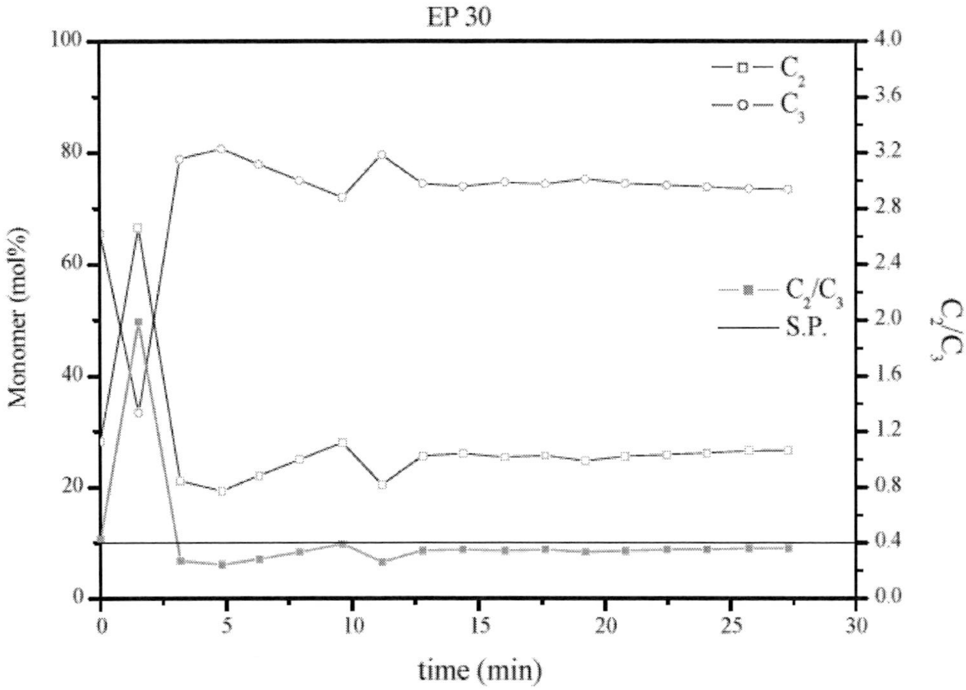

Figure 1. Gas phase ethylene/propylene composition ratio vs reaction time during the synthesis of EP 30.

BASIC CHARACTERIZATION OF COPOLYMERS

Basic properties considered for copolymers are the average molecular weight and polydispersity as well as the average comonomer concentration and its distribution along the chains (related to microstructure) because they are responsible of final copolymer properties.

Application of 13C nuclear magnetic resonance to polymers yields quantitative measurement of chain microstructure [42]. The overall composition of the copolymers is determined from the method of the triad distribution. The total ethylene content is determined by sum of the corresponding triads ([E] = [PEP]+[EEP]+[EEE]) according to the assignment of chemical shift given by Randall [43] and Kakugo [44]. Table 2 shows results obtained for the copolymer set listed in Table 1. Values for copolymer composition, determined by ^{13}C-NMR and given in weight (w%) and molar (mol%) percentage of ethylene are listed. As explained before the samples are named according to the molar composition.

Table 2. Basic characterization of polypropylene and polyethylene homopolymers, and ethylene/propylene copolymers obtained by ^{13}C-NMR and GPC

Sample	^{13}C-NMR				GPC	
	w(weight %)	w(mol %)	n_E	n_P	M_w	P.I.

PP	0	0				
EP 03	2.3	3.38	1.05	30.1	112186	2.45
EP 06	4.0	5.82	1.20	20.4	118757	2.44
EP 13	8.9	12.71	1.13	7.88	87690	2.49
EP20	14.6	20.46	1.13	4.51	69985	2.42
EP 30	22.6	30.44	1.24	2.85	61636	2.08
EP 59	49.0	59.03	2.08	1.41	61967	2.86
EP 81	73.5	80.66	4.70	1.14	54673	2.54
EP 87	81.3	86.7	6.68	1.12	163911	3.01
EP 94	91.0	93.84	15.6	1.10	268488	2.78
EP 99	98.6	99.03	102	1.00	260794	3.63
PE	100	100				

Additional relevant information can be obtained from NMR analysis, from the triad determination; it is possible to calculate the length of the monomeric blocks according to next expressions:

$$n_P = \frac{N_{EPE} + (N_{PPE} + N_{EPP}) + N_{PPP}}{N_{EPE} + \frac{1}{2}(N_{PPE} + N_{EPP})}$$
$$n_E = \frac{N_{PEP} + (N_{PEE} + N_{EEP}) + N_{EEE}}{N_{PEP} + \frac{1}{2}(N_{PEE} + N_{EEP})} \qquad (1)$$

where n_E and n_P represent the relative average number of ethylene and propylene, respectively, units in the copolymer chain and they give information about the distribution of comonomers along the copolymer chain. Obtained values are also listed in Table 2.

The molecular weight distribution is fundamental in determining how the material can be processed and what the physical properties of the final product will be [45]. Gel permeation chromatography with universal calibration is the most used technique to carry out that characterization. Molecular weight (M_w) and the polydispersity index determined by GPC are also presented in Table 2.

M_w values are dependent on comonomer content, showing the lowest values in the composition range from 30 to 80% molar ethylene percentage, because the high content of comonomer decreases the crystallinity of the polymer, favouring growth in solution, with features similar to those synthesized with homogeneous catalytic system. The polydispersity index is approximately constant, as expected for this single-site catalyst [46].

Molecular weight values for samples with high propylene content (EP 03) are lower than those with high ethylene content (EP 99). These differences are related to a lower probability of β-type termination reactions when the concentration of ethylene increases in the reaction medium because this reaction is more favourable for comonomers bigger than ethylene. This is in agreement with the results obtained when comparing average molecular weight of ethylene and propylene-rich copolymers [47, 48].

THERMODYNAMIC CHARACTERIZATION

Changes in the composition and structural parameters of the polymers affect its melting, glass transitions and crystallinity.

Calorimetric analysis are carried out by a differential scanning calorimetry (DSC), connected to a cooling system and calibrated with different standards. Samples are firstly heated from -70 to 190 °C at a heating rate and then cooled down to -50 °C at the same rate, followed by a second heating. The scanning rate used was 10 °C min^{-1}.

Figure 2 shows the results obtained by DSC for the glass transition temperature (Tg) and the second scan melting temperature (Tm_2) in function of the ethylene content. It is clear that values for these temperatures are lower for copolymers close to 50/50 ethylene/propylene ratio. This is related to the considerable decrease of crystallinity, since the crystal domains which are being formed are smaller because the increase of irregularities in the macromolecular chains.

Copolymers are studied from the point of view of crystallinity behaviour from DSC results. Values for the enthalpy of the second melting of the copolymers are obtained. It is possible to determine the percentages of crystallinity (f_c^{DSC}) by relating such values to those for the corresponding homopolymers (291.6 J/g for polyethylene [49] and 207.1 J/g for polypropylene [50]). The calculated values are plotted in Figure 1 and corroborate how the copolymers close to 50/50 ethylene/propylene ratio are amorphous materials and the crystallinity can not be determined.

The values previously calculated for n_E and n_P and listed in Table 1, can be related directly to the crystallinity obtained by DSC analysis. It can be noted that, for the heterogeneous copolymers obtained in the conditions of these reactions, when a chain contains at least 5 units of the same consecutive monomer, that sequence crystallizes.

During the past decade there has been a growing interest in the chemical heterogeneity characterization, i.e. all types of chemical composition distributions of synthetic polymers. This was due to increasing demand of polymer properties related to mechanical and chemical resistance and biodegradability, which required the development of tailor made products such as copolymers and terpolymers, block copolymers and polymer blends [51].

Deep investigations of CCD of semi-crystalline polyolefins have been carried out by means of Temperature Rising Elution Fractionation (TREF) since it was developed in 1980s, and by Crystallization Fractionation (CRYSTAF) method rather recently [52]. Chemical composition distributions are measured by TREF using o-dichlorobenzene (oDCB) as a solvent. The concentration of polymer was measured by a two-channel infrared detector. CRYSTAF analyses are carried in the temperature range 95 °C - 30 °C with a cooling rate of 0.1 °C/min.

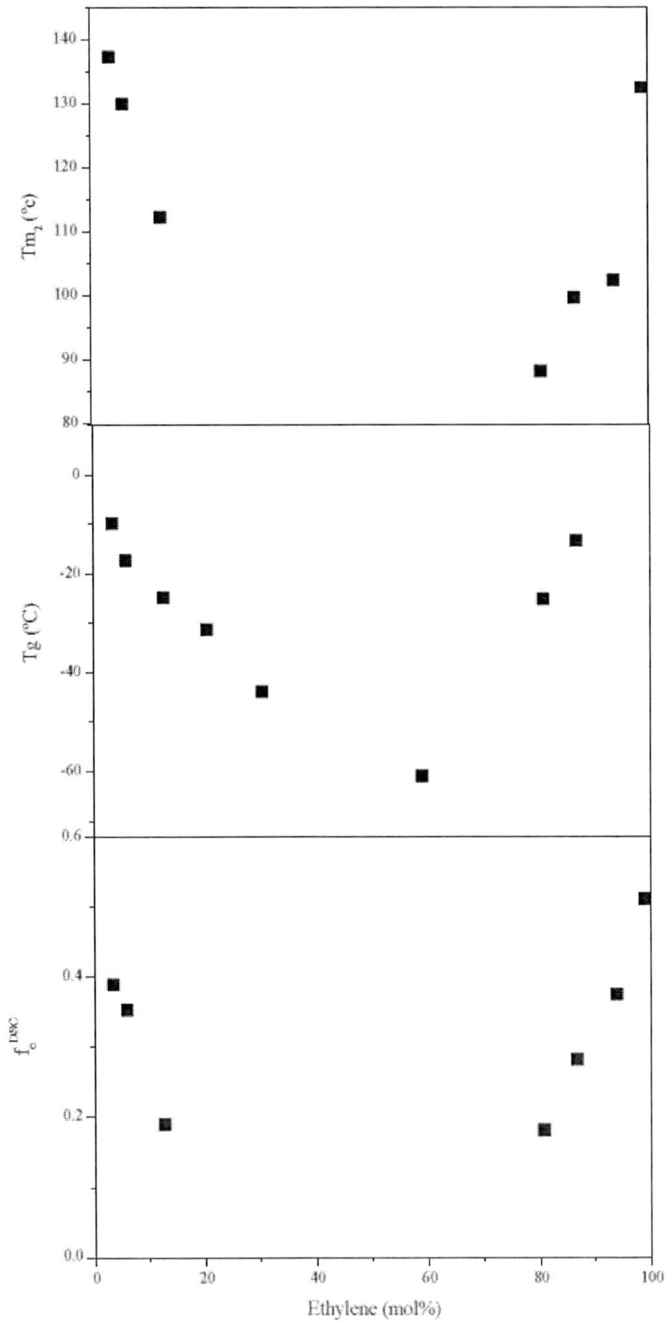

Figure 2. Glass transition temperature (Tg), second scan melting temperature (Tm_2) and crystallinity (f_c^{DSC}) for copolymers in function of the ethylene mol percentage.

The determined elution temperature (T_e) and the percentage of soluble fraction (SF%) obtained by TREF are also experimental properties related to crystallinity. Figure 3 shows the results obtained by TREF as a function of the ethylene comonomer content. This technique is limited to copolymers with very different comonomer contents, since in the middle range of comonomer content an amorphous copolymer is reached. In this case, the limit is lower than 10% molar of ethylene and higher than 80% molar of ethylene. For copolymers with low ethylene content (EP3 to EP13), values for Te decreases faster with ethylene molar percentage than in the high ethylene content (EP81 to EP99). This is due to the fact that the crystallinity of copolymers with high propylene content is influenced not only by the composition and sequence distribution but also tacticity. PP has three possible stereochemical configurations: isotactic (i-PP, with all methyls on the same side of the chain), syndiotactic (s-PP, with methyls on alternating sides of the chain), and atactic (a-PP, without any regular order) [53], this make the study of the chemical composition distribution of polypropylene copolymers more complicated.

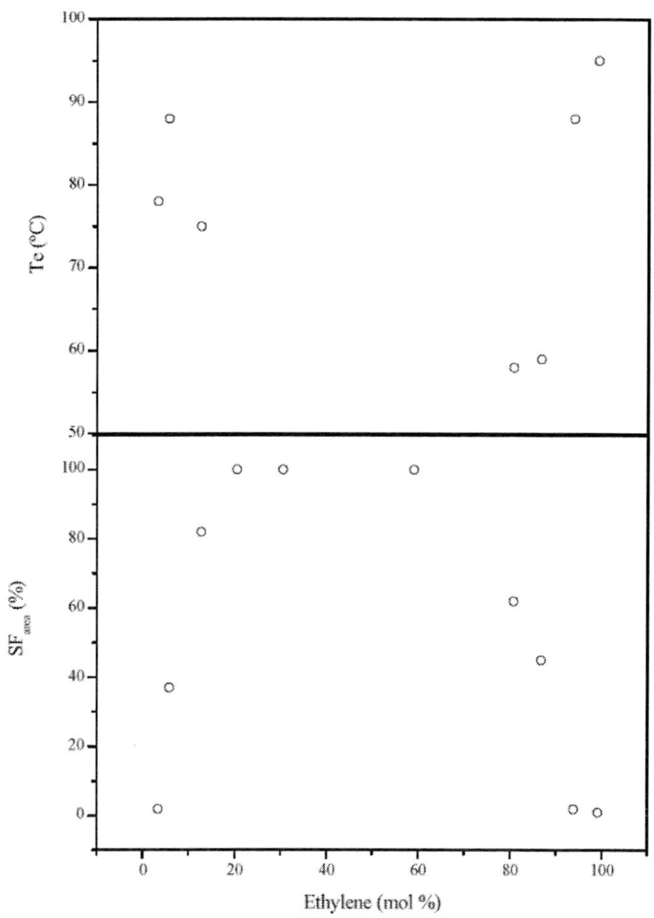

Figure 3. Influence of comonomer content on TREF elution temperature and percentage soluble fraction.

CRYSTAF separates the components present in a sample based on their crystallizability. The first derivative of the concentration profiles of these samples are shown in Figure 4 for each copolymer. As can be seen, first contribution observed due to the soluble fraction appears at low temperatures (25-30°C), this fraction is higher when monomer contents in the copolymer are similar, in agreement with TREF results (Figure 5). The main crystallization peak appears at variable temperature which decreases as comonomer content increases. It is important to underline that ethylene rich copolymers present higher crystallization temperature than propylene, due to the tacticity effect previously appointed. Some of these samples present two peaks of crystallinity which can be related with inhomogeneities in the copolymer chains related to heterogeneous ethylene distribution in spite the use of a metallocene catalyst.

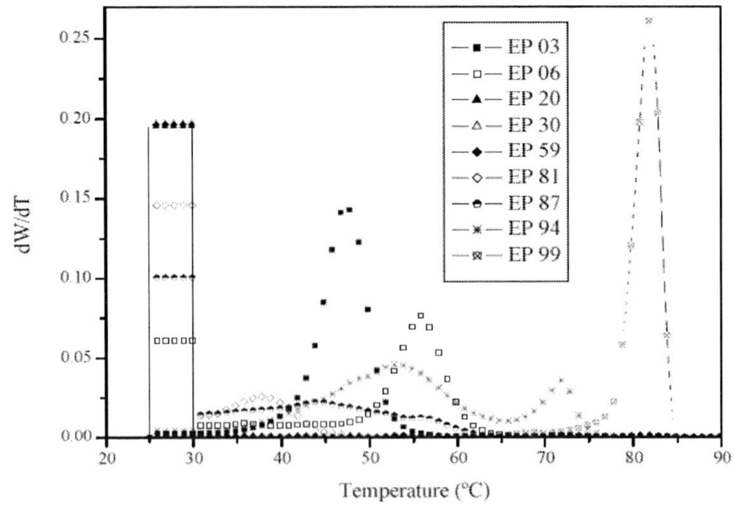

Figure 4. CRYSTAF profiles for ethylene/propylene copolymers in the whole composition range.

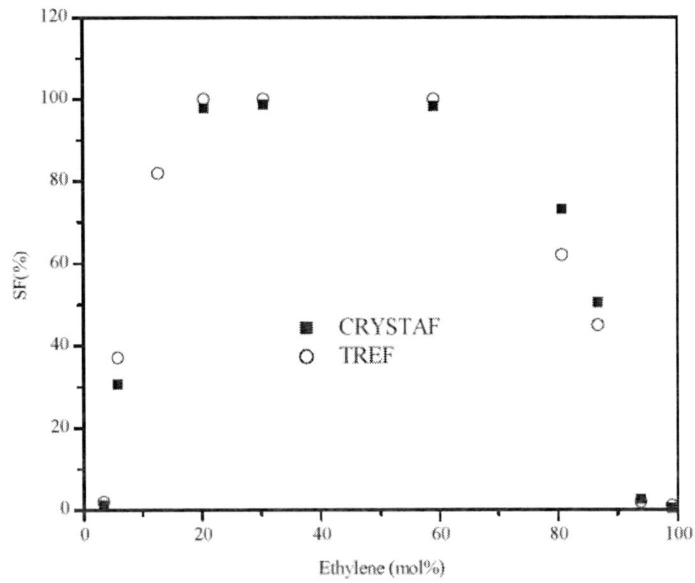

Figure 5. Percentage of soluble fraction determined by CRYSTAF and TREF.

CHEMICAL HETEROGENEITY CHARACTERIZATION

The properties of a copolymer are not only dependent on composition but they are dependent to a large extent on several structural features of the copolymer chains, as it was reported before: (a) the relative content of comonomer units in a polymer chains, the way that the comonomer units are distributed in the chain, and variation in the comonomer composition of different chains; (b) molecular weight and molecular weight distribution; and (c) relative content of normal head-to-tail addition vs. head-to-head and tail-to-tail additions.

This is the reason why, for copolymer samples, can be combined two kinds of molecular characterization with high interest: the molecular weight distribution (MWD) and the composition distribution. The MWD tells how much material of a certain molecular weight is present while the composition distribution tells how much material of a certain comonomer content is present. Comonomer incorporation gives rise to two different distributions: the monomer sequence length distribution (SLD) and the chemical composition distribution (CCD). The SLD describes how the monomers are distributed within the copolymer chains and reflects intramolecular heterogeneity. On the other hand, the CCD describes the distribution of the average comonomer content of the copolymer chains, thus reflecting intermolecular heterogeneity [54].

The method of GPC-IR (size exclusion chromatography coupled to IR detector) is used to measure the comonomer content on the molecular weight (M_w) and, consequently, to generate the comonomer distribution profile across the MWD of the resin. The GPC-IR method allows, in some extent, determining the comonomer content of the polymer molecules of similar sizes by means of the infrared detector. The final result of this method is the comonomer content for each molecular weight, i.e. the CCD. Actually, the average result obtained by this method (GPC-IR) can be compared with the percentage of comonomer incorporated determined by ^{13}C-NMR [55].

The GPC-IR studies are carried out by means of an IR5-MCT detector coupled to a GPC through a heated-transfer line controlled at 140°C. IR5-MCT determines the concentration of total CH_2, and the CH_2/CH_3 ratio and thus can be applied to the analysis of low number of short chain branches and to obtain the composition-molar mass dependence [56].

Figure 6 shows GPC-IR results for the samples synthesized with metallocene supported catalytic system, based on this plot, it can be seen that till EP30 and from EP94, the ethylene content is homogeneous along the whole molecular weight distribution, but EP59, EP81 and EP87 an uniform increase in the ethylene (mol%) value is shown with the molecular weight for these samples, i.e. the ethylene content is lower in the low molecular weight range and this effect can be check over the whole M_w range. When these results are fitted linearly the slope of the lines are plotted in Figure 7. As can be seen the behaviour is similar to those obtained when soluble fraction or crystallinity have been studied. Also, these samples have been previously characterized by CRYSTAF, in order to check their compositional uniformity. CRYSTAF results shown several peaks at different temperatures in response to inhomogeneities in the comonomer distribution and thus in agreement with GPC-IR results present here.

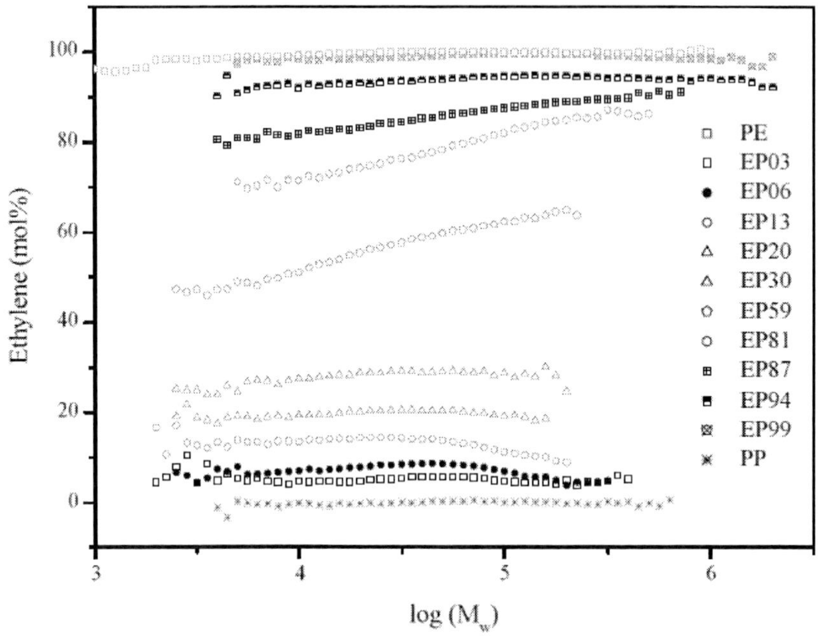

Figure 6. GPC-IR results of ethylene/propylene copolymers synthetized with heterogeneous metallocene catalytic system.

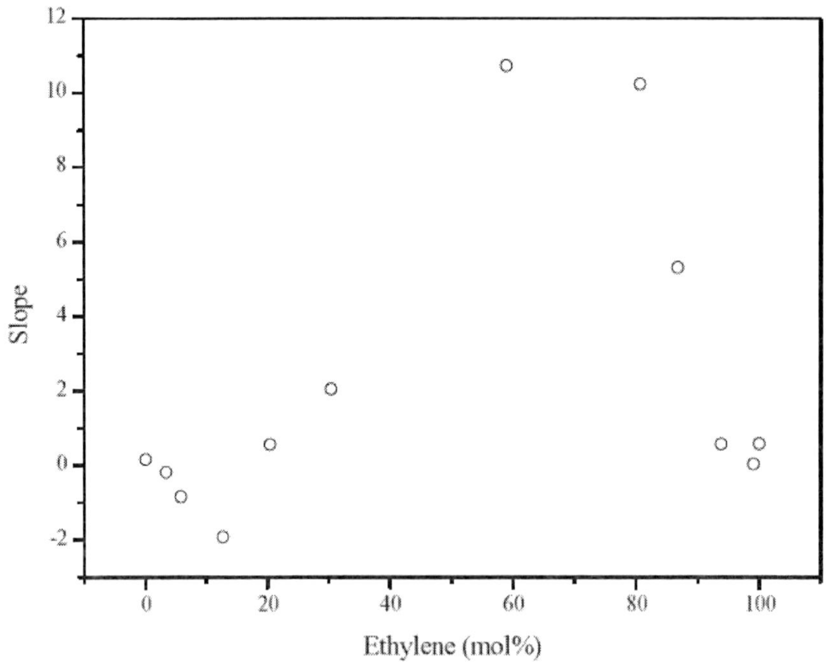

Figure 7. Slope of the fitted GPC-IR lines of ethylene/propylene copolymers.

CONFORMATIONAL CHARACTERIZATION

The method of GPC-MALS (multi-angle light scattering coupled to size exclusion chromatography) offers the possibility of analyzing a polymer sample in terms of the distribution of both molecular weight (M_w) and the radio of gyration (R_g) [57].

GPC-MALS studies are carried out in a gel permeation chromatograph combined with a multi-angle light scattering (MALS). MALS is equipped with a laser at 690 nm and 17 multi angle detectors. The software used, ASTRA V 5.1.9.1, from Wyatt Technology, allows on-line data acquisition of molecular weight and radius of gyration and the further calculation of distributions and averages.

The DAWN EOS system is calibrated with toluene to convert the measured voltage to intensity of scattered light. The normalization of the detectors is performed with a standard monodisperse PS of low molecular mass (M_w = 30000 g/mol) that did not show angular dependence on the light scattering signal. This standard polymer were also used to determinate the inter-detector volume [57].

The differential refractive index increments dn/dc of the polymers and copolymers are determined following a procedure previously described [58].

Molecular weight (M_w) and the radio of gyration (R_g) are measured in an absolute way [59, 60, 57]. The conformation of the macromolecular chains in solution was studied with the scaling law:

$$R_g = Q\, M_w^{\,q} \qquad (2)$$

where q is a shape parameter. Generally, q has a value of 0.33 for globular polymers and 0.5 for random coil polymers at theta conditions, but for random coil polymers in good solvents, the value of q could increase up to 0.6 [61].

Figure 8 shows the radius of gyration (R_g) against molecular weight for copolymers in whole composition range. Obtained linear results show that same scaling law given by Eq. (2) can be applied.

It is possible to check how copolymer R_g vs. M_w line is closer to the line for ethylene or propylene according to each component content. In addition, when the comonomer is propylene, the R_g values are similar to those for ethylene in the low molecular weight region but the comonomer influences the conformation of the chains in the higher molecular weight region.

Obtained shape parameter (q) values for copolymers are plotted in Figure 9. The bigger slope of such representations corresponds to both homopolymers where the same value, q = 0.54, was obtained both for PE and PP polymers, thus showing such random coil configuration. Copolymer q values are slightly lower than the homopolymers value, showing how the effect of the second monomer introduces an impediment to adopt the random coil distribution. q value present a trend with molar composition for the studied copolymers where the lowest value (q= 0.42) corresponds to copolymer with 30% of ethylene. These results are different from those observed previously by Sun et al. [62] where parallel power laws were found, indicating that q exponent was insensitive to the comonomer content.

In this work, the obtained Q values were 0.0426 and 0.0299 for PE and PP, respectively. The lowest value for PP imply a more compact molecule for the same molecular weight which can be directly related with the lower average length of the molecule due to the amount of short chain branches.

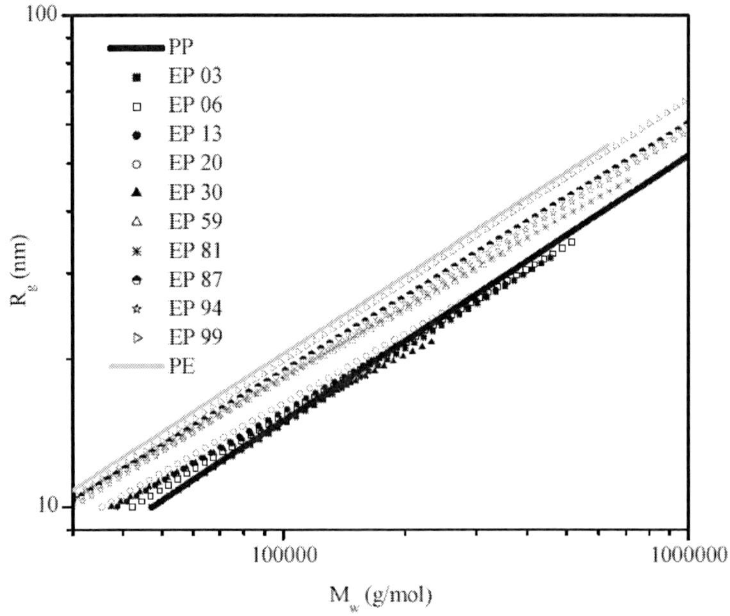

Figure 8. Radius of gyration vs. molecular weigh for EP copolymers.

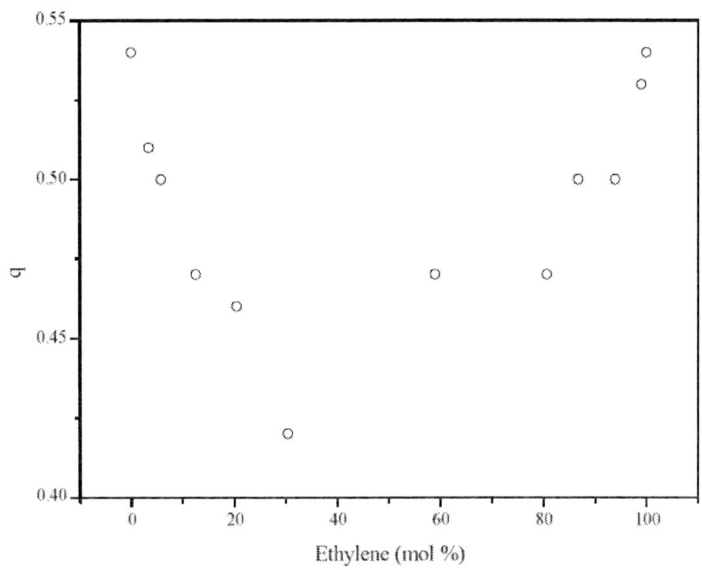

Figure 9. Plot of the q value vs. ethylene molar % for the studied copolymers.

Conclusion

Good control system for the gas phase ethylene/propylene ratio in the polymerization process allows obtaining copolymers in the whole composition range in a laboratory scale. The influence of copolymers composition allows observing similar behaviour in all properties independently of the technique used to be analyzed, because these results are directly related to the absence of crystallinity of samples when comonomer content are similar.

DSC analyses show, for copolymers with low comonomer content, similar behaviour to that of the corresponding homopolymer, but when the comonomer content increases, the melting temperature and glass transition temperature decrease and the copolymer becomes amorphous as concluded when crystallinity determined by DSC is analyzed. This tendency is corroborated when the case of soluble fraction by TREF and CRYSTAF are displayed.

The GPC-IR method allows the determination of the compositional distribution of the copolymers. In spite of these results, when the slopes of the GPC-IR results are analyzed, a similar behaviour to that obtained for thermodynamic properties is shown.

GPC-MALS allows studying the conformation of the polymers in solution. The q value calculated for both homopolymers (PE and PP) show a random coil configuration.

The q parameter determined for copolymers decreases with the composition, probably due to the effect of the second monomer, which introduces an impediment to adopt the random coil distribution. PP reaches a more compact structure than PE for the same molecular weight, which can be directly related with the lower average length of the molecule due to the amount of short chain branches.

From all these results it is possible to remark the high influence of the composition on different copolymers properties and the interest to determine this parameter.

References

[1] Akovali, G.; Atalay, A. *Polymer Testing.* 1997, 16, 165-171.

[2] Alamo, R. G.; Ghosal, A.; Chatterjee, J.; Thompson, K. L. *Polymer.* 2005, 46, 8774-8789.

[3] Anantawaraskul, S.; Soares, J. B. P.; Wood-Adams, P. M. *Macromol. Symp.* 2004, 206, 69–77.

[4] APME Plastics – the Facts 2010 An analysis of European plastics production, (demand and recovery for 2009).

[5] Arrowsmith, D.; Kaminsky, W.; Schauwienold, A. M.; Weingarten, U. *Journal of Molecular Catalysis A: Chemical* 2000, 160, 97-105.

[6] Asua, J. M. *Polymer reaction engineering*, Blackwell Publishing Ltd, OXFORD, pp.16.

[7] Bianchini, D.; Bichinho, K. M.; Dos Santos, J. H. Z. *Polymer.* 2002, 43(10), 2937-2943.

[8] Boor, J. Ziegler-Natta Catalysts and polymerizations, ACADEMIC PRESS: New York, 1979.

[9] Bovey, F. A.; Mirau, P. A. *NMR of polymers;* Academic press: California, 1996; pp.117.

[10] Caballero, M. J.; Suarez, I.; Coto, B.; Van Grieken, R.; Monrabal, B. *Macromol. Symp.* 2007, 257, 122–130.
[11] Carcuchinski Haag, M.; Krug, C.; Dupont, J.; Barrera de Galland, G.; Henrique Zimnoch dos Santos, J.; Uozumi, T.; Sano, T.; Soga, K. *Journal of Molecular Catalysis A: Chemical.* 2001, 169, 275-287.
[12] Carlson, E. C. Chemical Engineering Progress. 1996, 35-46.
[13] Cheng, H. N. *Structural Studies of polymers by Solution NMR*; Rapra Technology Limited, v. 11, 5, United Kingdom, 2001; pp.11.
[14] Chien, J. C. W.; He, D. *J. Polymer Sci. Polymer Chem.* 1991, 29, 1585-1593.
[15] Coto, B.; Escola, J. M.; Suarez, I.; Caballero, M. J. *Polymer Testing.* 2007, 26(5), 568-575.
[16] Dweik, H.; Al-Jabareen, A.; Marom, G.; Assouline, E. *Internacional Journal of*
[17] Faldi, A.; Soares, J. B. P. *Polymer.* 2001, 42, 3057-3066.
[18] Fink, G.; Steinmetz, B.; Zechlin, J.; Przybyla, C.; Tesche, B. *Chem. Rev.* 2000, 100, 1347-1376.
[19] Galimberti, M.; Piemontesi, F.; Fusco, O.; Camurati, I.; Destro, M. *Macromolecules.* 1998, 31, 3409-3416.
[20] Grigoryeva, O. P. ; Fainleib, A. M. ; Tolstov, A. L. ; Starostenko, O. M. ; Lievana, E. ; Karger-Kocsis, J. *J. Appl. Polym. Sci* 2005, 95, 659-671.
[21] Haaga, M. C.; Krug, C.; Dupont, J.; Barrera de Galland, G.; Zimnoch dos Santos, J. H.; Uozumib, T.; Sano, T.; Soga, K. *J. Mol. Catalysis A: Chemical* 2001, 169, 275-287.
[22] Hamielec, A. E.; Soares, J. B. P. *Prog. Polym. Sci.* 1996, 21,651-706.
[23] Hlatky, Gregory G. *Chem. Rev.* 2000, 100, 1347-1376.
[24] Holding, S. R.; Meehan, E. *Molecular weight characterisation of synthetic polymers;* vol. 7, 11, Rapra technology limitated: United Kingdom. 1997; pp. 3.
[25] Hosier, I. L.; Alamo, R. G.; Lin, J. S. *Polymer.* 2004, 45, 3441-3445.
[26] Huang, H. M.; Liu, I. C.; Chien-Chao, T. R. *Polymer.* 2005, 46, 955–963.
[27] Ioku, A.; Shiono, T.; Ikeda, T. *Appl. Catal. A.* 2002, 226, 15-22.
[28] Kakugo, M. *Macromolecules.* 1982, 15, 1150-1152.
[29] Kamfjord, T.; Wester, T. S.; Rytter, E. *Macromol. Rapid. Commun.* 1998, 19, 505-509.
[30] Kaminsky, W. *Catalysis Today* 2000, 62, 23–34.
[31] Karak, N. *Fundamentals of polymers: Raw materials to finish products*, 2009, PHI Learning Private Limited, New Delhi.
[32] Ko, Y. S.; Jeon, J. K.; Yim, J. H.; Park, Y. K. *Macromolecular Research.* 2009, 17, 296-300. *Polymeric Materials.* 2003, 52, 655-672.
[33] Kravchenko, R.; Waymouth, R. M. *Macromolecules.* 1998, 31 (1), 1-6.
[34] Laguna, M. T. R.; Medrano, R.; Plana, M. P.; Tarazona, M. P. *J. Chromatogr. A* 2001, 919, 13–19.
[35] Lethinen, C.; Löfgren, B. *European Polymer Journal.* 1997, 33, 115-120.
[36] Liu, S.; Meng, F.; Yu, G.; Huang, B. *Journal of Applied Polymer Science.* 1999, 71, 2253-2258.
[37] Philipsen, H. J. A. *Journal of Chromatography A* 2004, 1037, 329–350.
[38] Piel, C.; Karssenberg, F. G.; Kaminsky, W.; Mathot, V. B. F. *Macromolecules.* 2005, 38, 6789-6795.
[39] Pires, M.; Mauler, R.; Liberman, S. *J. Appl. Polym. Sci.* 2004, 92, 2155-2162.

[40] Quijada, A.; Rojas, R.; Alzamora, L.; Retuert, J.; Rabagliati, F. M. *Catal. Lett.* 1997, 46, 107-112.
[41] Randall, J. C. *J. Macrom. Chem. Phys.* 1989, 29, 201-317.
[42] Resconi, L.; Camurati, I.; Sudmeijer, O. *Topics in Catalysis.* 1999, 7, 145–163.
[43] Ribeiro, M. R.; Deffieux, A.; Protela, M. F. *Ind. Eng. Chem.* 1997, 36, 1224-1237.
[44] Sarzotti, D. M.; Marshman, D. J.; Ripmeester, W. E.; Soares, J. B. P. *J. Polymer Sci. Polymer Chem.* 2007, 45, 1677–1690.
[45] Singh, G.; Kothari, A. V.; Gupta, V. K. *Polym. Test.* 2009, 28, 475-479.
[46] Sinn, H.; Kaminsky, W. *Adv. Organomet. Chem.* 1980, 18, 99-149.
[47] Sinn, H.; Kaminsky, W., Vollmer, H. J. *Angew. Chem. Int. Ed. Engl.* 1980, 19, 390-392.
[48] Sinn, H.; Kaminsky, W.; Vollmer, H. J; Woldt, R. DE Patent 3007725, 29 Feb. 1980.
[49] Smit, M.; Zheng, X.; Loos, J. *J. Polym. Sci.* 2005, 43, 2734-2748.
[50] Soares, J. B. P.; Hamielec, A. E. *J. Polym. React. Eng.* 1995, 3, 131-200.
[51] Suarez, I.; Caballero, M. J.; Coto, B. *European Polymer Journal.* 2010, 46, 42–49.
[52] Suarez, I.; Caballero, M. J.; Coto, B. *Polym. Eng. Sci.* 2011, 51, 317–322.
[53] Sun, T.; Brant, P.; Chance, R.; Graessley, W. *Macromolecules.* 2001, 34, 6812-6820.
[54] Tynys, A.; Saarinen, T.; Hakala, K.; Helaja, T.; Vanne, T.; Lehmus, P.; Löfgren, B. *Macromol. Chem. Phys.* 2005, 206 (10), 1043-1056.
[55] Van Grieken, R.; Calleja, G.; Serrano, D.; Martos, C.; Melgares, A.; Suarez, I. *Polym. Reac. Eng.* 2003, 11, 17-32.
[56] Voegele, J.; Troll, C.; Rieger, B. *Macromol. Chem. Phys.* 2002, 203, 1918-1925.
[57] Voegele, J.; Troll, C.; Rieger, B. *Macromol. Chem. Phys.* 2002, 203, 1918-1925.
[58] Wild, L. *Adv. Polym. Sci.* 1990, 98,1-47.
[59] Yonggang, L.; Shuqin, B.; Yejuan, Z.; Wenhe, Z. *Polymer.* 2003, 44, 7209–7220.
[60] Youlu, Y.; DesLauriers, P. J.; Rohlfing, D. C. *Polymer.* 2005, 46, 5165–5182.
[61] Zambelli, A.; Grassi, A.; Galimberti, M.; Mazzocchi, R.; Piemontesi, F. *Makromolekulare Chemie Rapid Communications.* 1991, 12, 523-528.
[62] Ziegler, K.; Holzkamp, E.; Breil, H.; Martin, H. *Angew. Chem.* 1995, 57, 541.

In: Polymer Chains
Editors: Régine Penzkofer and Yasue Wu

ISBN: 978-1-62100-422-6

Chapter 9

STRUCTURE, PHYSICAL PROPERTIES AND INDUSTRIAL APPLICATION OF A PARTICULAR CLASS OF POLYMER CHAINS: THE PRECERAMIC POLYMERS

Samuel Bernard[1*], Mirna Bechelany[1,2], Laura Gottardo[3], Olivier Majoulet[2], Ting Ouyang[4], Christel Gervais[5], Corneliu Balan[4] and Philippe Miele[1]

[1] Institut Européen des Membranes,
IEM/Université Montpellier 2 Place E. Bataillon, France
[2] Laboratoire des Multimatériaux et Interfaces,
Université Claude Bernard Lyon 1, France
[3] Present address: Empa, Swiss Federal Laboratories for Materials Testing and Research, Laboratory for Advanced Fibers, Switzerland
[4] REOROM Laboratory, Hydraulics Department, "Politehnica" University of Bucharest, Bucharest, Romania
[5] Laboratoire de Chimie de la Matière Condensée,
UPMC, Collège de France, Paris, France

INTRODUCTION

Preceramic polymers are compounds with a skeletal structure that does not include only carbon atoms but also others atoms such as metals and metalloids (boron and silicon). We can distinguish homo-atomic polymers with one kind atom in the main chain such as polysilanes containing silicon-silicon and hetero-atomic polymers with more than one type of atom in the main chain, mostly two types of atoms alternate along the main chain such as polycarbosilanes (Si-C), polysilazanes (Si-N) [1-2], polymethylaminoboranes (B-N, Fig. 1

* Dr. Samuel Bernard: IEM, Université Montpellier 2; CC047, Place E. Bataillon, 34095 Montpellier Cedex 5, France; E-mail: samuel.bernard@iemm.univ-montp2.fr.

([3])) and others. They are available as liquid and solid (fusible or infusible) with a large range of viscoelastic properties to generate a large variety of shapes [4-9].

Figure 1. Structure of polymethylaminoborane [3].

The possibility to control the structure and physical properties of preceramic polymers is an interesting means to form "near net shape" polymers as reported in Fig. 2.

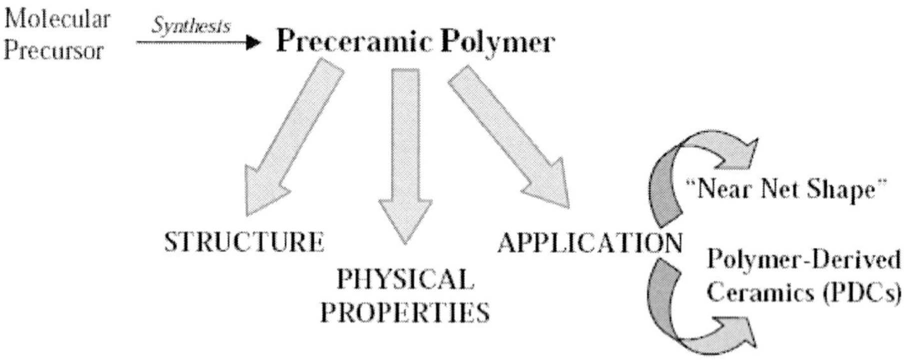

Figure 2. Structure, physical properties and application of preceramic polymer.

In addition, preceramic polymers are interesting candidates to prepare ceramics. The relatively recent *Polymer-Derived Ceramics (PDCs)* concept, which is based on the shaping then pyrolysis of preceramic polymers into, is a chemical route which offers original and new preparation opportunities in chemistry [1-2]. The PDCs route represents a synthetic approach in which the chemistry of molecular precursors and preceramic polymers is tailored at molecular scale to deliver complex ceramic compositions. This concept plays a major role in the preparation of materials endowed with designed performance properties that reach far beyond those of existing materials. The development of this scientific knowledge requires an interdisciplinary base that spans the fields of molecular chemistry, inorganic chemistry, polymer chemistry, synthesis, solid-state chemistry/physics, materials science and engineering.

In this sequence, detailed investigations of the structure and physical properties of the preceramic polymers are required to find the most appropriate applications.

This objective can be firstly reached by focusing on the molecular chemistry of inorganic polymers by solid-state NMR. Secondly, it is important to focus on the determination of their physcial properties by rheology and thermo-mechanical experiments to estimate their potentialities to be shaped into various kinds of shapes such as fibers, workpieces and porous materials. Thirdly, these polymers can be applied in industry as protective materials and used to generate ceramics by pyrolysis at high temperature.

STRUCTURE OF PRECERAMIC POLYMERS

In this part, we describe how the structure of preceramic polymers can be established. For that, we investigate the molecular structure of a particular class of preceramic polymers called poly[*B*-(methylamino)borazine]. These polymers are synthesized to form PDCs and in particular, boron nitride fibers. In our experiences, the same batch of a molecular precursor, i.e., *B*-tri(methylamino)borazine labeled MAB (Figure 3), is thermolyzed in argon between 150 and 200 °C forming solid polymeric networks, namely poly[*B*-(methylamino)borazine], with defined architectures, thermal properties and melt-spinnabilities. Here, our efforts focus on the characterization of the architecture and chemistry of a typical poly[*B*-(methylamino)borazine] labelled PMAB using a combination of characterization techniques including gas chromatography coupled with mass spectrometry (GC/MS), ^{15}N solid-state NMR, chemical analyses [10-11].

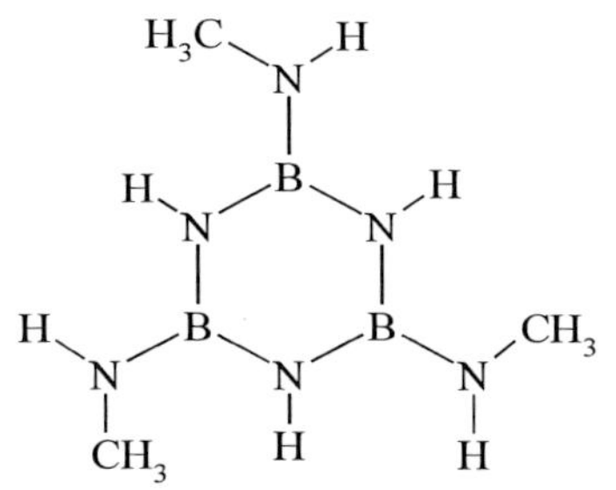

Figure 3. Schematic representation of the *B*-tri(methylamino)borazine.

According to GC/MS experiments, the transformation of MAB into the derived polymer PMAB occurs mainly with loss of methylamine (CH_3NH_2; $M/z = 31$) (Figure 4).

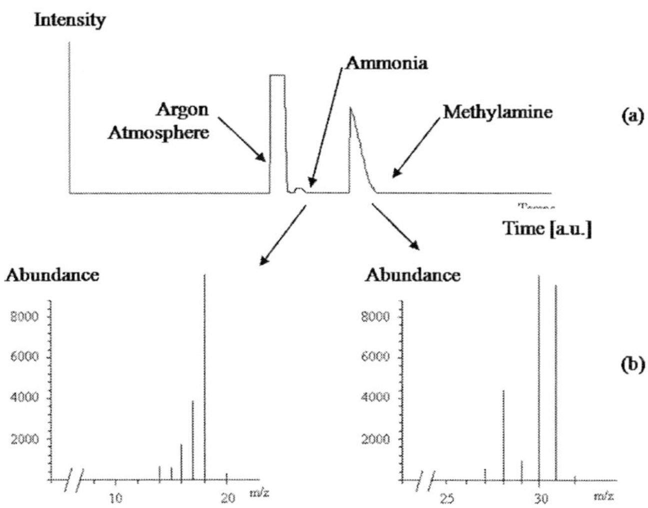

Figure 4. (a) Chromatogram and (b) corresponding molecular peaks of the gaseous species identified by GC/MS during the continuous polymerization process of MAB in flowing argon at 150 °C.

This should form -*N*(CH_3)- bridging borazine rings (δ_{iso} = -320 ppm) as depicted in Scheme 1.

Scheme 1. Condensation reaction of -N(H)CH_3 groups resulting in both the formation of bridged -N(CH_3)- units and generating methylamine.

Figure 5 includes the typical experimental and simulated ^{15}N MAS NMR spectra of PMAB with the assigned signals. During the polymerization of MAB leading to PMAB, an increase in the relative amounts of -*N*(CH_3)- bridges (δ_{iso} = -320 ppm) is observed, whereas the relative amounts of -*N*(H)CH_3 (δ_{iso} = -347 ppm) decrease. This agrees with the polycondensation mechanism proposed in Scheme 1. It is even shown that this mechanism is mainly effective at low temperature. A second route generating alkylamine species through polycondensation reactions including borazine N(H) units has been reported for *B*-tri(amino/alkylamino)borazines [10].

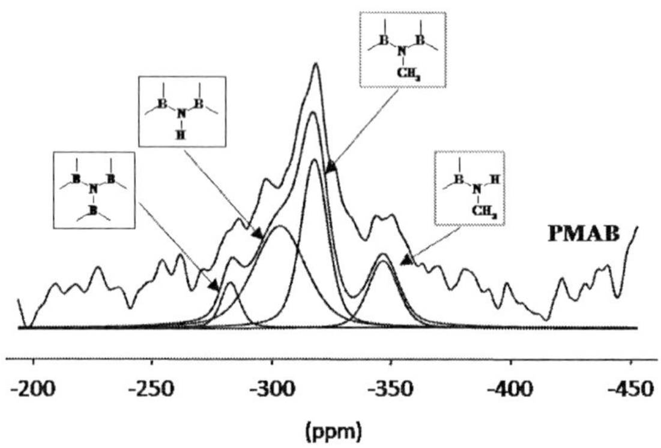

Figure 5. Selected experimental and simulated ^{15}N solid-state MAS NMR of a ^{15}N-enriched poly[*B*-(methylamino)borazine].

It consists in the formation of boron-nitrogen (B-N) bonds connecting two borazine rings. In our system, this mechanism occurs from the reaction of borazine N(H) units with -N(H)CH_3 groups generating methylamine (Scheme 2).

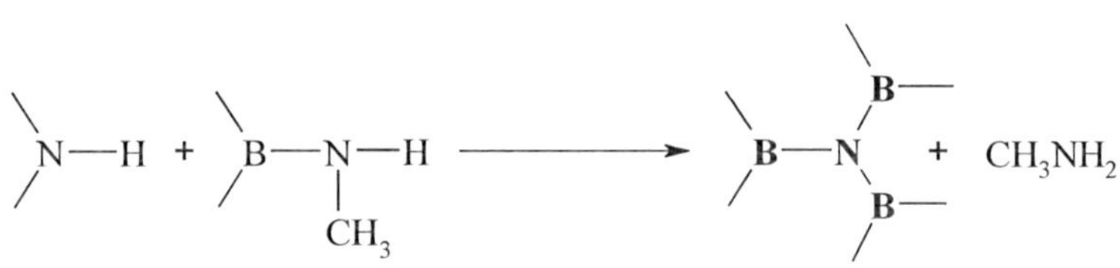

Scheme 2. Condensation reaction of -N(H)CH_3 groups with borazine N(H) units resulting in the formation of B-N bonds and generating methylamine.

The increase in the relative amounts of NB_3 sites associated with the disappearance of borazine N(H) units during polymerization is entirely consistent with the mechanism depicted in Scheme 2. By investigating the evolution of the -NB_3-to-N(H) ratio during polymerization of MAB, this second mechanism is seen to mainly occur above 160 °C. This is most probably attributed to the greater thermal stability of N-H bonds of borazine rings compared to the N-H bonds in -N(H)CH_3 groups. To our knowledge, any other reactions generating methylamine can be considered taking into account the functional groups, *i.e.*, borazine N(H) units and -N(H)CH_3 groups, which are present in the monomeric *B*-tri(methylamino)borazine (compound MAB).

Whereas the production of methylamine during polymerization of MAB was foreseeable according to the presence of -N(H)CH_3 groups, the identification of ammonia (NH_3; M/z = 17) as second gaseous by-product by GC/MS was not expected (See Figure 4). Further, the CH_3NH_2:NH_3 ratio decreases from 5000 at 120 °C to 9.5 at 200 °C based on the relative area of the corresponding GC signals. Therefore, production of ammonia increases with the progress of the polymerization. Evolution of ammonia suggests presence of -NH_2 groups in the polymeric network. Considering that such highly reactive groups are not identified by NMR and IR spectroscopies, it can be assumed that either their relative amount is too low or they quickly react to form more thermally stable species. On the basis of the actual knowledge of the chemistry of borazine-derived molecules, the only mechanism explaining the appearance of -NH_2 groups is a highly energetic ring-opening pathway. Such mechanisms historically proposed by Toeniskoetter *et al.* have been postulated to be activated when the borazine derivative contains -N(H)R (R = H, CH_3) units bonded to boron atoms [13]. Scheme 3 illustrates the most probable ring-opening mechanism occurring in our system which implies also the formation of a new $B_2N(CH_3)$ unit in the as-obtained structure.

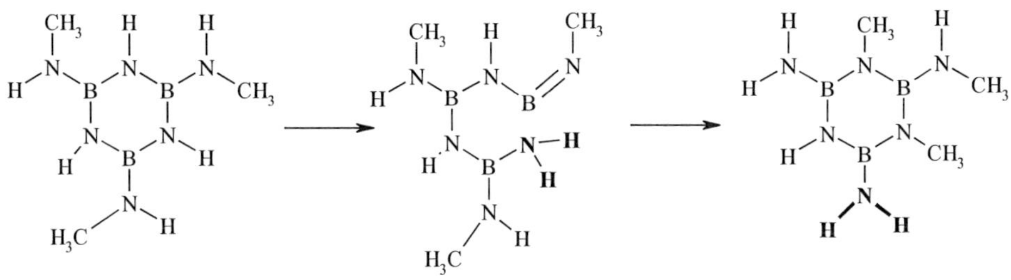

Scheme 3. Ring-opening mechanism leading to formation of -NH_2 groups.

According to their relatively high thermal reactivity, these -NH_2 groups should react with borazine N(H) units and/or -N(H)CH_3 groups with subsequent expulsion of ammonia and/or methylamine. Bridging -N(H)- units and inter-ring B-N bonds, *i.e.*, NB_3 motifs, should then result (Scheme 4).

According to the increase in ammonia production with the progress of the polymerization, the proportion of structural units formed must increase with the degree of polymerization of PMAB. However, the degree of polymerization does not have a strong impact on the polymer chemistry as illustrated with elemental analyses. The overall structure of the poly[*B*-(methylamino)borazine] established after synthesis at 150 °C is not significantly changed above that temperature. The typical empirical formula of ($[B_{3.0}N_{4.2}C_{1.9}H_{8.7}]_n$ clearly points to formation of cross-links connecting the borazine rings at several points of the same polymer chain in the high temperature regime of the polymerization.

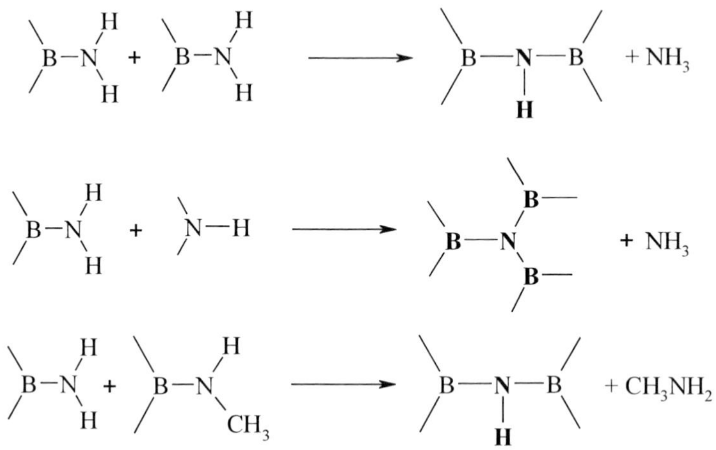

Scheme 4. Condensation reaction of $-NH_2$ and $-N(H)CH_3$ groups resulting in the formation of bridged -N(H)- and NB_3 sites and generating ammonia or methylamine.

As a consequence of the characterization results, it can be proposed that the basal structure of poly[*B*-(methylamino)borazine] composed of a majority of inter-ring $-N(CH_3)-$ bridges is shaped at $T_{Polymerization}$ = 150 °C. A relatively high amount of unreacted $-N(H)CH_3$ groups remains suggesting a poorly cross-linked network. Above that polymerization temperature, inter-ring B-N bonds, *i.e.*, NB_3 sites, are gradually formed in such a way that the B/N ratio increases in the derived polymer network synthesized up to 200 °C. In addition, the decrease in the relative amounts of $-N(H)CH_3$ groups associated with the poor changes in the polymer density in the temperature range 160 - 200 °C indicates that the poly[*B*-(methylamino)borazine] cross-links on itself in the high temperature regime of the polymerization. Lastly, in relation with the detection of ammonia, it should be mentioned that -N(H)- bridges/NH_2 groups are formed, but, despite a probable increase of their proportion with the progress of the polymerization, their influence on the architecture of poly[*B*-(methylamino)borazine] remains poor.

Physical Properties: The Rheology of Preceramic Polymers

Physical and chemical properties of preceramic polymers can be varied in a wide range by chemical modifications. Here, we have investigated as physical properties of preceramic polymers their rheology and in particular the rheological properties of the PMAB previously discussed. This helps us to correlate the effect of the polymer structure on their rheology behavior. PMAB are meltable preceramic polymers that provide fibers using a melt-spinning process [14-15].

Measurements of the viscosity and elasticity of molten polymers at their spinning temperature is important for controlling the melt-extrusion process and the subsequent drawing of the derived molten filament. The frequency dependence of the rheological parameters was determined during dynamic frequency sweep experiments within the LVE range giving the typical behavior of a melt-spinnable PMAB versus the oscillatory frequency (Figures 6). As-plotted curves showing the frequency dependence of G' (black square), G'' (black circle) and η^* (Empty symbols) for the different melts help us to predict the melt-spinnability of PMAB. Assuming the *Cox-Merz* relation [16] to be applied, it should be mentioned that we suppose that ω is related to the shear rate within the capillary. Therefore, we propose that the oscillatory frequency is proportional to a certain extent to the extrusion velocity.

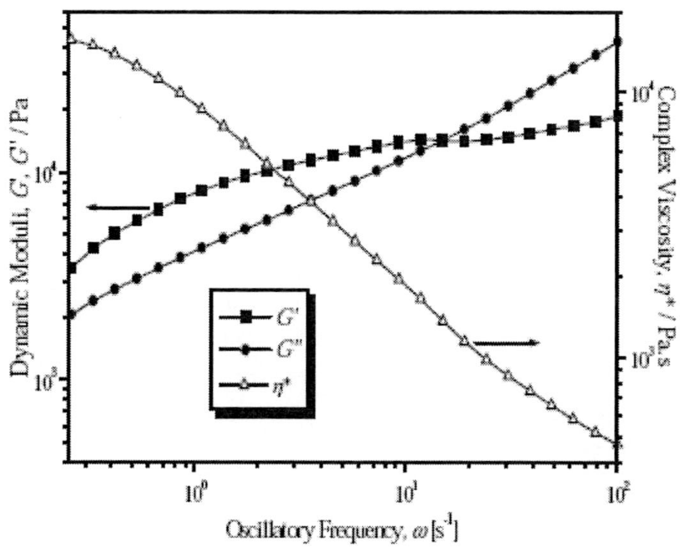

Figure 6. Frequency dependence of G', G'' and η^* for PMAB at T_{Test} = 165 °C [15].

As shown in the log-log plots in Figures 6, the complex viscosity η^* of PMAB is seen to continuously decrease in the oscillatory frequency range 0.25 - 100 s^{-1}, indicating clearly the shear-thinning behavior of PMAB melts. Therefore, PMAB investigated here are non-Newtonian compounds as it is usually required for melt-spinnable polymers. Regarding storage and loss moduli as a function of test frequency, it can be seen that polymer melts reveal an increase of the magnitude of G'' with increasing oscillatory frequency which is due to the higher energy required for molecular viscous response. The slight increase of the magnitude of G' for melts PMAB is attributed to a decrease in time available for molecular relaxation. It can be seen that its elastic behavior remains predominant (*i.e.*, G' is larger that G'') in the frequency range 0.25 - 16 s^{-1}. At ω = 16 s^{-1} occurs the crossover of G' and G'', and the viscous component (G'') becomes larger than the elastic component (G') above that oscillatory frequency. A slight stagnation of G' is observed between ω = 16 s^{-1} and ω = 100 s^{-1}, whereas G'' continues to increase in the oscillatory frequency range.

Rheological data provide data which are legitimate for the understanding of the melt-spinning bahavior of preceramic polymers. Fundamental assumptions for melt-spinning are given below. Based on the poor melt-spinnability of PMAB with high degree of polymerization it is reasonable to assume that an excess of G' over G'' is undesirable for reliable melt-spinning. Working on this assumption, we state that melt-spinning of PMAB is achieved above the crossover of G' and G'' at $\omega = 16\ s^{-1}$ (Figure 6) at which the viscosity component becomes predominant. This is in excellent agreement with spinning experiments, since this polymer displays excellent extrusion ability as melt at relatively high extrusion velocity. As a first pre-requisite condition for melt-extrusion, we therefore assume that the viscous component (G'') must be higher than the elastic one (G') in the frequency range investigated here. Based on this requirement, PMAB should be successfully extrudable from its molten state into fibers. Spinning experiments are in good agreement with such predictions, but a cohesive fracture occurred at high drawdown ratio ($\kappa > 93.9$) for PMAB with lower degree of polymerization. This spinning behavior is probably caused by the loss of elasticity under the high shearing level. As a first assumption for an appropriate filament drawing, we therefore suggest that a minimum level of elasticity is needed and/or the loss of elasticity must be restricted during rheology tests. From a rheological point of view, these established spinnability requirements are consistent with both phenomena occurring upon melt-spinning. First, the G''-to-G' ratio, *i.e.*, damping factor $\tan\delta$, is closely correlated to the melt-extrusion process, since the latter requires an appropriate viscosity-to-elasticity ratio allowing deformation of the molten polymer by flowing through the spinneret, and then retention of the filament shape at the capillary exit. Second, the elastic modulus is connected to the stretchability of the emerging molten filament which must exhibit a minimum level of elasticity to ensure cohesion of the material along the spinning line, and then retention of the fiber shape as it solidifies.

Industrial Application of Preceramic Polymers

There is an increasingly commercial development of preceramic polymers (KiON Ceraset® Polysilazane, PCS, PHPS [17-19]) to produce near-net-shapes in a way not known from other techniques.

Preceramic polymers are materials which can be processed or shaped using conventional polymer forming techniques such as polymer infiltration pyrolysis (PIP), injection molding, coating from solvent, extrusion, or resin transfer molding (RTM). Once formed, objects made from these preceramic polymers can then be used as-synthesized or be converted to ceramic components by heating to temperatures high enough to consolidate the elements contained in the polymer structure to a ceramic. Typically, such pyrolytic conversions occur in greater than about 80 wt% yield, meaning that less than 20 wt% of the mass of the original polymer is lost in the conversion to ceramic. Ideally, full shape retention is achieved during pyrolysis. Using techniques such as those listed above, a wide variety of composites, coatings, fibers, monoliths, and nano-phase material systems can be generated.

Probably, poly(carbo)silazanes and poly(carbo)silanes are the most produced and applied preceramic polymers in the world. Such polymers contain silicon, carbon (polycarbosilanes), and nitrogen (polycarbosilazanes) to be transformed into either silicon nitride or silicon carbide-containing ceramics. This is accomplished by performing the pyrolysis step in either an ammonia containing atmosphere or an inert atmosphere, respectively. Here, application of preceramic polymers is focused on polysilazanes.

Polymer Coatings

Taking advantage of their reaction with moisture and polar surfaces, polysilazanes are used as coating materials. Polysilazanes self condense in ambient conditions to give dense clear coats which strongly adhere to most surfaces from metal to glass, wood and plastics. For example, reaction of Si–N with OH leads to the formation of Si–O–M (M=Metal) bonds generating good adhesion of the coating to the substrate. Most of these applications are made by Clariant [17]

Applications are traditional techniques such as spraying, dipping or roll coating or just simply by wiping. Controlling the chemistry and the formulation of polysilazanes allows the obtention of clear and pigmented hybrid coatings with multifunctional properties such as antigraffiti, easy-to-clean, anti-corrosion, barrier effect against water vapor and oxygen. For example, the "free" surface of the coating can react with humidity thereby creating a siloxane-like structure with excellent "easy to clean" properties : tutoProm®, an organopolysilazane-based product, is used by Deutsche Bahn AG (German Railways) on their carriages as an anti-graffiti coating.

In addition, the properties of these coatings include thermal stability and resistance to acids, alkaline cleaners, UV radiation and corrosion. Ultra-thin polysilazane-based coatings protect metals such as aluminum, zinc, and steel against corrosion, erosion, weathering, and acids. The coating adheres very well to metallic materials and is not visible and can display no corrosion phenomena. The coating system is thus especially suitable for the automotive industry. Materials that are expected to gleam, while displaying high resistance to the effects of dirt and weather, are optimally protected with polysilazane-based layers. Aluminum wheels are a good example: here the coating protects against corrosion, its anti-adhesive properties prevent deposits of brake dust, and its easy-to-clean properties make washing the car a snap. Polysilazanes are especially effective in supporting the self-cleaning of coated objects with the aid of photocatalysis. As a result of the effect of UV irradiation together with humidity, radicals are formed during the photocatalytic process. These radicals degrade organic materials, e.g. dirt, but also plastics, paints, varnishes and lacquers. Polysilazanes form an effective protective layer against the undesired destruction of plastic or other organic carrier materials. Moreover, the photocatalytic process makes the surface superhydrophilic. It is thus completely wetted with water and the remaining dirt particles are easily floated off. Nanoscale titanium dioxide can be bound into the polysilazane matrix. Thus, unlike in other procedures, adhesion can be achieved through very gentle means.

Ceramics in Aerospace

Preceramic polymers are particularly attractive candidates to prepare ceramics for a variety of structural aerospace applications, including ceramic fibers, composite radomes, components made from carbon fiber-reinforced ceramics, particulate-reinforced SiC/SiC composites, exhaust nozzles, and coatings used to enhance the oxidative stability of a variety of refractory composite materials such as carbon-carbon.

Ceramics in Motorcycle and Automotive

The motorcycle and automotive markets also benefit from preceramic polymers. Driven by a desire to run engines hotter and hotter, ceramic materials are slowly finding their way into under-the-hood applications. Early work involving KiON's Ceraset polysilazanes [17] demonstrated that high performance silicon nitride engine rotors targeted for hybrid gas / electric vehicles could be manufactured by injection molding techniques. High performance ceramic coatings for mild steel exhaust manifolds made from ceramic particulate-reinforced Ceraset polymers passed aggressive 1000 hour dynamometer test protocols. In addition, Starfire Systems developed STARBlade CMC brake rotors fitted on a motorcycle based on silicon carbide prepared from polycarbosilanes [20].

Ceramics in Electronics

Preceramic polymers and polysilazanes are also suitable materials for use in electronics, where electrically insulating or passivating ceramic coatings are required, and in the computer industry, where ceramic coatings are used in multi-layer chip manufacture. In such applications, air pyrolysis results in silica-based ceramics which provide the desired electrical properties. Additionally, a variety of MEMS-related applications are currently being developed.

CONCLUSION

The objective of the present chapter was to discuss on the structure, physical properties and application of preceramic polymers. We described our strategy to identify the structure and characterize the physical properties (rheological properties) of a particular type of preceramic polymers called poly[*B*-(methylamino)borazine]. We demonstrated that we are able to control the synthesis and processability of preceramic polymers using solid-state NMR and shear rheology. This is important to prepare "near net shape" advanced polymers and derived ceramics, one of the applications of preceramic polymers. In addition, the latter can be used as-synthesized mainly as coatings. Such applications are described in the present chapter.

REFERENCES

[1] A. Staubitz, A. Presa Soto, I. Manners Angew. *Chem. Inter. Ed.* 2008, *47*, 6212.

[2] Duperrier, S. ; Bernard, S. ; Calin, A. ; Sigala, C. ; Chiriac, R. ; Miele P. ; Balan, C. *Macromolecules*, 2007, 40, 1028.

[3] Duperrier, S.; Calin, A.; Bernard, S.; Balan, C.; Miele, P. *Soft. Mater.*, 2006, 4, 123.

[4] Duperrier, S.; Gervais, C.; Bernard, S.; Cornu, D.; Babonneau F.; Miele, P. *J. Mater. Chem.*, 2006, 16, 3126.

[5] Duperrier, S.; Gervais, C.; S. Bernard, S.; Cornu, D.; Babonneau, F.; Balan C., Miele, P. *Macromolecules*, 2007, 40, 1018.

[6] J. Bill, F. Aldinger, *Adv. Mater.* 1995, *7*, 775.

[7] K. J. L. Paciorek, R. H. Kratzer, D. H. Harris, M. E. Smythe, *Polymer Preprints* 1984, *25*, 15. (a) R. H. Toeniskoetter, F. R. Hall, *J. Am. Chem. Soc.* 1963, *84*, 4619; (b) Y. Kimura, Y. Kubo, N. Hayashi, *J. Inorg. Organomet. Polym.* 1992, *2*, 231.

[8] K. J. Wynne, R. W. Rice, *Ann. Rev. Mater. Sci.* 1984, *14*, 297.

[9] M. Peuckert, T. Vaahs, M. Brück, *Adv. Mater.* 1990, *2*, 398.

[10] P. Colombo, G. D. Soraru, R. Riedel, H.-J. Kleebe, "*From Nano-structure to Applications*", DEStech Publications, Inc., 2009, p. 410.

[11] P. Colombo, G. Mera, R. Riedel, G. D. Soraru, *J. Am. Ceram. Soc.* 2010, *93*, 1805.

[12] P. Greil, *Adv. Engineer. Mater.* 2000, 339.

[13] R. Riedel, G. Mera, R. Hauser, A. Klonczynski, *J. Ceram. Soc. Jp.* 2006, *114*, 425.

[14] W. P. Cox, E. H. Merz, *J. Polym. Sci.* 1958, *28*, 619.

[15] Y. D. Blum, K. B. Schwartz, R. M. Laine, *J. Mater. Sci.* 1989, *24*, 1707.

KiON Ceraset® Polysilazane is a commercially available product of Clariant.
PCS is a commercially available product of Nippon Carbon Ltd; type S, Tokyo, Japan.
PHPS is a commercially available product of Tonen Co., N-N110, Tokyo, Japan.
Available at http://www.starfiresystems.com.

In: Polymer Chains
Editors: Régine Penzkofer and Yasue Wu

ISBN: 978-1-62100-422-6

Chapter 10

SYNTHESIS OF CARBON MATERIALS WITH HIGH POROUS ALUMINA AS TEMPLATE

O. K. Krasilnikova,* A. C. Pogosian, N. V. Serebryakova and T. Y. Grankina
A.N. Frumkin Institute of Physical Chemistry and Electrochemistry of Russian Academy of Science, Russia, Moscow

ABSTRACT

A novel method for template synthesis of carbon materials with homogeneous pores has been proposed on the basis of using a high porous monolithic material from alumina nanofilaments as a template matrix. Monolithic alumina with high surface area template was obtained by direct water vapor oxidation of aluminum plates with the purpose to apply it as a template matrix. The templates were impregnated by different carbon precursors – sucrose, furfuryl alcohol, polyacrylonitrile and petroleum pitch. After polymerization, carbonization of precursors and removing the template, the porous structures of initial template and the carbon materials were characterized by analyzing the low temperature nitrogen adsorption and water vapor isotherms. Also, the surface and structural characteristics of these nanostructured materials were investigated using FTIR and Raman spectroscopy, and electron microscopy methods. The results showed that the highly developed homogeneous porous structure of the carbon materials with narrow pore size distribution were the consequences of the replication of the high porous structure of the nanofibrous template. According to the spectroscopy data the carbon nanoporous materials are rather homogeneous, and their structural features are similar to nanomaterials produced by other methods. Their properties depend on precursor properties and impregnation conditions. The materials obtained from sucrose are similar to the double wall nanotubes (DWNTs). They have an average inner diameter in the range of 0.4 – 0.6 nm and the surface BET area equal to 400 m^2/g. It was concluded from the spectroscopic data that these materials had mainly semiconductor properties and showed a strong nonlinear response to an action of infrared radiation. Impregnation of the template by furfuryl alcohol resulted in the monolithic carbon materials, which occurred to be similar to the single wall nanotubes (SWNTs) with a diameter of about 1.8 nm and

* E-mail: krasilnikovaok@inbox.ru.

the BET surface equal to 580 m^2/g, exhibiting semiconductor and metallic properties. Uniformed supermicropores with the diameters exceeding 3.5 nm were found in the carbon materials obtained by using petroleum pitch as precursor.

1. INTRODUCTION

Nanoporous carbon materials offer unique opportunities for a variety of applications in almost all aspects of nanotechnologies as electronic, catalytic and hydrogen – storage systems productions. The most perspective carbon adsorbents are of high porosity with a narrow pore sizes distribution. An ideal model of the structure of the homogeneous carbon material can be the nanotubes with narrow pore size distribution. But these nanotubes with narrow diameter can't be isolated; they form disordered associates – bundles. The catalytic method extensively used for the synthesis of carbon nanotubes has several significant limitations. Purification of the tubes from catalyst impurities and isolation of the highly dispersed final product from the reaction mixture containing other carbon modifications occurs to be a difficult and costly technological problem. Its solution requires heavy investments, which substantially increase the prices of the final carbon nanomaterials. In addition, the catalytic method fails in the synthesis of monolithic carbon nanomaterials.

The template synthesis based on nanocasting and nanoreplication of carbon nanostructures inside the matrix of templates demands no cleaning procedure for carbon materials and has the significant preferences. The materials obtained by the template synthesis are structural replicas of a starting template. They are characterized by a large volume of pores and an ordered structure with a large specific surface area. In the template synthesis using rigid templates such as mesoporous silicates, the matrix can be dissolved, generally in hydrofluoric acid. The resulting carbon materials (as mesoporous carbon networks CMK-1, CMK-3, etc.) have the ordered and uniformed mesopores with diameters in the range of 2 – 9 nm and contained no impurities of catalyst [1, 2].

The template synthesis based on producing carbon materials inside a porous matrix structure can be used to obtain such materials as structurally ordered porous nanotubes, nanorods or nanowires.

Recently [3, 4, 5], an anodic oxidized aluminum has been proposed as the template for the synthesis of mesoporous carbon materials.

Later on, the oriented carbon materials having the shape of nanotubes were obtained by carbonization of large organic molecules within the pores of anodic alumina membranes [3]. The shape and the size of these nanotubes corresponded to the dimension of the template channels. These tubes were structurally oriented within the pores of alumina and formed a composite membrane.

After complete removal of the template, carbon nanotubes with a length of several micrometers, an outer diameter of about 100 nm and a wall thickness close to 20 nm were obtained. An undoubted advantage of this approach is a possibility of creating the monolithic mesoporous membranes, in contrast to the powdery carbon materials obtained with mesoporous silicate powders as template.

The recent development of this method is a preparation of mesoporous carbon nanofilaments with the diameter close to 5-9 nm as a replica from the silica based mesoporous nanowires deposited inside the channels of an anodic alumina membrane [4, 5].

The membrane formed by the ordered mesoporous carbon nanofilament was prepared by replication of a Fe-containing mesoporous nanofilament based on silica (Fe-SBA-15), prepared in the channels of anodic alumina membrane [4]. For this purpose, the Fe-SBA-15 mesoporous silica was pre-grown within the large pores (170– 250 nm in diameter) of alumina membrane. Then, the carbon precursor was polymerized in the pores (9 nm in diameter) of the resulting mesoporous silica to give a three-component composite of a “matreshka” type. With the silicate structure of SBA-15 (with 9-nm pores) as the template matrix in the pores of alumina, the carbon tubes with the diameter of 5–9 nm were grown by pyrolysis of carbon precursor. Thus, after dissolution of the matrix, the resulting carbon tubes had the structurally oriented mesopores with the outer diameter equal to the diameter of the pores in the SBA-15 matrix. The mesophase carbon nanofilaments formed the ordered structure of macroscopic membranes. This structure was hierarchically ordered due to arrangement of pores of the anodic alumina membrane and to the silica mesophase, as well. The diameter of pores depended on the size of pores of the mesoctructured silica template and could be equal to 5–9 nm.

A long one-dimentional (1D) mesoporous carbon nanofiber was synthesized by the template method using an Al doped mesoporous silica nanofiber template within the porous alumina template and furfuryl alcohol as the carbon precursor [5]. The typical diameter and the length of the mesoporous carbon nanofibers are of 150-300 nm and 10-30 μm, respectively. The resulting mesoporous carbon nanofibers showed a unique mesoporous structure of circularly wound nanochannel alignment perpendicular to the long fiber axis and had the high surface area with the pore diameter near 5 nm.

The carbon nanotubes and nanorods which were obtained using the hybrid template of mesoporous silica nanofibers within the pores of anodic alumina were mesoporous adsorbents. But the carbon structure with the narrower pore sizes as micro- and supermicropores, can’t be synthesized using such a mesoporous template. For this aim it is necessary to use a matrix with a structure different from the porous anodic alumina with the mesoporous silica template inside it.

The nanowire structure based on porous alumina [6] can be also considered as the matrix for the template synthesis (imprinting) of carbon materials. Carbon precursor covered the nanowires, and after following polymerization a polymer film was formed around the template filament of alumina matrix. After carbonization the template was removed from the carbon-template composite by dissolving in NaOH solution and after water washing by the second step – the flushing by HF. A principal scheme of this template synthesis is demonstrated on Figure 1.

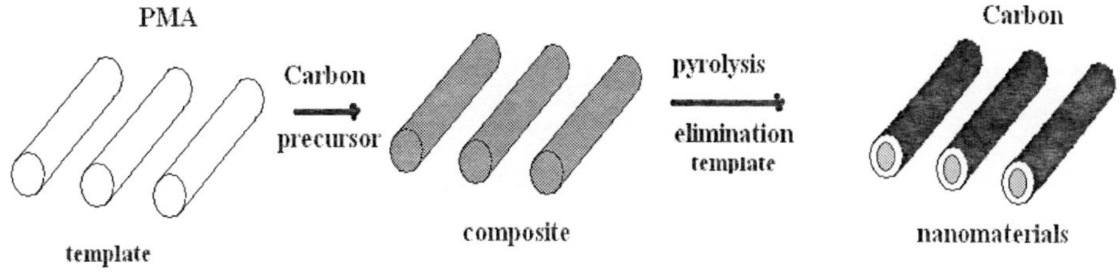

Figure 1. The scheme of the template synthesis.

The obtained nanocarbon materials with uniformity in pore sizes and high surface areas are the replica of monolithic alumina template. Unlike the carbon nanotubes the obtained materials do not require any accessory purification from other reaction products.

Here we describe the method of producing the carbon materials within a porous space of a specially prepared porous monolithic alumina template (PMA) with the finer porous structure and the higher BET surface compared to the anodic alumina. This template consists of the intertwined alumina nanofilaments that give the possibilities to use simple molecules as the carbon precursors to obtain nanorods or filaments, or tubes without modifying the porous voids as it has been done for the anodic alumina by silica. The formation of graphene plates was realized without metal containing particles due to the surface properties of alumina nanofibrous template. Prepared from sucrose, furfuryl alcohol and polyacrylonitrile these carbon materials have a microporous structure with narrow distribution of pore sizes. The pore sizes of microporous materials depend on a nature of carbon precursor. The carbon materials made out of petroleum pitch have the wider uniform micropores or so-called supermicropores. The specific feature of these new materials is the homogeneity in the pore sizes.

2. EXPERIMENTAL

2.1. The Synthesis of High Porous Monolithic Alumina

The template from high porous monolithic alumina (PMA) was prepared by special oxidizing the aluminum plate using the modified method by Vignes [7]. After removing the natural oxidization layer covering the aluminum plates (99,999 % purity), the metal surface was treated by mercury compounds to obtain a surface amalgam of Hg-Al. Then the aluminum plate was oxidized by water vapor to prepare the monolithic sample of aluminum oxide, which was grown up perpendicularly to the metal surface (Figure 2).

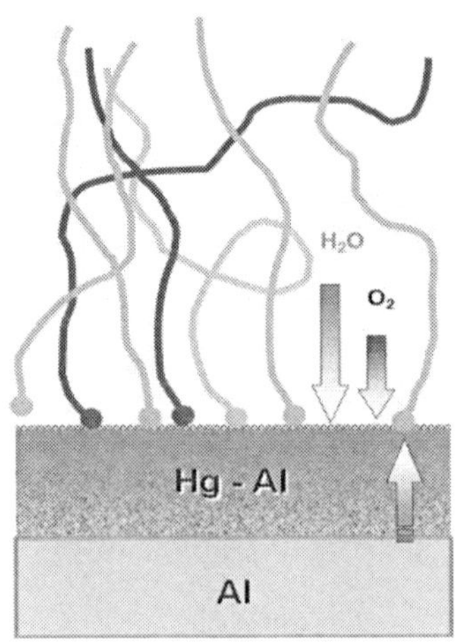

Figure 2. Principal scheme of growth of alumina nanofilament.

The fine filaments of alumina have the high specific surface reaching the value of 400 m^2/g. The material is amorphous. It transforms into γ-modification of alumina by heating up to 870 °C and to α-alumina by heating above 1200 °C. Prepared in this way the alumina monoliths have high porous structure with low density of about 0.004 g/cm^3. The PMA samples took the form like the aluminum plate used as precursors in this synthesis. The growth rates (or oxidizing rate) varied with the experimental condition such as humidity and temperature in the range of 2 – 7 mm/hour. The diameter of alumina nanofilaments in the PMA sample was equal to 5.0 nm (Figure 3). It occurred to be a suitable material for using as template (see Figure4).

2.2. Synthesis of Carbon Replica

The preparation of carbon replica of templates was founded on the formation of polymer film of various carbon precursors around the alumina nanofilaments. The PMA filaments were impregnated by different carbon precursors as sucrose (samples S, Table 2) or furfuryl alcohol (samples F, 2F, Table 1), polyacrylonitrile (samples PAN, Table 1) and petroleum pitch (samples PP, Table 1).

After impregnation by sucrose (3.2 g/g), the PMA sample was dried and carbonized in an inert atmosphere by slowly increasing the temperature up to 350 °C at a heating rate of about 1 °C/min.

In the case of using furfuryl alcohol (FA) as the carbon precursor, the PMA sample was pre-impregnated with 5% TsOH in ethanol as a polymerization initiator. Then the sample was dried at room temperature, heated and kept at 80 °C, impregnated with fresh distilled FA, and gradually heated up to 50 °C. This procedure was repeated several times and then the sample was kept in an oven at 80 °C for 8 h. The resulting composite was carbonized in the inert atmosphere at 350 °C as described above. After the gas evolution was over, slow heating was continued until gaseous products ceased to be generated with an increase in the temperature. After that the Al_2O_3/C composite resulted from the low temperature carbonization was chemically treated step-by-step by 1 N alcoholic solution of NaOH and diluted HF to dissolve selectively the Al_2O_3 matrix. The shrunk sample was washed to neutral reaction and dried. Washing was repeated after drying at 120 °C.

The other type of precursor – 1% solution of polyacrylonitrile in DMSO was used to impregnate the alumina template. After drying the template-polymer composite was treated and carbonized as described above. The template from the composite was removed by dissolution. Two analogous carbon samples 7 and 8 in Table 1 and 3 were obtained by consecutive dissolution.

The melted petroleum pitch (PP) soaked the template to obtain the composite PMA-PP which was also treated by low temperature carbonization as described above.

The sample 1FG was prepared by deposition from argon flow the FA vapors of furfuryl alcohol at temperature reaction up to 300 °C. The carbonized FA-PMA composite before dissolution of template is designated as 1FG sample, while the carbon material after dissolution – as a 1FGC sample and are represented in Tables 1 and 3.

Table 1.Carbon materials

No	name	precursor	Synthesis condition
1	2S-0	Sucrose S	Impregnation of PMA 1 stage dissolution of template
2	2S	sucrose	Impregnation of PMA 2 stages dissolution of template
3	1F-0	furfuryl alcohol (FA)	Impregnation of PMA 1 stage dissolution of template
4	1F	furfuryl alcohol (FA)-	Impregnation of PMA 2 stages dissolution of template
5	2F	furfuryl alcohol (FA)	Impregnation of PMA 2 stages dissolution of template
6	4F	furfuryl alcohol (FA)	Impregnation of PMA 2 stages dissolution of template
7	PAN	polyacrylonitrile (PAN)	Impregnation of PMA 1stage dissolution of template
8	PAN	polyacrylonitrile (PAN)	Impregnation of PMA 2 stages dissolution of template
9	5PP	petroleum pitch (PP)	Impregnation of PMA 2 stages dissolution of template
10	1PP	petroleum pitch (PP)	Impregnation of PMA 2 stages dissolution of template
11	1FG	furfuryl alcohol (FA)	Deposition from flow of FA, carbonization, before dissolution of template
12	1FGC	furfuryl alcohol (FA)	Deposition from flow of FA carbonization after dissolution of template

2.3. Porous Structure of the PMA and Composite Samples Using the Microscopy Data

The PMA template sample and the carbon composite products were characterized by scanning and transmission electron microscopy micrographs with a transmission electron microscope LEO Supra 50 VP microanalysis system INCA Energy + and Tecnai G^2 20 TWIN with an accelerating voltage of 200 kV and scanning electron microscope Jeol 6490 LV with X-ray microanalysis.

Exterior views of the PMA template and the carbon materials are shown on Figures 3 and 4.

The data of microanalysis showed that the PMA template and the replicated carbon composite had no impurity substances. Replicated carbon composites included imperceptible traces of petroleum pitch.

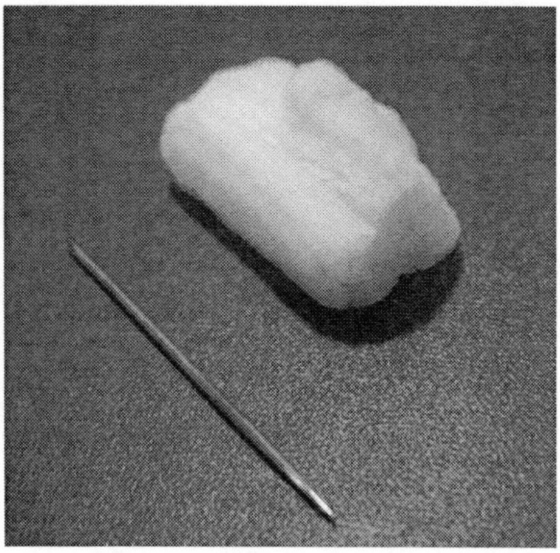

Figure 3. The external view of the PMA template sample.

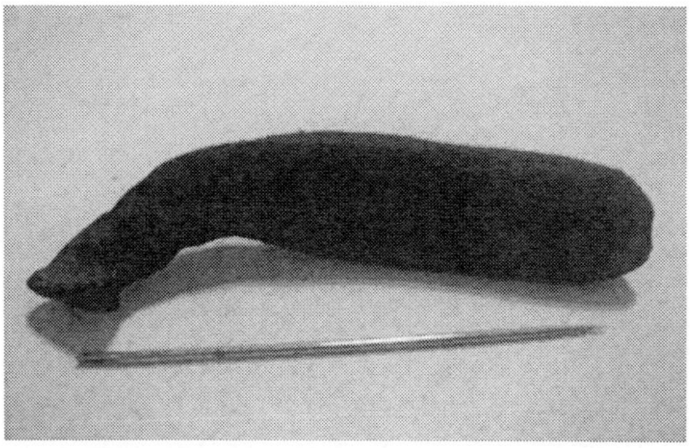

Figure 4. The external view of the carbon material replicated from the PMA template.

a)

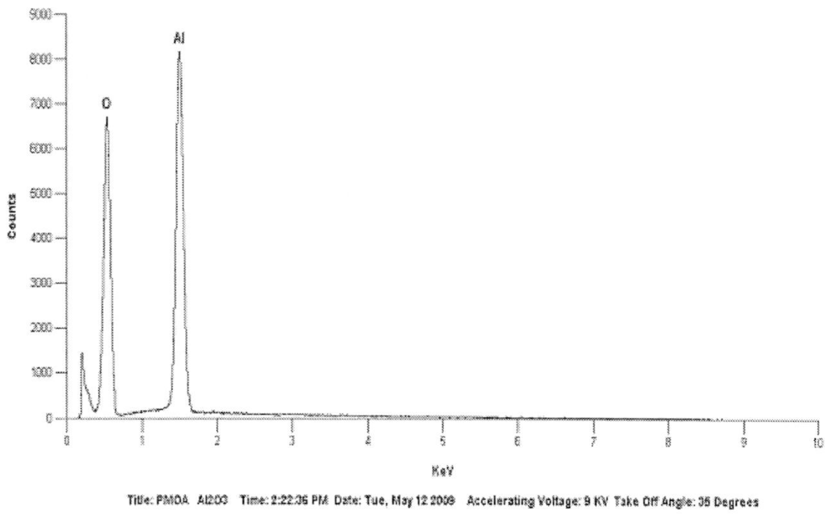

Figure 5. (Continued).

b)

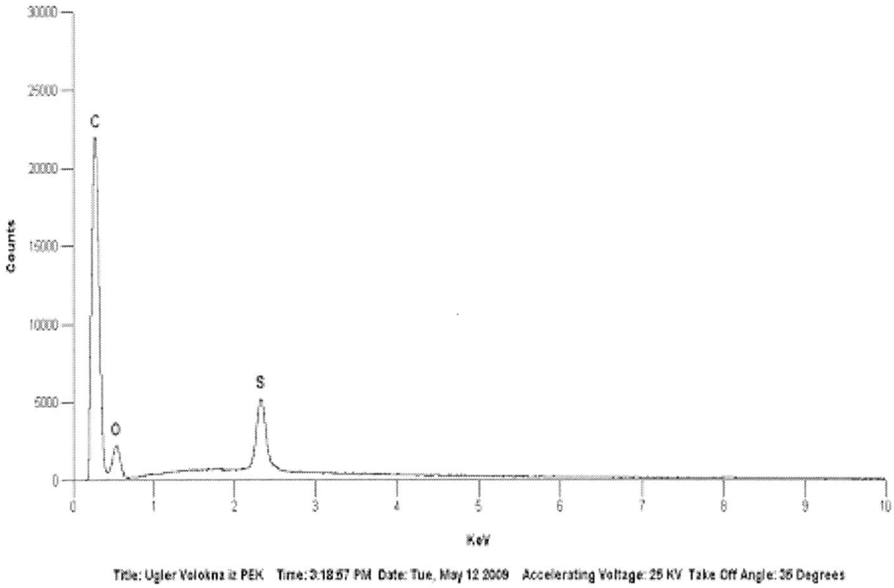

Figure 5. a – X-ray microanalysis data for the PMA template. b – X-ray microanalysis data for the carbon material based on petroleum pitch.

2.4. Determination of Parameters of Porous Structure

The low temperature nitrogen vapor adsorption and water vapor adsorption measurements were performed in the carbon replicas as well as in alumina template.

The porous structure parameters of the prepared materials were determined from the isotherms of nitrogen adsorption at 77 K measured using a volumetric adsorption device ASAP 2020 MP (Micromeritics, USA) and a GRAVIMAT-4303 Sartorius NETZSCH (Germany) automatic vacuum adsorption microbalance with sensitivity of 1 μg for a weighed portion of 100 mg within a range of relative pressures from 10^{-4} to 0.9.

The total specific area was determined by the BET method [8], the volume of micropores, W_0, and characteristic energy of adsorption, E_0, were calculated by the equations of the micropore volume filling theory using the Dubinin–Radushkevich equation (in a form proposed by Stoeckli) [9-12], the surface of mesopores, S_{ME}, was found using the comparative plots according to the MP method (S_{MP}) [8, 13]. The half-width of micropores, x_0, was calculated by the generalized Stoeckli equation, which related E_0 and x_0 [12]. The averaged isotherm of nitrogen vapors adsorption at 77 K on the nonporous carbon blacks was used as a basis in the comparative MP method [8, 9, 13].

Water vapor adsorption isotherms were studied using a vacuum weight Mac-Ben quartz spring balance with sensitivity of 10 μg with weight sample up to 100 mg. The quantity of primary active centers of water adsorption (PAC) for the samples under investigation were determined using the comparative plots based on the isotherm water vapors adsorption on a Vulcan carbon-black, graphitized at 2800 °C [14].

2.5. Spectroscopy Investigations of Structure and Surface Features of PMA and Carbon Materials

The structural characteristics and surface properties of prepared samples were studied by different techniques of molecular spectroscopy. The FTIR spectrometer "Perkin-Elmer 2000" has been explored for optical absorption spectra measurements in the region of 2700 – 15000 cm^{-1}. Raman spectra in the region of 100 – 3500 cm^{-1} have been obtained by using Fourier IR-Raman spectrometer "Perkin-Elmer" having a laser with an excitation energy of 1064 nm (9400 cm^{-1} or 1.175 eV)

Raman spectra were recorded in a wide spectral range of 150-3500 cm^{-1} using a laser Raman spectrometer from Nogo LabRAM firm Jobin Yvon, equipped with a microscope, TV camera and a cooled CCD detector. The laser line λ= 632.8 nm from a He-Ne laser was used as an exciting line. A laser beam power was less than 1 mW. Collection of scattered light was carried out on the reflection (180 °). Spectral slit width was equal to 2 cm^{-1}.

3. Results and Dicussion

3.1. Structure and Properties of Alumina Template

3.1.1. Adsorption Nitrogen Vapors on PMA Samples

The fresh-made PMA sample contains more than 40% water and has a composition of $Al_2O_3{\cdot}4H_2O$ [7]. The growth of alumina nanowires is an exothermic process and occurs in humid air. The fresh-made PMA seemed as a transparent amorphous monolithic aerogel. It has the density varying within 0.004 – 0.06 g/cm^3 and is stable in air. By variation of conditions of the synthesis (humidity and temperature) the aerogels with the density reaching 0.1 g/cm^3 can be obtained.

Optimization of parameters of the PMA synthesis (humidity 70% and temperature near 25 °C) allowed producing the stable amorphous aerogel with the considerably lower density of 0.004g/cm^3 as compared with the previous data [7]. The SEM and TEM micrographs of the fresh-made PMA samples show the structure formed by inosculated and interlaced filaments of alumina, which have the diameter of about 5 nm (Fig 6b).

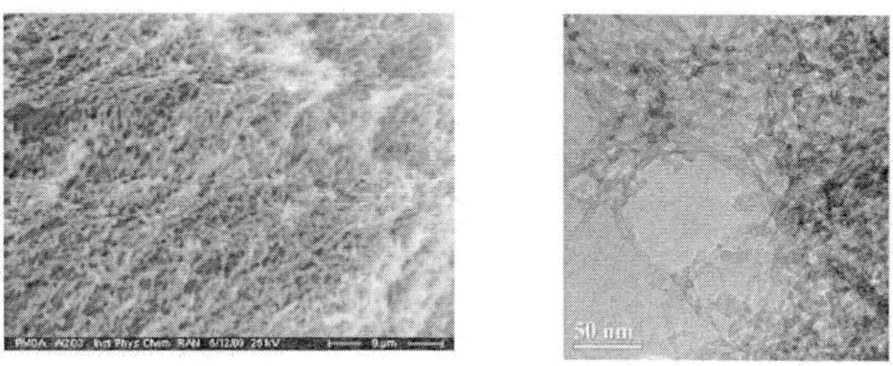

a

b

Figure 6. SEM (a) and TEM (b) micrographs of the as-grown PMA sample.

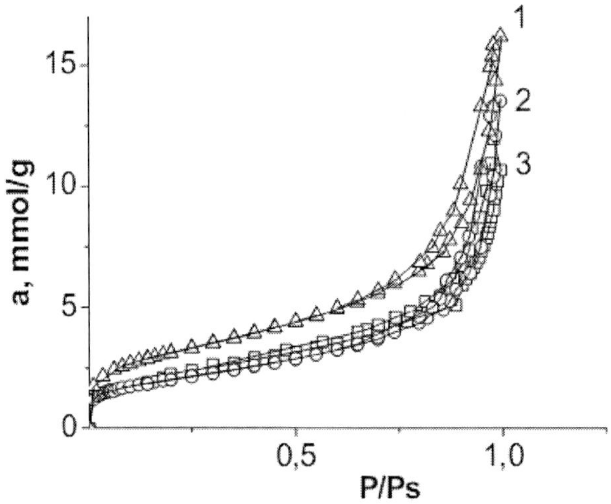

a

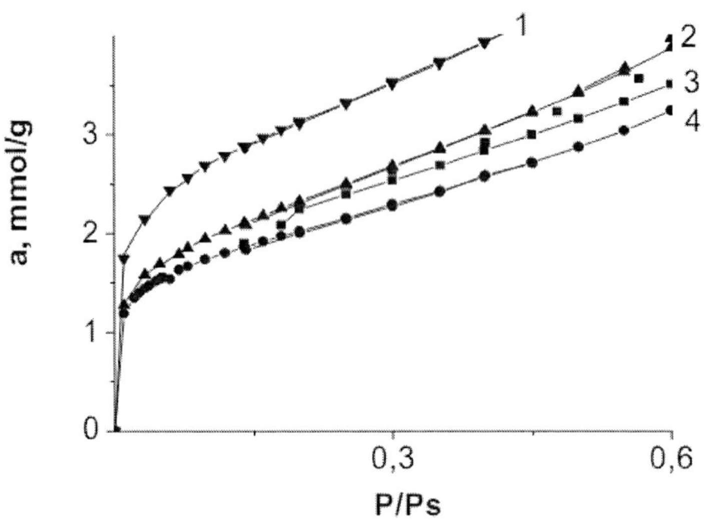

Figure 7. Adsorption isotherms of nitrogen at 77 K for the samples: a – PMA 3 (curve 1), PMA 4 (curve 2) and PMA 5 (curve 3) depending on the temperature of dehydration (Table 2), b – initial region of the adsorption isotherm of nitrogen at 77 K for the samples PMA 7 (curve 1), PMA 6 (curve 2), PMA 3 (curve 3) and PMA 4 (curve 4) depending on the increase of dehydration temperature.

Figure 7 shows the equilibrium adsorption isotherms of nitrogen vapors at 77 K by the alumina aerogels. They exhibited the specific mesoporous type of isotherm with a well developed plateau. The isotherms are reversible and have a hysteresis in a capillary condensation range. After dehydration by high temperature treatment it may be seen the decrease of the BET surface and the adsorption activity due to the contraction of length and thickening the alumina filaments that was also observed by the electron microscopy experiments [7].

In contrast to the range of low relative pressures, when $P/P_S > 0.6$, the isotherms indicated that a heat treatment leads to the decrease both of the adsorption capacity and hysteresis loop connected with the compaction of PMA structure due to dehydration. The initial ranges of isotherms shown on Figure 7b give more explicit evidence of the heat treatment effect on the porous structure of PMA samples. In particular, the increase of the dehydration temperature resulted in the distinct decrease of the BET surface areas listed in Table 2. The samples produced with identical condition such as the PMA 5 and PMA 7 samples have equal values of BET surface area. Some differences in surface areas for fresh-made samples may be explained by the differences in temperature and humidity in reactor. The high humidity led to reducing the BET surface area and to thickening nanofibers (sample PMA 6).

The comparative plots of isotherms of adsorption nitrogen vapors by the samples of fresh-made PMA (PMA 5) and PMA heated at 900 °C (PMA 3, PMA 4) and after that degassed at 30°C (PMA 3) and 200 °C (PMA 4) are shown on Figure 8.

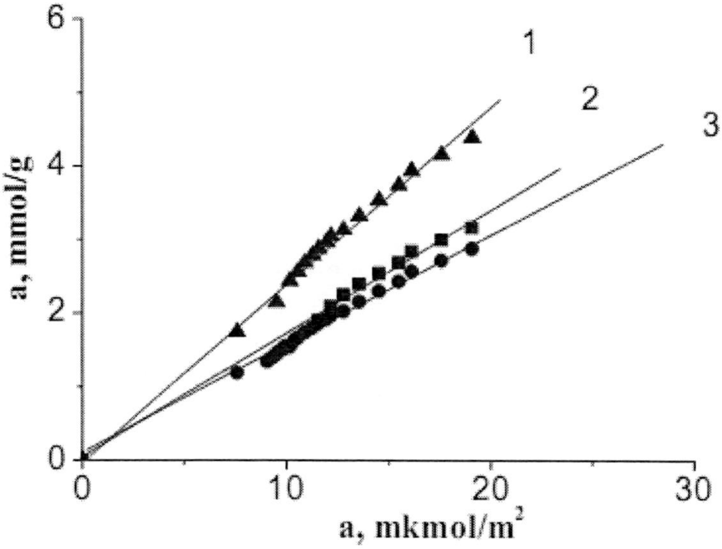

Figure 8. Comparative graphs of the nitrogen adsorption isotherms on the samples: fresh-made PMA 5 (curve 1), PMA 3 (curve 2) and PMA 4 (curve 3) (see Table 2).

Table 2. Porous structure of the PMA samples

№	Symbol	S_{BET}, $m^2 \cdot g^{-1}$	Comment
1	PMA 1	320	Fresh-made sample*
2	PMA 2	400	Fresh-made sample*
3	PMA 3	185	Heat treatment at 900 °C in nitrogen flow and degassing at 30 C in vacuum
4	PMA 4	152	Heat treatment at 900 °C in nitrogen flow and degassing at 200 °C in vacuum in vacuum
5	PMA 5	253	Fresh-made sample*
6	PMA 6	191	Fresh-made sample*
7	PMA 7	256	Fresh-made sample*

* The fresh-made PMA was degassed at 30 °C up to 10^{-6} torr in vacuum.

The comparative graphs are plotted relative to the averaged isotherm of nitrogen vapor adsorption on nonporous carbon adsorbents at the same values of relative pressures [8, 13]. There is a linear relationship between the adsorption values for PMA samples and the adsorption values defined from the averaged isotherm for nonporous carbon. The linearity of the comparative graphs indicated to the nonporous or mesoporous structure of PMA samples. The structure of PMA nanofilaments didn't changed after the heating up to 900 °C. According to the MP method [8, 13], the slopes of straight lines of the comparative plots passed through the coordinate origin are proportional to the total surface area, S_{ME}. For these PMA samples the S_{ME} values are practically equal that of S_{BET}.

3.1.2. Adsorption of Water Vapors by the PMA Samples

The adsorption of water vapors at 293 K by the alumina samples was investigated for comparative estimating the amount of surface functional groups and the amount of primary adsorption centers, PAC, on the alumina surface. The equilibrium water vapors isotherms at 293 K for the fresh-made sample designated as PMA 5 and that after the heat treatment at 900 °C, PMA 3, are shown on Figure 9. These concave isotherms are of the characteristic type IV and occurred to be reversible. There is a hysteresis in a capillary condensation range. The proximity of water vapors isotherms of fresh and dehydrated by heat treatment aerogels testified that the structural changes of PMA caused by thermal dehydration had no considerable influence on the amount of adsorption centers in these aerogels. The maximum of the water adsorption capacity for aerogels was equal to 20 mmol/g (0.36 g/g). The initial part of water vapors isotherm adsorption indicated that the fresh-made sample PMA 5 has more adsorption capacity then the sample PMA 3 obtained by dehydration at 900 °C.

The comparative plots based on the isotherms of water vapors adsorption by the fresh-made PMA and PMA after heat treatment at 900 °C are shown on Figure 10.

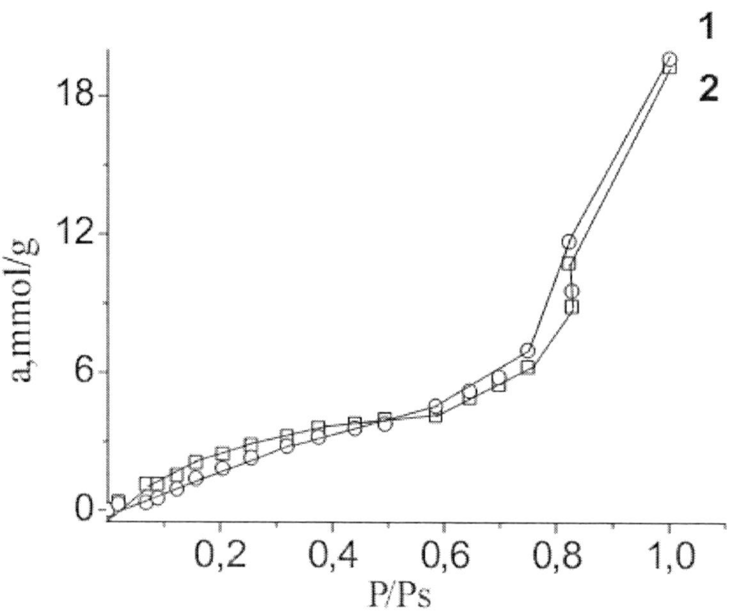

Figure 9. Adsorption isotherms of water vapor at 293 K for the samples: initial PMA 5 (curve 2) and PMA 3 obtained by heat treatment at 900 °C (curve 1) (Table. 2).

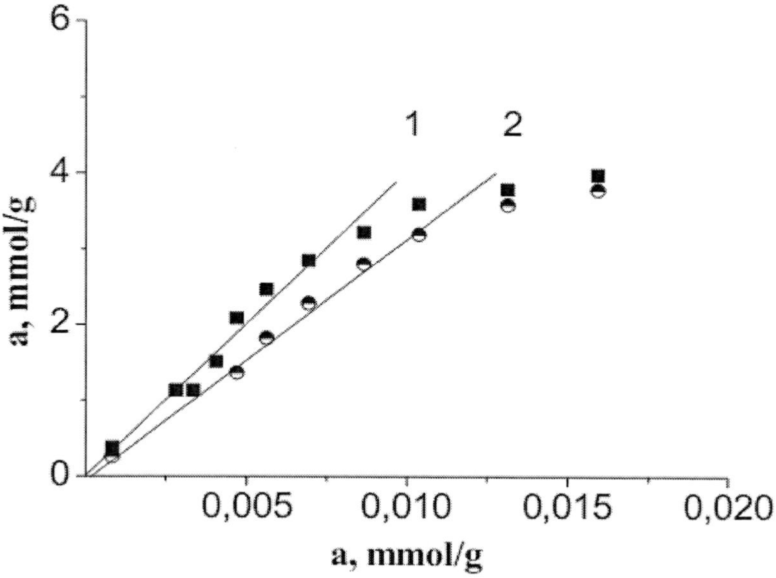

Figure 10. Comparative graphs built by using the adsorption isotherms of water vapor for the fresh-made samples PMA 5 (curve 1) and after heat treatment at 900 °C – PMA 3 (curve 2), (Table 2).

The water adsorption values for PMA samples under study are plotted as a function of the adsorption value of water vapors on the surface of a standard nonporous carbon adsorbent – the carbon black Vulkan-7H, graphitized at 2800 K, at the same values of relative pressures. In the initial range of adsorption all curves seemed as straight lines passing through the origin. In this case the mechanism of the water adsorption is based on the formation of clusters of water molecules on the surface of alumina. The initial water adsorption is caused by the interaction of water molecules with the polar groups on the alumina surface which may act as the primary adsorption centers, PAC. The PACs amounts were calculated from the comparative plots on Figure 10 for the PMA samples employing the cluster model [14].

There is a linear relationship between the water adsorption values for both of PMA samples and that on the nonporous carbon black observed at the same relative pressure.

It is known that for nonporous and mesoporous samples with the surfaces of similar chemical nature the comparative plots are the straight lines starting from the coordinate origin. The ordinates at a/a_m=1 correspond to the amount of PAC per gram of adsorbent [14] where a_m is equal to 0.004 mmol/g. In the low pore filling water molecules interacted with PAC of adsorbent. While in the range of low pressures, where the comparative plot is the straight line starting from the coordinate origin, when $P/P_s > 0.6$ there is an inflection downward in the comparative plots of water adsorption that is a characteristic feature for mesoporous and nonporous adsorbents. This inflection of the comparative plots is observed for the PMA samples and denotes an absence of micropores, but an existence of mesopores in alumina filaments. In the range of higher adsorption values water molecules are adsorbed on water molecules arranged on PAC and form the clusters. The interaction between the neighboring water clusters formed on the external surface of alumina leads to merging of clusters into the water film on the alumina filaments.

The cluster model was employed for estimation of the PACs amounts in the PMA samples from the comparative plots shown on Figure 10. Note that the cluster model fits the water adsorption data in the samples under study, when the pore filling are low, and the comparative plots are linear. According to the data calculated by the comparative method, the PACs amounts for water are larger for the fresh alumina (1.56 mmol/g) as compared to that obtained in the sample after the heat treatment at 900 °C (1.21 mmol/g).

Apparently, the nature of PAC in the fresh-made and the heated PMA samples is identical, and there are no essential changes in their features after the heat treatment.

Consequently it may be suggested that the dehydration of alumina template up to a phase transition had no effect on the chemical state of alumina surface, but the decreasing the value of BET surface area calculated from the low temperature nitrogen vapors adsorption isotherms denotes the contraction and swelling of alumina filaments due to heat treatment.

3.2. Porous Structure of Carbon Materials

3.2.1. Electron Microscopy Data for Carbon Materials

According to the electron microscopy data, the carbon composites resulted from the template synthesis are porous. In the photomicrograph made by the scanning electron microscope (see Figure 11), the three different types of pores can be observed. The carbon materials have a hierarchical structure of different pores, including the uniform micropores produced due to structural arrangement, and some amount of meso- and macropores formed by interconnection and overlapping the porous carbon nanofibers.

There are, firstly, the macropores with the diameters of 0.1–10 nm replicating the structure of initial PMA template; secondly, the mesopores with the sizes of more than 2 nm; and, finally, the pores with the sizes of 0.5–2 nm. The walls of macropores formed by accreted bundles are pierced with small pores. The pores, constituting the structure of the 2S samples, have small sizes, but such samples as 2F and 5P (Table 3) have more large pores or supermicropores. Thus, the character of porous structure depends on the nature of carbon precursor. On the other hand, the dissolution of the template-matrix leads to a considerable contraction of carbon structure. These drastic changes resulted in the formation of opened micropores, but not mesopores. The diameters of the template filaments equal to 5 nm appreciably exceed the diameters of micropore observed in the final carbon materials (0.7 – 3.7 nm).

Scanning and transmission electron microscopy micrographs and the electron-microscope images of the synthesized samples are presented in Figure 11. The SEM and TEM images were obtained after complete removal of PMA template. The image of bulk carbon material was presented above on Figure 3.

Due to removing the template the samples contracted, and the volume shrinkage over about 30% was observed. The SEM image of bulk carbon (Figure 11 a) presents the cellular structure of porous carbon, which was generated by nano replication process from the amorphous alumina structure. This structure was produced by agglutinating carbon bundles (Figure 11 b).

The TEM image (Figure 6) showed that the alumina nanofibers used as template in this study formed irregular structure. But the final carbon materials obtained using these alumina nanofibers differed by more ordered quasi-regular geometric shape from the parent alumina. After carbonization and complete removal of the template the final replicated carbon nanomaterial from PP carbonization is composed of long fibers, which are vertically assembled, as shown in Figures 11 c, d, e. The long fibers can be aggregated into the ordered structure (Figure 11 a). The typical diameter and the length of nanocarbon fibers bundles are equal to 100 nm and 10-30 μm, respectively.

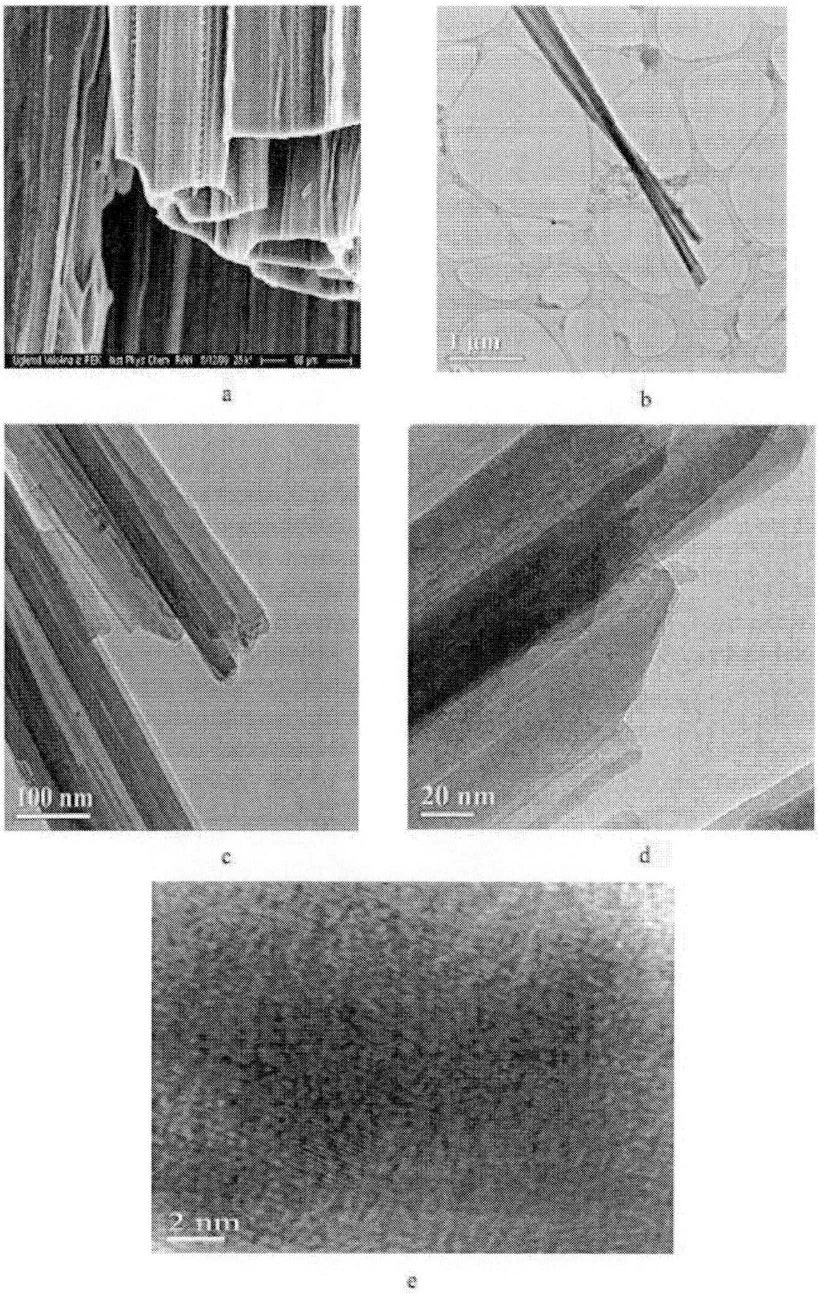

Figure 11. SEM micrographs (a, b) and TEM images (c, d) of the carbon materials based on carbonization of PP in PMA, e - TEM images for the carbon material based on FA (2F sample, see Table 1).

The nanofibers obtained from PP represented the bundles of fibers which composed a turbostratic carbon replicated alumina structure. This indicates that the alumina as well as the carbon material experienced both the structure transformation and rearrangement of porous, permolecular and supramolecular structures due to replication. The nanofibers of the final carbon material consist mainly of aggregated carbon nanoclusters within the nanofibers, which were templated by PMA.

Carbon nanomaterials resulted from FA carbonization show the internal structure with very narrow nanochannels, as shown in Figure 11 e, which are similar to the structure obtained in zeolite templating [15].

3.2.2. Low Temperature Adsorption of Nitrogen Vapors by Carbon Materials

The parameters of porous structure of carbon materials were analyzed by means of the adsorption isotherms of nitrogen vapors at 77 K. Nitrogen adsorption-desorption isotherms were measured for the nanoporous carbon materials with different carbon precursors (sucrose, FA, PAN and PP), templated from the PMA matrix (Table 3). It was shown that after pyrolysis, but before dissolution of the template, all of the carbonized composites were practically nonporous. But double dissolution of template in alkali and HF led to appearance of porous structure. Whereas, the isotherms measured for the samples prepared by impregnation of sucrose and FA, (see Figure 12 a) indicate that one stage dissolution by alkaline solution in alcohol and washing in distilled water are not effective. Two stage dissolution with intermediate thermal treatment at 120 °C, washing by dilute HF and multiple flushing by distilled water gave the necessary results. The porous structure of the final nanocarbon materials from PP displayed after overall dissolution of template (Figure 12 b).

The parameters of porous structure of the carbon materials were calculated from the low temperature nitrogen adsorption data and are summarized in Table 3. It was shown that the BET surface area of the carbon based on sucrose was equal to 415 m^2/g, including the surface area of mesopores equal to 40 m^2/g. It is a typical microporous material with low amount of mesopores. The carbon materials from sucrose have small micropores, so-called ultramicropores, and their diameter was calculated to be equal to 0.6 nm with the high adsorption energy of $E_0 = 29$ kJ/g.

The porous structure of the samples, prepared by FA carbonization (for example, 1F, 2F and 4F) is quiet different from other materials. These samples as 1F and 2F have the wider pores compared to micropores but more narrow than mesopores, usually named as supermicropores, and more amount of mesopores with the surface area from 35 up to 220 m^2/g (for sample 2F: $S_{ME} = 220$ m^2/g). The increase in the mesoporous surface area observed in the samples from FA has been attributed to more loose arrangement of the carbon tubes bundles than for the samples from sucrose that was caused by the influence of template properties.

The nitrogen adsorption-desorption isotherm observed on the carbon materials resulted from the carbonization of PP, which have an exclusively supermicroporous structure without mesopores is plotted in Figure 12 b. This isotherm exhibits a clear hysteresis loop located in the range of relative pressures of 0.5-0.9. But the analysis of the comparative plot for this sample denotes the lack of mesoporous structure. The absence of closing the hysteresis loop observed in low relative pressure region may be connected with deformation, caused by nitrogen adsorption.

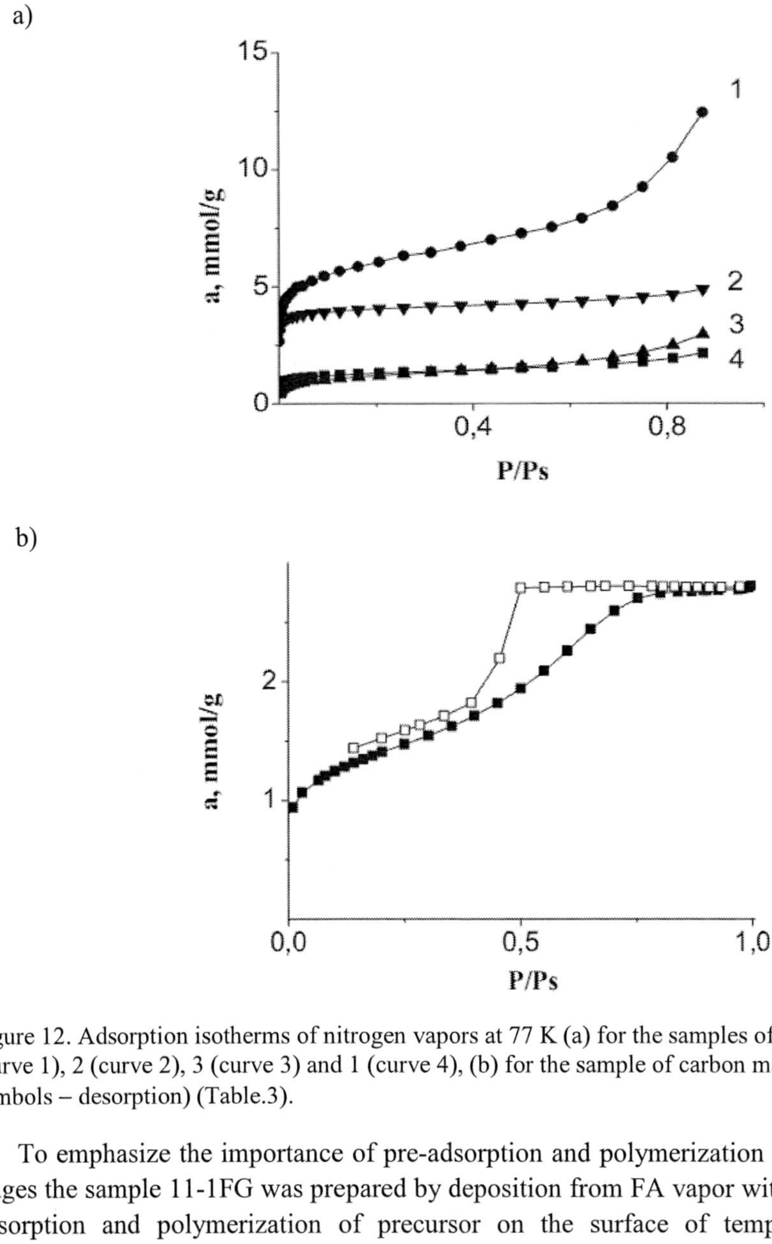

Figure 12. Adsorption isotherms of nitrogen vapors at 77 K (a) for the samples of carbon materials 5 (curve 1), 2 (curve 2), 3 (curve 3) and 1 (curve 4), (b) for the sample of carbon material 5P (open symbols – desorption) (Table.3).

To emphasize the importance of pre-adsorption and polymerization of carbon precursors stages the sample 11-1FG was prepared by deposition from FA vapor without the preliminary adsorption and polymerization of precursor on the surface of template. In process of application of FA from the flow of inert gas which contained FA, the PMA template slightly changed their volume. The carbonized composite FGC had the surface area equal to 290 m^2/g, close to that of parent template. But after the dissolving alumina template, a typical carbon black (sample 1FG) with the surface area of 85 m^2/g was obtained. It was shown that in the absence of doping metals or preliminary polymerization of precursor on the surface of nanofilaments of template no nanomaterials could be generated.

On the other hand, both the pre-adsorption of precursor and interaction between precursor molecules and alumina surface play a significant part in developing the carbon porous structure. For this purpose polymer film based on PAN was deposited on the alumina filaments from solution in DMSO. After pyrolysis and dissolution of template-matrix the carbon materials resulted from PAN carbonization (samples 7 and 8 in Table 3) with the surface area of about 190 m^2/g were obtained.

Thereby, it is apparent that the adsorption and subsequent polymerization of carbon precursors inside the porous structure of template from silica as well as the polymerization on the surface of PMA nanofibers are of fundamental importance for the formation of self-ordered polymer layer of precursor. The following carbonization of this layer leads to the formation of nanoporous carbon material with rolled-up graphene layers, which is similar to the carbon nanotubes. As it will be mentioned below in the section considering the spectroscopy data, there are some graphene layers in sample under investigation, which are similar to the carbon nanotubes.

The comparative plots for the isotherms of nitrogen adsorption at 77 K by carbon nanomaterials are shown in Figure 13. The comparative graphs are plotted relative to the average isotherm of nitrogen vapors adsorption on nonporous carbon adsorbents (carbon blacks) [8, 13].The analysis of the comparative plots for nitrogen vapors adsorption indicates the differences in porous structure of the samples with differing carbon precursor.

The slopes of straight lines in the comparative plots in the region of polymolecular adsorption (up to ~ 13-15 μmol/ m^2) are proportional to the mesopores surface area, S_{ME}, and the intercepts of the ordinate axis – to the micropores volume, W_0. When the amount of mesopores increased, the slope of straight line in the region of polymolecular adsorption also increases, corresponding to the increased value of $S_{ME.}$ At the same time the section of the comparative plot for the sample obtained from FA carbonization shifted to the region of low filling and transformed into a straight line passing though the coordinate origin. Such peculiarities of the comparative plots are typical for supermicroporous adsorbents and serves as a basis of the applying the MP method for determining a micropore size distribution [8-13]. As it was mentioned above the slope of initial section of the comparative plot are proportional to the total surface area of supermicro- and mesopores (S_{MP}). The comparative plot for the supermicroporous adsorbent resulted from FA carbonization is demonstrated by curve 1 in Figure 13a. In the region of polymolecular adsorption the comparative plot indicates the existence of some amount of mesopores. The comparative plot for microporous adsorbent from sucrose (curve 2, Figure 13a) distinguished from curve 1 and gives the possibility to estimate both values of micro- and the mesopores area.

a)

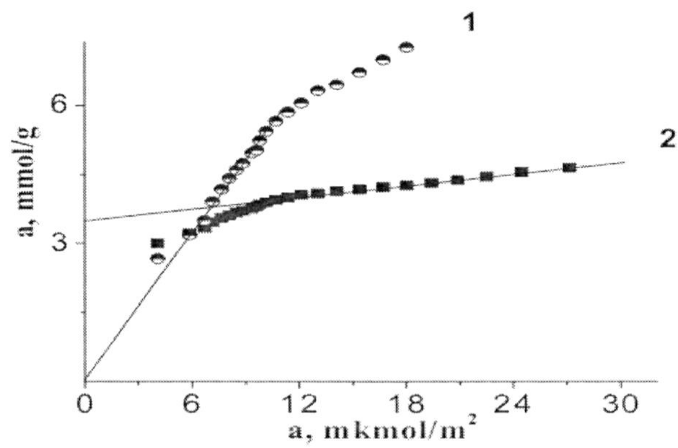

Figure 13. (Continued).

b)

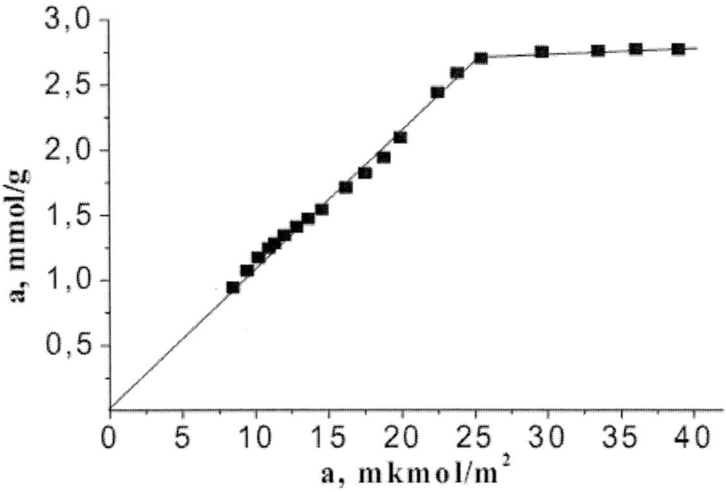

Figure 13. Comparative graphs of the adsorption isotherms of nitrogen at 77 K (a) for samples of carbon materials 5 (curve 1) and 2 (curve 2), (b) for sample of carbon material 9 (5P) (see Table 3.).

Table 3. Parameters of porous structure of carbon materials

No	name	S_{BET}, m^2/g	W_0, cm^3/g	E_0, kJ/mol	x_0, nm	S_{ME}, m^2/g	S_{MP}, m^2/g
1	2S-0	85		-		90	90
2	2S	415	0.14	29.2	0.30	40	400
3	1F-0	125	0.03	20.9	0.57	35	120
4	1F	500	0.20	22.6	0.52	90	400
5	2F	580	0.19	17.6	0.87	220	500
6	4F	520	0.25	14.4	0.93	219	540
7	PAN	195	-	-	–	–	185
8	PAN	190	-	-	–	–	170
9	5PP	110		–	1.75*	–	110
10	1PP	191	0.07	15.8	1.20	13	187
11	1FG	290	–	–	–	–	270
12	1FGC	85	–	–	–	–	80

* Data obtained by the MP method.

The adsorbents obtained from PP (see Figure 13b) have a unique uniform supermicroporous structure, which has neither mesopores, nor micropores. The pore size distribution is very narrow. Only in the case of 5PP sample, the comparative plot is parallel to abscissa axis in the region of polymolecular adsorption.

The comparative plots represented in Figure 13 are divided in several groups: the plot for microporous carbon adsorbents (2 – 2S), and for supermicroporous (9 – 5PP) and the supermicro+mesoporous adsorbent (5 – 2F).

3.2.3. Water Vapors Adsorption by the Nanocarbon Materials

The significant difference in porous structure of the resulted carbon materials was confirmed by the results of the study of water vapors adsorption. The water adsorption isotherms (Figure 14, 15) and the comparative plots (Figure 16) for the carbons materials occurred to be characteristic to the isotherms of water clusters adsorption on carbonaceous surface.

The concave water vapors isotherms showed on Figure 16 are specific for carbon materials. Though for the sample 9 obtained from PP carbonization (Figure 15) in the range of low filling there is a sharp increase of adsorption unlike the isotherm for the supermicroporous sample 2F. The isotherm for the microporous sample 2S from sucrose has more pronounced concave form.

The isotherm for sample 9 is of IV type having a hysteresis loop closing at P/Ps of 0.5-0.9.

The comparative graphs for water vapors plotted the relationship between the experimental isotherm measured in the samples under study and the averages isotherm of water vapor adsorption on the nonporous carbon black (Vulcan 7H 2800) at fixed relative pressures P/P_S. Here water adsorption on the carbon black was presented as a quantity of water molecules on primary active center. The amounts of PAC per 1g of adsorbent were evaluated from the slope of comparative plots (for the carbon materials from FA – 0.5 mmol/g and for carbon from sucrose – 0.4 mmol/g).

Also the formation of unconnected clusters of water in nanopores was found by the analysis of the comparative plots of water adsorption (Figure 16). The straight line for ultramicropororous sample 2S indicated to narrow micropores, but the plot for 2F sample turned down to abscissa proving the formation of continuous water film on the walls of pores at high adsorption values.

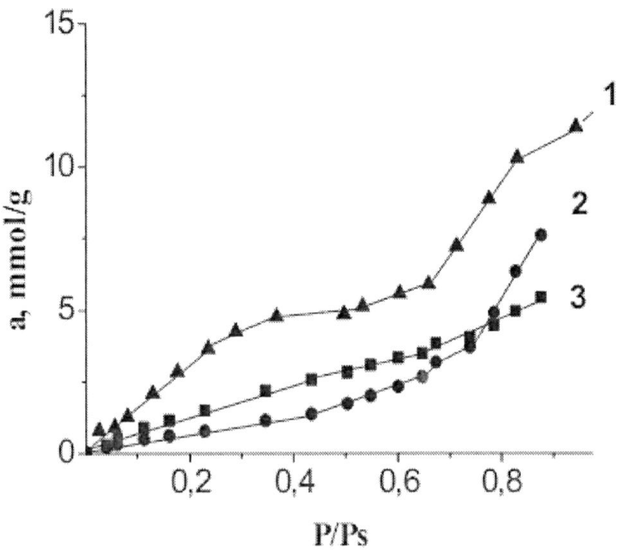

Figure 14. Adsorption isotherms of water vapors at 293 K for the samples of carbon materials from carbonization of PP – sample 9 (curve1), of sucrose - sample 2 (curve 2) and of FA – sample 5 (curve 3), see Tables 1 and 3.

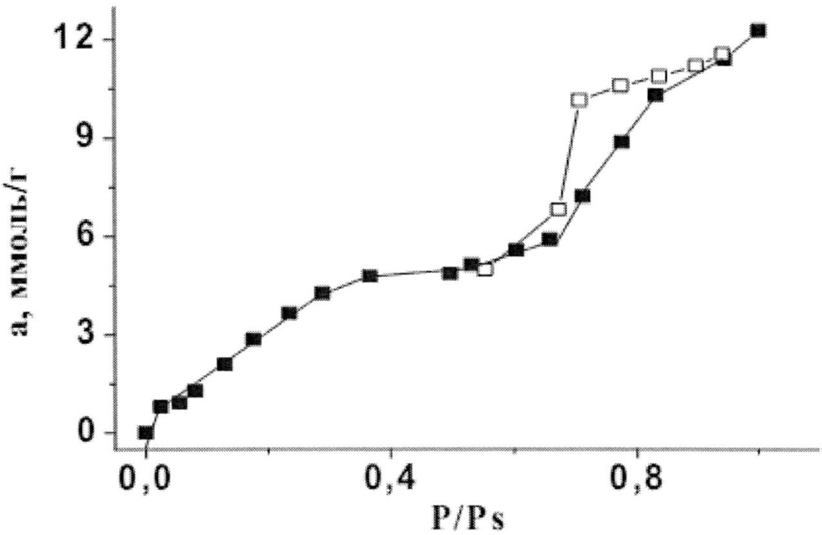

Figure 15. Adsorption isotherm of water vapor at 293 K for the sample of carbon material 5PP (non painted icons - desorption) (Table 3).

The comparative graph for the sample 2S is typical for the formation of two-dimensional clusters in the slit-like ultramicropores [14]. An analysis of the comparative graphs shows that a clusters (for 2S sample) consist of about 4-5 water molecules. This sample virtually does not contain micropores, which is evident from the nonlinearity of the comparative graph at high filling values. The absence of micropores and very few mesopores suggest that the sample 2S is an aggregate of almost homogeneous carbon turbostratic nanotubes with small pores diameters. In this case, linear clusters of water coalesced producing a continuous chain in a very narrow pore like spheres in a fine tube.

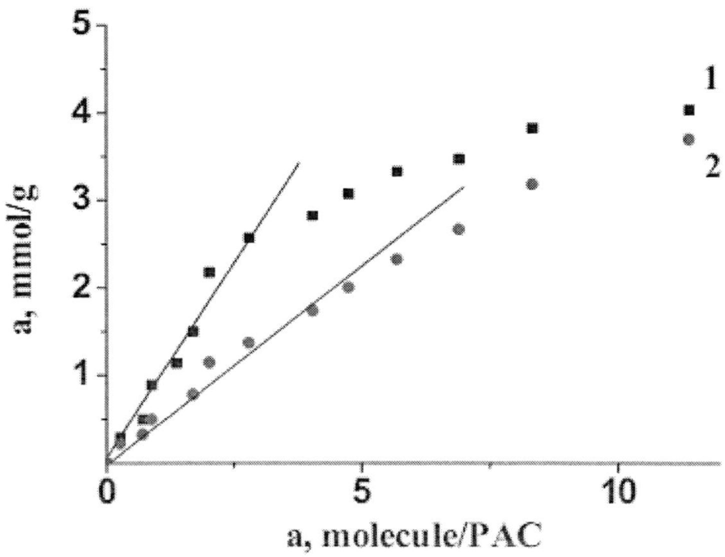

Figure16. Comparative graph of the adsorption isotherms of water vapor at 293 K for sample of carbon material 2F (curve 1) and 2S (curve 2) (see Table 3).

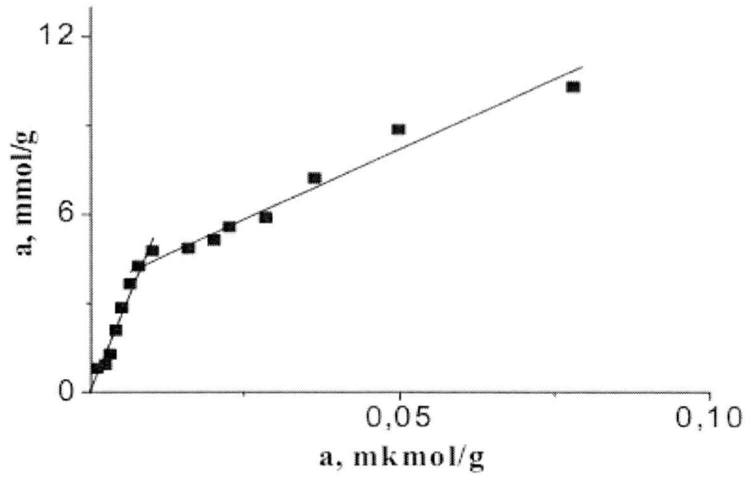

Figure 17. Comparative graph of the adsorption isotherms of water vapor at 293 K for the sample of carbon material 5PP from petroleum pitch (see Table 3.).

The sample 2F, containing wider micropores (called as supermicropores), also adsorbs water molecules according to the cluster mechanism. In contrast, the plot for the supermicroporous sample 2F (curve 1) turned down to abscissa confirmed the formation of continuous water film on pore walls at high adsorption values in the same way as the comparative plot for the supermicroporous sample 9 or 5PP (see Figure 16, 17). The clusters for supermicroporous samples 2F and 5PP, which have more wide micropores, contain only 3 water molecules per PAC.

3. 3. Data of Molecular Spectroscopy Investigations

Molecular spectroscopy techniques have been used for understanding both the nature and structure of obtained carbon materials.

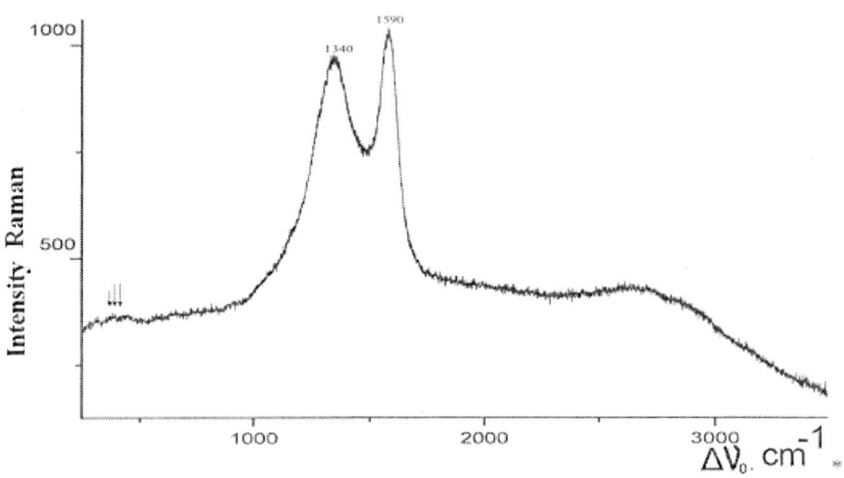

Figure 18. Raman spectrum of the sample 2F.

Figure 19 shows two peaks with maxima at 1590 (G-band) and 1340 cm^{-1}(D-band) which are characteristic for carbon structures with sp^2 hybridization including carbon nanotubes. However, the bands in the region 100-500 cm^{-1} (marked by arrows) belong only to the structures of tubular type (so-called breathing modes in carbon nanotubes (CNT)) as other types of sp^2-carbon have no any absorption at these frequencies [16, 17].

Thus, the data of adsorption, electronic microscopy and Raman-spectra investigations allow to conclude that an initial stage of solving the problem of the synthesis of consolidated carbon adsorbents with nanochannels using the PMA template has been carried out. The nanochannels in the synthesized materials apparently are formed by the aggregates and bundles of nanotubes.

More complete information concerning to the structure of the obtained materials was received from the IR-absorption and resonant Raman spectra analysis.

It is known that the breathing mode frequency (the frequency of radial vibration in CNT) is inversely proportional to nanotube diameter [17]. Thus, measuring the frequencies of breathing modes in Raman spectra gives the possibility to calculate the carbon nanotubes diameter.

As it follows from the Raman spectrum of the sample 2F (Figure18) the intensity of breathing vibrations bands is rather weak at excitation energy of 632.8 nm. A number of theoretical and experimental studies showed the possibilities to observe a very intensive spectrum of breathing modes. For this purpose it is necessary to use an advantage of a resonance and to select the laser with the excitation energy close to energy of one of electronic transition of CNT's [18, 19].

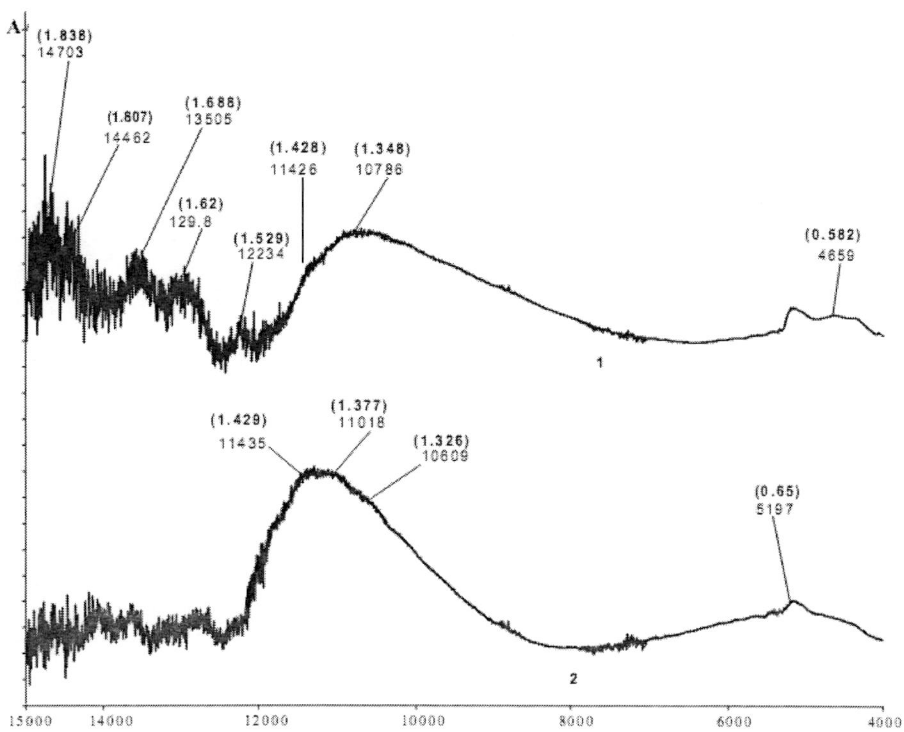

Figure 19. Near Infrared Absorption of the carbon materials: 1 — sample 2F; 2 — sample 2S.

To identify the structure of electronic transitions in the synthesized materials, the absorption spectra were registered in the region of 15000 —2700 cm^{-1} (Figure 20). All samples were outgassed by vacuum pumping during several days up to a constant weigh before the measurements of the spectra for removing the adsorbed water, CO_2 and other adsorbed gases. As it is known the doping CNT by different adsorbed substances resulted in vanishing the bands of electronic transitions in the CNT spectra [20].

Figure 19 shows two main bands at about 1.4 eV and about 0.65 eV in the optical absorption spectrum 2 for the sample 2S. Observing the bands at 1.348 eV and 0.58 eV in the spectrum 1 for the sample 2F indicates that the absorption takes place at the energies above 1.5 eV.

Among all carbon materials, such type of spectra is characteristic for carbon nanotubes where these bands are interpreted as van Hove singularities (VHSs) and attributed to electronic quantum transitions in 1D systems [18, 19]. The optical peculiarities at 0.6–1.3 eV and 1.3–2.2 eV correspond to the allowed transitions between S_{11} and S_{22} VHSs in the semiconducting CNTs, respectively. The M_{11} transitions in the metallic CNTs are disposed in the region with higher energy, 1.6 – 2 eV for the sample 2F. Thus, the sample 2F, the product of FA carbonization, reveals the metallic properties (with semiconducting CNTs), but the sample 2S, product of sucrose carbonization, is a semiconductor [19].

After the irradiation of these two samples with the laser energy 1094 nm, the spectra of resonant scattering were recorded as the laser energy corresponds to VHSs S_{22} (see Figure20). These spectra are presented on Figures 21 and 22.

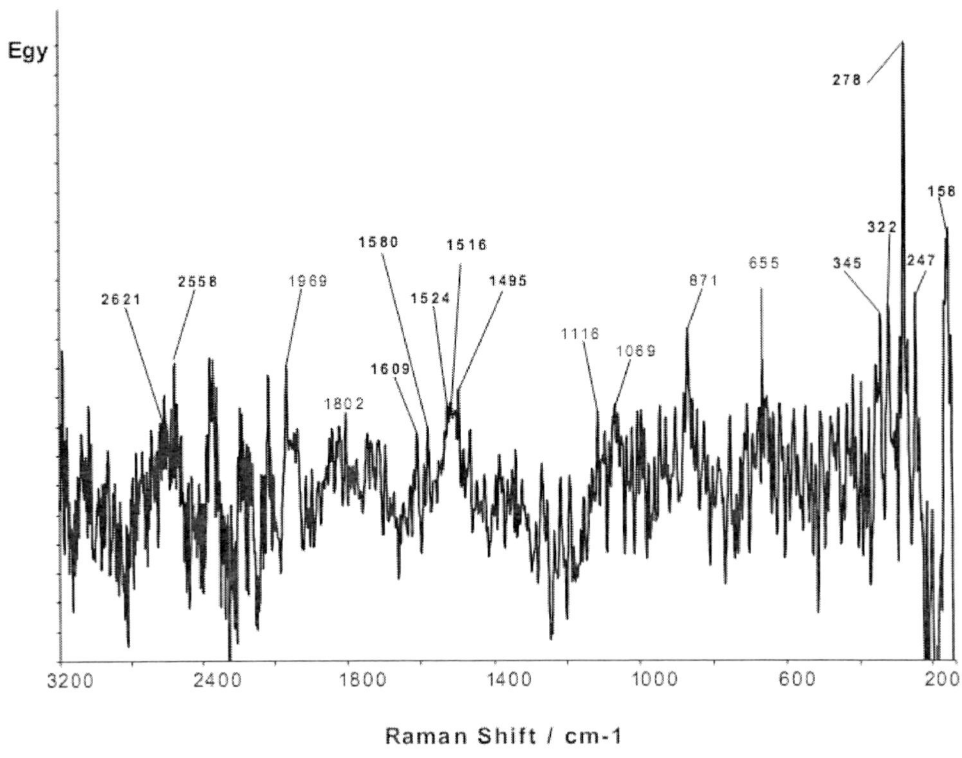

Figure 20. The resonance Raman spectrum of carbon material produced from FA carbonization – the 2F sample.

Figure 20 displays that the spectrum of 2F sample reveals the absorption bands at 1609, 1580, 1524, 1516 and 1450 cm^{-1} characteristic for longitudinal oscillatory modes in the SWCNT, i.e. G-band [17–20]. The relevant breathing modes include the adsorption bands at 345, 322, 278, 247 and 159 cm^{-1}.

It is significant to note that while the tangential modes with the frequencies lower than 1600, 1580 cm^{-1} are inherent to the nanotubes with semiconducting properties, the appearance of tangential modes with the frequencies of 1524, 1516 and 1450 cm^{-1} is a characteristic feature of the metal nanotubes [21].

The obtained values of frequencies allowed estimating the diameters of nanotubes from which this sample consists of.

Thus, the observation of the frequencies of breathing modes is the evidence that in the sample obtained by FA carbonization contains the SWNTs: for the mode 345 $см^{-1}$ (8,8), (10,10) metallic and or (8,0) semiconducting; for the mode 322 $см^{-1}$ (9,0) metallic and /or (11,11) and (9,1) semiconducting; for a mode 278 $см^{-1}$ (10,1) metallic and/or (10,1) semiconducting; for the mode 247 $см^{-1}$ (10,1) metallic and/or (11,1) and (11,0) semiconducting; for the mode 159 $см^{-1}$ (10,10) and/or (11,11) [18–22]. The relevant NT diameters equal to 0.627 nm are for (8,0); for the mode 278 $см^{-1}$ the diameter of semiconducting (10 ,0) NT is equal to 0.829 nm and for metallic (10,1) NT — 0.825 nm; metallic (10,1) NT with breathing band at 247 nm has the diameter of 0.95 nm and semiconducting (11,1) and (11,0) NT's — diameter of 0.903 nm; 159 $см^{-1}$ belong to metallic (10,10) or (11,11) NT's with diameter of 1.1 nm [18– 23].

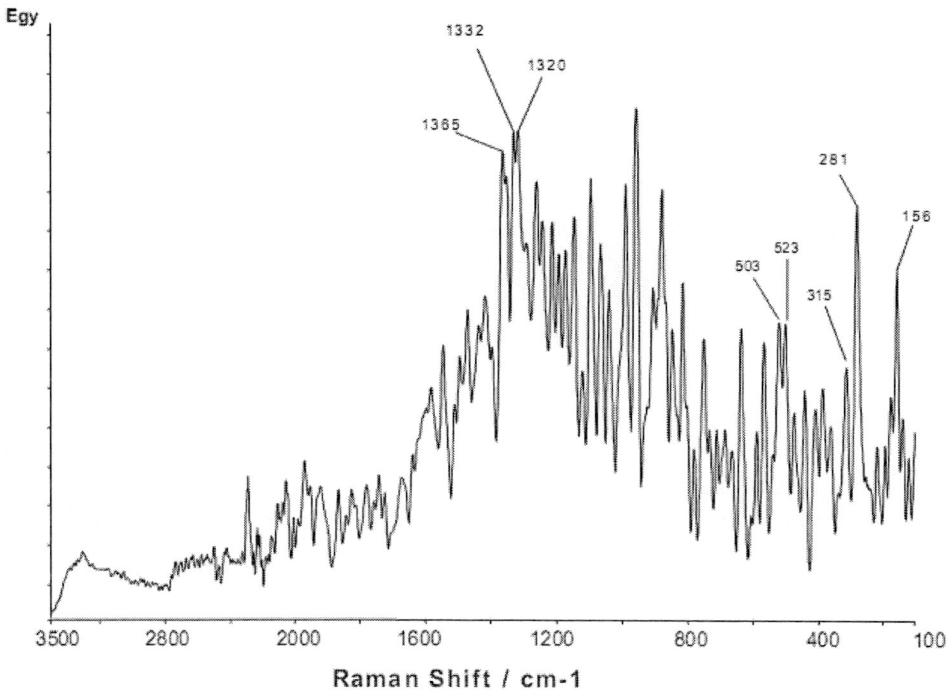

Figure 21. Resonant Raman spectrum of carbon material resulted from sucrose carbonization, 2S sample.

Another singularity of the considered spectrum is a presence of several intensive bands above 1600 cm^{-1}. As previously was established, there are bands in this spectral region that are attributed to the vacancies and topological imperfections in NT's walls, which are caused by the bending of NTs or some other reasons. So, the bands at 1813, 1948 and 2024 cm^{-1} can be attributed to pentagons, and the bands at 1494, 1458 cm^{-1} — to heptagons, respectively. A pair "pentagon — heptagon" is characterized by the bands at 1505 and 1706 cm^{-1} [24]. In addition, the concatenation of several SWCNTs of different diameters also can be carried out by formation of heptagons, pentagons or pairs 'pentagon – heptagon' in the places of junction which will be exhibited in occurrence of relevant bands in the Raman spectrum [24].

Figure 21 showed the spectrum of the sample 2S resulted from sucrose carbonization that has many essential differences from the spectrum of SWNT observed for 2F sample, product of FA carbonization (see Figure 20). Instead of absorption at 1600-1450 cm^{-1} found for 2F sample, the intensive bands for 2S sample are observed at 1365, 1328 and 1319 cm^{-1} (D-band). In addition, the absorption bands attributed to imperfections are weaker for sample 2S. Moreover, in low-frequency region at 523, 503, 281 and 157 cm^{-1}, there are the bands characteristic for CNT breathing modes.

As it was is established recently, the bands at 1307 cm^{-1} may belong to the tangential modes of DWCNTs [25-28]. From the comparison of the frequencies of breathing modes at 523, 503, 281 and 157 cm^{-1}, based on above mentioned references it is possible to conclude that the sample resulted from sucrose carbonization contains the DWCNTs of type (4,8)and(16,5) and - or (2,9)and(19,0) with the diameter of an inner tube of about 0.4 nm [25] which are aggregated into the monolith sample.

The morphological distinctions of these two samples are the following:

Monolithic adsorbents, produced by FA carbonization, are composed of the SWNTs with the average pore diameter no more than 2 nm with metallic and semiconducting properties. The bundles of NTs are strongly curved; apparently, they consist of the loose connected CNT's of different diameters.

Carbon materials, produced by sucrose carbonization, are semiconductors and, probably are composed of the DWNTs bundles. Apparently, the arrangement of DWNTs in this adsorbent is rather dense so that there are no opportunities for bending and deforming, therefore it found little imperfections.

CONCLUSION

The microporous structures formed by the bundles of CNTs with narrow inner diameter have been synthesized by the template method applying the nanofibers of monolithic alumina as a template. The synthesized nanotubes have the surface area dependent on the nature of carbon precursor. The carbon materials produced using sucrose as the precursor are the microporous adsorbents with the surface area of about 500 m^2/g, that is equal to the surface area of the template. This carbon structure is formed by the DWCNTs bundles and has the properties of semiconductor. The usage of furfuryl alcohol as the precursor led to the formation of the supermicroporous adsorbent with wider pores and the surface area reaching 580 m^2/g. It is composed of SWCNT and displays both semiconducting and metallic properties. The carbon material, produced with petroleum pitch as the precursor, has a unique uniform supermicroporous structure with homogeneous distribution of pore sizes. Investigations of the nature and structure of the obtained carbon materials by the methods of molecular spectroscopy and electron microscopy showed that their supramolecular structure is composed of the aggregates of CNTs bundles.

REFERENCES

[1] Ryoo, R.; Hoon, S.; Jun, S. *J. Phys. Chem. B.* 1999, vol.103 (37), 7743–7746.
[2] Hoon, S.; Jun, S. *J. Am. Chem. Soc.,* 2000, vol.122 (43), 10712–10713.
[3] Cott, D.J.; Petkov, N.; Morris, M.A.; Platschek, B.; Bein, T.; Holmes, J.D. *J. Am. Chem. Soc.* 2006, vol.128, 3920–3921.
[4] Lu, Q.; Gao, F.; Komarneni, S. Mallouk, T. E. *J. Am. Chem. Soc.* 2004, vol.126, 8650-8651.
[5] Chae, W-S.; An, M-J.; Lee, S-W.; Son, M-S.; Yoo, K-H.; Kim, Y-R. *J. Phys. Chem. B.* 2006, vol.110, 6447-6450.
[6] Krasilnikova, O.K.; Pogosyan, A.S.; Serebryakova, N.V.; Grankina, T.Yu, Khodan, A.N. *Protection of Metals and Physical Chemistry of Surfaces*, 2008, vol.44, 362-366.
[7] Vignes J.-L.; Frappart C.; Di Costanzo T.; Rouchaud J.-C.; Mazerolles L.; Michel D. *J. Mater. Sci.* 2008, vol.43, 1234-1240.
[8] Gregg S.J.; Sing K.S.W. *Adsorption, Surface Area, and Porosity*. 2nd edition, Academic Press, New York, USA, 1982, pp.306.
[9] Dubinin, M.M. *Carbon.* 1989, vol.27, 457–467.
[10] Dubinin, M.M. *Progr. Surf. Membr. Sci.* 1975, vol.9, 1-70.
[11] Dubinin, M.M. In *Chemistry and Physics of Carbon*; Ed. Walker P.L.; Marcel Dekker: N.Y., USA, 1966. Vol. 2, pp.51-120.
[12] Stoeckli, H.F.; Rebstein, P.; Ballerini, L. *Carbon.* 1990, vol.28, 907-909.
[13] Voloshchuk, A. M.; Dubinin, M. M.; Moskovskaya, T. A.; Larionova A. O.; Shilkova, L. A. *Russ. Chem. Bull.* 1989, vol.37, 1979-1983.
[14] Vartapetyan, R .Sh.; Voloshchuk, A.M. *Russ. Chem. Rev.* 1995, vol.64, 985-1001.
[15] Paredes, J. I.; Martnez-Alonso, A.; Yamazaki, T.; Matsuoka, K.; Tascn, J. M. D.; Kyotani, T. *Langmuir.* 2005. vol.21, 8817-8823.
[16] Eklund, P.C.; Holden, J.M.; Jishi, R.A. *Carbon.* 1995, vol.13, 959-972.
[17] Jishi, R.A.; Venkataraman, L.; Dresselhaus, M.S.; Dresselhaus. G. *Chem. Phys. Letters.* 1993, vol.209, 77-82.
[18] Rao, A.M.; Richter, E.; Bandow, S.; Chase, B.; Eklund, P.C.; Williams, K.A.; Fang, S.; Subbwasvamy, K.R.; Menon, M.; Thess, A.; Smalley, R.E.; Dresselhaus, G.; Dresselhaus, M.S. S*CIENCE.* 1995, vol.275, 187-191.
[19] Kataura, H.; Kumazawa, Y.; Maniva, Y.; Umezu, I.; Sudzuki, S.; Ohtsuka, Y.; Achiba, Y. *Synth. Met.* 1999, vol.103, 2555-2562.
[20] Kazaoui, S.; Minami, N.; Jacquemin, R.; Kataura, H.; Axhiba, Y. *Phys. Rev.* B. 1999, vol.60, 13339-13342.

[21] Pimenta, M.A.; Marucci, A.; Empedocles, S.A.; Bawendi, M.G.; Hanlon, E.B.; Rao, A.M.; Eklund, P.C.; Smalley, R.E.; Dresselhaus, G.; Dresselhaus, M.S. *Phys. Rev. B.* 1998, vol.58, R16016-R16019.

[22] Maultzsch, J.; Telg, H.; Reich, S.; Thomsen, C. *Phys. Rev. B.* 2005, vol.72, 205438 - 205456.

[23] Popov, V.N.; Lambin, *Ph. Phys. Rev. B.* 2006, vol.73, 085407(1-9).

[24] Wu, G.; Dong, J. *Phys. Rev. B.* 2006, vol.73, 245414(1-9).

[25] Xia, M.; Zhang, S.; Zuo, X.; Zhang, E.; Zhao, S.; Li, J.; Zhang, L.; Liu, Y.; Liang, R. *Phys. Rev. B.* 2004, vol.70, 205428(1-8).

[26] Ren, W.; Li, F.; Tan, P.; Cheng, H.-M. *Phys. Rev. B*, 2006, vol.73, 115430(1-6).

[27] Pfeiffer, R.; Simon, F.; Kuzmany, H.; Popov. V.N. *Phys. Rev. B.* 2005, vol.72, 161404(1-4).

[28] Ye, J.T., Tang, Z.K. *Phys. Rev. B.* 2005, vol.72, 045414(1-4).

INDEX

A

B

C

D

E

F

G

H

I

J

K

L

M

Q

R

S

T

U

V

W

X

Y

Z